THE NUTRITION OF HERBIVORES

Second International Symposium on the Nutrition of Herbivores

International Organizing Committee

J. H. TERNOUTH
Chairman

D. J. MINSON
Deputy chairman

D. G. ARMSTRONG	R. F. BARNES
K. F. DOWSETT	A. R. EGAN
R. ELLIOTT	R. J. W. GARTNER
J. B. HACKER	H. S. HOFMEYR
J. P. HOGAN	P. HOPKINS
I. D. HUME	N. P. McMENIMAN
R. G. MOIR	G. M. MURPHY
D. B. PURSER	D. ROBERTSHAW
M. E. ROSE	R. W. SUTHERST
A. R. SYKES	R. F. THORNTON

Symposium organisers

Uniquest Conference Systems, University of Queensland.

The symposium was organized under the auspices of the Australian Society of Animal Production Inc.

G. E. ROBARDS Federal President	**T. F. LECHE** Federal Secretary	**B. G. LUKINS** Federal Treasurer

THE NUTRITION OF HERBIVORES

Edited by

J. B. HACKER

CSIRO Division of Tropical Crops and Pastures, St Lucia

and

J. H. TERNOUTH

Department of Animal Sciences and Production,
University of Queensland, St Lucia

ACADEMIC PRESS

(Harcourt Brace Jovanovich, Publishers)

Sydney Orlando San Diego New York Austin
London Montreal Tokyo Toronto

ACADEMIC PRESS AUSTRALIA
30–52 Smidmore Street,
Marrickville, NSW 2204

United States Edition published by
ACADEMIC PRESS INC.
Orlando, Florida 32887

United Kingdom Edition published by
ACADEMIC PRESS, INC. (LONDON) LTD.
24/28 Oval Road, London NW1 7DX

Printed in Australia

National Library of Australia Cataloguing-in-Publication Data

---

Nutrition of herbivores.

Bibliography.
Includes index.
ISBN 0 12 750052 9.

1. Animal nutrition. 2. Herbivora – Food. I. Hacker, J. B. (John Bryan), date– . II. Ternouth, J. H. (John Hardy), date– .

636.08'5

Library of Congress Catalog Card Number: 87–70846

Academic Press Rapid Manuscript Reproduction

Contents

Contents

Introduction

Nutrition is a complex quantitative science, the complexity of which is probably greatest in the case of the grazing animal. Because of the difficulty of studying the complete range of factors influencing the nutrition of grazing herbivores, the Organising Committee of the Second International Symposium on the Nutrition of Herbivores, held at the University of Queensland between 6th and 10th of July, 1987, approached the discipline in a mechanistic manner. First, the factors affecting the growth of plants, the concentration of nutrients and secondary compounds in the aerial portion of the plant are considered, then how much and how readily the plants are consumed by herbivores, and finally how effectively the ingested nutrients are digested, absorbed, partitioned and utilised by the animal for survival and growth. As man has modified the natural habitat of animals, one chapter has been devoted to understanding the migration of animals and three to understanding the diversity of herbivorous animals in the world today. Herbivores have developed a special symbiotic relationship with their gastro-intestinal microflora, but other unicellular and multicellular animals are parasitic on herbivores and effect their nutritional status. Thus chapters have been included which discuss the effects of endoparasites and ectoparasites on the nutrition of their unwilling hosts. Finally, estimating the nutrient status of animals is discussed as supplementation is becoming an increasingly important quantitative science. Two chapters are devoted to supplementing the diets of herbivores.

The annual University of Queensland D. O. Butler Memorial Lecture has been included in this volume. The 1987 lecture discusses the genetic basis for the survival and growth of herbivores. Research papers presented at the Second International Symposium will be published as an Occasional Publication of the Australian Society of Animal Production. The annual Harry Stobbs Memorial Lecture organised by the University of Queensland and the Tropical Grassland Society of Australia is to be presented by Dr R. J. Jones during the symposium and is to be published in the journal *Tropical Grasslands*.

As it is intended that this volume be available for the participants at the Symposium, time constraints have been placed upon both the authors and the editors in their separate endeavours. Thus the order of the chapters does not coincide with that originally intended, nor does it follow the order of the programme for the symposium itself. In addition, although every effort has been made to remove errors in the text, it is likely that some will have escaped the editors' attention.

The editing of this work would have been impossible without the skilled and dedicated assistance of Douglas Bodero, who converted the diversity of micro-computer discs into usable files with a common format and style, Dianne Burling for her technical assistance, and Janice Niven for creating the many drafts and the final document.

The Organizing Committee wishes to thank the Australian Wool Corporation for sponsoring some of the plenary lecturers.

J. B. HACKER
J. H. TERNOUTH

Authors

D. F. Balph	College of Natural Resources, Utah State University. Logan, Utah 84322. USA.
T. N. Barry	Department of Animal Science, Massey University. Palmerston North, New Zealand.
B. Blaney	Animal Research Institute, Queensland Department of Primary Industries. Yeerongpilly, Queensland 4105. Australia.
C. Devendra	Agriculture, Food and Nutrition Sciences, Institute Developmental Research Centre. Tanglin, P. O. Box 101, Singapore.
P. T. Doyle	Department of Agriculture and Forestry, University of Melbourne. Parkville, Victoria 3052. Australia.
R. Elliott	Department of Agriculture, University of Queensland. St Lucia, Queensland 4067. Australia.
D. F. Holleman	Institute of Arctic Biology, University of Alaska. Fairbanks, Alaska 99701. USA.
M. E. Hubbert	Institute of Arctic Biology, University of Alaska. Fairbanks, Alaska 99701. USA.
I. D. Hume	School of Biological Sciences, University of Sydney. Sydney, NSW 2006. Australia.
P. A. Janssens	Department of Zoology, Australian National University, Canberra, ACT 2601. Australia.

D. I. H. Jones	Welsh Plant Breeding Station, The University College of Wales, Abertstwyth, Dyfed AY23 3EB. Wales.
R. N. B. Kay	Rowett Research Institute. Bucksburn, Aberdeen AB2 9SB. Scotland.
J. P. Langlands	Division of Animal Production, CSIRO, Pastoral Research Laboratory. Armidale, NSW 2350. Australia.
N. P. McMeniman	Department of Animal Sciences and Production, University of Queensland. St Lucia, Queensland 4067. Australia.
S. J. McNaughton	Biological Research Institute, Syracuse University. Syracuse, New York 13244–1220. USA.
R. I. Mackie	Department of Rumen Biochemistry, Animal and Dairy Science Research Institute. Irene, 1675. Republic of South Africa.
J. C. Malechek	College of Natural Resources, Utah State University. Logan, Utah 84322. USA.
J. I. Murray	Institute of Animal Physiology and Genetic Research Edinburgh, EH9 3JQ. Scotland.
E. R. Ørskov	Rowett Research Institute. Bucksburn, Aberdeen AB2 9SB. Scotland.
D. P. Poppi	Animal Sciences Department, Lincoln College, Canterbury. New Zealand.
B. R. Roberts	School of Applied Sciences, Darling Downs Institute of Advanced Education. Toowoomba, Queensland 4350. Australia.
H. Staaland	Department of Zoology, Agricultural University of Norway. 1432 Aas–NLH. Norway.
R. W. Sutherst	Division of Entomology, CSIRO, Long Pocket Laboratory. Indooroopilly, Queensland 4068. Australia.

N. F. Suttle — Moredun Research Institute.
408 Gilmerton Road,
Edinburgh, EH17 7JH. Scotland.

A. R. Sykes — Animal Sciences Department,
Lincoln College,
Canterbury. New Zealand.

St C. S. Taylor — Institute of Animal Physiology and Genetic Reseach.
Edinburgh Research Station,
Edinburgh, EH9 3JQ. Scotland.

J. H. Ternouth — Department of Animal Sciences and Production,
University of Queensland.
St Lucia, Queensland 4067. Australia.

R. F. Thornton — Meat Research Institute,
CSIRO, Division of Food Research.
Cannon hill, Queensland 4170. Australia.

R. H. Weston — Division of Animal Production,
CSIRO, Ian Clunies Ross Animal Research Laboratory.
Blacktown, NSW 2148. Australia.

R. G. White — Institute of Arctic Biology,
University of Alaska.
Fairbanks, Alaska 99701. USA.

B. A. Young — Department of Animal Science,
University of Alberta.
Edmonton, Alberta. Canada.

1

NATIVE AND INTRODUCED HERBIVORES IN AUSTRALIA

by I.D. Hume

Apart from the marine mammal the dugong (*Dugong dugon*) all native herbivorous mammals in Australia are marsupials. These include terrestrial grazers and browsers such as kangaroos and wallabies (family Macropodidae) and wombats (Vombatidae) and arboreal folivores such as tree-kangaroos (Macropodidae), the koala (Phascolarctidae) greater glider and ringtail possum (Pseudocheiridae) and brushtail possum (Phalangeridae).

The arboreal folivores are absolutely dependent on forests for their survival, and suffer from competition by Man for these forests. This competition is direct, being in the form of removal of trees for timber, modification of the structure of the remaining forest, and replacement of native forests by other forms of land use, principally exotic pine plantations and agriculture.

The human competition faced by the native terrestrial grazers and browsers is less direct. It comes not only from removal and modification of their habitat, but also from competition for food resources by a wide range of introduced herbivores, both domestic and feral. All domestic herbivores have been introduced and include cattle, sheep, goats, deer and horses. Feral herbivores, which have escaped or have been released from captivity, include water buffalo, cattle, goats, deer, camels, horses and donkeys, rabbits, hares and pigs (which in the feral state are largely herbivorous (Pavlov 1980)).

In common with many native and feral herbivores, domestic cattle, sheep, goats and horses must often survive and reproduce in environments characterized by extreme climatic unpredictability. Some of the adaptations of herbivores to such environments, and the problems inherent in increasing, or even maintaining, animal production in these environments are discussed in detail in subsequent chapters of this volume.

THE ENVIRONMENT

In general, higher rainfall areas of Australia are restricted to coastal regions and with increasing distance from the coast total rainfall decreases. The annual distribution of rainfall is most even in a small area of the south east of the continent (Fig. 1), although even here high summer temperatures reduce effective rainfall to such an extent that plant growth is often restricted during the summer. With increasing distance north and west from this region rainfall becomes more seasonal, with distinct summer peaks in the monsoonal north and

NUTRITION OF HERBIVORES
ISBN 0 12 750052 9

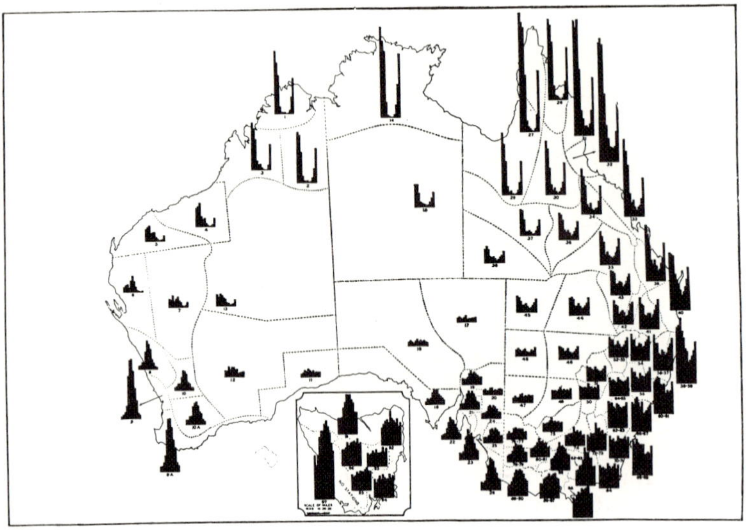

Figure 1. Monthly distribution of rainfall in various districts of Australia. Columns for the months January to December run from left to right (Commonwealth of Australia Year Book No. 56, 1970 reproduced by permission of the Bureau of Meteorology).

distinct winter peaks in the Mediterranean climates of the south west of Western Australia, southern South Australia and parts of Victoria and southern New South Wales.

Just as importantly, rainfall variability around the annual average also increases from less than 15% in the south east and south west of the continent to a maximum of more than 50% in central Australia (Fig. 2).

These patterns of total rainfall, its seasonal distribution and its variability result in a characteristic distribution of natural vegetation in Australia (Fig. 3). Forests are restricted to the south west, south east and east coast of the continent. Woodlands are found inland of the forested regions as well as in the tropical monsoonal regions of northern Australia. Inland of the woodlands are the shrublands and grasslands of the semi-arid and arid zones.

Within each broad vegetational zone there are more specialized environments. These include alpine heaths in the highest parts of the Great Dividing Range in the south east, and rainforest and rainforest remnants along the east coast and on the eastern escarpment of the Great Dividing Range.

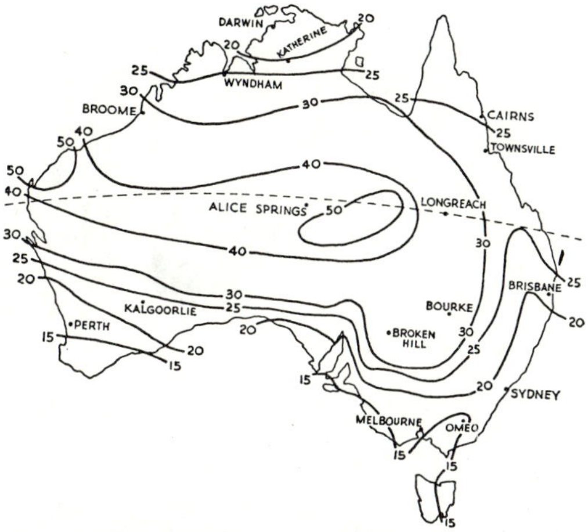

Figure 2. Rainfall variability in Australia expressed as a percentage of total rainfall. All values calculated over at least 40 years (Wadham *et al.* 1964).

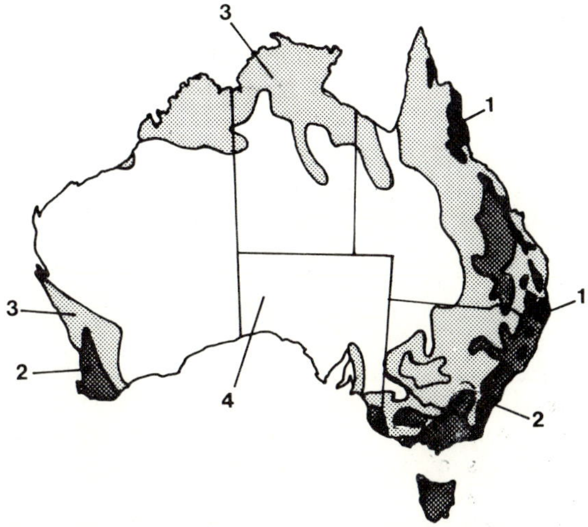

Figure 3. Vegetation of Australia (adapted from Moore 1970). 1: rainforest; 2: wet and dry sclerophyll forest; 3: sub-humid and semi-arid woodland; 4: semi-arid shrubland and grassland.

High ambient temperatures for a greater part of the year often accompany the highly variable rainfall pattern of much of the shrubland and grassland vegetational zones, particularly north of the Tropic of Capricorn. At the other extreme, low ambient temperatures restrict pasture growth in the highlands of south eastern Australia and impose considerable thermal stress on native and introduced herbivores alike (Fitzpatrick and Nix 1970).

INTRODUCED HERBIVORES

Greatest densities of beef cattle are found in the wetter coastal and subcoastal regions, but their distribution includes all vegetational and climatic zones of the continent. Lowest densities are within the arid zone. Dairy cattle are clustered around centres of human population, including all capital cities except Darwin in the north. Greatest numbers are found in the previously forested vegetational zones. Sheep densities are highest in the woodland and shrubland vegetational zones, with few sheep found north of the Tropic of Capricorn or in the forested vegetational zones.

Apart from the south east and south west the first limiting factor to beef and wool production is the unpredictability of rainfall and thus feed supplies. This unpredictability results in huge deficiencies in plant production seasonally and annually, with droughts an expected component of the environment. This theme recurs throughout this volume, with emphasis on management strategies and research priorities aimed at ameliorating this variability in plant and thus animal production.

The replacement of native pastures with introduced and improved pasture species has been one of the principal approaches to reducing these productivity gaps (Fig. 4). However, these are generally suitable only for higher rainfall zones. Thus improved tropical pastures are largely restricted to areas with annual average rainfalls of at least 500 mm, with only Mediterranean annual pastures being used successfully in areas with as little as 300 mm rainfall.

The variability and unpredictability of rainfall throughout the pastoral zone (Fig. 4) and in many parts of the improved pasture regions means that the land systems are often overstocked and overgrazed. This abuse of the environment is manifested in numerous ways. For example, the dieback of *Eucalyptus* spp. on the New England Tablelands of northern New South Wales, although complex in its aetiology, has been attributed to continuous overgrazing of the Tablelands by cattle and sheep for at least 100 years, resulting in a distinct lack of replacement seedling trees (Breckwoldt 1986). The remaining over-mature trees are less resilient to defoliation, and in the face of repeated attack by leaf-eating insects eventually succumb. The result is progressive loss of the native woodland vegetation, loss of stock shelter, and therefore of animal production. The effects of

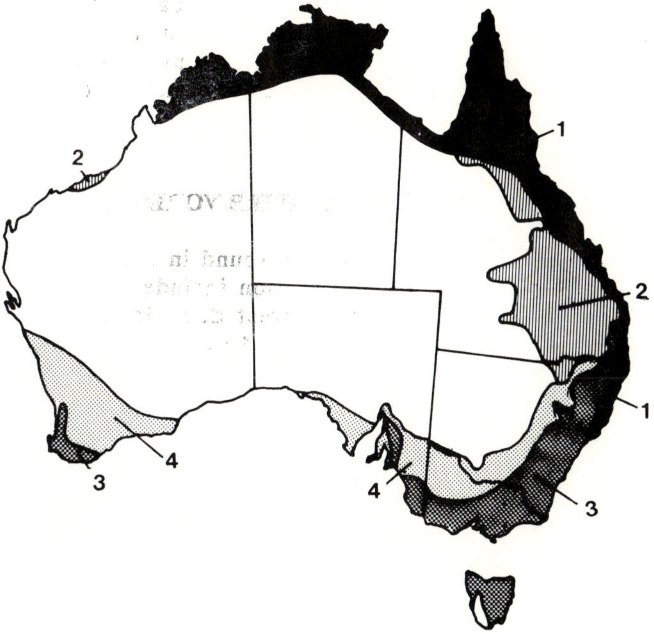

Figure 4. Pastures of Australia (adapted from Moore 1970). 1: tropical perennial pastures (*Digitaria, Setaria, Panicum, Paspalum, Pennisetum, Desmodium, Macroptilium, Neonotonia, Stylosanthes* spp.); 2: xerophytic perennial pastures (*Cenchrus ciliaris, Chloris gayana, Panicum maximum* var. *trichoglume, Sorghum almum*); 3: temperate perennial pastures (*Lolium perenne, Trifolium repens*) and temperate perennial grass – annual legume pastures (*Phalaris aquatica, Lolium perenne, Trifolium subterraneum*); 4: Mediterranean annual pastures (*Lolium rigidum, Bromus rigidus, Hordeum leporinum, Trifolium subterraneum, Medicago* spp.). Area in Tasmania not shaded is too mountainous for sown pastures, and in central Australia too arid.

artificial and natural windbreaks in this environment in reducing soil water losses, and increasing pasture and animal production have been clearly demonstrated by Lynch *et al.* (1980) and Lynch and Donnelly (1980).

The effects of overgrazing in the semi-arid and arid rangelands are seen in the loss of palatable shrubs and the encroachment over large areas by dense inedible shrubs, especially *Eremophila* spp., *Cassia* spp. and *Dodonaea* spp. The result is loss of grazing land as well as a drought reserve in the form of edible shrubs.

Other effects of overgrazing include permanent degradation of the rangelands by loss of the grass cover, increased erosion and silting of waterholes in creeks and riverbeds (Noble *et al.* 1984).

FERAL HERBIVORES

The detrimental effects of overgrazing by domestic stock are exacerbated by feral herbivores, which are found in nearly all vegetational zones of Australia (Fig. 5). In the more closely settled areas of the south east the main feral herbivores are rabbits and hares (Fig. 5f, g). Away from the more populated areas larger feral herbivores predominate, including horses, donkeys and camels (Fig. 5c, d, e).

The most easily recognized effects of feral herbivores are those associated with competition with domestic stock. These include overgrazing and subsequent erosion, damage to fencing and watering points, and other environmental damage (e.g. spread of inedible shrubs, fouling of swamps and lagoons). However, there is a less visible problem associated with the potential of feral animals to act as reservoirs of several livestock diseases such as tuberculosis, brucellosis, leptospirosis, rabies, rinderpest, blue tongue and foot-and-mouth.

Table 1. Estimates of total populations of feral herbivores in the Northern Territory based on aerial surveys (P. Whitehead, unpublished).

Species	Year	Population
Buffalo	1985	340,000
Cattle*	1985	140,000
Camels	1984	25,000 - 30,000
Horses	1985	250,000 - 300,000
Donkeys	1985	120,000 - 140,000

* Assuming 10% of total herd is feral.

The most feared disease is foot-and-mouth, because of the constant risk of introduction by a variety of means. An outbreak of the disease, which is not present in Australia, would be economically devastating for the pastoral industries. A localised outbreak in the Northern Territory, for instance, would probably result in an embargo by major importing countries of livestock, meat and animal products from the whole of Australia. The presence of large numbers of feral water buffalo, cattle and pigs in the Northern Territory (Table 1) would greatly complicate eradication of foot-and-mouth disease.

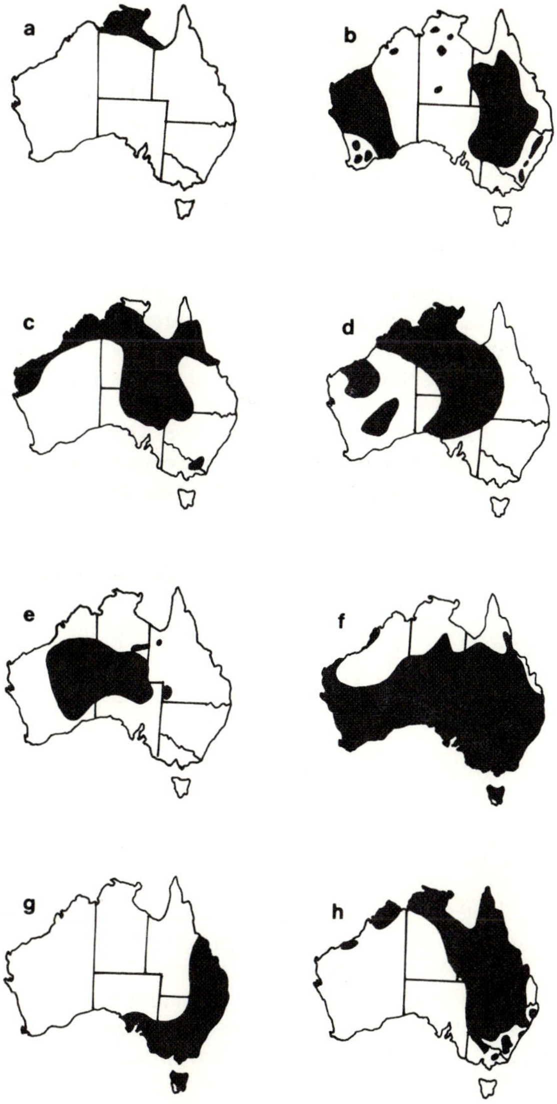

Figure 5. Distributions of eight feral herbivores in Australia (from Strahan 1983). (a) Water buffalo, (b) Goats, (c) Horses, (d) Donkeys, (e) Camels, (f) Rabbits, (g) Hares, (h) Pigs.

Water buffalo (*Bubalus bubalis*)
The water buffalo was introduced into the Northern Territory from Indonesia between 1825 and 1843, and is now firmly established in the "top end" of the Territory in rainfall areas above 1,000 mm per annum (Fig. 5a). Its distribution is limited by water; during the annual wet season water buffalo spread out over the whole of their range but in the dry season they are restricted to the plains around rivers and swamps. Some areas are now managed specifically for water buffalo meat production, most of which is exported overseas. In 1980–81, this industry earned $A7 million. However, in terms of beef production feral herbivores in the Northern Territory, including water buffalo, represented an estimated net loss of over $A12 million in 1979 (Letts *et al.* 1979). This loss did not account for environmental damage (e.g. spread of noxious weeds such as *Mimosa pigra* due to overgrazing along river frontages or fouling of swamps and lagoons) and disease costs (e.g. tuberculosis, which is present at a level of 1–2%, but in some localised areas at 10% or more of water buffalo).

Feral cattle
Feral cattle exist in all areas of the Northern Territory and Queensland where poor management or uncontrolled areas make their mustering difficult or impossible. There is a problem of definition, since these animals would normally be regarded as stock by many pastoralists, but in the event of an exotic disease outbreak they would more likely be regarded as feral (Mitchell *et al.* 1982). On northern Queensland properties about 8% of cattle are usually missed at muster. On this basis up to 50,000 cattle could be regarded as feral in the true sense of the word, but if exotic disease made immediate mustering necessary this number could increase 3- to 5-fold (Mitchell *et al.* 1982).

Goats (*Capra hircus*)
Feral goats are found throughout much of the pastoral zone (Fig. 5b). Although in certain areas they are harvested for use in the pet meat industry and as game meat for export to Europe, and thus are a useful supplementary source of income to land-holders in these areas, feral goats are undoubtedly major competitors of both domestic and native herbivores for food resources. Because of their gregarious nature feral goats can reach locally high population densities. This, together with their browsing habits, often leads to serious effects on vegetation cover and species balance in fragile plant communities. Goats can strip leaves and bark off valuable fodder shrubs, and perennial pasture species often fail to recover from overgrazing, being replaced by less desirable annuals and weeds.

One of the native herbivores directly affected by competition from feral goats for food resources is the yellow-footed rock-wallaby (*Petrogale xanthopus*). This rare rock-wallaby is a mixed feeder (Hume 1982) with a preferred habitat of rocky hills and escarpments. In this habitat the rock-wallabies are largely free of competition from cattle and sheep, but feral goats are able to occupy all parts of the rock-wallabies' range. The result has been removal of a significant proportion of the rock-wallabies' food resources, especially browse, jeopardising the continued existence of these small macropods.

Numbers of feral goats are highly variable depending on, amongst other factors, the extent of harvesting in different areas for pet food and for export of carcasses to Europe. As with water buffalo and feral cattle, feral goats would undoubtedly play a major role in the spread of exotic diseases such as foot-and-mouth, rinderpest and blue-tongue.

Deer

Six species of deer have been introduced into Australia, from Europe and Asia. Deer farming has increased in recent years, but feral populations of deer have never exploded in the same way that other feral ungulates have. All six species are confined to coastal and forested vegetational zones. The sambar (*Cervus unicolor*) appears to be the only deer species in Australia capable of extending its range without assistance from Man. Its population in Victoria and southern New South Wales seems to be controlled by recreational hunting.

Horses and donkeys

Except for a small area in the Southern Tablelands in south eastern Australia feral horses (*Equus caballus*) are generally found in the pastoral zone (Fig. 5c). They are pastoral pests, particularly in the Northern Territory where the population in 1985 was estimated to be 250,000–300,000 (Table 1). In such large numbers, feral horses are responsible for destruction of fences, fouling of watering points and direct competition with domestic stock for pasture.

Feral donkeys (*E. asinus*) were imported from Europe for use as pack animals. The donkey is well adapted to arid regions and is found in large numbers in Western Australia, South Australia and the Northern Territory (Fig. 5d). In 1978 the Western Australian population was estimated at 0.5–1.0 million, with a total Australian population of at least 1.5 million (Strahan 1983). Donkeys can be found in mobs as large as 500 animals and compete with domestic stock for water and pasture. Donkeys will congregate around a water source and deny stock access to it, at the same time degrading pastures and initiating erosion, both sheet and gully.

Camels (*Camelus dromedarius*)

Australia is the only place in the world where *C. dromedarius* is found in the wild state. Its range is principally the sandy ridge-dunes of central and central-western Australia (Fig. 5e). The total population is estimated to be at least 25,000 (Strahan 1983) and probably more (P. Whitehead unpublished). There is a small export trade to the Middle East, but apart from this their economic, and environmental, impact is negative. Being tall and predominantly browsers, local concentrations of camels can have a devastating effect on the shrub layer, and it is likely that this has reduced the shelter available for small native mammals in the arid zone. Economic damage to fences and watering points, though, is more obvious and direct (Strahan 1983).

Rabbits and hares

The European rabbit (*Oryctolagus cuniculus*) spread from introductions in the 1850s near Melbourne and Adelaide to reach epidemic proportions across much of temperate Australia by the turn of the century (Fig. 5f). The introduction of the viral disease myxomatosis in the 1950s caused a major decline (80–90%) in rabbit numbers (Frith 1970). Myxomatosis is now endemic in rabbit populations, and mortality has fallen from over 99% to less than 40%. The rabbit flea, the principal vector of the disease in Europe, was introduced in the early 1970s to assist the spread of the virus. This has been effective in areas of higher rainfall, but survival of the flea in the arid zone has been poor. It is in the semi-arid and arid zones that environmental damage caused at least partly by rabbits has been greatest. This includes denudation of land surfaces through permanent degradation of grass and shrub components of the delicately balanced native vegetation, physical displacement of native burrowing animals, and removal of food sources and refuges of small native herbivores.

The brown hare (*Lepus capensis* = *L. europaeus*) first established itself in Victoria in 1862 and is now widespread in grasslands, woodlands and some urban habitats in south eastern Australia (Fig. 5g). It has occurred in plague proportions in Victoria, and is responsible for significant damage to crops and young tree plantations.

Pigs (*Sus scrofa*)

The feral pig is an opportunistic omnivore, and is a predator on both native and introduced fauna. The plant component of its diet includes roots and shoots of grasses and other forbs, as well as irrigated grain crops and pastures. The distribution of feral pigs is limited by water, but nevertheless includes most climatic regions of Australia (Fig. 5h). In the pastoral zone feral pigs are common along water courses, which are often fouled by mud wallows. Other environmental effects of feral pigs are less well documented, but damage to vulnerable plant communities such as rainforests was suggested as being significant by Letts *et al.* (1979). However, its reputation as the most significant mammalian pest of Australian agriculture (Strahan 1983) is based partly on its role as a reservoir for brucellosis, tuberculosis, leptospirosis and Murray Valley encephalitis, and a potential reservoir for swine fever and foot-and-mouth disease should these devastating diseases reach Australia.

NATIVE HERBIVORES

Interactions with domestic and feral herbivores in the arid zone

The initial impression is that grazing of the native pastures of the Australian arid zone by introduced herbivores (both domestic and feral) has had two distinct effects on the native herbivores. Tall grass tussocks are used as shelter by small marsupials such as wallabies, hare-wallabies and bandicoots. The replacement of tall tussocks with heavily grazed short grasses has resulted in a sharp decline of populations of small native mammals, some to extinction (Figs. 6, 7). On the other hand, the same vegetational change has been cited as the reason for the present relative abundance of the large grazing kangaroos.

However, Newsome (1975) has suggested that the response of the two groups of native herbivores to competition from introduced herbivores may have been similar, only on different time scales. That is, the initial response of both groups was possibly an increase in population size, followed by a decline. The time scale for the large native herbivores (i.e. the kangaroos) may be longer than 50 years, but for the small herbivores less than 25 years (Newsome 1975).

Small native herbivores. Initially, patchy grazing of the natural grasslands by introduced herbivores probably improved the habitat for small wallabies by creating "edge effects"; that is there was an increase in food supply and quality in the form of heavily grazed short grasses (mainly *Eragrostis setifolia*) while adequate shelter remained close by in the form of a reduced number of tall grass tussocks (*Aristida* spp., *Danthonia* spp., *Astrebla* spp.). The new shoots of the short grasses would have been lower in cell wall content than the tall tussocks. Small herbivores, because of their relatively high mass-specific nutrient requirements, must select plant material lower in cell walls and higher in cell contents than large herbivores (Demment and Van Soest 1983).

Under these conditions the numbers of small wallabies and hare-wallabies probably increased (Newsome 1975). However, continued grazing by increasing numbers of sheep and cattle, together with feral goats, horses, donkeys and camels, eventually removed the shelter provided by the tall grass tussocks; the grass tussocks would have given protection from environmental extremes and from predators. This habitat destruction led to the decline of small native herbivores such as the rufous hare-wallaby (*Lagorchestes hirsutus*), bridled nailtail wallaby (*Onychogalea fraenata*) and brush-tailed bettong (*Bettongia penicillata*) (Fig. 6) and bandicoots such as the rabbit-eared bandicoot or greater bilby (*Macrotis lagotis*) and golden bandicoot (*Isoodon auratus*) (Fig. 6). For some species this decline was to extinction (e.g. eastern hare wallaby (*L. leporides*), crescent nailtail wallaby (*O. lunata*), desert bandicoot (*Perameles eremiana*), pig-footed bandicoot (*Chaeropus ecaudatus*) (Fig. 7).

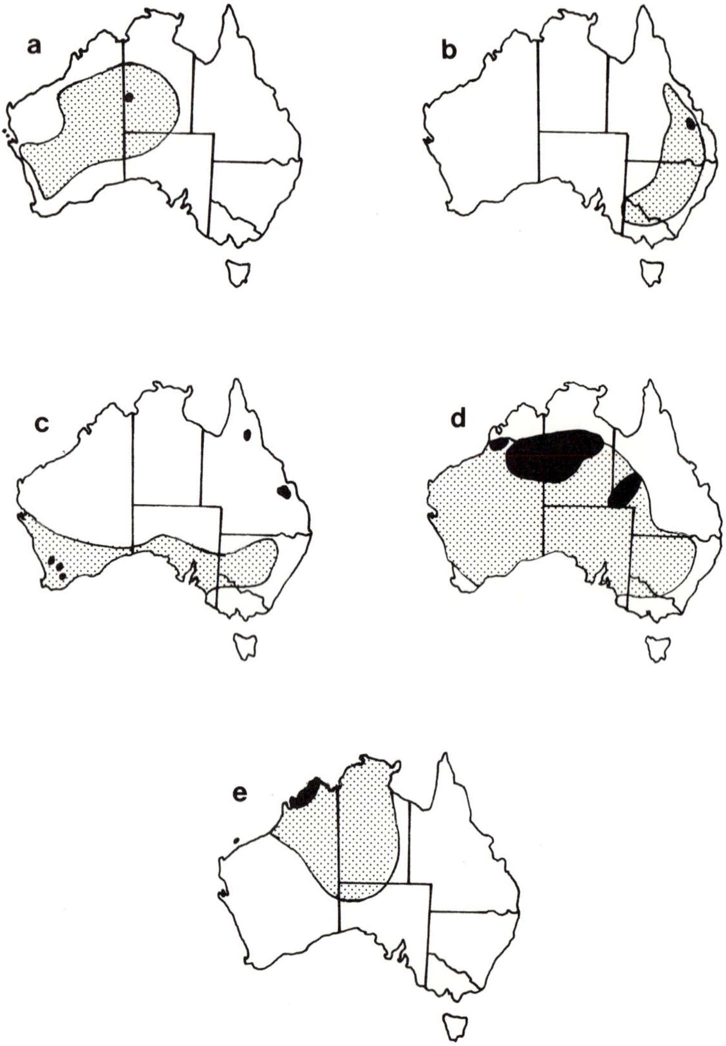

Figure 6. Present and past distributions of five small marsupials in Australia (from Strahan 1983). (a) Rufous hare-wallaby (*Lagorchestes hirsutus*), (b) Bridled nailtail wallaby (*Onychogalea fraenata*), (c) Brushtailed bettong (*Bettongia penicillata*), (d) Greater bilby (*Macrotis lagotis*), (e) Golden bandicoot (*Isoodon auratus*).

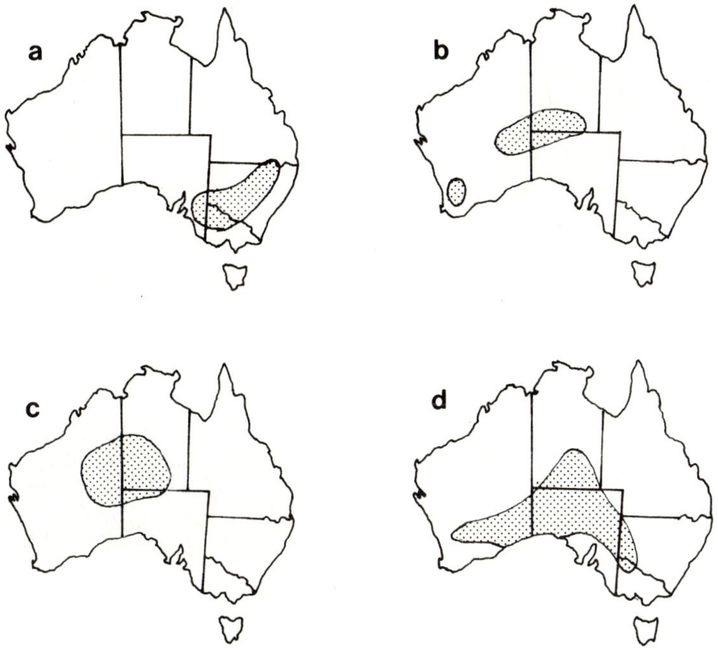

Figure 7. Past distributions of four extinct small marsupials in Australia (from Strahan 1983). (a) Eastern hare-wallaby (*Lagorchestes leporides*), (b) Crescent nailtail wallaby (*Onychogalea lunata*), (c) Desert Bandicoot (*Perameles eremiana*), (d) Pig-footed bandicoot (*Chaeropus ecaudatus*).

Large native herbivores. The grazing kangaroos of the arid zone are not dependent on tall grass tussocks as refuges. The virtual complete replacement of these tussock grasslands by short grasses therefore continued to improve their food supply. Thus present populations of the red kangaroo (*Macropus rufus*), grey kangaroos (*M. giganteus* and *M. fuliginosus*) and wallaroos and euros (*M. robustus*) are probably much higher than they were before the introduction of exotic herbivores to the arid zone. All four species are widely distributed (Fig. 8) and are seen by graziers as competitors of domestic stock for food, and are harvested for pet food and for hides on a regulated commercial basis (Sinclair 1977).

However, it is possible that, with increased grazing by introduced herbivores, both domestic and feral, widespread permanent degradation of the rangelands of the arid zone will occur (by removal of palatable grasses, erosion and loss of soil and thus vegetation cover,

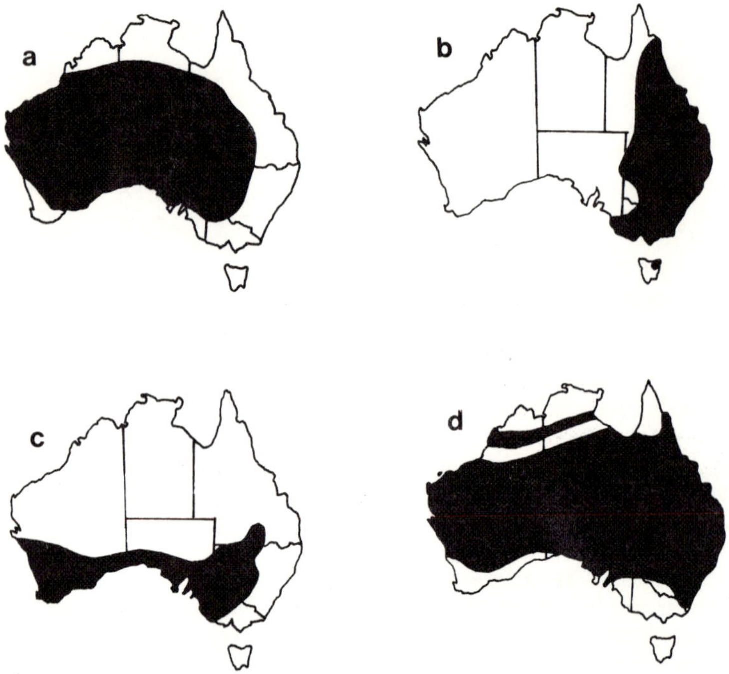

Figure 8. Distributions of four large kangaroos in Australia (from Strahan 1983). (a) Red kangaroo (*Macropus rufus*), (b) Eastern grey kangaroo (*M. giganteus*), (c) Western grey kangaroo (*M. fuliginosus*), (d) Euros and wallaroos (*M. robustus*).

and encroachment by unpalatable shrubs and grasses). If this happens, populations of the large native herbivores in the arid zone can be expected to follow the same response curve of the smaller species, that is, an irreversible decline.

This has already occurred for the red kangaroo in the Pilbara region of Western Australia. Here overgrazing has resulted in replacement of palatable short grasses by xeric spinifex (*Triodia pungens*) which is not eaten by either red kangaroos or sheep. As a result numbers of both herbivores have declined since the 1930s (Newsome 1975).

Dietary overlap between kangaroos and domestic stock. Several studies have been made of the dietary habits of the large kangaroos and domestic stock (mainly sheep) in the arid zone. The extent to which kangaroos and stock compete is still not clear, mainly because all the studies have been short term, not long enough to span both good and drought years. Direct comparisons between studies are also limited by the different techniques used (Hume 1982). Nevertheless, it does appear that the red kangaroo, eastern grey kangaroo (*Macropus giganteus*) and euro (*M. robustus erubescens*) are all more dependent on grasses than are sheep or cattle which eat a broader diet of grasses, forbs and shrubs. The overlap in species eaten is greatest during good seasons when sheep and cattle eat more grass, and least during drought years when sheep and cattle eat more browse (Newsome and Corbett 1977; Hume 1982). Dietary overlap is further reduced in central Australia by the spatial separation of cattle and red kangaroos in their utilization of open and wooded communities, especially during drought (Dudzinski *et al*. 1982). As forage conditions deteriorate cattle tend to leave the open communities for more wooded areas, while red kangaroos mostly move out of the wooded communities, used in good forage conditions, towards the open areas.

Interactions with European man in forested zones

Wallabies and wombats. The edge effect in the arid zone created by the initial grazing activities of introduced herbivores in the tussock grasslands (Newsome 1975) has a parallel in the wetter forested zones of Australia. In the latter zones the edge effect has been created by the clearing of forest for agriculture by European man. Initial patchy clearing maximised the interface between forest (the refuge area of native herbivores) and cleared fields (their new food resource). Establishment of improved pastures enhanced this food resource. This probably maximised the numbers of wombats, wallabies and the smaller rat-kangaroos. The greatest diversity of macropodoid marsupials (10 species) presently known is at Wallaby Creek in the catchment of the Clarence River in northern New South Wales (Calaby 1966). This area was patchily cleared about 130 years ago, and has remained as a "partly improved" beef cattle property surrounded on three sides by forest since that time.

In other areas clearing of the forest has continued, with a consequent reduction in the edge effect and removal of refuge areas. The result has been a decline in diversity of native herbivores, particularly the smaller wallabies and rat-kangaroos, as well as the omnivorous bandicoots. Complete removal of the forest has led to local extinction of some native herbivores in the same way that the loss of tall grass tussocks in the arid zone led to extinction of several small native herbivores there.

Arboreal folivores. This group of marsupials is dependent on forests not only for refuge but also for their food resource. Thus local abundance depends on the extent to which clearing for agriculture or forestry has led to complete removal of native forests. Some species such as the koala (*Phascolarctos cinereus*) and greater glider (*Petauroides volans*) are absolutely dependent on *Eucalyptus* spp. for food. Replacement of native mixed-species eucalypt forests with exotic pines (*Pinus* spp.) or single-species eucalypt plantations is as effective as clear-felling for wood chips in removing the habitat of these folivores.

Populations of koalas were reduced, often to the extent of local extinction, earlier this century by unregulated harvesting for their skins (furs). Total protection since 1974 has resulted in a gradual recovery of their numbers. This is being assisted by translocation of excess animals from locally overpopulated areas to other areas which appear to be appropriate habitats but from which the species is absent, possibly as a long-term effect of earlier local extinction.

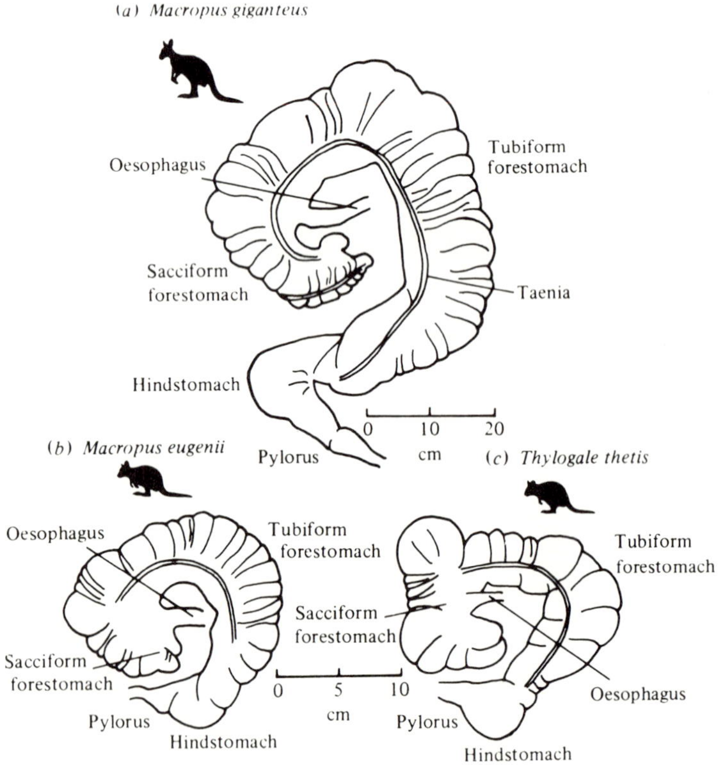

Figure 9. Stomach of three macropod marsupials (Hume 1982). (a) Eastern grey kangaroo (*Macropus giganteus*), (b) Tammar wallaby (*M. eugenii*), (c) Red-necked pademelon (*Thylogale thetis*).

Nutritional adaptations of herbivorous marsupials

Study of the digestive physiology of Australian marsupial herbivores has revealed some remarkable examples of parallel evolution with herbivores from other continents. For example, like ruminants the macropodid marsupials (kangaroos and wallabies) are foregut fermenters. However, in contrast to ruminants the macropod stomach is basically a tubular "colon-like" organ (Fig. 9) rather than a large diverticulum like the rumen (Hume 1982). As a result of the tubular nature of digesta flow through the macropod stomach the more readily fermentable components of the diet disappear early, and there is thus a progressive change in the nature of the substrate fermented along the length of the forestomach (Fig. 10). However, despite the differences in the pattern of digesta flow and microbial fermentation between the two foregut fermenters, the efficiency of the fermentation in terms of volatile fatty acid production and microbial protein synthesis per unit of digestible organic matter is similar in kangaroos and ruminants (Dellow *et al.* 1983).

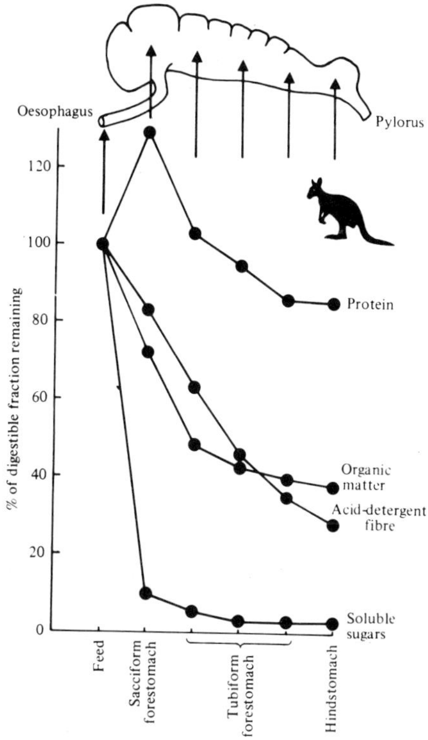

Figure 10. Disappearance of digestible components of chopped lucerne hay along the stomach of the eastern grey kangaroo (*Macropus giganteus*) (Dellow and Hume 1982).

A further consequence of the "colon-like" morphology of the macropod stomach is that digesta passage is faster and thus the extent of fibre digestion is usually less than in the ruminant. However, digesta passage rate in the kangaroos is also much less affected by the fibre level in the diet. Consequently the large kangaroos are able to maintain feed intake as the fibre level in drought-affected grasses increases better than can ruminants (Hume 1984). This is analagous to the response of eutherian equids, which are colon fermenters (Hume and Warner 1980), to highly fibrous feeds (Janis 1976).

The ability of large kangaroos to utilize highly fibrous feeds is pertinent to the success of the euro (*Macropus robustus erubescens*) in the Pilbara region of Western Australia where the more palatable and nutritious grasses have been replaced by the spiny grass *Triodia pungens* (spinifex) in response to continuous overgrazing by sheep (Newsome 1975). Spinifex is now often the principal component of the euro's diet. The capacity to maintain intake of such a fibrous grass, together with a low maintenance nitrogen requirement (Brown and Main 1967), helps to explain why the euro can survive on a diet dominated by spinifex.

The low maintenance nitrogen requirement of the euro is shared by several other macropods, and is related to the generally low basal metabolic rate of macropod marsupials (about 30% below that of eutherian herbivores such as the sheep). Low basal metabolic rates appear to be linked to low turnover rates and therefore low requirements not only of protein but also of energy and water (Hume 1982). This is of great survival value to marsupial herbivores such as kangaroos, wallabies and wombats in the Australian arid zone.

In contrast, two forest wallabies, the red-necked pademelon (*Thylogale thetis*) and parma wallaby (*M. parma*) have been found to have maintenance nitrogen requirements similar to those of sheep (Hume 1986). These relatively high nitrogen requirements restrict the range of habitats which these small wallabies can exploit, but within the forested vegetational zones this is not a great problem.

In the forested zones there is an apparently abundant food resource available to four arboreal marsupials (the koala, greater glider, brushtail possum (*Trichosurus vulpecula*) and ringtail possum (*Pseudocheirus peregrinus*), in the form of *Eucalyptus* foliage. However this material is uniformly low in available protein and high in lignified fibre. It also contains appreciable levels of secondary compounds (allelochemicals) including essential oils, tannins and other phenolics (Hume 1982).

The four folivores mentioned above have been found to have several adaptations in common. These include low maintenance requirements for energy and thus low voluntary feed intakes (Hume *et al*. 1984). This minimises the intake of allelochemicals. Low feed intakes, coupled with a large hindgut (i.e. caecum and colon) (Fig. 11) also result in very slow digesta passage, which maximises fibre digestion

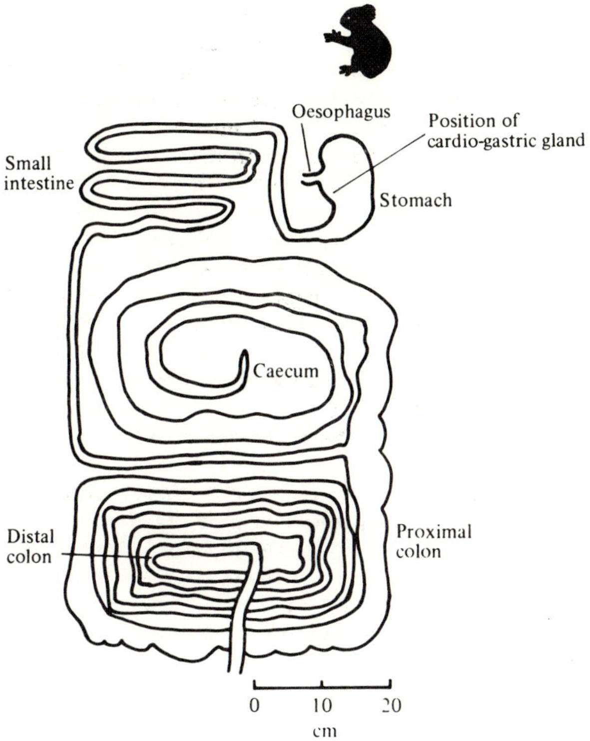

Figure 11. Digestive tract of the koala (*Phascolarctos cinereus*) (Hume 1982).

and the reversal of tannin-protein complexes in the gut. In three of the four species there is also selective retention of fluid digesta, along with fine particles, primarily in the caecum. This concentrates digestive effort on the potentially more fermentable fractions of the digesta, the coarse particles being eliminated relatively rapidly.

In the smallest species, the ringtail possum, selective digesta retention is coupled with caecotrophy, as it is in the unrelated eutherian rabbit. Caecotrophy is a form of coprophagy (ingestion of faeces); caecotrophy is the ingestion of soft faeces (caecotrophes) of high nutritive value derived from the caecum. Caecotrophy enables the ringtail to salvage microbial protein and B-vitamins synthesised in the caecum and which would otherwise be lost in the faeces. By this means the ringtail possum can survive on poor quality eucalypt leaves despite its high mass-specific nutrient requirements consequent upon its small body size (Chilcott and Hume 1985).

CONCLUSION

Despite their many adaptations to the often harsh Australian environment our native herbivores have evolved in the absence of competition from eutherian herbivores except for the last 150–200 years. Similarly the Australian vegetation has evolved in the absence of grazing by large ungulates. The impact of introduced herbivores, domestic and feral, on the vegetation has been dramatic, often resulting in its complete removal. The impact on many native herbivores has been devastating (Figs. 6, 7).

Domestic livestock are now affecting their own numbers. In the pastoral areas numbers have declined from a peak which lived on a climax vegetation with abundant reserves of forage. As rangelands became degraded stock numbers decreased. This effect was exacerbated by periodic droughts.

The challenge facing the Australian grazing industry now is to develop safe ways of utilizing our rangelands in the arid zone and the tropics so that present productivity can at least be maintained and hopefully improved, and soil and pasture degradation halted. To achieve this, the management of domestic stock must be geared to the production of forage. This means that properties must be very large and managed to ensure continual productivity of the range. Range must be destocked during drought to an extent determined by the condition of the range, and the frequency and extent of destocking during drought must increase with decreasing property size. If destocking is not followed, stock must be reduced to nucleus herds or flocks retained on well-watered "sacrifice areas" and fed for survival on imported feed (Newsome and Corbett 1977). There must also be control of feral herbivores, perhaps by extension of present commercial harvesting for export markets (e.g. camels, horses, goats).

In the forested and improved pasture zones the challenge is to improve the long-term economics of animal production and to retain forests in the face of increasing pressures from an expanding human population. In both cases the basic aim must be to increase habitat diversity. In the improved pasture zone the result is likely to include a halt to the loss of productive areas through salination resulting from overclearing and rising water tables, an improvement in pasture production through better soil water-plant relations, and reduced dependence on chemicals for the control of pasture and crop pests.

The conservation of native herbivores would be automatically enhanced by such ecologically-based approaches to long-term improvement of domestic herbivore production in both the arid zone and improved pasture zones.

Breckwoldt, R. (1986). "The Last Stand. Managing Australia's Remnant Forests and Woodlands." (Aust. Govt. Printing Service: Canberra.)

Brown, G.D., and Main, A.R. (1967). *Aust. J. Zool.* **15**, 7–27.

Calaby, J.H. (1966). Tech. Paper Div. Wildl. Survey, CSIRO Aust. **10**, 1–55.

Chilcott, M.J., and Hume, I.D. (1985). *Aust. J. Zool.* **33**, 1–15.

Dellow, D.W., and Hume, I.D. (1982). *Aust. J. Zool.* **30**, 767–77.

Dellow, D.W., Nolan, J.V., and Hume, I.D. (1983). *Aust. J. Zool.* **31**, 433–43.

Demment, M.W., and Van Soest, P.J. (1983). "Body Size, Digestive Capacity, and Feeding Strategies of Herbivores." (Winrock Int.: Morriton, Arkansas.)

Dudzinski, M.L., Lowe, W.A., Muller, W.J., and Low, B.S. (1982). *Aust. J. Ecol.* **7**, 69–74.

Fitzpatrick, E.A., and Nix, H.A. (1970). In "Australian Grasslands." (Ed. R.M. Moore.) pp. 3–26. (Aust. Nat. Univ.: Canberra.)

Frith, H.J. (1970). In "Australian Grasslands." (Ed. R.M. Moore.) pp. 74–83. (Aust. Nat. Univ.: Canberra.)

Hume, I.D. (1982). "Digestive Physiology and Nutrition of Marsupials." (Cambridge Univ.: Cambridge.)

Hume, I.D. (1984). *Proc. Nutr. Soc. Aust.* **9**, 76–81.

Hume, I.D. (1986). *Aust. J. Zool.* **34**, 147–55.

Hume, I.D., Foley, W.J., and Chilcott, M.J. (1984). In "Possums and Gliders." (Eds A.P. Smith and I.D. Hume.) pp. 247–51. (Aust. Mammal Soc.: Sydney.)

Hume, I.D., and Warner, A.C.I. (1980). In "Digestive Physiology and Metabolism in Ruminants." (Eds Y. Ruckebusch and P. Thivend.) pp. 665–84. (MTP: Lancaster, UK.)

Janis, C. (1976). *Evolution* **302**, 757–74.

Letts, G.A., Bassingthwaighte, A., and de Vos, W.E.L. (1979). "Feral Animals in the Northern Territory." (North. Terr. Govt.: Darwin.)

Lynch, J.J. and Donnelly, J.B. (1980). *Aust. J. Agric. Res.* **31**, 967–79.

Lynch, J.J., Elwin, R.L., and Mottershead, B.E. (1980). *Aust. J. Exp. Agric. Anim. Husb.* **20**, 170–4.

Mitchell, J., Merrell, P., and Allen, L. (1982). "Vertebrate Pests of Queensland." (Dept. Lands: Brisbane.)

Moore, R.M. (1970). (Ed.) "Australian Grasslands." (Aust. Nat. Univ.: Canberra.)

Newsome, A.E. (1975). *Quart. Rev. Biol.* **50**, 389–424.

Newsome, A.E., and Corbett, L.K. (1977). Proc. Second U.S./Aust. Rangeland Panel, Adelaide, pp. 331–56.

Noble, J.C., Cunningham, G.M., and Mulham, W.E. (1984). In "Management of Australia's Rangelands." (Eds G.N. Harrington, A.D. Wilson and M.D. Young.) pp. 171–86. (CSIRO: East Melbourne.)

Pavlov, P.M. (1980). M.Sc. Thesis, Monash Univ.: Melbourne.

Sinclair, R.G. (1977). *Aust. Wildl. Res.* **4**, 207–18.

Strahan, R. (1983). (Ed.) "Complete Book of Australian Mammals." (Aust. Museum: Sydney.)

Wadham, S.M., Wilson, R.K., and Wood, J.L. (1964). "Land Utilization in Australia." 4th ed. (Melbourne Univ.: Melbourne.)

2

HERBIVORES IN THE ARID AND WET TROPICS

by C. Devendra

The herbivores in the climatic extremes of the arid and wet tropics constitute very important animal genetic resources. Their importance relates to two particular issues. Firstly, there exist specific adaptational peculiarities to the climatic extremes that are coupled to various functional values. Secondly, their adaptation and ability to thrive in difficult environments is principally due to the interaction of components of the environment and the genetic potential of the animals. The magnitude of this expression is dependent on the severity or conduciveness of the environmental factors of which the climatic components (mainly temperature, radiation, humidity and rainfall) are especially important. Of the non-climatic factors, the availability of feeds is especially critical.

The arid and wet tropics are unique in that the regions present contrasting extremes of the effects of climatic components, and therefore differential responses in animals. These bioclimates range from tropical desert climates characterised by temperature extremes (0° to $52^{o}C$), insignificant rainfall and sparse vegetation, to the wet tropics with high temperature, humidity and rainfall (3000 to 5500 mm) and abundant vegetative cover. In both situations, herbivores play an important role in agriculture and render valuable service to man. These range from bedouins with their camels, sheep and goats practising shifting agriculture in the arid regions of northern Africa, to small farmers with water buffaloes (*Bubalus bubalis*) and cattle in complex crop-livestock systems in monsoonal Asia. In this region, meat, milk and draught are vital contributions by these species.

With each species, inherent biological characteristics peculiar to individual herbivores are apparent which enable adaptation to the environment. These peculiarities are extremely variable, but constitute important expressions of survival and function which are not adequately understood at the present time. A major constraint in this context is the sparse base of knowledge on comparative nutrition. If the biological characteristics are to be identified with economic animal production (mainly meat, milk and draught), much more research is needed on the type of response in these herbivores and also their potential performance in specific climates, when individual species are managed alone or in combination. For example, the efficient utilisation of rangelands by sheep, cattle, goats or native kangaroos in Australia is dependent on knowledge of various behavioural and production characteristics of these species (Wilson 1978). Thus, increased contribution from these species is consistent with more efficient

NUTRITION OF HERBIVORES
ISBN 0 12 750052 9

utilisation of the available production resources, increased resource allocation and the choice of species appropriate to specific situations.

The total number of species of ungulates is 210, of which 16 are odd-toed and 194 are even-toed (Yousef 1976). For reasons of brevity and space, this chapter will only focus on ruminants (water buffaloes, cattle, goats, sheep and camels) and equines (asses, horses and mules). This presentation aims to review and discuss current understanding of the mechanisms of adaptation (anatomy and physiology, heat loss, feeding behaviour and metabolism, water economy and turnover rate), as they affect the performance and the nutrient requirements of each species. In addition, this presentation will allude to the significance of these features with respect to the search for increased productivity and efficiency in the use of herbivores.

RUMINANT AND EQUINE RESOURCES

The population of ruminant and equine animal resources in developing countries and their contribution to food production is presented in Table 1. Almost all the water buffalo and camel populations, and 0.94 to 0.97 of the goat, mule and ass populations are found in the developing countries. Thus almost all water buffalo meat and milk is produced in developing countries. Goats in developing countries contributed about 0.93 of the world goat meat and about 0.55 of the world goat milk production. Unfortunately no statistics are available relating to the contribution of camels, mules and asses.

Table 1. The contribution of ruminant and equine animal resources in developing countries to world food production (FAO, 1984).

Species	Number (10^6)	Percentage of world Population	Meat	Milk
Water buffaloes	125.4	99.4	98.8	99.7
Cattle	844.4	66.4	27.0	16.1
Goats	432.7	94.1	92.8	54.8
Sheep	595.3	52.2	-	-
Camels	16.9	98.5	-	-
Horses	41.8	65.5	49.0	-
Mules	14.8	96.7	-	-
Asses	38.1	95.6	-	-

DISTRIBUTION OF RUMINANT ANIMALS

Table 2 sets out the distribution of water buffaloes, cattle, goats, sheep and camels in the arid/semi arid, sub-humid and humid zones. These together account for 0.63 of the total world populations of water buffalo, 0.47 of the cattle, 0.59 of the goats, 0.19 of the sheep and 0.79 of camels. Of these, the water buffaloes are exclusively found in the sub-humid and humid zones. In addition, the zones contains 0.87, 0.62 and 0.66 of the cattle, goat and sheep population respectively. By comparison, 0.89 of the camel population is found in the arid and semi-arid regions, with none in the humid zone.

Table 2. Distribution of ruminant animals in the arid and humid tropics.

Climatic zone	Distribution (%)				
	Water Buffaloes	Cattle	Goats	Sheep	Camels
Arid/Semi-arid	0	13	28	34	89
Sub-humid	85	77	62	63	11
Humid	15	10	10	3	0
Total population					
$x10^6$	78.8	543.0	270.7	213.1	13.6
% of world pop.	62.5	46.6	58.9	18.7	79.2

+ Adapted from UNESCO (1979) and FAO (1984).

VALUE TO MAN

Table 3 summarises the value of each species to man. Many species have primary, secondary and miscellaneous identifiable functions which need to be kept in perspective. Water buffaloes, cattle and goats, and to a lesser extent sheep are valued for meat and milk production. Goats are useful for fibre (mohair and pashmina (cashmere)) and sheep for wool. The equines are especially useful for draught. In all cases, the miscellaneous functions include value in dung production, recreation and culture.

ADAPTATION

Anatomy and morphology

Individual ruminants and equines have distinctive anatomical and morphological adaptational characteristics. While ruminants are found in both arid and humid climates, buffaloes (swamp and water types) are specific to the humid tropics and the one-humped camel (dromedary) to

Table 3. The value of ruminants and equines to man.

Species	Primary	Secondary	Miscellaneous
Milch buffaloes (river)	Milk	Draught	Dung, skin, recreational and cultural
Water buffaloes	Draught	Meat	Dung, skin, (swamp) recreational
Cattle	Meat/Milk	Draught	Dung, skin, recreational and cultural
Goats	Meat	Milk/Fibre+	Skins, hair, dung, recreation
Sheep	Mutton/ Wool	Milk	Skin, dung, recreational and cultural
Camels	Draught	Meat/Milk	Dung, hair
Asses	Draught		Dung
Horses	Draught	Meat	Dung
Mules	Draught		Dung

+Mohair, pashmina (cashmere) and coarse wool

the arid regions. Equines are widely distributed. Table 4 summarises the more important features of these species. In the semi-arid and arid regions, relatively taller animals are apparent, with loose skins and appendages. The larger size is an advantage in these environments because of smaller relative surface area and this advantage is maximal where there is much exposure to direct radiant heat. Goats and camels are particularly adapted to this environment which is characterized by very high temperatures, very low rainfall and sandy soils. The situation is well illustrated by larger (30 - 50kg live weight) and taller (70 - 100 cm height at withers) goats found here, compared to smaller (10 - 25 kg live weight) and shorter (50 - 65 cm height at withers) goats found in the humid regions.

Goats in arid regions absorb low amounts of solar radiation, can withstand dehydration and show greater capacity for survival in this environment than do sheep.

The fat tail of some sheep breeds such as the Awassi in Israel, and the Barki, Ossimi and Rahmani in Egypt is an important adaptive mechanism of significance in enabling grazing for long distances in the dry season. The storage of fat in the tail provides an important source of energy during periods of feed scarcity such as droughts. Fat tail breeds thus assume greater significance in arid and semi-arid regions.

Table 4. Adaptation of ruminants and equines.

Species	Arid/Semi-arid zone	Wet zone
Water buffaloes (*Bubalus bubalis*)		Larger size (400–600kg) Stocky Low hair density Limited sweat glands (140–150/cm^2) Light to dark colour Grey skin to reduce solar and radiant heat Very indented epidermis Strong pigmentation
Cattle (*Bos indicus*)	Taller Large size (350–450kg) Loose skin and appendages Able to walk long distances	Shorter, compact body Medium size (300–500kg) Thicker skin Higher hair density Sparse pigmentation More sweat glands (800–1500/cm^2)
Goats (*Capra hircus*)	Large size (30–50kg) Tall (70–100cm height) Long legs and ears Soft spongy hooves White, black or brown colour, shiny skin Short hair covering Prehensile lips Fat reserves less than sheep (mainly visceral) Survivability greater than sheep	Small size Short (50–65cm height) Short legs and small ears Mainly black or brown coat Less shiny skin
Sheep (*Ovis aries*)	Large size (30–50kg) Tall (60–80cm height at withers) Long legs Thick hooves Cream to brown colour Fibre covering High fat reserves	Small size (10–25kg) Short legs and small ears Coarse wool more matted
Camels (*Camelus dromedarius*)	Long legs Long neck Distinct single hump Split upper lip Nostrils almost closed	
Equine (*Equus asinus*)	Generally small in size Relatively smaller stomach	Smaller

Camels in the arid regions are distinctively one-humped, which together with the long neck and legs enable adaptation to a very harsh environment. In addition, the camel has long eye lashes and nostrils that are almost closed to protect it against the sand in the desert.

In the humid tropics, water buffaloes are particularly well adapted to take advantage of the relatively lower temperatures, high rainfall and high humidity. The skin thickness is about two to three times greater than that of cattle, is indented, heavily pigmented and provides protection from radiation. Fischer (1975) has reported that the epidermis of the buffalo hide is much stronger than that of cattle, and that the papillary bodies are strongly developed. An adaptational factor in water buffaloes is the relatively small number of sweat glands (140-150 /cm^2 *vs* 800-1500 /cm^2 in cattle). These glands show a tendency to atrophy which further limits the water buffaloes' capacity to loose heat (Prusty 1973); it is not clear when the glands atrophy but this characteristic contributes to their relatively low heat tolerance compared with cattle. When exposed to direct sunlight, water buffalo significantly increase their rectal temperature, respiratory rate and heart rate (Moran 1973; Chikamune 1983; Loypetjra *et al.* 1984). Nevertheless, water buffaloes are capable of activity in open sunlight and in situations where there are no wallows (Shute 1966). In these situations they attempt to dig holes to remain cool. Their capacity to work in open sunlight is probably related to a decline in metabolic heat production during prolonged exposure to heat (Rao *et al.* 1979).

Recently, it has been reported that zebu cattle (*Bos indicus*) in Bangladesh fed untreated rice straw had significantly greater rumen and total gut contents than those given urea-treated rice straw. Expressed as a percentage of live weight, both groups had gut contents and tissues far greater than European cattle (*Bos taurus*) (Mould *et al.* 1982). This could be genetic or due to adaptation, but is an advantage to these animals in utilising coarse roughages.

In the humid tropics and notably in West Africa, dwarf goats of the Fouta Djallon type are common. The precise reason for dwarfism is not known, but it has been suggested that this is an adaptation to poor nutrition and a low metabolic rate (Devendra and Burns 1983).

Heat loss

The avenues of heat loss in ruminants and equines are the same as in all mammals and include non-evaporative and evaporative mechanisms (Table 5). There are species differences in the pattern of panting. Some species pant through the nose and others pant with an open mouth and protruding tongue. Panting, sweating and cooling are common in cattle, goats and sheep in both arid/semi-arid and the humid/sub-humid regions to varying degrees. With camels and equines, sweating is the most common.

The camel, unlike most mammals, exhibits fluctuations in body temperature especially during periods of water restriction. When

Table 5. Heat loss in ruminants and equines

Species	Arid/Semi-arid zone	Wet zone
Water Buffaloes	–	Sweating Wallowing Salivating
Cattle	Panting Sweating Cooling	Reduced panting Sweating Cooling
Goats	Panting Sweating Cooling	Reduced panting Sweating Cooling
Sheep	Panting Sweating Cooling	Reduced panting Sweating Cooling
Camels	Sweating	–
Equines[+]	Sweating	Reduced sweating

[+] Asses, horses and mules.

ambient temperature is low, the camel's body temperature is low (34°C) whereas in the afternoon it is *ca* 41°C. These changes are beneficial to the camel in that, firstly, when the animal's body temperature is low, less heat is emitted by the body by radiation, and secondly, that no water is expended for evaporative cooling (Yagil 1986).

Cutaneous evaporative cooling occurs by diffusion of vapour through the skin or by sweating. This depends primarily on simple diffusion of water. Under heat stress conditions, evaporative cooling depends on sweating. Three patterns of sweating are identified: cyclical discharge such as in the ass (Robertshaw and Taylor 1969), continuous but stepwise rise in discharge and continuous but immediate rise in discharge such as in the water buffalo (Robertshaw and Taylor 1969).

Water buffaloes loose heat by sweating and salivation. Covering of the body trunks with white cloth has been shown to decrease the sweating rate (Chikamune and Shimizu 1985). However, heat loss is mainly achieved by wallowing in water and in mud. Studies in the dry zone of Sri Lanka during the hottest part of the year (25.8 to 31°C) with Lanka swamp buffaloes, Indian Surti and Murrah milch buffaloes indicated that wallowing occurred 7.5 times daily over a duration of 12.9 minutes each time. Rectal temperature decreased an average by 1.8°C and it was also calculated that the daily heat lost was 137.2 kJ /kg $^{0.75}$ (Ranawana *et al.* 1985).

Brout *et al.* (1979) reported that panting was the main avenue of heat loss in Black Bedouin goats in climatic chambers, but under field conditions, heat loss by sweating was twice as great as that lost in the respiratory system, due to solar radiation. Isabirye (1972) however, did not detect any cutaneous moisture loss in white goats in equatorial Kenya. Dmi'el *et al.* (1979) demonstrated that the difference was due to the sweat rate in the Black Bedouin goats which was 5 to 10 times more than in bovine species. This high sweat rate was caused by high density of sweat glands and mean secretion rate of the glands which was 4 to 5 times that of bovine species.

The horns of goats probably have a thermo-regulatory function (Taylor 1966). The blood vessels of the horns dilate in response to heat stress and there is a linear relationship between ambient temperature and heat loss per unit area of horn surface. It has been suggested that this may lead to selective control of the temperature of the brain by the arteries and capillary meshes.

Feeding behaviour and metabolism

There are considerable differences in the feeding behaviour and metabolism of ruminants and equines (Table 6).

In the arid and semi-arid region, animals generally walk long distances in search of feed and water, much more than in the humid and sub-humid region where animals are more sedentary. The longest distances walked are by camels, with progressively lesser distances by goats, sheep and cattle. This pattern is also consistent with the fact that goats and camels are very much better adapted to heat stress, can withstand dehydration and survive on very low planes of nutrition, utilising desert shrubs and coarse roughages. Their browsing habits are facilitated by the possession of split upper lips and, in the case of camels, long necks.

Goats are browsers and have a bi-pedal stance (Malechek and Provenza 1981). They have a competitive advantage over sheep in woodland and shrubland, are generally more active, selective, walk longer distances in search of feed and relish variety in feeds (Devendra 1987). Thus they are natural leaders of mixed goat and sheep flocks in many developing countries. Sheep, by comparison, are less selective and essentially grazing animals. Another feature of the feeding behaviour of goats is their discerning ability to taste. Goats can distinguish between bitter, sweet, salty and sour tastes, and show a higher tolerance for bitter taste than do sheep and cattle (Bell 1959; Goatcher and Church 1970). Additionally, some desert goats such as the Egyptian Zaraiby are known to have lower resting metabolic rate than would be predicted from the known equations relating body weight, metabolic rate and surface area (Salem *et al.* 1982). Such an adaptation is advantageous to life in arid zones where water and feed are at a premium.

Table 6. Feeding behaviour and metabolism of ruminants and equines

Species	Arid/Semi-arid zone	Wet zone
Water buffaloes		Grazer (3–5 km/day) Slower rumen movements Use coarse roughages more efficiently Superior N retention
Cattle	Grazer (3–5 km/day) More active Able to walk long distances Dietary N less well digested	Utilises forages efficiently With coarse roughages, longer retention time in *Bos indicus* cattle
Goats	Browser (8–12 km/day) Bi-pedal stance More active Selective feeding More discerning taste Relishes variety in feeds Utilises coarse roughages more efficiently Recycling of urea greater	Daily consumption of dry matter up to 6% BW
Sheep	Grazer (3–5 km/day) More sedentary Less discerning taste Recycling of urea less Use coarse roughages less efficiently	Preference is for grass Daily consumption of dry matter up to 3% BW
Camels	Grazer and browser (15–25 km/day) High digestive efficiency for cellulose Less selective Higher efficiency of N utilisation Very tolerant to salt Feeding and digestion unimpeded by water restriction	
Asses Horses Mules	Grazer Lower digestive efficiency Digestion mainly in the colon No rumination or eructation Faster rate of passage	

Cattle, like sheep, are essentially grazing animals and thrive on a variety of grasses, forages or agro-industrial by-products. Their digestive efficiency is comparable to that of sheep and hence sheep are often used to evaluate the nutritive value of feeds. By comparison, goats and water buffaloes appear to have the ability to utilise coarse fibrous feeds more efficiently than do cattle or sheep. Precise reasons for this are not clear, but are associated with feed particle size, amount of salivary secretion, rumination, concentration of cellulose splitting micro-organisms, fermentation rate, absorptive capacity, water turnover, recycling of urea, rate of passage and retention time (Devendra and Burns 1983). Differences in saliva secretion between goats and sheep are evident when given the same feed, with production rates of 848 ml and 502 ml /day respectively (Seth *et al.* 1976). Differences in saliva secretion are likely to influence urea recycling, an important factor in fibre digestion. There appear to be important breed differences in urea recycling; Black Bedoiun goats recycled 0.18 g N-urea /kg liveweight which is twice that produced by Saanen goats fed the same diet (Silanikove *et al.* 1980).

Recently, attention has been drawn to the higher rumen NH_3-N in goats which has been attributed to greater rumen protein degradation as a result of a longer retention time of digesta in the rumen (Watson and Norton 1982). Since goats drink less water than sheep per unit dry matter intake (Gihad 1976; Gihad *et al.* 1980; Owen and Ndosa 1982; and Alam *et al.* 1983), it has further been suggested that the lower water intake may be the cause of the higher rumen NH_3-N concentration (Alam *et al.* 1984).

Water buffaloes have certain metabolic features that are characteristic of the species. They have an ability to utilise low quality forages more efficiently, but evidence for an inherent species ability remains to be established (Devendra 1983). This ability is probably associated with slower rumen movements (Bhattacharya and Mullick 1965) and the longer time that food remains in the gastro-intestinal tract (Ponappa *et al.* 1971; Kumar and Raghavan 1974).

Water buffaloes consistently exhibit a greater ability to utilise nitrogen than cattle. On low quality roughages, water buffaloes have been found to have a higher plasma urea concentration and therefore a greater absorption of urea than zebu cattle (Norton *et al.* 1979). Both the swamp buffalo (Devendra 1983; 1985) and the Murrah buffalo (Mohan and Raghavan 1974) have been shown to have higher nitrogen retention than do zebu cattle, resulting from a lower urinary nitrogen output and efficient glomerular filtration (Moran 1983; Devendra 1985; Table 7). Buffaloes have also been reported to have a higher concentration of bacterial nitrogen than zebu cattle (Singh *et al.* 1986; Mohan and Raghavan 1974; Reddy and Raghavan 1976) and total nitrogen content (Sharma and Mudgal 1975; Sharma and Mudgal 1979; Aranas and Lopez 1976) of rumen contents. Dry matter degradabilities of different sources of protein fed together with untreated rice straw

Table 7. Mean nitrogen utilisation of swamp buffaloes and Kedah-Kelantan cattle when fed nine different diets containing 60-220g CP and 10.48 MJ ME /kg DM (Devendra 1985).

Parameter	Swamp Buffalo[+]	Kedah-Kelantan Cattle[+]
Total N intake (g /day)	128.8	89.5 (109.9)[++]
Total N intake (g /$W^{0.75}$kg /day)	1.24	1.18
Total N in faeces (g /day)	50.1	34.1 (42.3)
Total N in urine (g /day)	27.4	35.4 (53.5)
Total N output (g /day)	77.5	69.5 (95.8)
Total N balance (g /day)	52.5	19.8 (14.1)
N balance as % N intake (%)	38.0	12.5 (12.9)
Total apparently digestible N (g /day)	78.7	55.5 (67.6)
Apparent digestibility of N (%)	59.0	62.0 (61.5)
True N digestibility (%)	88.5	89.3

[+] Nine animals per dietary treatment, all fed *ad libitum*.
[++] Values adjusted to the same total N intake as swamp buffalo.

and urea-treated rice straw have been shown to give higher values for water buffaloes compared to cattle (Wanapat 1985).

The camel also possesses certain inherent features concerning feeding and metabolism. It has a lower resting metabolic rate per unit weight than other ungulates (Yousef 1976), which is of adaptive significance since this enables the camel to require less food and water. There is a preference and adaptation to high salt diets which appears to ensure that the stomach and intestines are rich in electrolytes (Maloiy and Clemens 1980). The electrolytes and water are slowly utilised even if there is restriction in water, unlike cattle or sheep which stop eating after three days without water (Farid *et al.* 1980). Maloiy (1972) has reported a greater rumeno-reticular outflow rate and a faster rate of passage compared to zebu cattle, probably due to a more rapid frequency of fore-stomach contraction and rumination cycle (Vallenas and Stevens 1973). Hintz *et al.* (1973) have reported higher overall digestive efficiency of camels compared to sheep and horses. The camel, like the buffalo, appears to utilise nitrogen more efficiently than do cattle (Farid *et al.* 1985). This is reflected in low blood urea-nitrogen values of 0.4 - 0.8 mg /kg compared to 0.6 - 2.7 /kg for cattle and 0.8 - 2.0 /kg for sheep (Bhatia 1986). The ability to utilise N efficiently became more significant when water was restricted in comparative studies between sheep and cattle (Emmanuel 1980). This feature is similar to that in goats (Alam *et al.* 1984).

Asses appear to require less feed than do horses (Yousef 1985). Intermittent heat loads have no effect on feed consumption and slightly increase dry mater digestibility (Maloiy 1970). However, a 0.15 dehydration level and simulated desert conditions (22 to 40°C) depressed feed intake but increased the apparent digestibility of dry matter. When the animals lost 0.20 to 0.22 of body weight, they stopped eating. Asses deprived of feed and water at temperatures of 35 to 45°C selected hay first and then water when the increase in plasma osmolality was less than 0.10. At 0.17 increase, water was chosen first (Dill *et al.* 1980). It was concluded that the ability of this species to utilise dry hay after long periods of dehydration is essential for their survival where feed and water are constantly limiting.

Horses neither ruminate nor eructate. Because of the small stomach size, the rate of passage of digesta from it to the large intestine is rapid. The feed is exposed to enzyme digestion in the small intestine prior to fermentation rather than after fermentation. Also, microbial cell bodies escape enzymic digestion since they are found posterior to the small intestine rather than anterior to it (Robinson 1974). The digestive function in horses appears to be more efficient when the horses are exercised than when they are penned (Orton *et al.* 1984), but this difference is not evident in cattle (Lawrence 1985).

The effect of high temperature on the ruminant digestive system is not well documented. Attebery and Johnson (1969) reported that the frequency of rumen contraction in cows was markedly reduced after five days of exposure to a temperature of 38°C compared to that recorded at 18°C; the response was opposite to that due to low temperature. Rumen contractions of sheep maintained at temperatures between O and 5°C or 20° showed a 0.21 increased frequency in favour of the latter (Westra and Christopherson 1976); similar results were obtained with cattle (Christopherson and Kennedy 1985). Rumination activity is also affected by environmental temperature. Appleman and Delouche (1958) showed that the rate of rumination chewing in goats increased by a factor of 0.22 during exposure to 0°C compared with activity at 20 to 30°C. The effect is likely to be conspicuous where temperature changes are marked. In continuously high temperatures, it is unlikely that there will be a limiting influence on rumination.

High temperature also results in longer retention times and increased digestibilities. Warren *et al.* (1974) showed that retention times of steers fed forage diets increased from 36.2 to 43.2 h when the environmental temperature increased from 18 to 32°C with associated increased digestibility of dry matter, cellulose, acid detergent fibre (ADF) and neutral detergent fibre (NDF) of the diet. It is not clear if thermal stress affects some or all parts of the digestive tract, but it is likely that physical form of the diet is implicated in the rate of passage of digesta. The exposure to heat has been shown to result in increased rumen ammonia concentration (Mishra *et al.* 1970) with the likelihood of reduced flow of dietary nitrogen to the intestines. Since nitrogen is often the main limiting factor in the diet of ruminants in many parts of the tropics (Devendra 1986), a reduced availability due to excessive degradation in the rumen will necessitate strategic supplementation.

High fibre diets are common for ruminants in the climatic extremes and although water buffaloes and goats appear to be able to utilise these more efficiently, the deleterious effects of high temperatures on the animals remain. Tsai *et al.* (1967), found that Red Sindhi cows in the Philippines fed high roughage diets (crude fibre content of 185 g /kg) gave significantly reduced) milk yields. These changes were associated with significant increases in acetic acid production, rectal temperature, pulse rate and respiration rate. Thus, there needs to be a balance between adequate dietary fibre and other nutrients that can ensure production. Where this is not possible, such as in the extensive production systems in arid and semi-arid environments, strategic supplementary feeding is essential.

WATER ECONOMY AND TURNOVER RATE

Water economy and turnover rate are important indices of adaptation to various environments by a species and even breeds within a species. Water turnover is influenced mainly by species, age, environmental temperature, feed supply and lactation. The water and energy economies in the animal body are closely associated (Macfarlane and Howard 1972).

In semi-arid and arid climates, water supply is minimal and dehydration is inevitable. Water is conserved by ability to resist dehydration and by producing concentrated urine, dry faeces and reducing evaporative water loss. Goats, camels and asses show tolerance of dehydration up to 0.20 of their body weight and this is associated with their ability to maintain constant plasma volume and hence maintain circulation. Yousef and Johnson (1985) have calculated that the limits to tolerance of dehydration and survival time for camels and asses without water were 23 and 9 days respectively. Such information is of considerable practical significance in the management of these species.

Table 8 summarises the salient features of the water economy and turnover rates of ruminants and equines. Camels have a very low water turnover rate and this enables them to retain their physiological integrity even after 50 days without drinking water (Hassan 1971). By comparison, water buffaloes in the humid tropics have a high water turnover rate. Like camels, goats have an efficient mechanism to conserve water through resistance to dehydration and reduced water loss in faeces and urine. In both species the urine is concentrated. In the camel this ability enables it to eat saltbushes and to drink brackish water (Yagil and Berlyne 1978). Bedouin goats can replenish all their water loss in a few minutes by consuming up to 40% of their liveweight (Maltz and Shkolnik 1980). Camels can rehydrate rapidly following losses exceeding one quarter of liveweight (Yagil and Etzion 1986).

Table 8. Water economy and turnover rate of ruminants and equines.

Species	Arid/Semi-arid zone	Wet zone
Water buffaloes		High water turnover rate Low urine concentration
Cattle	Water turnover rate higher than buffaloes Normal urine concentration Evaporative water loss comparable to sheep	Evaporative water loss less due to humidity
Goats	Resists dehydration Efficient water economy Reduced water loss in faeces and urine More concentrated urine Evaporative water loss less	Water turnover rate less
Sheep	Less efficient water economy Water loss in faeces, and urine greater Less concentrated urine[+] Evaporative water loss greater	Water turnover rate economy
Camels	Efficient water economy Very low water turnover rate Very concentrated urine	
Asses Horses Mules	Intermediate rates of water turnover rate Normal renal concentrating ability	

[+] With the exception of the Merino breed

Awassi sheep and the German Mutton Merino sheep breeds have been shown to reabsorb water from the alimentary tract at about the same rate. However, the Awassi excreted more concentrated urine and retained 0.84 of their plasma volume compared to 0.75 in the Merino. It was also reported that the Awassi retained 0.97 of the

Table 9. Water turnover of adult ruminants and equines in the arid and humid environments.

Species and breed	Body Weight (kg)	Temperature (^{o}C)	Water turnover (ml/kg$^{0.82}$ day)	Reference
Banteng	72	29	367	Siebert & Macfarlane (1969)
Boran	425	21	260	Macfarlane & Howard (1970)
Brahman	532	29	382	Siebert & Macfarlane (1969)
Santa Gertrudis	523	29	385	Siebert & Macfarlane (1969)
Shorthorn	332	25	286	Macfarlane & Howard (1970)
	278	23	315	Siebert & Macfarlane (1969)
Water buffalo	354	29	574	Siebert & Macfarlane (1969)
Water buffalo				
(shade)	234	23–32	418	Dollah *et al.* (1986)
(pasture)	237	25–34	918	Dollah *et al.* (1986)
Javanese goat	16	25	212	Terrill (1968)
Katjang crossbred goat	15	24	189	Tuen *et al.* (1984)
Somali goat	22	25	193	Terrill (1968)
Awassi sheep	40	19–30	275	Degen (1977a)
Masai sheep	29	25	193	Terrill (1968)
Merino sheep	41	25	133	Macfarlane & Howard (1970)
Priangan sheep	20	25	272	Terrill (1968)
Camel	656	25	93	Macfarlane & Howard (1969)
Ass	236	20–27	184	Yousef (1971)

water consumed in a single drink compared with 0.64 in the Merino (Degen 1977a; 1977b).

Maloiy *et al.* (1971) have reviewed osmoregulation in large herbivores and classified them into groups adapted to three main biomes:

(i) Wet tropical or wet temperate region animals, with high rates of water and energy use and poor urine concentrating ability – water buffalo, cattle, pig, elephant, eland (*Taurotragus oryx*), waterbuck (*Cobus defassa*), horse, moose (*Alces malchis*) and reindeer (*Rangifer tarandus*).

(ii) Warm dry savanna and semi-arid zones with intermediate rates of water and energy use and good renal concentrating ability – sheep, wildebeest (*Connochaetes taurinus*), kongoni (*Alcephalus* sp.) and ass (*Equus asinus*).

(iii) Arid zone animals with low rates of energy and water turnover and with medium to high urine concentrating ability – camel, goats, oryx (*Oryx* spp.) and gazelles (*Gazella* spp.).

Table 9 presents a summary of water turnover rate of ruminants and equines in the arid and humid environments. The camel has the slowest rate while water buffaloes have the fastest. Breed differences are apparent but it is not possible to ascertain whether these are due to genotype, lactation, stage of pregnancy or other factors such as quantity and quality of the feeds. Lactating ewes (Degen 1977a), goats (Maltz and Shkolnik 1980) and camels (Macfarlane *et al.* 1963) have a faster water turnover rate than non-lactating animals. Similarly, pregnancy increased the water turnover rate in Awassi and German Mutton Merino sheep (Degen 1977a).

MILK PRODUCTION

Production of large quantities of milk is an important attribute of many herbivores, notably water buffaloes, cattle, goats, sheep and camels. Species differences are apparent in qualitative and quantitative terms. Table 10 provides a comparison between the different types of milks. Total solids, protein and ash contents of water buffalo, goat, sheep and camel milks are higher than are present in cow milk. Water buffalo milk has the highest content of total solids, solids-not-fat and fat. Milk production in goats compared to water buffaloes and cattle in the humid tropics has been reviewed (Devendra 1980).

Goat, cow and human milks are approximately isocaloric and supply about 3.10 MJ /kg of energy, but there are differences in the proportions of the energy derived from lactose and protein. In goat and cow milk, fat, protein and lactose account for about 0.50, 0.25 and 0.25 of the energy, but in human milk they furnish 0.55, 0.07 and 0.38 respectively. Calculations of the nutritional adequacy of goat milk for human infants (Jenness 1980) demonstrated that the supply of protein, calcium, phosphorus, vitamin A, thiamin, riboflavin, and pantothenate were in excess. On the other hand, it was deficient in iron and vitamins C, D and B_{12}. Goat's milk, like cow's milk, has a satisfactory balance of essential amino acids equalling or exceeding current requirements.

Three special attributes of goat milk are worthy of mention:

(a) The fat globules are small in size. While the range of size of the fat globules is the same as in cow milk (1 – 10 μm diameter), the content of smaller globules is greater (Fahmi *et al.* 1956). Fat globules up to a diameter of 4.5 μm represent 0.86 of those found in the milk in sheep, 0.83 in goats, 0.62 in cow and 0.41 in water buffalo milks.

(b) The fat and protein contents are more easily digestible. Tubercle bacillus is rare and there are also anti-allergy properties.

Table 10. Average composition (g /kg) of water buffalo, goat, cattle, sheep and camel milk in the tropics.

Constituent	Murrah buffalo	Zebu cattle	Goat	Sheep	Camel
Total solids	187	135	139	182	137
Fat	65	48	48	64	45
Protein	37	28	37	55	32
Solids-not-fat	97	81	91	118	92
Ash	8.6	7.4	8.5	9.2	7.8
Lactose	48	46	50	46	49

Goat milk can replace cow milk for those allergic to the latter.

(c) The vitamin A is readily available in the absence of caretenoid pigments and the precursors.

The Black Bedouin goats have been reported to produce 0.8–0.9 of their body weight in milk per lactation which was associated with the water turnover rate being twice as high as in non-lactating goats. This was achieved by the goats increasing their total body water by 0.35 over the non-lactating period. During dehydration, milk yield was only affected 48 hrs post-dehydration for which reason these goats are only watered every 2 to 4 days (Maltz and Shkolnik 1980). During dehydration, plasma volume decreased in proportion to total body fluids indicating ability to balance both gain and loss of body fluids. Repeated periods of water deprivation had no effect on the Bedouin goat's ability to resume full milk production, which within three days of rehydration returned to normal.

Diusembin and Rahimberdiew (1975) give fat globule size of Bactrian camel as 0.18–0.34 µm. Nutritionally, the high concentration of small fat globules is an advantage, but for butter making, the milk fat of goat milk is more difficult to separate than that of cow or water buffalo milk.

When compared to water buffaloes, Sahiwal cattle and Friesian x Sahiwal crossbred cattle, camels excel in total milk yield, average daily milk yield, total fat and protein produced and total proteins as percentage of liveweight. When expressed in terms of liveweight, Sahiwal cows and the crossbreds produced more total milk fat and total milk solids and calories per 100 kg liveweight (Table 11). It has been reported that the daily protein requirements of an adult man are contained in 1.8 kg of camel milk and 2.5 kg of the milk can also provide the calorie requirements (Dahl and Hjort 1980). The concentration of vitamin C in camel milk is up to 100 mg /kg (Knoess 1980) and is three times that of cow's milk. This supply is significant in the arid zones where fruits, vegetables and therefore vitamin C are scarce.

Table 11. Milk yield and percentage of fat in the milk of different species in Pakistan.

Animal	Camels	Water Buffaloes	Sahiwal Cattle	Friesian x Sahiwal crosses
Liveweight(kg)	573.6	630.0	400.0	400.0
Daily milk yield				
(kg)	18.68	6.80	11.10	11.97
(% of liveweight)	3.26	1.08	2.78	2.99
Total yield				
(kg /365 d)	6,687.8			
(kg /305 d)	5,695.5	2,065.0	3,385.0	3,652.0
Fat content (g /kg)				
305 days	29.3	55.9	44.7	41.7
Total fat				
(kg /305 d)	167.1	115.5	151.3	152.3

NUTRIENT REQUIREMENTS

The nutrient requirements of individual species are quite different, varying in relation to the environment and more particularly function. Kearl (1982) has reviewed the available information concerning water buffaloes, cattle, goats and sheep in the developing countries. Water buffaloes appear to be more efficient in utilising the energy in low protein fibrous diets for work (Ffoulkes and Bamualim 1986). Lawrence (1985) has recently reviewed the nutrient requirements of draught cattle. Recommendations are available for the nutrient requirements of meat, milk and Angora goats (NRC 1981). The available information is in general inadequate and attention is drawn especially to the following major gaps in the literature:

(i) With water buffaloes, inadequate data exist for the nutrient requirements (energy, protein, minerals and vitamins) for growth, reproduction, lactation and draught capacity.

(ii) The requirements for water buffaloes need to address differences in the swamp (draught and meat), the Murrah (milk and draught) and the triple purpose animals (draught, meat and milk).

(iii) For goats, there is a paucity of information on nutrient requirements (energy, protein, minerals and vitamins) under extensive conditions. The requirements for fibre (pashmina) production also remain to be determined.

(iv) The nutrient requirements (energy, protein, minerals and vitamins) for draught capacity in cattle need to be determined.

(v) Within each species, the stresses of the environments, and especially that of the climatic components that affect nutrient requirements need to be considered.

The requirements for asses and mules appear to be similar to those published for horses (Morrison 1950; NRC 1966). Estimated energy requirements for horses and oxen of different weights working at various rates for 4, 6 and 8 hours have been reported (Goe and McDowell 1980). The same authors have reported that horses require more energy when travelling at a speed of 4 km /hr than at lower speeds, given selected working intensities.

Heat stress is likely to increase the nutrient requirements of horses by increasing the amount of nutrients lost in sweat. Thus the total excretion of calcium and phosphorus in polo horses was 80 - 145 mg Ca and 11 - 17 mg P after a 20 minute exercise (Schryver *et al.* 1978). Horses ridden on a hot day for 80 km had reduced plasma potassium and calcium values and it was concluded that intravenous administration of balanced electrolytes could improve performance (Lucke and Hall 1980). Temperature has a negligible effect on digestibility in horses (Hintz and Schryver 1985).

The nutrient requirements of camels have also been inadequately investigated. Limited research in Pakistan suggests that the nutrient consumption of lactating camels is approximately the same as *Bos taurus* cattle, and also meat goats, sheep and swamp buffaloes (Knoess *et al.* 1986). Lactating goats and buffaloes on the other hand have higher dry matter intakes as percentage of body weight (Devendra and Burns 1983).

HERBIVORE NUTRITION AND FUTURE IMPORTANCE

In this review of herbivore nutrition it is evident that information is increasingly sparse in the order cattle, sheep, water buffaloes, goats, camels and equines. With cattle and sheep, the scientific understanding of their value in general, and of nutrition in particular, is quite comprehensive. By comparison, the situation with regard to water buffaloes, goats and especially camels and equines within the climatic extremes of the arid and humid tropics is one of extreme paucity. The reason for this is that water buffaloes, goats and camels and also large populations of equines are mainly found in developing countries, where resource constraints have not yet enabled objective research and development efforts to demonstrate the potential value of these species. This trend needs to be corrected.

There is a need to study more fully the inherent biological attributes of herbivores and identify these with economic production in contrasting environments. Their value needs to be addressed very much more than it has been in the past and also the selection of species for individual situations. In this respect, there is a major challenge to increase our understanding of the significance of nutrition on productivity in individual species. Comparative nutrition provides a special motivation that should result in much more rapid progress than when one species is studied in isolation. This is apparent, for example, in the contrasting pattern of utilisation of energy and protein between cattle and water buffaloes for meat and milk production, and similarly, the utilisation of coarse fibrous feedstuffs by ruminants and equines for draught. Throughout the climatic extremes of the developing countries, the large ruminants, camels and equines are very important for draught; however, the nutritional requirements of draught animals remains largely unknown. These efforts merit high priority, and there is every justification for increased resources allocation and facilities for pragmatic research and development, especially on water buffaloes, goats, camels and equines, commensurate with their current contribution, potential future importance and search for efficiency in the use of available resources in the developing countries.

Alam, M.R., Poppi, D.P., and Sykes, A.R. (1983). *Proc. N.Z. Soc. Anim. Prod.* **43**, 119–21.

Alam, M.R., Borens, F., Poppi, D.P., and Sykes, A.R. (1984). In "Ruminant Physiology – Concepts and Consequences." (Eds S.K. Baker and D.B. Purser.) p. 184. (Univ. West. Aust.: Nedlands.)

Allen, T.E., and Bligh, J. (1961). *Comp. Biochem. Physiol.* **31**, 347–63.

Appleman, R.D., and Delouche, J.C. (1958). *J. Anim. Sci.* **17**, 326–35.

Aranas, T.J., and Lopez, O.L. (1976). *Phil. J. Vet. Anim. Sci.* **2**, 111–20.

Attebery, J.T., and Johnson, H.D. (1969). *J. Anim. Sci.* **29**, 734–8.

Bell, F.R. (1959). *J. Agric. Sci.* **52**, 125–8.

Bhatia, J.S. (1986). *Farm Animals* **1**, 15–29.

Bhattacharya, N.K., and Mullick, J.N. (1965). *Indian J. Exp. Biol.* **3**, 255–9.

Brout, A., Dmi'el, R., and Shkolnik, A. (1979). *Physiol. Zool.* **52**. 105–112.

Chikamune, T. (1983). In "Current Development and Problems in Swamp Buffalo Production." (Ed. H. Shimizu.) Preconf. Symp., Vth Wrld Conf. Anim. Prod., Tsukuba, Japan. pp. 107–27.

Chikamune, T., and Shimizu, H. (1985). *Buffalo J.* 1, 19–27.

Christopherson, R.J., and Kennedy, P.M. (1983). *Can. J. Anim. Sci.* **63**, 477–82.

Dahl, G., and Hjort, A. (1980). Proc. Workshop on Camels, International Foundation for Science, Stockholm, Sweden. pp. 215–40.

Degen, A.A. (1977a). *J. Agric. Sci.* **88**, 699–704.

Degen, A.A. (1977b). *Physiol. Zool.* **50**, 284–93.

Devendra, C. (1980). *J. Dairy Sci.*, **63**, 1755–67.

Devendra, C. (1983). In "Current Development and Problems in Swamp Buffalo Production." (Ed. H. Shimizu.) Proc. Preconf. Symp., Vth Wrld. Conf. Anim. Prod., Tsukuba, Japan. pp. 107–27.

Devendra, C. (1985). Proc. 3rd Asian-Australasian Anim. Sci. Congr., Seoul, Korea, Vol. **2**, 873–5.

Devendra, C. (1987). Proc. IVth Int. Conf. on Goats, Brasilia, Brazil (In press).
Devendra, C., and Burns, M. (1983). "Goat Production in the Tropics." (Commonwealth Agricultural Bureaux: Farnham Royal, UK.)
Dill, D.B., Yousef, M.K., Cox C.R., and Barton, R.G. (1980). *Physiol. Behav.* **24** 975-81.
Diusembian, Kh.D., and Rahimberdiev, S.A. (1975). *Trudy Inst. Fiziol. Akad. Nauk Kaz. SSR.* pp. 122-7.
Dmi'el, R., Robertshaw, D., and Choshniak, I. (1979). *Physiol. Zool.* **52**, 558-64.
Doolah, M.A., Marson, S., Kamaruddin, M., and Liang, J.B. (1986). Proc. Vth Int. Conf. on Livestock Prod. and Diseases in the Tropics. Serdang, Malaysia. pp. 118-20.
Emmanual, B. (1980). Proc. Workshop on Camels, International Foundation for Science, Stockholm, Sweden. pp. 322-46.
Fahmi, A.H., Sirry, I., and Safwat, A. (1956). *Indian J. Dairy Sci.* **9**, 80-5.
Farid, M.F.A., Shawket, S.M., and Abdul-Rahman, M.H.A. (1980). Proc. Workshop on Camels, International Foundation for Science, Stockholm, Sweden. pp. 214-40.
Farid, M.F.A., Sooud, A.O., and Hassan, N.I. (1985). Proc. 3rd Asian-Australasian Anim. Sci. Congr., Seoul, Korea, **2**, 781-3.
Ffoulkes, D., and Bamualim, A. (1986). *Buffalo Bull.* **5**, 3-5.
Fischer, H. (1975). *Anim. Res. Dev.* **2**, 118-30.
Food and Agricultural Organisation (1984). Production Yearbook. Vol. 38. (FAO: Rome.)
Gihad, E.A. (1976). *J. Anim. Sci.*, **43**, 879-83.
Gihad, E.A., El-Badawy, T.M., and Mehrez, A.Z. (1980). *J. Dairy Sci.* **63**, 1701-6.
Goatcher, W.D., and Church, D.C. (1970). *J. Anim. Sci.* **31**, 373-82.
Goe, M.R., and McDowell, R.E. (1980). Cornell Int. Agric. Mimeo No. 81. Suppl.
Hassan, Y.M. (1971). *Sudan J. Vet Sci. Anim. Husb.* **12**, 111-2.
Hintz, H.F., Schryer, H.F., and Halbert, M. (1973). *Anim. Prod.* **16**, 303-5.
Hintz, H.F., and Schryver, H.F. (1985). In "Stress Physiology in Livestock." Vol., 1 : Basic Principles. (Ed. M.K. Yousef.) pp. 181-7. (CRC: Boca Raton, Florida.)
Isabirye, J.S.J. (1972). M. Sc. Thesis, Univ. of Nairobi, Kenya.
Jenness, R. (1980). *J. Dairy Sci.* **63**, 1605-30.
Kearl, L.C. (1982). Int. Feedstuffs Institute, Utah State Univ., Logan, USA.
Knoess, K.H. (1980). *Comp. Biochem. Physiol.* **50**, 201-14.
Knoess, K.H., Makhudum, A.J., Rafiq, M., and Hafeez, M. (1986). *World. Anim. Rev.* **57**, 11-21.
Kumar, B.A., and Raghavan, G.V. (1974). *Indian J. Anim. Sci.* **44**, 953-8.
Lawrence, P. (1985). Proc. Draught Animal Power for Production, ACIAR Proc. Series **10**, 59-63.

Loypetjra, P., Chaiyaputi N., Usanakornkul, S., Pichaichanarong, and Djurdjeric, D. (1984). *Acta Vet.*, Belgrade **31**, 13–9.

Lucke, J.N., and Hall, G.N. (1980). *Equine Vet. J.* **12**, 189–94.

Macfarlane, W.V., Morris, R.J.H., and Howard. B. (1963). *Nature*, London **197**, 270–1.

Macfarlane, W.V., and Howard, B. (1972). *Symp. Zool. Soc. London* **31**, 261–96.

Malechek, J.C., and Provenza, F.D. (1981). Proc. Int. Symp. on Nutrition and Systems of Goat Feeding, Tours, France, **1**, pp. 411–28.

Maloiy, G.M.O. (1970). *Am. J. Physiol.* **219**, 5122–7.

Maloiy, G.M.O. (1972). *Symp. Zool. Soc. London* **31**, 243–59.

Maloiy, G.M.O., and Clemens, E.T. (1980). *Comp. Biochem. Physiol.* **67A**, 21–5.

Maloiy, G.M.O., Macfarlane, W.V., and Shkolnik, A. (1979). In "Comparative Physiology of Osmoregulation in Animals." **2**. pp. 185–209. (Academic: London.)

Maltz, E. and Shkolnik, A. (1980). *Physiol. Zool.* **53**, 12–8.

Mishra, M., Martz, F.A., Stanley, R.W., Johnson, H.D., Campbell, J.R., and Hilderbrand, E. (1970). *J. Anim. Sci.* **31**, 1023–7.

Mohan, K.D.V., and Raghavan, G.V. (1974). *Indian J. Anim. Sci.* **44**, 510–9.

Moran, J.B. (1973). *Aust. J. Agric. Res.* **24**, 775–82.

Moran, J.B. (1983). *J. Agric. Sci.* **100**, 13–23.

Morrison, F.B. (1950) "Feeds and Feeding." (Morrison: Ithaca, New York.)

Mould, F.L., Saadullah, M., Haque, M., Davis, D., Dolberg, F., and Orskov, E.R. (1982). *Trop. Anim. Prod.* **7**, 174–81.

Norton, B.W., Moran, J.B., and Nolan, L.J.V. (1979). *Aust. J. Agric. Res.* **30**, 341–51.

NRC (1966). "Nutrient Requirements of Horses." (National Research Council: Washington DC, USA.)

NRC (1981). "Nutrient Requirements of Goats : Angora, Dairy, and Meat Goats in Temperate and Tropical Countries." (National Research Council: Washington DC, USA.)

Orton, R.K., Hume, I.D., and Leng, R.A. (1984) *Proc. Aust. Soc. Anim. Prod.* **15**, 729.

Owen, E., and Ndosa, J.E.M. (1982). Proc. 3rd Int. Conf. Goat Prod. and Disease, Tucson, Arizona, 362 (Abstr.)

Ponappa, C.G., Uddin, M.N., and Raghavan, G.V. (1971). *Indian J. Anim. Sci.* **41**, 1026–31.

Prusty, J.N. (1973). *Indian J. Anim. Hlth* **12**, 33–7.

Ranawana, S.S.E., Rajaratne, A.A.J., and Tilakaratne, N. (1985). Proc. 3rd Asian-Australasian Anim. Sci. Congr., Seoul, Korea, **1**, 894–6.

Rao, A.V.N., Sukla, D.C., Sarma, P.V., and Opadhyay, R.C. (1979). *Indian J. Anim. Res.* **13**, 113–4.

Reddy, J.T., and Raghavan, G.V. (1976). *Indian Vet. J.* **53**, 106–13.

Robertshaw, D., and Taylor, C.R. (1969). *J. Physiol.* (London) **203**, 135–53.

Robinson, D.W. (1974). Univ. of Nottingham Nutr. Conf. for Feed Manufacturers. pp. 185–99. (Butterworth : London.)

Salem, M.H., Yousef, El-Sherbiny, A.A., and Khalil, M.H. (1982). In "Animal Production in the Tropics." (Ed. M.K. Yousef.) pp. 148–57. (Praeger Scientific: New York.)

Schryver, H.F., Hintz, H.F., and Lowe, J.E. (1978). *Am. J. Vet. Res.* **39**, 245–8.

Seth, D., Dai, G.S., Yadav, P.C., and Pandey, M.D. (1976). *Indian J. Anim. Sci.* **46**, 660–3.

Sharma, D.D., and Mudgal, V.D. (1975). *Indian J. Anim. Sci.* **45**, 332–8.

Sharma, D.D., and Mudgal, V.D. (1979). *Indian J. Nutr. Diet.* **16**, 17–28.

Shute, D.J. (1966). DTA Report (Unpublished.) Univ. of the West Indies, Trinidad.

Siebert, B.D., and Macfarlane, W.V. (1969). *Aust. J. Agric. Res.* **20**, 613–22.

Silanikove, N., Tagri, H., and Shkolnik, A. (1980). *Comp. Biochem. Physiol.* **67A**, 215–8.

Singh, S., Langar, P.N., Sidhu, G.S., Kochar, A.S., and Bhatia, I.S. (1968). *Indian J. Vet. Sci.* **38**, 674–81.

Taylor, C.R. (1966). *Physiol. Zool*, **39**, 129–39.

Terrill, C.E. (1968). In "Adaptation of Domestic Animals." (Ed. E.S.E. Hafez.) (Lea and Febiger : Philadelphia.)

Tsai,Y.C., Castillo, L.S., Hardison, W.A., and Payne, W.J.A. (1967). *J. Dairy Sci.* **50**, 1126–31.

Tuen, A.A., Alimon, A.R. and Jelan, Z.A. (1984). Proc. 8th Conf. Malays. Soc. Anim. Prod., Serdang, Malaysia. pp. 168–72.

UNESCO (1979). *National Resources Res.* No. 16.

Vallenas, A., and Stevens, C.E. (1973). *Am. J. Physiol.* **220**, 275–82.

Wanapat, M. (1985). Proc. Microbial Utilization of Renewable Resources, Vol. **4**. pp. 475–8.

Warren, W.P., Martz, F.A., Asay. K.H., Hilderbrand, E.S., Payne, C.G., and Vogt, J.R. (1974). *J. Anim. Sci.* **39**, 93–8.

Watson, C., and Norton, B.W. (1982) *Proc. Aust. Soc. Anim. Prod.* **14**, 467–70.

Westra, R., and Christopherson, R.J. (1976). *Can. J. Anim. Sci.* **56**, 699–704.

Wilson, A.D. (1978). Proc. 1st Int. Rangeland Congr. pp. 444–6.

Yagil, R., and Berlyne, G.M. (1978). *Renal Physiol.* **1**, 104–12.

Yagil, R. (1986). *World Anim. Rev.* 57, 2–10.

Yagil, R., and Etzion, Z. (1987). In "Adaptive Physiology to Stressful Environments." (Eds S. Samueloff and M.K. Yousef.) (CRC: Boca Raton, Florida.)

Yousef, M.K. (1971). In "Tritium." (Eds A.A. Moghissi and M.W. Carter.) pp. 333–41. (Messenger Graphics : Las Vegas, Nevada.)

Yousef, M.K. (1976). In "Progress in Animal Biometeorology." Vol. 1. (Ed. H.D. Johnson.) pp. 109–22. (Swets and Zeitlinger : Amsterdam.)

Yousef, M.K. (1985). In "Stress Physiology in Livestock." Vol. 2. Ungulates (Ed. M.K. Yousef.) pp. 81–97. (CRC: Boca Raton, Florida.)

Yousef, M.K. and Johnson, H.D. (1985). In "Stress Physiology in Livestock." Vol. 1, pp. 189–201. (CRC: Boca Raton, Florida.)

3

THE AVAILABILITY OF HERBAGE

by B.R. Roberts

The primary constraints to the production of edible forage may be considered to be moisture and temperature as they are the basic determinants of the climatic regions of the world. The secondary constraints upon herbage production are soil and topography as they both have vital effects upon the nutrition and growth of plants in the various climatic regions.

The major part of herbivore production, both in the form of domestic animal production and in the sustenance of grazing wildlife, occurs in the middle latitudes of the world where rainfall and topography are conducive to a reasonable animal carrying capacity on either native or improved pastures. In particular, herbivores are grazed in conjunction with arable agriculture in Europe, eastern and southern Asia, eastern USA, south eastern South America, southern Africa and eastern Australia. Ranching and extensive herbivore production are practised in western USA, central Asia, in the inland south east of South America, central Africa and all but the eastern and south western regions of Australia. The more extreme climatic zones which result in biomes of desert, tundra, coniferous forest and rainforest are used only to a very limited extent for the production of herbivores for use by man.

THE DEVELOPMENT OF ANIMAL HUSBANDRY AND FODDER CROPPING

There is evidence that about a million years ago man occupied a range of biomes including woodlands, grasslands and arid steppes. The first indications of some form of regionally adapted agricultural technology appeared in the late Pleistocene, about 125,000 years before the present day (BP). In Asia and Australasia the hunting society was associated with fire at least 50,000 years ago. Fire was utilised to open the woodland cover to encourage grass growth, which, in turn, attracted wildlife. The relative lateness of the development of agriculture in Africa, about 4000 BP, may have been the result of animal and human disease (Desowitz 1980) and the abundance of food in the open woodlands (Jewell 1980). Two of the world's great religions, namely Hinduism (1200 BC) and Buddhism (550 BC) have always forbidden the killing of cattle and the felling of trees. The policy of "holy cows" over a period of at least 1200 years has had a significant effect on the animal stocking rate in many countries where these beliefs have dominated. As animals were more completely domesticated, there was

NUTRITION OF HERBIVORES
ISBN 0 12 750052 9

development of fodder crops and less dependence was placed on the seasonally variable natural pastures for animal feed supply. For 2,000 years Man has sought to improve grain and fodder crop production. In addition, in recent times, he has consciously applied various animal breeding programmes to improve production from pasture.

CLIMATIC CONSTRAINTS TO DRY MATTER PRODUCTION

In general terms, herbage production is limited by low rainfall or low temperature as influenced by latitude, altitude and topography. The amount of dry material produced in each biome of the globe largely determines the land use alternative selected by man in each agricultural region. However, while production of herbage dry matter is subject to climatic control, it may be significantly increased by improving soil conditions through inputs such as fertilizers and irrigation. These inputs are limited in their effect by the length of the growing season which is determined in turn by latitude, altitude, the effects of ocean currents and wind systems and proximity to the sea.

The biological productivity map of the world therefore reflects a direct correlation between production of dry matter and the combined effects of temperature and moisture availability. The world herbivore population is concentrated in those regions of the middle latitudes which are characterised by moderate temperatures, low altitude and high rainfall.

Table 1. Relationship between rainfall and production of herbage dry matter (DM) in subtropical Australia (Mott *et al.* 1985).

Vegetation type	Dominant grass	Annual rainfall (mm)	Annual DM production (g /m^2)
Monsoon tallgrass	*Themeda triandra*	750-1100	112-270
Tropical tallgrass	*Heteropogon contortus*	600-1460	320-420
Subtropical tallgrass	*Heteropogon contortus*	700-1150	180-350
Midgrass on clay soils	*Dichanthium* spp.	500-900	200-300
Tussock grassland	*Astrebla* spp.	380-1800	100-230
Acacia	*Enneapogon* spp. *Thyridolepis* spp.	180-500	70-120

In sub-tropical Australia, the general relationships between dry matter yields and rainfall across a range of native pastures (Table 1) have been described by Mott *et al.*(1985). Particularly in the lower rainfall regions, the variability of rainfall, both within and between years, is as important as average rainfall in influencing dry matter yield and animal production. This is important in Australia where the variability of annual rainfall is 10–30% greater than in comparable homoclimates in other countries (Leeper 1973).

Within climatic regions, important local differences in herbage production occur as a result of topographic differences. The effects of slope, aspect and altitude on dry matter production act through their effects on soil temperature and water runoff. The microclimatic effect of slope on soil temperature is similar to the temperature effect of moving towards the equator. For instance, a 5°C slope toward the south in the northern hemisphere results in a similar change in microclimate to moving to a site 402 km nearer the equator (Alter 1912). This is similar to the effect of altitude upon plant growth, a reduction in soil temperature of 1°C occurs for every 152 m increase in altitude. The direct relationship between the degree of slope and water runoff has been well established. Increased run-off is associated with increased annual soil loss (Capelin and Truong 1985; Table 2). The sequential series of effects of topography upon animal carrying capacity, and hence animal production, are shown in Figure 1.

Table 2. Relationship between the degree of slope and soil erosion (Capelin and Truong 1985) on cultivated soils contour-planted with pineapples near Brisbane, Australia.

Land Slope (%)	Annual soil loss (t /ha)
11	7.0
14	15.0
17	36.4

On a global scale, animal production is affected more by drought than by any other single factor. The recent dramatic and widespread effects of continental drought caused by marine temperature changes (El Niño) have resulted in stock losses of enormous socio-economic significance. Donaldson (1967) and Keep (1973) have demonstrated the varying degrees of drought tolerance displayed by a range of herbage plants. Of particular interest from an animal production viewpoint are studies of the relative drought mortalities of adapted plant and animal species. Scholes (1985) has reported that, of the wild mammal species in Southern Africa, omnivores such as the warthog (*Phacochoerus aethiopicus*) had highest drought mortality, less susceptible were grazers with high intake requirements like the Cape buffalo (*Syncerus caffer*) and the impala (*Aepyceros melampus*), with the efficient grazers, for instance, the zebra (*Equus burchelli*), the least affected. In addition, in those animals that survived, the severe drought led to almost total breeding failure, indicating that drought and the consequential critical shortage of herbage had a significant effect on reproduction.

EDAPHIC CONSTRAINTS TO DRY MATTER PRODUCTION

The general effects of soil conditions on natural grazing and animal feed production have been known since man first domesticated animals. Few definitive studies exist on the association between individual herbaceous species and measured edaphic attributes, although some grasses and legumes and their cultivars are associated with particular soil types. Soil affects plant growth and dry matter production through the influences of texture, structure, depth and fertility on plant nutrition and moisture availability. These effects on dry matter production are modified by slope and drainage, as both factors influence water runoff and soil saturation.

Drainage schemes and the provision of irrigation have increased dry matter production in many regions of the world. However, improvement of soil fertility has been the major factor responsible for raising animal carrying capacity through improved plant growth. The increases in agricultural production over the last decade in both India and China are noteworthy and have contributed significantly to the food self-sufficiency of these most populous countries (FAO 1982).

Soil type has a direct influence on dry matter production through mineral nutrition, especially when soil moisture is adequate. For example, in South America correction of several mineral deficiencies is required to maximise the dry matter production of the pasture legumes *Stylosanthes guianensis* and *Desmodium ovalifolium*. Deficiencies in phosphate and calcium occur in almost half the soils potentially suitable for intensive cattle production in South America. Both elements applied as single superphosphate have given significant yield increases in a range of fodder plants (Ritchey *et al.* 1980). Potash deficiency is even more widespread, while magnesium and sulphur levels are also low over much of the South American continent (McClung and Martens de Freitas 1959).

In some high rainfall zones, excessive leaching, especially on lateritic soils, is associated with levels of iron and aluminium which may be toxic to plants (Howeler 1973). Such toxicity depresses plant growth and carrying capacity and can generally be remedied by the application of lime (Cochrane 1980).

It has been shown (Roberts 1971) that both physiographic factors such as altitude, slope and aspect, together with soil characteristics such as texture, structure and pH, can influence the natural distribution of many of the more important pasture grasses. Analyses of grasses for protein and phosphorus (Roberts and Silcock 1982) exhibit important differences in the value of the herbage produced by different species under similar soil and climatic conditions. Studies in southern Africa (Roberts *et al.* 1975) reflect similar differences between species and other studies expose significant differences within species (e.g. Hacker *et al.* 1985).

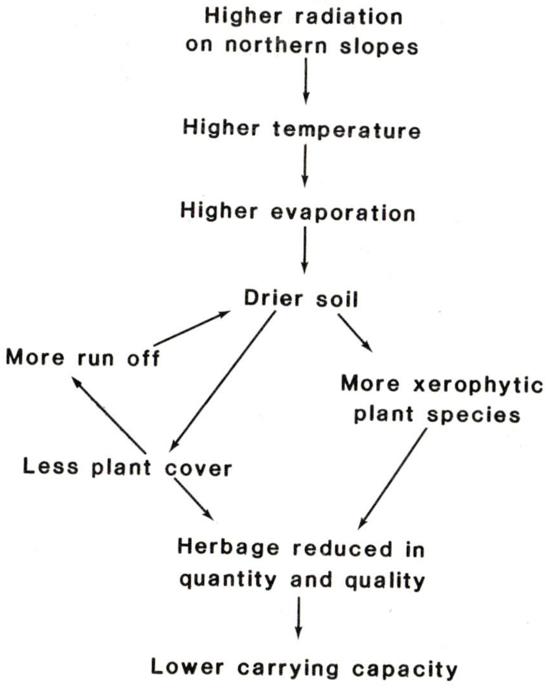

Fig. 1. Ecological processes affected by northern aspects in the southern hemisphere (Roberts 1975).

RELATIVE PRODUCTIVITY OF PLANT SPECIES

The evaluation of herbaceous species can be based on both yield and quality criteria and it has been shown that an understanding of species differences (Table 3) and the differential response of species to grazing, can form a useful basis for pasture and animal management decisions (Roberts *et al.* 1975). The productivity of the natural sward is thus a function of plant species, soil fertility, available moisture and length of the growing season.

From an animal nutrition viewpoint, given similar seasonal and maturity conditions, the most important predictors of dry matter production are plant species and density of basal cover (Roberts 1972). These criteria are widely used as the basis for reporting natural pasture condition and from such reports both condition and trend over time can be estimated (Roberts 1972).

Table 3. The digestibile organic matter (DOM) and crude protein (CP) contents (g /kg DM) of various grass species in the Northern Cape of South Africa (Roberts *et al*. 1975).

Herbage Species	Leafy Stage		Mature Stage	
	DOM	CP	DOM	CP
Anthephora pubescens	700	80	550	40
Cymbopogon plurinodis	560	50	410	30
Eragrostis lehmanniana	550	60	330	30
Eragrostis superba	670	60	480	40
Fingerhuthia africana	670	70	420	40
Heteropogon contortus	560	60	440	40
Panicum stapfianum	600	80	450	60
Schmidtia pappophoroides	650	70	460	40
Sporobolus fimbriatus	630	70	480	40
Stipagrostis uniplumis	470	50	340	30
Themeda triandra	570	50	440	30

Plant species is a predominant factor in determining both the quantity and quality of feed production. While this review is concerned with the availability of dry matter, it should be stressed that climate, soil, species and management have significant effects on both dry matter availability and on herbage quality. The relationship between yield and percentage digestible nutrients is thus of critical importance in evaluating how pasture management may achieve maximum yield of digestible nutrients as compared to maximizing dry matter production. The interrelationships between soil, climate and management and their effect upon carrying capacity are shown in Fig. 2 (Roberts 1973). A typical example from the subhumid tropics is given by Powell and Waters-Bayer (1985) who reported the relationships between dry matter, digestibility and protein content for the widely used fodders sorghum and millet (Table 4).

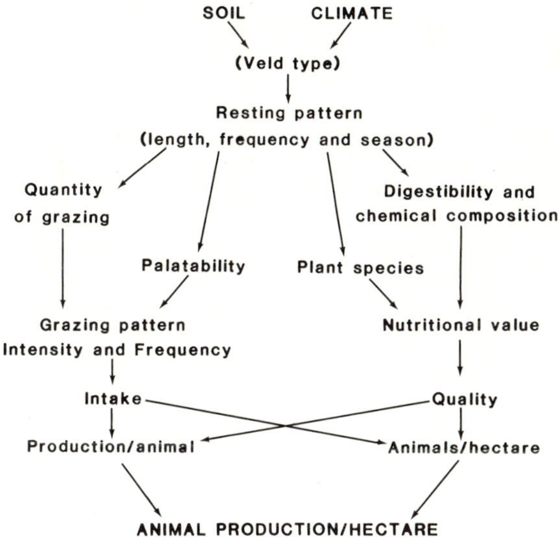

Fig. 2. Relationships between factors affecting carrying capacity (Roberts 1973).

Table 4. Relationship between amount of herbage and quality of sorghum and millet in Nigeria (Powell and Waters-Bayer 1985).

Herbage type	DM Yield kg /ha	Digestibility (g /kg DM)	Crude Protein (g /kg DM)
Native grasses	2400	350	20
Immature panicles:			
sorghum	40	500	70
millet	66	530	130
Upper leaves:			
sorghum	370	570	70
millet	290	590	110
Lower leaves:			
sorghum	530	460	30
millet	380	560	70
Stalks:			
sorghum	3110	330	10
millet	2440	360	20

The volume of research on the contribution of the various plant species to dry matter yields of native and improved pastures, is constrained by the laborious nature of determining the individual contributions of each species to both the sward as a whole and to differential dry matter yields in particular. A computer package named BOTANAL has been developed for use in field measurements of botanical composition (% species contribution) and dry matter yield of individual species (Jones and Tothill 1985). This method has considerable advantages over the use of the dry weight rank method of Mannetje and Haydock (1963) which has been used successfully for botanical composition determination in many countries.

THE INFLUENCE OF ANIMAL SPECIES ON HERBAGE UTILISATION

Many studies have shown that potential animal production cannot be reliably predicted from dry matter yields, in the absence of information on the type and mix of grazing and browsing animals to harvest the material available (Owen-Smith 1985).

The basic differences in the mode of intake, selectivity and height of grazing by cattle, sheep and equines have been well established. Sheep graze shorter and more selectively than do cattle and equines generally consume much courser material than either cattle or sheep (Tribe 1950). The complementary browsing behaviour of goats and wild herbivores has more recently indicated how significant increases in total carrying capacity can be attained through an understanding of the needs of grazers and browsers in mixed vegetation types.

Improvements in the application and interpretation of results from the oesophageal fistula technique have given impetus to intake studies in recent years. Alternative techniques of measuring material removed in the field are laborious and time-consuming. The work of Gammon and Roberts (1980) has demonstrated how species selection, spatial selection and plant-part selection by cattle, can result in animals selecting the most nutritious of feed available at different seasons. Du Toit (1972) and Field (1978) have reported on the effects of sheep and goats on African grazinglands, through their preferential intake of dry matter from grasses and shrubs respectively.

Dry matter intake studies in wildlife have revealed a complex selection of diet from different strata of the vegetation (Talbot and Talbot 1962; Meissner 1982; Novellie 1983; Pellew 1984). The most recent studies (Owen-Smith 1985) reveal how an understanding of dietary habits of each species of herbivore in a particular vegetation type can lead to improved dry matter intake and carrying capacity.

THE EFFECTS OF BURNING

Grazing habits and thus dry matter useage are directly influenced by biotic factors such as fire. Natural fires are the first link in a chain of events in many ecosystems in which herbivores are the prime consumers. Large herbivores are generally attracted to newly burnt areas where fresh grass and new shrub growth provide a high protein diet (Frost 1984). This has led to management policies based on the fire-ecology of African game reserves (Brynard 1971), and to grazingland management aimed at preventing shrub dominance of natural pastures (Trollope 1974). The use of prescribed burning as an ecological management tool has been reviewed by Roberts (1979). Fire can be used to achieve one or more of the following effects beneficial to herbivores (West 1965):

1. To remove old grass and improve pasture quality.
2. To stimulate out of season growth.
3. To control woody weeds and timber regrowth.
4. To attract animals to otherwise unused areas.
5. To reduce the hazard of hot wildfires.
6. To control ticks and diseases.

The scientific use of fire was pioneered in the USA and Southern Africa and has yet to be fully implemented in many other countries where the plant biomass tends to develop toward unpalatable woody species.

INCREASING HERBAGE PRODUCTION

Dry matter production on a global scale is limited by moisture in drier regions and by soil fertility in moister regions. Sandford (1983) has classed techniques for increasing rangeland productivity into six groups:

1. Mechanical - contour banking, pitting, water spreading, clearing, haymaking.
2. Planting improved species.
3. Burning.
4. Chemicals - herbicides, fertilizers.
5. Grazing management, fences and watering.
6. Animal numbers and type.

Although this overview is concerned with available dry matter, i.e. that component of plant biomes available above ground to the herbivore, production methods must consider not only maximizing plant growth but minimizing waste, namely physical and chemical reduction in dry matter yield through senescence, disease and insects.

The introduction of legumes and improved pasture grasses has increased available dry matter substantially, notably in Australia, South America and the USA. The search for more efficient pasture plants in terms of moisture use and soil nutrient uptake continues to produce "two blades of grass where one grew before".

In many cases the introduction of improved pasture species requires the addition of fertilizer, notably nitrogen for grasses and phosphate for legumes. Unfortunately the cost of such inputs has limited the use of this new technology in most third world countries.

The combination of improved pasture species and greater soil fertility has produced spectacular improvement in subtropical animal performance in extensive ranching areas. Edye and Gillard (1985) report up to ten-fold carrying capacity increases from *Stylosanthes* spp. plus superphosphate in North Queensland, combined with weight gains per head of twice those achieved on native pastures (in other words, a 20-fold increase in animal production). Investigations in Indonesia have shown the potential of vast areas (more than 40,000 km^2) of blady grass (*Imperata cylindrica*) for beef production (Roberts 1985). These areas could be more extensively grazed if herbage quality was raised. Ibrahim *et al.* (1985) have shown how a combination of legumes (*Stylosanthes guianensis*, *Calopogonium mucunoides* and *Centrosema pubescens*) and improved grasses (*Brachiaria mutica*) in Indonesia resulted in significantly higher carrying capacity being sustainable with improved botanical composition (Table 5). Despite the early success of this work, persistence of legumes remains an important problem in many regions where dry matter production potential has as yet not been exploited.

Table 5. Effects of stocking rate upon the botanical composition of an Indonesian pasture (Ibrahim *et al.* 1985).

Pasture Treatment	% Botanical Composition			
	Legume	*Imperata*	*Other grass*	Weed
Oversown pasture				
2 AU* /ha	41.5	34.1	23.8	1.2
3 AU /ha	54.9	18.5	24.6	2.3
Native pasture				
0.5 AU /ha	8.1	51.8	30.5	10.0
1 AU /ha	8.7	41.8	40.2	9.2

*AU Animal units

The opportunities for increasing dry matter production on a large scale are probably greater on the South American continent than elsewhere. Some 750,000 km^2 of well drained open rolling country is available in Brazil, Venezuela, Colombia and Bolivia, for improved pastures based on legumes. The native pastures (llanos, pampas and cerrados) produce liveweight gains of the order of 20 kg /ha (Thomas *et al.* 1983) although dry matter yields are approximately 4 t /ha (San

Jose and Medina 1975). Legume-based pastures given appropriate fertilizer, produce liveweight gains of 86–580 kg /ha in the same region (Thomas *et al.* 1983). The potential of tropical regions to produce greater yields has been reviewed by Cochrane (1981), who considered that South America, Central America, Northern Australia and upland regions of several Asian countries hold most potential for the introduction of legume-based pastures.

In addition to the criteria for increasing herbage production listed above can be added animal and plant breeding, the provision of mineral supplements, the grazing of mixed animal species and biological control of weeds and pests. In practice, under ranching conditions, the manager can combine animal groups, alter stocking rates and use mineral licks strategically to modify both intake and grazing behaviour (Roberts 1972). It has been demonstrated (Gammon and Roberts 1980) that in mixed natural swards, the patterns of selective grazing under different management systems are not significantly different (Table 6). After reviewing the literature, Gammon (1976) found no predictable relationship between grazing system and animal production within a particular season.

Table 6. A comparison of the time spent on continuous and rotational grazing of three plant communities (Gammon and Roberts 1980).

	Hyperthelia dessoluta		*Themeda triandra*		*Sporobolus iocladus*	
	min	%	min	%	min	%
Continuous	266*	60.0	101	22.8	76	17.3
Rotational	219	58.6	91	24.4	64	17.1

LSD (P = 0.05) Continuous grazing = 53 min
Rotational grazing = 43 min

* Mean day-grazing times for each of three veld types.

Despite isolated successes, especially in the less populated areas of the world, the general level of improvement of animal production in modern times is unimpressive. For instance, in Africa during the last 20 years there has been only a 2% increase in animal and crop production; agricultural production *per capita* has fallen and human food imports have soared. There has been a simultaneous five-fold increase in desertification in some areas, notably in Mali (Haywood 1980), in the period from 1953 to 1976. Similarly in India, carrying capacity in Rajasthan has decreased from 2.5 to less than 0.7 cattle /ha and production has dropped from 2 t /ha to 0.5 t /ha (Ahuja 1975), of animal fodder on a dry matter basis.

The critical factor affecting both dry matter production and the stability of many landscapes used for animal production, is excessively high stocking rates. This is evident in Africa and Asia, and also in some fragile ecosystems of Australia and the western USA (Roberts 1972).

Drege (1983) has summarized the causes and extent of desertification on a global scale. The greatest expansion of deserts is occurring in semi-arid regions where the combined effects of overgrazing, overbrowsing and the removal of firewood are exposing the soil to water and wind erosion. In most cases this process is occurring under tribal or communal systems of grazing in societies where human population growth is high. The current annual increase of 54,000 km^2 in area under desert represents the greatest source of loss of dry matter production in world history.

At the International Savanna Symposium dealing with maintaining productivity, Roberts (1985) referred to the universal application of Hardin's (1968) classic case study entitled "The Tragedy of the Commons". In this scenario, Hardin explains the conflict between personal gain through increasing an individual owner's stock numbers on commonage grazing land, and the eventual permanent production loss to the individual, the community and their descendants. This has resulted in serious denudation in many of the world's grazing lands where communal grazing is practised.

In some regions, increased dry matter production has resulted from improved animal management through fencing, watering points and rotational grazing (Gammon 1976). When the use of supplements such as urea, molasses and mineral licks is combined with improved physical layout of grazing paddocks, significant increases in the intake of the low quality dry matter have been achieved (Roberts 1972). Intake is increased as a result of stimulation of the rumen flora and thus of appetite for mature herbage.

The effects of differing defoliation patterns (height and frequency) on dry matter yield of herbage plants, vary widely according to the plant/climate complex involved. Conflicting experimental evidence exists on the advantages of grazing systems which theoretically lead to differential defoliation patterns. The compounding effects of season, stocking rate and type of animal have made interpretation of data on gains/head and gain/ha very complex. As a result, the predicted increased yield from managed stock movements cannot be demonstrated in most experiments (Gammon 1976).

Much has been written on the merits of alternative grazing systems (Roberts 1969). Measurements of the herbage on offer and the patterns of defoliation under continuous and rotational grazing (Gammon and Roberts 1978a) reveal differences which are at variance with the theoretical advantages of the systems compared. As an example the data in Table 7 indicate that the degree of utilisation of neither the most common grassland type (*Hyperthelia*) nor of the most nutritious

Table 7. Percentage utilisation of available herbage under two grazing systems (Gammon and Roberts 1978).

Veld type	Study 1 Continuous grazing 1971–72	Study 2 Rotational grazing 1973–74
Hyperthelia dessoluta	63.2	61.7
Themeda triandra	65.2	70.5
Sporobolus iocladus	96.7	90.4

grassland (*Themeda*) is increased by rotational grazing compared with continuous grazing of a mixed grassland.

The effects of stocking rate, as reflected in intensity of defoliation, markedly influence dry matter and crude protein yields in tufted grasses (Opperman *et al.* 1969). Similarly, season of grazing has a direct influence on both DM yield and on infiltration capacity of the topsoil, through the effect of differential defoliation on surface organic matter and trampling (Van den Berg *et al.* 1976). The processes involved relate to the direct detrimental effect of intensive defoliation, during the active plant growth season, on the perenniality of the more nutritious species. The effect of heavy summer grazing in the summer rainfall region is thus much more severe physiologically than winter grazing, and results in decreased plant density (cover) which in turn causes greater runoff of rainfall. The percentage basal cover of climax grasses is directly related to the percentage total plant cover.

Riewe's (1961) experimental work allowed for a direct relationship between liveweight gains and stocking rate to be established. Humphreys (1966) has reviewed the effects of defoliation intensity and frequency on herbage yields and Raymond (1965) has summarized the important relationship between grazing intensity and both output per animal and per unit area. Although the slopes of the trends suggested by Raymond differ markedly with the intensity of grazing applied, it is clear that in many countries the limitation on feed intake, as a result of overstocking, prevents both cattle and sheep from producing at or near their genetic potential.

Vast areas of the world are limited in their herbage production by an abundance of trees. Where plant succession progresses beyond the grassland or savanna stage, yields of herbage dry matter available to herbivores are markedly reduced through competition from woody species for moisture and sunlight. The modification of woodland through clearing, thinning, selective herbicides and fire, has given marked increases in carrying capacity in many humid climates and in some semi-arid regions. Australia and South America offer good examples of how tree removal, combined with the introduction of legumes, has produced impressive gains in stocking rate and liveweight (Edye and Gilland 1985). In some cases, such as in Australian landscapes developed from sedimentary geological formations, tree removal has induced salinity problems which have ultimately reduced dry matter production (Peck 1980).

RESEARCH DIRECTIONS FOR INCREASED HERBAGE PRODUCTION

The overall picture of herbage production on a global scale is one of unused technological expertise – unused for socio-economic reasons rather than for lack of demonstrated advantages. In general, there are several alternative methods of improving dry matter production in all but the most arid or arctic environments. Research will be required for many years to come, on selection of improved fodder species for particular environments and on the alleviation of nutrient deficiencies in both plants and animals. However, the two prime constraints to current fodder production are a lack of:

1. Implementation of land use policies which ensure maintenance of land stability principles.
2. Socio-political action to control human population growth rate as the prime means of limiting pressures for more intensive land use.

The situation in large parts of Africa and Asia is presently dominated by human population pressure and religious constraints on efficient animal production. In South America these pressures are lower but a lack of organization and finance is delaying the large scale adoption of improved fodder production techniques. In the developed western nations, the level of input into fodder production is constrained only by economic cost/benefit considerations, complicated by significant levels of subsidy in Europe and the United States. In young developing countries such as Australia, technological advances in land clearing have outstripped the development of stewardship of the land and an appreciation of the importance of stable production systems.

Ahuja, L.D. (1975). In: Range Management. Mimeo, New Delhi.

Alter, J.C. (1912). USDA Yearbook, 309–19.

Brynard, A.M. (1971). *Proc. Tall Timbers Fire Ecol. Conf.* **11**, 219–31.

Capelin, M.A., and Truong, P.N. (1985). Proc. Fourth Aust. Soil Con. Conf., Maroochydore.

Cochrane, T.T. (1980). In: IV International Symposium of Machine Processing of Remotely Sensed Data. pp. 227–33. (Purdue Univ.: Lafayette, Indiana.)

Cochrane, T.T. (1981). Food and Climate Review. pp.18–28. (Boulder: Colorado.)

Desowitz, R.S. (1980). In "Human ecology in Savanna Environments." (Ed. D.R. Harris.) pp. 457–77. (Academic: London.)

Donaldson, C.H. (1967). *Proc. Grassl. Soc. S. Afr.* **2**, 137–41.

Dregne, H.E. (1983). "Desertification of Arid Lands." (Harwood: New York) Du Toit, P.F. (1972). *Proc. Grassl. Soc. S. Afr.* **7**, 44–50.

Edye, L.A., and Gillard, P. (1985). In "Ecology and Management of the World's Savannas." (Eds J.C. Tothill and J. J. Mott.) pp. 303–9. (Aust. Acad. Sci.: Canberra.)

FAO, (1982) "Soil Conservation for Sustained Agricultural Production." (UNO: Rome.)

Field, A.C. (1978). ODM–IPAL sheep and goats project: preliminary report on the impact of sheep and goats on the vegetation of the arid zone of northern Kenya. UNEP–MAB Integrated Project on Arid Lands.

Frost, P.G.H. (1984). In "Ecological effects of Fire in South African Ecosystems." (Eds P. de V. Booysen and N.M. Tainton.) pp. 273–309. (Springer–Verlag: Berlin.)

Gammon, D.M., (1976). D.Sc. Thesis, University of the Orange Free State.

Gammon, D.M., and Roberts, B.R. (1978a). *Rhod. J. Agric. Res.* **16**, 133–46.

Gammon, D.M., and Roberts, B.R. (1978b). *Rhod. J. Agric. Res.* **16**, 3–22.

Gammon, D.M., and Roberts, B.R. (1980). *Zimbabwe J. Agric. Res.* **18**, 13–28.

Hacker, J.B., Strickland, R.W., and Basford, K.E. (1985). *Aust. J. Agric. Res.* **36**, 201–12.

Hardin, G. (1968). *Science* **162**, 1243.

Howeler, R.H. (1973). *Soil Sci. Soc. Am. Proc.* **37**, 898–903.

Haywood, M. (1980). ILCA Working Document 3, International Livestock Centre for Africa: Addis Ababa.)

Humphreys, L.R. (1966). *J. Aust. Inst. Agric. Sci.* **32**, 93–105.

Ibrahim, T., Bulo, D., Rudolf, W., Hunt, M., and Ivory, D. (1985). In "Ecology and Management of the World's Savannas." (Eds J.C. Tothill and J.J. Mott.) pp. 291–4. (Aust. Acad. Sci.: Canberra.)

Jewell, P.A. (1980). In "Human Ecology in Savanna Environments." (Ed. D.R. Harris.) pp. 353–81. (Academic: London.)

Jones, R.M., and Tothill, J.C. (1985). In "Ecology and Management of the World's Savannas." (Eds J.C. Tothill and J.J. Mott.) pp. 318–20. (Aust. Acad. Sci.: Canberra.)

Keep, M.E. (1973). *Lammergeyer* **19**, 16–23.

Leeper, G.W. (1973). "The Australian Environment." (Melbourne University: Melbourne.)

Mannetje, L.'t, and Haydock, K.P. (1963). *J. Brit. Grassl. Soc.* **18**, 268–75.

McClung, A.C. (1959). *Ecology* **40**, 315–7.

Meissner, H.H. (1982). *S. Afr. J. Wild. Res.* **12**, 41–7

Mott, J.J., Williams, J., Andrew, M.H., and Gillison, A.N. (1985). In "Ecology and Management of the World's Savannas." (Eds J.C. Tothill and J.J. Mott.) pp. 56–82. (Aust. Acad. Sci.: Canberra.)

Novellie, P.A. (1983). D.Sc. thesis, University of Pretoria.

Owen-Smith, N. (1985). In "Ecology and Management of the World's Savannas." (Eds J.C. Tothill and J.J. Mott.) pp. 345–9. (Aust. Acad. Sci.: Canberra.)

Pellew, R.A. (1984). *J. Zool.* **202**, 57–82.

Powell, J.M., and Waters-Bayer, A. (1985). In "Ecology and Management of the World's Savannas." (Eds J.C. Tothill and J.J. Mott.) pp. 252–5. (Aust. Acad. Sci.: Canberra.)

Raymond, W.F. (1965). *J. Brit. Grassl. Soc.* **19**, 81–9.

Riewe, M.E. (1961). *Agron. J.* **53**, 310–3.

Ritchey, K.D., Souza, D.M.G., Lobato, E., and Correa, O. (1980). *Agron. J.* **72**, 40–4.

Roberts, B.R. (1964). Veld Management Conference Proceedings, Bulawayo, Rhodesia.

Roberts, B.R. (1971). *Proc. Grassl. Soc. S. Afr.* **6**, 44–9.

Roberts, B.R. (1972). Ecological studies on pasture condition in semi-arid Queensland. (Department Primary Industries: Brisbane.)

Roberts, B.R. (1973). *S. Afr. Wool Grower*, May, 1–4.

Roberts, B.R. (1975). *Proc. Grassl. Soc. South. Afr.* ***10***, 49–55.

Roberts, B.R. (1979). Ecol. Soc. Aus. Open Forum, Proc., May, Brisbane.

Roberts, B.R. (1984). Report on range management in south Sulawesi, Indonesia. (Australian Universities International Development Program: Canberra.)

Roberts, B.R. (1985). In "Ecology and Management of the World's Savannas." (Eds. J.C. Tothill and J.J. Mott.) pp. 195–96. (Aust. Acad. Sci.: Canberra.)

Roberts, B.R., Anderson, E.R., and Fourie, J.H. (1975). *Proc. Grassl. Soc. S. Afr.* **10**, 133–40.

Sandford, S. (1983). "Management of Pastoral Development in the Third World." (Wiley: New York.)

San Jose, J.J. and Medina, E. (1975). In "Tropical Ecological Systems. Ecological Studies II." (Eds F.B. Golley and E. Medina.) pp. 251–64. (Springer-Verlag: Berlin.)

Scholes, R.J. (1985). In "Ecology and Management of the World's Savannas." (Eds J.C. Tothill and J. J. Mott.) pp. 350–3. (Aust. Acad. Sci.: Canberra.)

Talbot, L.M., and Talbot, M.H. (1962). *East Afr. Agric. For. J.* ***27***, 131–8.

Tribe, D.E. (1950). *J. Brit. Grassl. Soc.* 5, 209–24.

Thomas, D., de Andrade, R.P., Couto, W., Moore, C.P., and da Rocha, C.M.C. (1983). *World Rev. Anim. Prod.* **19**, 38–44.

Trollope, W.S.W. (1974). *Proc. Grassl. Soc. S. Afr.* **9**, 67–72.

Van den Berg, J.A., Roberts, B.R., and Vorster, L.F. (1976). *Hand. Weidingsveren. S. Afr.* **11**, 91–5.

West, C. (1965). "Fire in Vegetation and its Use in Pasture Management." (Commonwealth Bureau of Pastures and Field Crops: Hurley, UK.)

4

NUTRITIVE QUALITY OF FORAGE

by D.I.H. Jones and A.D. Wilson

A wide diversity of herbivores are sustained by the grasslands of the world. Many are dependent on natural grassland species for their survival, often having to adapt their feeding patterns to the available forage by utilising browse plants and migration. To make farming economic it has often been necessary to improve pastures by the introduction of sown pastures. Norton (1982), however, points out that only 40 of the 10,000 grass species and 30 of the 11,000 legume species are presently used to any significant degree as sown pasture species.

In many situations the supply of available green forage is an overriding consideration and overshadows the importance of nutrient content and supply. Mannetje (1982) regards the main limitations of tropical pastures in seasonally dry regions to be the lack of green feed for at least half the year, coupled with the low nutritive quality of herbage during most of the period of active pasture growth. The problems of temperate grasslands are very different. These pastures have an inherent advantage of higher nutritional quality, in terms of digestibility and intake potential, enabling high levels of animal production to be attained (Reid and Jung 1982). In recent years the tendency in the temperate pasture regions of the United States and western Europe is for concentrate feeds to substitute for pasture herbage as a source of animal production. There are indications, however, of this trend reversing, partly due to the need to reduce the surpluses of animal products by reducing the current levels of animal production. The need to reduce feed costs has, therefore, become paramount and more emphasis is being given to obtaining production from more efficient use of pasture herbage.

The main factors influencing the chemical composition and nutritive quality of forages have been the subject of several extensive reviews including those compiled by Butler and Bailey (1973) and those of Raymond (1969), Minson (1982), Norton (1982), and Wilson (1982). There are, additionally, innumerable reports of the nutrient content of native and improved forages from most countries in the world which would be impossible to incorporate into a general review. This paper, therefore, aims at discussing the principles determining nutrient content and the significance of variation in nutrients on intake and digestibility. The contribution that browse plants can make to the herbivores' nutrient supply is given special attention in view of its importance in arid and semi-arid environments. Chemical constituents which are toxic or exert anti-nutritive effects are the subject of following contribution to this symposium.

NUTRITION OF HERBIVORES
ISBN 0 12 750052 9

PASTURE HERBAGE AS A SOURCE OF NUTRIENTS

Herbage contains a wide diversity of chemical constituents (Butler and Bailey 1973) many of which can be considerd to be nutrients for the herbivore. To be so considered, a constituent must either be utilised as an energy source or satisfy a specific nutrient requirement. The position is, however, complicated, since all herbivores have well developed and specialised fermentation organs in their gut, which enable them to derive energy and nutrients from substrates not utilisable by non-herbivores. Thus constituents such as cellulose and hemicellulose are nutrients for the herbivore provided they are accessible to microbial attack in the rumen or other fermentative organ. The herbivore derives most of its energy from the end product of microbial fermentation, in other words, volatile fatty acids. Particularly in the foregut fermenters, protein quality and vitamin content of the diet are of limited importance as the absorbed amino acids and vitamins are of microbial origin.

Table 1. Potential digestibility of major constituents (adapted from Minson 1976a).

	Cell contents	Cell wall
Completely digestible	Sugars, reserve carbohydrates Protein, non-protein N Lipid Organic acids Soluble minerals	Unlignified cellulose, hemicellulose, pectin
Partially digestible		Partially lignified cellulose and hemicellulose
Indigestible	Pigments	Lignin Highly lignified cellulose and hemicellulose Silica Cuticle

The chemical constituents of herbage can be conveniently divided into those making up the structure of the plant (the cell wall constituents) and those contained within the cell structure (the cell contents). This division has virtue in dividing the constituents in a way which is nutritionally meaningful and lends itself to simple routine chemical analysis by extraction in neutral and acid detergent (van Soest and Moore 1965). In terms of major constituents, the cell contents comprise the sugars and 'reserve' carbohydrates (fructosan or starch), protein and non-protein nitrogenous constituents, lipids, organic acids, vitamins and minerals. The cell wall components are mainly cellulose, hemicellulose, pectin and lignin.

The availability or digestibility of these constituents varies from the completely digestible sugars to the largely indigestible lignin. The digestibility of others, such as cellulose or hemicellulose, is influenced by the degree of lignification and also by the conditions and duration of fermentation and their potential digestibility, i.e. the maximum digestibility attainable, when fermentation conditions and time are not limiting, may not be achieved in practice (Wilkins 1969). This concept has been further developed by Minson (1976a) who proposed a summation model of potential digestibility. Table 1 is an adaptation of this model and suggests that virtually all cell constituents can be regarded as nutrients together with a variable proportion of the cell wall polysaccharide. The efficacy of herbage as a nutrient source for the herbivore is thus clearly a function of its content of non-structural constituents and the extent to which the potential nutrients of the cell wall can be released by fermentation.

Non-structural constituents

Non-structural carbohydrates. The main non-structural carbohydrates found in grasses and legumes are the sugars (glucose, fructose and sucrose), together with the storage polysaccharides (starch and fructosan). The content of sugars does not normally vary greatly and most of the variation found in total non-structural carbohydrates between species or due to environment can be attributed to differences in the content of storage polysaccharides (Smith 1973).

Legumes, whether of temperate or tropical origin, accumulate starch as their reserve polysaccharide. Grasses of tropical origin have a C_4 photosynthetic pathway characterised by a specialised leaf anatomy, higher growth rates, a higher nitrogen use efficiency (Norton 1982) and accumulate starch as reserve polysaccharide. Temperate origin grasses have a C_3 photosynthetic pathway and characteristically accumulate fructosan as storage polysaccharides (Smith 1973). Significant levels of starch can, however, be found in temperate grasses after active photosynthesis in cool conditions (Wilson and Bailey 1971) but the levels are generally very low in comparison with fructosan. Since fructosan is water soluble, the estimation of water soluble carbohydrates gives a reasonable approximation of total non-structural carbohydrates for temperate grasses. A similar approximation for tropical grasses or legumes requires an analytical procedure to include starch, which is not soluble in cold water and only partially so in hot water (Smith 1971).

Considerable variation has been found in the water soluble carbohydrate content of temperate grass species. Cocksfoot (*Dactylis glomerata*) generally shows the lowest content and ryegrass (*Lolium* spp.) the highest with fescue (*Festuca* spp.) and timothy (*Phleum* spp.) intermediate (Waite and Boyd 1953; Dent and Aldrick 1963). Less information appears to be available for legumes or tropical grasses, possibly due to the analytical difficulties involved. Smith (1972), however, found relatively little variation (57-88 g /kg DM) among a range of seven tropical species at anthesis. Higher contents and more variation (61-193 g /kg DM) were found for six temperate grasses grown on the same site. Smith did not include Italian (*Lolium multiflorum*) or perennial ryegrass (*Lolium perenne*) in his comparison but values of up to 500 g /kg DM have been recorded in these species (Bugge 1978).

Legumes are generally regarded as being higher in soluble carbohydrates than grasses but Smith (1973) points out that this belief may rest on comparisons of water soluble carbohydrate content i.e. excluding starch. Smith (1972) shows the total non-structural carbohydrate content of grasses and legumes to be similar, or not consistently different, when grown on the same site.

Considerable variation has also been shown in water soluble carbohydrates between varieties of temperate species (Dent and Aldrich 1963; Cooper 1973), a consistent feature being the higher content of water soluble carbohydrates in ryegrass species than other species. It has proved possible to exploit this variation to develop high and low selection lines in temperate (Cooper and Breese 1971) and tropical species (Hutton 1970).

Table 2. The effect of monthly nitrogen fertiliser (90 kg /ha) on the fructosan content (g /kg DM) of monthly cuts of perennial ryegrass (*Lolium perenne*) and cocksfoot (*Dactylis glomerata*) (Jones *et al.* 1965).

	Perennial ryegrass		Cocksfoot	
	-N	+N	-N	+N
April	69	20	27	6
May	132	28	53	3
June	127	88	53	16
July	75	74	21	4
August	68	43	30	6
September	44	38	10	6

Since stem contains higher levels of fructosan than leaf (Waite and Boyd 1953) the percentage water soluble carbohydrate content of temperate origin grasses is greatly influenced by the ratio of stem to leaf. Concentration generally increases with age reaching a peak at around the time of inflorescence emergence. There appears to be no comparable trend with advancing maturity in the non-structural carbohydrate concentration of legumes (Smith 1973).

A number of studies have shown the water soluble carbohydrate content of temperate grasses to decline markedly when nitrogen fertiliser is applied (Table 2; Smith 1973). The effect may be attributed to an accelerated growth promoted by increased nitrogen supply and hence less opportunity for accumulation of reserve polysaccharide. Consequently, the reduction is more marked in periods of more active growth and stem elongation and is particularly evident in low soluble carbohydrate species such as cocksfoot.

Less information is available on grasses of tropical origin but the data summarised by Smith (1973) indicates similar trends to temperate grasses but with a less marked reponse to nitrogen fertiliser.

Environmental conditions favourable to photosynthesis, but not rapid growth, increase the accumulation of non-structural carbohydrate in grasses (Deinum 1966) and legumes (Lee and Smith 1972). The decrease in content at higher temperatures is more marked in the storage fraction (fructosan or starch) than in sugars (Smith 1970). Most of the evidence on environmental effects on non-structural carbohydrate content has been derived from studies in controlled environments. Experience from field experiments, however, supports the conclusion that soluble carbohydrate levels in temperate grasses are lower in wet and overcast summers.

The non-structural or soluble carbohydrate fraction of temperate grasses normally lies in the range 50–300 g /kg DM. Although higher values have been reported under rapid growing conditions such as in New Zealand, values tend to be near the bottom of the range. Since the soluble carbohydrates are completely digested (Table 1) they represent a significant contribution to the herbivore's nutrient supply. They may also be significant in relation to palatability (Bland and Dent 1964) and conservation as silage (Jones 1970a). Variation in non-structural constituents in legume and tropical grasses is probably of lesser significance to the herbivore since their content and variation is less than in temperate grasses.

Protein and non-protein nitrogen constituents. Pasture herbage contains a diversity of nitrogenous constituents including proteins, peptides, amino-acids, amides, ureides, nitrates, ammonia and nucleic acids. Herbage proteins are almost entirely in the form of metabolic protein, in other words, they are concerned with the growth and biochemical processes of the plant. About half of the total protein is soluble and of the soluble protein up to half is a single homogeneous protein (fraction 1, RuDP carboxylase), the remainder (fraction 2) is a mixture of diverse proteins. The major part of the insoluble protein is associated with lipid (chloroplast membrane protein) and can be extracted with detergent. A small amount (1–2%) of the total protein is closely associated with the cell wall. The amino-acid composition of herbage protein appears to vary very little between species or as a result of nitrogen fertiliser application; it has a high biological value with methionine as the first limiting amino acid (Lyttleton 1973; Brady 1976; Mangan 1982).

The true protein content generally amounts to around 750 g /kg of the crude protein (N x 6.25) depending on the level of nitrogen fertiliser applied and growth stage. This may fall to 600-650 g /kg when ammonium fertilisers are applied or as low as 420-460 g /kg after nitrate fertilisation (Nowakowski *et al.* 1965). Hegarty and Peterson (1973) have reviewed the non-protein nitrogenous constituents of forage and the chemical diversity involved can be judged from the fact that over 200 non-protein amino acids have been isolated and characterised. Plant amines are almost as diverse but generally occur in lower quantities than the free amino acids. Nitrate can be a major constituent in grasses when heavily fertilised with nitrogen (ap Griffith 1960).

The crude protein content of herbage is greatly influenced by the growing conditions, notably the level of nitrogen fertiliser applied and the stage of growth. General trends are, however, evident in comparisons between species and particularly between grasses and legumes. Minson (1976b) has compared the frequency distribution of protein content in tropical and temperate grasses and legumes from data reported in the literature. His analysis shows a considerably higher median for the legumes (150-180 g CP /kg DM) than the grasses (60-90 g CP /kg DM) although the overall range was similar. The tropical grasses showed skewed distribution with 22% of the values below 60 g CP /kg DM. The lower protein content of tropical grasses has been related to the higher efficiency with which C_4 plants use nitrogen for growth. Norton (1982) concludes that the low protein content often found in tropical grasses, even when fertilised with nitrogen, is an inherent characteristic of C_4 plants closely related to their survival under conditions of low fertility.

Table 3. Effect of age on the protein content (g/kg DM) of tropical grasses (from Gomide *et al.* 1971).

Species	Age in weeks		
	4	12	24
Napier grass (*Pennisetum purpureum*)	24	10	7
Guinea grass (*Panicum maximum*)	23	12	9
Pangola grass (*Digitaria decumbens*)	13	7	6
Kikuyu grass (*Pennisetum clandestinum*)	22	15	14
Bermuda grass (*Cynodon dactylon*)	18	10	9

Legumes are generally higher in protein than grasses (Lyttleton 1973) and significant genotypic variation has been reported in lucerne (Heinrichs 1970). Genotypic variation has also been found in temperate and tropical grasses (Cooper 1973). A high protein content is, however, often associated with lower production of dry matter and a lower content of soluble carbohydrate (Rogers 1970).

Protein content is strongly influenced by growth stage (Table 3; Lyttleton 1973; Norton 1982). Young vegetative growth is high in protein but the content rapidly falls as the proportion of leaf decreases. There is a slower decline in the protein content of the leaf than of the stem with advancing age reflecting the decrease in the proportion of cell contents in the stem as it lignifies and the cell walls thicken. The decline in protein with age is similar in temperate and tropical grasses but some species appear to decline less rapidly. Kikuyu grass (*Pennisetum clandestinum*), for example, has been found by several workers to maintain its protein content at higher levels than other grasses (Milford and Haydock 1965; Gomide *et al.* 1969).

Water stress is a contributory factor to the low protein content of tropical grasses at a mature stage of growth (Lyttleton 1973). There is also evidence of nitrogen content being decreased by high temperatures (Colman and Lazenby 1970). The main factor influencing protein content is, however, nitrogen supply and grasses generally respond to applied nitrogen fertiliser through increased growth and a higher protein content. The extent of the increase depends on the timing of the applications in relation to growth stage, yield response and harvest date but crude protein content can often be doubled by heavy applications of N fertiliser (Henderson *et al.* 1962). Some species such as Italian ryegrass appear to give a very rapid and marked response resulting in crude protein contents as high as 300 g /kg DM (Wilman 1970).

Pasture management will clearly have a marked effect on the protein content of pasture herbage reflecting the effects of growth stage, leaf content and nitrogen supply. Management which maintains herbage at a young vegetative stage by frequent grazing will result in a high leaf and protein content provided the nitrogen supply is adequate. The proportion of legumes, inherently higher in protein content than grasses, is also important and fertiliser and management regimes favouring legume growth will increase protein content.

The protein and non-protein nitrogen constituents of pasture herbage are important nutrients for the herbivore and deficiencies of protein are a major limitation to the intake and utilisation of tropical forages (Milford and Minson 1965). Protein content is seldom limiting in temperate pastures especially under grazing but the high degradability of the protein results in a high proportion being inefficiently used by the ruminant (Filmer 1982; Beever 1982). The balance of soluble carbohydrate and nitrogen constituents may thus be of greater significance than total protein content.

Lipids. Hawke (1973) has reviewed the composition of the lipid fraction of pasture herbage and factors influencing lipid content. Lipids constitute 30–109 g /kg DM of the leaf and most is located in the chloroplasts; lipids comprise around 30% of the dry weight of chloroplasts and 50% of the chloroplast lamellae. The lipid fraction comprises neutral lipids (glycerides, sterols etc.) phospholipids, glycolipids, sterol and acyl sterol glycosides and sulpholipid. Hawke points out that much of the data on pasture species is fragmentary but lipid content appears to be as much attributable to age as to species.

The fatty acid composition of the total lipid appears to be relatively constant probably reflecting the important role of lipids as structural constituents of chloroplast membranes. Leaf lipid is generally rich in linolenic and linoleic acids, linolenic acid usually accounting for 60–75% of the total fatty acids.

The external surface of leaves comprises a cuticular layer covered with epicuticular waxes. These waxes include hydrocarbons, esters of long-chain alcohols with long-chain fatty acids, primary alcohols, free fatty acids and a variety of minor constituents (Hawke 1973). They are important in relation to water conservation and amount to 2.0–4.5 g /kg DM depending on species, growth stage and other factors. The distribution of wax constituents has significance in chemotaxonomy and characteristic differences have been shown between grasses and legumes and between grass species (Eglinton and Hamilton 1967).

These variations in lipid content of pasture herbage can be sufficient to have an effect on the composition of depot or milk fat of grazing animals (Hawke 1963) and may also constitute a useful source of energy. High lipid levels may, however, depress the digestibility of cellulose and protein in herbage diets.

Organic acids. Relatively little attention appears to have been given to organic acids in pasture herbage as a nutrient source to the herbivore. Dijkshoorn (1973) has reviewed factors influencing their accumulation and quotes a total content in the range 7.3–72.2 g /kg DM for temperate grasses during the growing season. The total content in the leaf can be as high as 38–45 g /kg DM in fescue species and 103–107 g /kg DM in lucerne (*Medicago sativa).* High levels of oxalic acid are found in many tropical grasses such as *Setaria sphacelata* and *Cenchrus ciliaris* (Barry and Blaney 1987) and high levels of transaconitic acid have been reported in some grass species.

Vitamins. Butler (1973) has reviewed the vitamin content of pasture herbage pointing out that the need of herbivores for vitamins is greatly modified by the capacity of microbial fermentation to synthesise a range of vitamins. Herbage contains abundant amounts of carotene, the precursor of vitamin A, and deficiency only occurs when animals exist on mature dry herbage in drought conditions. Aitken and Hankin (1970) have tabulated levels of vitamins in livestock feeds. Fresh green herbage is high in many vitamins, for example vitmain C and riboflavin, but most vitamins can be synthesised in adequate amounts in the rumen or gut and herbage content is thus of little significance.

Minerals. The present evidence suggests that over twenty mineral elements are essential for animals. Of these, economically important deficiencies in herbivores may be attributed to phosphorus, magnesium, cobalt, copper, iodine and selenium (Russel and Duncan 1956; Underwood 1981; Little 1982). There are significant differences between the mineral requirements of plants and animals. Pasture herbage, for example, has no requirement for cobalt, selenium or iodine and a low requirement for copper. Widespread deficiencies of these minerals occur in herbivores despite normal herbage growth. Other elements such as potassium and iron are required in higher amounts by plants than animals; deficiencies of these elements do not therefore occur in grazing animals.

Most mineral deficiencies in grazing animals occur on infertile soils under extensive systems of grazing management, for example, unimproved hill or rangeland, and are often accompanied by deficiencies of energy and protein. Other deficiencies are related to intensive systems of production, for example magnesium, where the nutrient supply from herbage fails to meet the very high requirements of animals. Pasture composition may serve as a guide to likely deficiencies in comparison with the estimated requirements of animals. The availability of elements such as magnesium and copper to the animal is, however, greatly modified by the content of other dietary constituents and deficiency cannot be simply related to the levels found in herbage (Butler and Jones 1973).

The factors influencing the mineral content of pasture herbage have been reviewed by Whitehead (1966), Fleming (1973), Reid and Horvath (1980) and others. The content of some elements in pasture herbage, such as cobalt and selenium, is well correlated with their content in the soil parent material. For most elements, however, herbage mineral content is determined by the availability of soil minerals as influenced by factors such as soil pH and water content. In fertilised swards the content of calcium, phosphorus and potassium is mainly influenced by fertiliser applications.

Marked species differences are evident in mineral content. Legumes and herbs contain more of the major and trace elements than grasses with the exception of sodium and manganese. The inclusion of legumes and herbs in a pasture will, therefore, increase the mineral content of pasture herbage.

Norton (1982) has compared published data on the mineral content of tropical and temperate legumes and grasses. The data clearly show the higher mineral content of legumes compared to grasses and also temperate grasses and legumes were higher in phosphorus, calcium and sodium, but lower in magnesium than those of tropical origin. Norton concluded that magnesium deficiency appears more likely in animals grazing temperate than tropical pastures.

Grass species differ widely in mineral content (Fleming 1973), differences in sodium content are particularly striking and consistent (ap Griffith and Walters 1966; Hacker *et al*. 1985). Considerable variation between and within varieties has also been shown; Cooper (1973) and Hacker (1982) suggest that sufficient variation exists between plants for improvement of mineral content by breeding to be feasible. Selection for magnesium content in Italian ryegrass has been used to develop high magnesium populations (Hides and Thomas 1981).

Plants take up minerals rapidly during early growth but, as growth accelerates, dry matter accumulates more rapidly than minerals are taken up and the content of most minerals therefore falls with advancing maturity. Leaf is almost invariably higher in minerals than the stem and this also contributes to the decline with age as leaf percentage decreases. Frequent defoliation, for the same reason, tends to maintain pasture mineral levels (Fleming 1973). The presence of the grazing animal is important since the return of excreta maintains an efficient cycle and hence herbage mineral levels are maintained (Wilkinson and Lowrey 1973).

The effect of added fertilisers on mineral content is often difficult to interpret. The application of a nutrient may increase growth but not necessarily mineral content if the soil is deficient in that nutrient. Further applications may increase both growth and content or only content. Nitrogen fertiliser generally increases mineral content (Whitehead 1970) and potassium fertilisers reduce the content of most other cations (Reid and Horvath 1980). Lime greatly influences the uptake of calcium and trace elements due to its effect on soil pH, the availability of soil trace elements (except molybdenum) being greatly reduced as pH increases.

Structural constituents of pasture herbage

The structural constituents of herbage consist largely of polysaccharides with lesser amounts of lignin and some protein. According to Bailey (1973) they can be divided into the matrix and fibre polysaccharides. The latter are largely crystalline and held together by various bands in a cement of mainly amorphous matrix polysaccharide, lignin and some protein. The matrix polysaccharides are composed of hemicellulose (β 1–4 linked xylopyranoses, β 1–4 linked glucose and mannose) and pectic substances (α 1–4 linked galacturonic acids, galactans and arabans). The fibre polysaccharide is cellulose (β 1–4 linked glucose). Lignin is a polymer derived mainly from three closely related phenylpropanoid monomers, p-coumaryl, coniferyl and sinapyl alcohols, interconnected in varying proportions and random sequences. Lignin cannot be depolymerised by straightforward hydrolysis with chemical agents or enzymes due to the strong carbon–carbon or ether linkages interlinking its constituents; its detailed structure is therefore largely unknown (Harkin 1973).

Much of the data available on structural constituents refers to crude fibre and is often difficult to interpret. Systems of analysis are, however, now available which allow partitioning of the cell wall by acid hydrolysis (Bailey 1964) or detergent extraction (van Soest 1963a, b).

The data summarised by Bailey (1973) and Harkin (1973) show the stem fraction to be higher in structural polysaccharides and lignin than the leaf. A strong association thus exists with growth stage, structural constituents increasing as the plant matures and the proportion of stem increases.

This may be illustrated by the date of Jarrige (1961) and Jarrige and Minson (1964) who found the hemicellulse and cellulose content of ryegrass and cocksfoot to double during growth to maturity; the differences were much less marked under a frequent cutting regime. Jones (1970b) reported similar trends in varieties of perennial ryegrass, cocksfoot and timothy.

Table 4. Changes in structural constituents (g /kg DM) of perennial ryegrass and cocksfoot with maturity (Jones 1970b).

Species	Growth stage	Xylan	Cellulose	Lignin	DMD*
Perennial ryegrass	Vegetative	60	156	39	0.83
(*Lolium perenne*)	Stem elongation	79	171	41	0.79
	Ear emergence	93	196	43	0.71
	Flowering	125	249	59	0.64
Cocksfoot	Vegetative	59	138	57	0.77
(*Dactylis*	Stem elongation	80	171	68	0.71
glomerata)	Ear emergence	100	219	76	0.63
	Flowering	103	218	87	0.55

**In vitro* DM digestibility coefficient (Tilley and Terry 1963)

Tropical grasses have generally also been found to show an increased content of structural constituents with maturity (Reid and Jung 1965; Gomide *et al.* 1969) although some grasses such as Kikuyu (*Pennisetum clandestinum*) and coastal Bermuda grass (*Cynodon dactylon*) showed remarkably little seasonal change. There are, however, important differences in the behaviour of tropical and temperate grasses attributable to the high proportion of vascular tissues associated with the specialised anatomy of C_4 plants (Wilson and Minson 1980; Norton 1982). Tropical grasses appear to have a higher leaf cellulose content and a higher proportion of hemicellulose to cellulose than do temperate grasses (Bailey 1973). Tropical grasses are generally regarded as being higher in structural constituents (Norton 1982) but this appears to be related to the growing temperature. Wilson and Ford (1971) showed the cell wall content of tropical species to be similar to that of perennial ryegrass at high temperature but to be 10–15% higher than perennial ryegrass at the optimum growth temperature for the species. It was subsequently shown that the cell wall content of tropical grasses decreased with increasing temperature while that of temperate grasses increased. At high temperature there was, therefore, little difference between the two types. Temperate grasses have been reported to have higher leaf lignin contents than tropical grasses (Ford *et al.* 1979).

Temperate legumes contain less cellulose and hemicellulose than do temperate grasses (Jarrige 1961) and the composition of the hemicelluloses of legumes differs significantly from that of grasses (Gaillard 1965). Pectin is an important polysaccharide in legumes averaging 40–70 g /kg DM compared to 10–40 g /kg DM in grasses (Bailey 1973). In some tropical legume species much higher levels of pectin have been reported, for example, 140–200 g /kg in DM of *Stylosanthes* spp. (Pangway and Richards 1971). According to Bailey *et al.* (1971) levels of pectin and hemicellulose varied little during growth in legumes but there was a marked rise in stem cellulose content with advancing maturity. Legumes, whether whole plants, leaves or stems, tend to have higher lignin levels than do temperate grasses. Levels in the leaves and petioles are much less than in the stem and the content in all tissues increases with maturity (Harkin 1973).

There is considerable information available on the variation in cell wall constituents between grasses, particularly those of temperate origin (Bailey 1973). The interpretation of differences is, however, often difficult, due to the dominating effect of growth stage and leaf content. When species and varieties are compared at the same growth stage, differences are nevertheless apparent, as illustrated in Table 5.

Table 5. Differences in cell wall constituents of temperate grasses at ear emergence (g /kg dry matter) (Jones 1970b).

Species	Variety	Xylan	Cellulose	Lignin	DMD*
Perennial ryegrass (*Lolium perenne*)	Early flowering	85	177	48	0.73
	Late flowering	93	196	43	0.71
Cocksfoot (*Dactylis glomerata*)	Early flowering	79	149	65	0.75
	Late flowering	100	219	76	0.63
Timothy (*Phleum pratense*)	Early flowering	113	230	64	0.65
	Late flowering	125	235	79	0.64

**In vitro* DM digestibility coefficient (Tilley and Terry 1963)

Perennial ryegrass varieties are lower in lignin than cocksfoot and timothy while earlier flowering varieties are lower in structural constituents (and higher in digestibility) than later flowering types. Similar results have been reported more recently by Morrison (1980) who showed the lignin and hemicellulose content of ten temperate grasses to increase with maturity; perennial ryegrass was lower in both constituents than cocksfoot or timothy and tetraploid lower than diploid varieties. Less information appears to be available for tropical species but considerable variation has been shown in whole plant, stem and leaf structural constituents (Bailey 1973).

Significant genotypic variation has also been shown. Cooper (1973) has reviewed the data available on temperate and tropical species; Wilson (1965), for example, has shown genetic variation in the cellulose content of ryegrass and that cellulose content is strongly heritable. Significant genotypic variation has also been shown in the cell wall constituents of legumes (Gil *et al.* 1967).

Variation in the total content of structural constituents or their components is of lesser significance in the nutrition of the herbivore than the interrelations between the constituents. The polysaccharides of the cell wall of herbage are totally digestible once the lignin is removed (Bailey and Jones 1971). Their association with lignin, however, hinders attack by microbial enzymes to a varying extent depending primarily on the degree of lignification. Harkin (1973) emphasises the importance of covalent lignin-carbohydrate bonds and the resultant tight binding of lignin to the polysaccharide fraction. This has the effect of preventing the fibres swelling to allow penetration by microbial enzymes while the cross-linking and substitution inhibits enzyme attack. Bacon (1979) visualises a matrix material of the cell wall based on the highly substituted xylan chain having covalent bonds to lignin, enzymes could act on this chain until halted by points of attachment or by substituents. Lignin is thus seen as impeding digestion by its linkage to specific points of the polysaccharide chain. Richards (1976), however, sees lignin as a substance inhibiting digestion by preventing the physical attachment of bacteria to the cell walls. Minor constituents of the cell wall can also be of importance in relation to the degradability of the polysaccharides by enzymes, for example acetyl groups (Bacon *et al.* 1975) and various phenolic components (Hartley 1973; Hartley and Jones 1977). Silica has been shown to depress cell wall digestibility in temperate grasses (van Soest and Jones 1968) but Minson (1971) failed to show a similar effect in *Panicum* spp..

In view of the importance of lignification it is not surprising that differences in accessibility to rumen and fungal cellulases has been shown for different cell types and different anatomical characteristics. Akin and Amos (1975) and Selim *et al.* (1975) have shown the mesophyll cells are the first to be digested, followed by the phloem and epidermis, the most resistant cell walls being those of the xylem while the cutin layer of the epidermis is not digested. The characteristic specialised leaf anatomy of tropical, C_4, grasses is clearly important in determining the resistance offered to mechanical and microbial degradation (Norton 1982). Recent work by Wilson and Hattersley (1983), Wilson *et al.* (1983) and Akin *et al.* (1983) has related ease of digestion of bundle sheaths of leaf of *Panicum* spp. with cell wall structure and photosynthetic type. Leaf digestibility was higher for C_3 than C_4 types, the latter having a higher proportion of bundle sheath, vascular tissue, epidermis and schlerenchyma. The differing anatomical features of temperate grasses and legumes have been shown to result in a different particle geometry in the rumen during digestion (Moseley and Jones 1984).

HERBAGE CONSTITUENTS AS AN ENERGY SOURCE

Herbivores are able to derive a considerable, and often a major, proportion of their energy needs from the complex polysaccharides of the cell wall by virtue of fermentation by microbial organisms in specialised organs within their gut. This energy is additional to that derived from non-structural carbohydrates, protein and lipids etc. all of which are highly digestible (Minson 1976). The end products of fermentation are mainly volatile fatty acids and these are used as an energy source by the herbivore. The proportion of the various acids formed is influenced by herbage composition and affects the efficiency of utilisation of energy for maintenance and production (Armstrong 1982).

Since the gross energy content of different herbages of varying maturity is relatively constant (Raymond 1969; Minson 1980) digestible energy content is closely related to the digestibility of dry matter or organic matter. Assessment of the digestibility of these is, therefore, regarded as an acceptable measurement of the metabolisable energy of herbage since the energy losses in urine and fermentation gases also tend to remain relatively constant. Potential digestibility may not, however, always be realised if the rate of passage in the gut is accelerated or the diet is deficient in essential nutrients, such as nitrogen and sulphur, required for efficient microbial fermentation (Campling *et al.* 1962; Rees *et al.* 1974).

Forage digestibility is primarily determined by its cell wall content and its accessibility to enzyme attack during microbial fermentation. Techniques taking account of these factors, for example *in vitro* digestibility using rumen inoculum (Tilley and Terry 1963) or enzyme solubility (Jones and Hayward 1975), have thus been found to be more successful in predicting *in vivo* digestibility than those based simply on fibre content or fibre constituents such as cellulose or lignin.

While considerable information is available on the *in vivo* digestibility of herbage, the advent of precise laboratory techniques has greatly increased knowledge of the factors influencing digestibility and particularly the digestibility of plant parts and their significance.

These factors have been the subject of several major reviews (Raymond 1969; Hacker and Minson 1981).

Digestibility is very much influenced by advancing maturity since the proportion of stem increases and this is less digestible than leaf. At the same time the digestibility of the stem, and to a lesser extent the leaf, decreases as the tissues become more lignified. There is, therefore, a progressive decline in digestibility with age, values remaining high during vegetative growth and the early stages of stem elongation but falling rapidly after ear emergence (Minson *et al.* 1964). The decline in stem digestibility is particularly marked, Hacker and Minson (1982) quote a mean daily decline of 0.5–0.8 percentage units compared to 0.1–0.4 for leaf in a range of temperate grasses. Similar trends have been reported in red clover (*Trifolium pratense*) and lucerne but the decline is less obvious in white clover (*Trifolium repens*) since only the leaf and petioles are harvested (Cooper 1973). Differences are also less pronounced in regrowth or under frequent cutting or grazing. Walters *et al.* (1967) found differences in the digestibility of a range of grasses to be closely associated with leaf content and growth stage in primary growth but not in regrowth cuts.

Care is therefore needed in the interpretation of differences between species since in many instances these may merely reflect differences in maturity at the harvest date. Characteristic differences have, nevertheless, been established between species. Minson *et al.* (1964) showed clear differences in the *in vivo* digestibility of a range of temperate grasses, notably the low digestibility of cocksfoot compared to perennial ryegrass at the same growth stage, a difference independent of leaf content. Considerable variation has also been found between varieties within species, early flowering varieties generally being more digestible than late flowering ones at the same growth stage (Dent and Aldrich 1963; Raymond 1969). Tropical grasses are generally less digestible than temperate (Minson and McLeod 1970), as are the leaf and stem fractions of tropical compared to temperate grasses (Hacker and Minson 1981). These characteristics are related to the higher cell wall content and specialised anatomy of the C_4 grasses discussed previously. A survey of the published literature by Minson and Wilson (1980) showed the mean digestibility of temperate legumes and grasses to be 0.61 and 0.68 respectively compared to 0.57 and 0.55 for tropical legumes and grasses. The digestibility of tropical grasses also continuously declines with advancing maturity and significant differences have been noted between species of grasses and legumes (Reid *et al.* 1973). Some species show a relatively slow decline in digestibility with age and are thus of particular value in tropical pastures (Norton 1982).

Considerable genotypic variation in digestibility has been found within tropical and temperate species and there is clearly a potential for improving digestibility by breeding (Cooper 1973; Hacker 1982). Burton *et al.* (1967) were able, for example, to introduce higher digestibility to the standard coastal bermuda grass (*Cynodon dactylon*).

There are difficulties, however, if selection results in morphological changes. Cooper and Breese (1980) reported selection for digestibility as single plants in cocksfoot to result in marked morphological changes incompatible with survival in swards; selection may therefore need to be directed at specific tissues rather than whole plants.

The effects of various environmental factors on digestibility have been reviewed in detail by Wilson (1982). The main effect appears to be associated with growth temperatures (Dirven and Deinum 1977), a higher temperature resulting in decreased digestibility due to an increase in cell wall content. The effects of variation in other environmental factors, such as light, water supply and fertiliser applications, are either insignificant or difficult to interpret due to contradictory findings by different workers.

The digestibility of forage is also an important determinant of voluntary intake (Balch and Campling 1962; Raymond 1969) although important differences in intake have been noted between grasses at the same level of digestibility (Minson *et al.* 1964; Walters 1971). The chemical constituents of forage have an important influence on voluntary intake (Minson 1982) as have the availability of feed, diet selection and animal factors. These aspects are considered in detail by other contributors to this symposium.

TREES AND SHRUBS AS A SOURCE OF NUTRIENTS

The leaves of trees and shrubs (browse) are a component of most natural pastures and are common in cold mountainous areas and semi-arid regions. The species present and their relative density vary widely with climate, aspect, soil type and rainfall. The number of species is, therefore, large and only selected species that illustrate particular points or relationships will be considered here.

The amount of leaf eaten by herbivores is dependent on a wide range of both plant and animal factors. Browse is usually fibrous and may also contain tannins, essential oils or other aromatic compounds that decrease palatability. Browse species are, therefore, almost without exception, of low palatability and are eaten in the dry (or cold) season when herbaceous forage becomes either sparse or unpalatable. Animal species also differ in their preference for browse. Animals such as sheep, cattle, kongoni (*Alcelaphus* sp.), wildebeest (*Connochaetes* sp.) and kangaroos eat little browse, at least while herbaceous species are available. However, browsing animals such as Grant's gazelle (*Gazella granti*) and giraffe (*Giraffa camelopardalis*) may live predominantly on browse. Other animals, such as deer, elk and goats, are intermediate. Nevertheless, browsing animals still exercise widely differing preferences for individual plant species. Browse makes the greatest contribution to forage supplies where it occurs as a low shrubland, such as in the South African Karoo. Heather (*Calluna vulgaris*) may provide half the energy intake of sheep in the Scottish hill country (Herriot *et al.* 1981).

Browse as a source of energy

The energy available in browse is usually low, most species having digestibility coefficients within the range 0.45–0.55 (Walker 1980). The major contrast with herbaceous forages is that there is little seasonal variation in digestibility. Leigh *et al.* (1978) found essentially no seasonal changes in the *in vitro* digestibility of semi-arid fodder trees, and Bamualim *et al.* (1982) found a similar lack of variation in tropical browse plants. In the colder Scottish uplands, seasonal change in the digestibility of heather (*Calluna vulgaris*) was found, but limited to the range 0.42–0.56 (Milne 1974). As may be expected, the leaves of most browse species are high in both cell walls and lignin. They are important as forage, despite their low quality, because of the presence of foliage in the dry (cold) season. Animals such as giraffe (Pellew 1980) and elk (*Alces americana*; Hobbs *et al.* 1981) depend on browse at these times. A few browse species are of unusually high energy availability (for example *Leucaena leucocephala, Aspilia africana*). However, it is not unusual to find that such species are eaten only sparingly (with the notable exception of *L. leucocephala*) because of the presence of anti-nutritional factors (see below). In fact, nutritious and palatable shrubs are relatively uncommon because they lack the natural defense mechanisms which are required for persistence in a world of browsing predators.

Browse as a source of protein

The crude protein of browse is also relatively constant throughout the year and usually much higher than that of grasses during the dry season. Browse is therefore often viewed as a protein supplement for both livestock (Varma *et al.* 1982) and wildlife such as elk (*Alces alces*; Hobbs *et al.* 1981). An important characteristic of *Acacia* species in the Serengeti is that they produce a flush of new leaves at the end of the dry season and up to one month before the opening rains. The flush growth has a crude protein content of approximately 230 g /kg DM for the first 10 days and this provides an important supplement for the giraffe (Pellew 1980). Specific species and food niche situations are often important in defining the role of browse in the maintenance of herbivore populations. Surprisingly, leguminous browse species do not seem to be much higher in crude protein (mean 150 g /kg DM) than non-leguminous species (mean 130 g /kg DM) although there are many leguminous species that are high in protein (Lamprey *et al.* 1980).

Many browse species contain condensed tannins and other phenolic compounds which reduce the availability of protein and fibre, and reduce palatability. Although their role is not fully understood, they are nutritionally important components of browse. The magnitude of their effect can be gauged from an experiment by D. Pritchard (unpublished) in which sheep were fed on the leaves of mulga (*Acacia aneura*). Daily drenching with polyethylene glycol, which preferentially binds with the tannins, led to a 2-3 fold increase in food intake. The same treatment of eucalypts gave rise to increases in both intake and cell wall digestibility in the brushtail possum (*Trichosurus vulpecula*) (Cork 1984).

The true digestibility of protein in many browse species is therefore much less than that in herbaceous species. Wilson (1977) found the true nitrogen digestibility of four Australian tree species was 0.66-0.83 in sheep whilst Milne (1974) reported values of 0.55-0.75 for heather (*Calluna vulgaris*). Similarly, Mould and Robbins (1981) noted that ingested phenolics increased the faecal nitrogen excretion of elk when feeding on maple (*Acer* spp.). Diagayete (1981) found four West African browse species with tannin contents of 200 g /kg DM and eleven with tannin contents of 50 g /kg DM. He also found a logarithmic relationship between tannin content and *in vitro* protein digestibility, although much variation remains unexplained (r = 0.63). The effect is dependent on the protein - precipitating capacity of the tannins rather than total phenolic content (Kumar 1983). Nevertheless some families of browse shrubs are free of such problems. Saltbushes (*Atriplex* spp.), for instance, have true digestibilities of nitrogen in excess of 0.90 (Wilson 1977).

Tannins may also be important palatability factors and it is of interest that tannin content may increase after defoliation. Grazed shoots of blackbush (*Coleogyne ramosissima*) have a higher tannin content than ungrazed shoots and are subsequently avoided by goats (Provenza and Malechek 1984). Tannin content is therefore not necessarily a fixed characteristic of a species, but may vary with recent grazing history and with soil type (Barry and Forss 1983).

Assessment of nutritive value of browse

The development of general relationships between chemical composition and nutritive value would be of considerable value for future research on browse; these relationships appear to be more complex in browse than in pasture grasses and legumes although useful relationships may be found within restricted groups of species. Thus Bamualim *et al.* (1980) reports a negative correlation (r = -0.92) between digestibility and lignin content in tropical browse legume shrubs. However, this does not hold over a wider array of species. For instance, the shrub *Atriplex nummularia* has an apparent digestibility of organic matter of 0.69 compared with 0.40 for the tree *Heterodendrum oleifolium*, despite similar cell wall and lignin contents (Wilson 1977). The role of lignin in particular is ambiguous; Sidahmed (1981) reports an apparent negative digestibility of lignin in chapparal shrubs (*Quercus* and other species) by goats, which is attributed to the polymerization of tannins, whilst lignin in *Eucalyptus* leaves is partially digested (0.09-0.26) by marsupial folivores (Cork 1984). Species variation in cell wall structure and the distribution of lignin is likely, although analytical problems are also implicated. Reed (1986) concluded that the behaviour of phenolics and tannins in the detergent system is highly variable, with some fractions soluble and others insoluble in neutral detergent.

A similar problem with the interpretation of crude protein values arising from complexing by tannins, was mentioned earlier. These complexes are formed within the digestive tract so *in vitro* techniques applied to freeze dried samples may be required to assess the availability of nitrogen. *In vitro* techniques remain the most reliable methods for assessing digestibility, but these are subject to considerable bias in individual species (Milne 1977) and require *in vivo* standards. Older tables of crude protein and crude fibre, which are still the common source of information on individual species, are of little value.

The assessment of nutritive value is also hampered by the lack of relationship between intake and digestibility in browse. For instance, the intake of *Casuarina cristata* (OM digestibility = 0.29) by sheep is greater than the intake of *Atriplex nummularia* (OM digestibility = 0.69). It is not an uncommon observation that a species with a high nutritive value (e.g. *A. nummularia, Geijera parviflora*), has a low intake because of its content of other anti-nutritional factors. This does not detract from its value as a protein supplement.

A further complication for assessment of nutritive value is differences between herbivore species. Such differences are difficult to confirm because of variation arising from differences in relative intake and differences between individuals. Nevertheless, somewhat greater digestibility of NDF and ADF by goats than sheep is frequently reported (Louca *et al.* 1982). Frinton (1982) speculates on the presence of different metabolic strategies to overcome insoluble complexes of condensed tannin. Whilst there is no evidence to support this, it is probable that browsing animals such as giraffe and elk do digest more of the cell walls of browse than do sheep and cattle. Marsupial folivores obtain their full nutrient requirements from *Eucalyptus* leaves and, as mentioned earlier, digest significant amounts of lignin. There is evidence that detanning of leaf protein by microbial action occurs in the caecum and lower colon of possums and that this subsequently becomes available to the animal through coprophagy (O'Brien *et al.* 1986).

Usefulness of browse as a nutrient source

There are a few individual species of shrub that have high nutritive value. For example, the shrub *Leucaena leucocephala* is planted in tropical regions and *Atriplex nummularia* is planted in the semi-arid areas of South Africa. The vast majority are, however, of low nutritive value.

On occasions they provide a much needed protein supplement but more commonly their value lies in maintaining animals through periods of forage shortage. They are supplements or reserve forages so that nutritive value is comparative rather than absolute, that is comparative with alternative species that may grow on the same land unit under different management regimes.

They contain an array of anti-nutritional factors which affect palatability, digestibility and intake and these complicate the assessment of nutritive value. Analytical relationships derived from pasture plants cannot therefore be applied with confidence. Many opportunities remain for defining new predictive relationships in the many families of browse species and for defining differences in digestion of cell walls and protein between herbivore species.

Akin, D.E., and Amos, H.E. (1975) *Appl. Microbiol.* **29**, 692-701.

Akin, D.E., Wilson, J.R., and Windham, W.R. (1983). *Crop Sci.* **23**, 147-55.

Aitken, F.C., and Hankin, R.G. (1970). Vitamins in feeds for livestock. Technical Communication No.25. (Commw. Agric. Bureaux: Farnham Royal, UK.)

Armstrong, D.G. (1982). In "Nutritional Limits to Animal Production from Pastures." (Ed. J.B. Hacker.) pp. 225-44. (Commw. Agric. Bureaux: Farnham Royal, UK.)

ap Griffith, G. (1960). *Nature* (London) **185**, 627-8.

ap Griffith, G., and Walters, R.J.K. (1966). *J. Agric. Sci.* **67**, 81-9.

Bacon, J.S.D. (1979). Rep. Rowett Res. Inst. **35**, 99-108.

Bacon, J.S.D., Gordon, A.H., and Morris, E.J. (1975). *Biochem. J.* **149**, 485-7.

Bailey, R.W. (1964). *N.Z. J. Agric. Res.* 7, 496-507.

Bailey, R.W. (1973). In "Chemistry and Biochemistry of Herbage." Vol. 1. (Eds G.W. Butler and R.W. Bailey.) pp. 157-212. (Academic: London.)

Bailey, R.W., and Jones, D.I.H. (1971). *N.Z. J. Agric. Res.* **14**, 847-57.

Bailey, R.W., Allison, R.M., and O'Connor, K.F. (1971). *Proc. N.Z. Grassld Assoc.* **32**, 127-36.

Balch, C.C., and Campling, R.C. (1962). *Nutr. Abstr. Rev.* **32**, 669-86.

Bamualim, A., Jones, R.J., and Murray, R.M. (1980). *Proc. Aust. Soc. Anim. Prod.* **13**, 229-32.

Bamualim, A., Jones, R.J., and Murray, R.M. (1982). *Proc. Aust. Soc. Anim. Prod.* **14**, 677.

Barry, T.N., and Forss, D.A. (1983). *J. Sci. Fd Agric.* **34**, 1047-56.

Beever, D.E. (1982). In "Forage Protein Conservation and Utilisation." (Eds T.W. Griffiths and M.F. Maguire.) pp. 99-119. (Commission of the European Communities.)

Bland, B.F., and Dent, J.W. (1964). *J. Br. Grassld Soc.* **19**, 306-15.

Brady, C.J. (1976). In "From Plant to Animal Protein." Reviews in Rural Science II. (Eds T.H. Sutherland, J.R. McWilliam and R.A. Leng.) pp. 13-16. (Univ. New England: Armidale, NSW.)

Bugge, G. (1978). *Z. PflZucht.* **81**, 235-40.

Burton, G.W., Hart, R.H., and Lowrey, R.S. (1967). *Crop Sci.* **7**, 329-32.

Butler, G.W. (1973). In "Chemistry and Biochemistry of Herbage." Vol. 2. (Eds G.W. Butler and R.W. Bailey.) pp. 1–12. (Academic: London.)

Butler, G.W., and Bailey, R.W. (1973). (Eds.) "Chemistry and Biochemistry of Herbage." (Academic: London.)

Butler, G.W., and Jones, D.I.H. (1973). In "Chemistry and Biochemistry of Herbage." Vol. 2. (Eds G.W. Butler and R.W. Bailey.) pp. 127–62. (Academic: London.)

Campling, R.C., Freer, M., and Balch, C.C. (1962). *Br. J. Nutr.* **16**, 115–24.

Colman, R.L., and Lazenby, A. (1970). Proc. 11th Int. Grassld Congress, Surfers Paradise, Queensland. pp. 392–7.

Cooper, J.P. (1973). In "Chemistry and Biochemistry of Herbage." Vol. 2. (Eds G.W. Butler and R.W. Bailey.) pp. 379–417. (Academic: London.)

Cooper, J.P., and Breese, E.L. (1971). In "Potential Crop Production." (Eds P.F. Wareing and J.P. Cooper.) pp. 295–318. (Heinemann: London.)

Cooper, J.P., and Breese, E.L. (1980). *Proc. Nutr. Soc.* **39**, 281–6.

Cork, S.J. (1984). *Proc. Nutr. Soc. Aust.* **9**, 88–97.

Deinum, B. (1966). Proc. 10th Int. Grassld Congress, Helsinki. pp. 415–8.

Dent, J.W., and Aldrich, D.T.A. (1963). *J. Nat. Inst. Agric. Bot.* **9**, 261–81.

Diagayete, M. (1981). *Landwirtsch. Forsch.* **37**, 416–26.

Dijkshoorn, W. (1973). In "Chemistry and Biochemistry of Herbage." Vol. 2. (Eds G.W. Butler and R.W. Bailey.) pp. 163–88. (Academic: London.)

Dirven, J.G.P., and Deinum, B. (1977). *Forage Res.* **3**, 1–17.

Eglinton, G., and Hamilton, R.J. (1967). *Science* (New York) **156**, 1322–35.

Filmer, D.G. (1982). In "Forage protein in ruminant animal production." *Occ. Publ. Br. Soc. Anim. Prod.* **6**, 129–38.

Fleming, G.A. (1973). In "Chemistry and Biochemistry of Herbage." Vol. 1., (Eds G.W. Butler and R.W. Bailey.) pp. 529–66. (Academic: London.)

Ford, C.W., Morrison, I.M., and Wilson, J.R. (1979). *Aust. J. Agric. Res.* **30**, 621–33.

Frinton, P.A.J. (1982). Proc. 3rd Int. Conf. Goat Prod. Dis., Tucson, Arizona, p. 279.

Gaillard, B.D.E. (1965). *Phytochem.* **4**, 631–4.

Gil, H.C., Davis, R.L., and Barnes, R.F. (1967). *Crop Sci.* 7, 19–21.

Gomide, J.A., Noller, C.H., Mott, G.O., Conrad, J.H., and Hill, D.L. (1969). *Agron. J.* **61**, 116–20.

Hacker, J.B. (1982). In "Nutritional Limits to Animal Production from Pastures." (Ed. J.B. Hacker.) pp. 305–26. (Commw. Agric. Bureaux: Farnham Royal, UK.)

Hacker, J.B., and Minson, D.J. (1981). *Herb. Abstr.* **51**, 459–82.

Hacker, J.B., Strickland, R.W., and Basford, K.E. (1985). *Aust. J. Agric. Res.* **36**, 201–12.

Harkin, J.M. (1973). In "Chemistry and Biochemistry of Herbage." Vol. 1 (Eds G.W. Butler and R.W. Bailey.) pp. 323–74. (Academic: London.)

Hartley, R.D. (1973). *Phytochem.* **12**, 661–5.

Hartley, R.D., and Jones, E.C. (1977). *Phytochem.* **16**, 1531–4.

Hawke, J.C. (1963). *J. Dairy Res.* **30**, 67–75.

Hawke, J.C. (1973). In "Chemistry and Biochemistry of Herbage." Vol. 1. (Eds G.W. Butler and R.W. Bailey.) pp. 213–64. (Academic: London.)

Hegarty, M.P., and Peterson, P.J. (1973). In "Chemistry and Biochemistry of Herbage." Vol. 1. (Eds G.W. Butler and R.W. Bailey.) pp. 1–62. (Academic: London.)

Heinrichs, D.H. (1970). Proc. 11th Int. Grassld Congr., Surfers Paradise, Queensland. pp. 267–70.

Henderson, J.L., Edwards, R.S., and Hammerton, J.L. (1962). *Agric. Sci.* **59**, 199–206.

Herriott, J.B.D., Edwards, R.A., and Gray, A. (1981). *Occ. Symp. Br. Grassld Soc.* **12**, 174–5.

Hides, D.H., and Thomas, T.A. (1981). *J. Sci. Fd Agric.* **32**, 990–1.

Hobbs, N.T., Baker, D.L., Ellis, J.E., and Swift, D.M. (1981). *J. Wildl. Manage.* **45**, 156–71.

Hutton, E.M. (1970). *Adv. Agron.* **22**, 1–74.

Jarrige, R. (1961). *Annls Biol. Anim. Biochim.* **1**, 163–212.

Jarrige, R., and Minson, D.J. (1964). *Annls Zootech.* **13**, 117–50.

Jones, D.I.H. (1970a). *J. Agric. Sci.* **75**, 517–21.

Jones, D.I.H. (1970b). *J. Sci. Fd Agric.* **21**, 559–62.

Jones, D.I.H., and Hayward, M.V. (1975). *J. Sci. Fd Agric.* **26**, 711–8.

Jones, D.I.H., ap Griffith, G., and Walters, R.J.K. (1965). *J. Agric. Sci.* **64**, 323–8.

Kumar, R. (1983). *J. Agric. Fd Chem.* **31**, 1364–6.

Lamprey, H.F., Herlocker, D.J., and Field, C.R. (1980). In "Browse in Africa: The Current State of Knowledge." (Ed. H.N. LeHouerou.) pp. 33–54. (ILCA: Addis Ababa.)

Lee, C., and Smith, D. (1972). *Agron. J.* **64**, 527–30.

Leigh, J.H., Wilson, A.D., and Mulham, W.E. (1978). *Aust. Rangel. J.* **1**, 137–41.

Little, D.A. (1982). In "Nutritional Limits to Animal Production from Pastures." (Ed. J.B. Hacker.) pp. 259–83. (Commw. Agric. Bureaux: Farnham Royal, UK.)

Louca, A., Antoniou, T., and Hatzipanayiotou, H. (1982). Proc. 3rd Inter. Conf. Goat Prod. Dis., Tucson, Arizona. pp. 122–32.

Lyttleton, J.W. (1973). In "Chemistry and Biochemistry of Herbage." Vol. 1. (Eds G.W. Butler and R.W. Bailey.) pp. 63–103. (Academic: London.)

Mangan, J.L. (1982). *Occ. Publ. Brit. Soc. Anim. Prod.* **6**, 25–40.

Mannetje, L.'t (1982). In "Nutritional Limits to Animal Production from Pastures." (Ed. J.B. Hacker.) pp. 67-85. (Commw. Agric. Bureaux: Farnham Royal, UK.)

Milford, R., and Haydock, K.P. (1965). *Aust. J. Exp. Agric. Anim. Husb.* **5**, 13-7.

Milford, R., and Minson, D.J. (1965). Proc. 9th Int. Grassld Congr., Sao Paulo. pp. 815-22.

Milne, J.A. (1974). *J. Agric. Sci.* **83**, 281-8.

Milne, J.A. (1977). *J. Br. Grassld Soc.* **32**, 141-7.

Minson, D.J. (1971). *Aust. J. Agric. Res.* **22**, 589-98.

Minson, D.J. (1976a). In "Carbohydrate Research in Plants and Animals." Misc. Papers 12, Landbouwhoges School, Wageningen, The Netherlands. pp. 101-14.

Minson, D.J. (1976b). In "From Plant to Animal Protein." Reviews in Rural Science II. (Eds T.M. Sutherland, J.R. McWilliam and R.A. Leng.) pp. 27-30. (Univ. of New England: Armidale, NSW.)

Minson, D.J. (1980). In: "Standardization of Analytical Methodology for Feeds." (Eds W.J. Pigden, C.C. Balch and M. Graham.) pp. 72-78. (Int. Dev. Res. Centre: Ottawa.)

Minson, D.J. (1982). In "Nutritional Limits to Animal Production from Pastures." (Ed. J.B. Hacker.) pp. 167-82. (Commw. Agric. Bureaux: Farnham Royal, UK.)

Minson, D.J., and McLeod, M.N. (1970). Proc. 11th Int. Grassld Congr., Surfers Paradise, Queensland. pp. 719-22.

Minson, D.J., Harris, C.E., Raymond, W.F., and Milford, R. (1964). *J. Br. Grassld Soc.* **19**, 298-305.

Minson, D.J., and Wilson, J.R. (1980). *J. Aust. Inst. Agric. Sci.* **46**, 247-9.

Morrison, I.M. (1980). *Grass and Forage Science* **35**, 287-93.

Moseley, G., and Jones, J.R. (1984). *Br. J. Nutr.* **52**, 381-90.

Mould, E.D., and Robbins, C.T. (1981). *J. Wildl. Manage.* **45**, 323-34.

Norton, B.W. (1982). In "Nutritional Limits to Animal Production from Pastures." (Ed. J.B. Hacker.) pp. 89-110. (Commw. Agric. Bureaux: Farnham Royal, UK.)

Nowakowski, T.Z., Cunningham, R.K., Nielsen, K.F. (1965). *J. Sci. Fd Agric.* **16**, 124-34.

O'Brien, T.P., Lomdahl, A., and Sanson, G. (1986). *Aust. J. Zool.* **34**, 157-76.

Pangway, C., and Richards, G.N. (1971). *Aust. J. Chem.* **24**, 1041-8.

Pellew, R.A. (1982). In "Browse in Africa: The Current State of Knowledge." (Ed. H.N. LeHouerou.) pp. 223-31. (ILCA: Addis Ababa.)

Provenza, F.D., and Malechek, J.C. (1984). *J. Appl. Ecol.* **21**, 831-41.

Raymond, W.F. (1969). *Adv. Agron.* **21**, 1-108.

Reed, J.D. (1986). J. *Range Manage.* **39**, 5-7.

Rees, M.C., Minson, D.J., and Smith, F.W. (1974). *J. Agric. Sci.* **82**, 419-22.

Reid, R.L., and Horvath, D.J. (1980). *Anim. Feed Sci. Technol.* **5**, 95–167.

Reid, R.L., and Jung, G.A. (1965). *J. Anim. Sci.* **24**, 615–25.

Reid, R.L., and Jung, G.A. (1982). In "Nutritional Limits to Animal Production from Pastures." (Ed. J.B. Hacker.) pp. 21–43. (Commw. Agric. Bureaux: Farnham Royal, UK.)

Reid, R.L., Post, A.J., Olsen, F.J. and Mugerwa, J.S. (1973). *Trop. Agric.* (Trin.) **50**, 1–15.

Richards, G.N. (1976). In "Carbohydrate Research in Plants and Animals." Misc. Papers 12, Landbouwhoges School, Wageningen, The Netherlands. pp. 129–35.

Rogers, H.H. (1970). Proc. 11th Int. Grassld Congress, Surfers Paradise, Queensland. pp. A70–A77.

Russell, F.C., and Duncan, D.L. (1956). Minerals in pastures: deficiencies and excesses in relation to animal health. Tech. Communication No. 15. (Commw. Agric. Bureaux: Farnham Royal, UK.)

Selim, O.I., Wilson, D., and Jones, D.I.H. (1975). *J. Agric. Sci.* **85**, 297–9.

Sidahmed, A.E., Morris, J.G. Koong, L.J., and Radosevich, S.R. (1981). *J. Anim. Sci.* **53**, 1391–1400.

Smith, D. (1970). Proc. 11th Int. Grassld. Congress, Surfers Paradise, Queensland. pp. 510–4.

Smith, D. (1971). *J. Sci. Fd Agric.* **22**, 445–7.

Smith, D. (1972). *Agron. J.* **64**, 705–6.

Smith, D. (1973). In "Chemistry and Biochemistry of Herbage." Vol 1. (Eds G.W. Butler and R.W. Bailey.) pp. 105–55. (Academic: London.)

Tilley, J.M.A., and Terry, R.A. (1963). *J. Br. Grassld Soc.* **18**, 104–11.

Underwood, E.J. (1981). "The Mineral Nutrition of Livestock." 2nd ed. (Commw. Agric. Bureaux: Farnham Royal, UK.)

van Soest, P.J. (1963a). *J. Assoc. Official Agric. Chem.* **46**, 825–9.

van Soest, P.J. (1963b). *J. Assoc. Official Agric. Chem.* **46**, 829–35.

van Soest, P.J., and Jones, L.H.P. (1968). *J. Dairy Sci.* **51**, 1644–8.

van Soest, P.J., and Moore, L.A. (1965). Proc. 11th Int. Grassld. Congr., Sao Paulo, Vol. 1, pp. 783–9.

Varma, A., Yadav, B.P.S., Sampath, K.T., and Roy, D.J. (1982). *Ind. J. Anim. Sci.* **52**, 859–65.

Waite, R., and Boyd, J. (1953). *J. Sci. Fd Agric.* **4**, 197–204.

Walker, B.H. (1980). In "Browse in Africa: The Current State of Knowledge." (Ed. H.N. LeHouerou.) pp. 7–24. (ILCA: Addis Ababa.)

Walters, R.J.K., ap Griffith, G., Hughes, R., and Jones, D.I.H. (1967). *J. Br. Grassld Soc.* **22**, 112–6.

Walters, R.J.K. (1971). *J. Agric. Sci.* **76**, 243–52.

Whitehead, D.C. (1966). "Nutrient Minerals in Grassland Herbage." Mimeo Publ. 1/1966. (Commw. Agric. Bureaux: Farnham Royal, UK.)

Whitehead, D.C. (1970). "The Role of Nitrogen in Grassland Productivity." Bulletin 48. (Commw. Agric. Bureaux: Farnham Royal, UK.)

Wilkins, R.J. (1969). *J. Agric. Sci.* **73**, 57–64.

Wilkinson, S.R., and Lowrey, R.W. (1973). In "Chemistry and Biochemistry of Herbage." Vol. 2. (Eds G.W. Butler and R.W. Bailey.) pp. 248–316. (Academic: London.)

Wilman, D. (1970). *J. Br. Grassld Soc.* **25**, 242–5.

Wilson, A.D. (1977). *Aust. J. Agric. Res.* **28**, 501–8.

Wilson, D. (1965). *J. Agric. Sci.* **65**, 285–92.

Wilson, D., and Bailey, R.W. (1971). *J. Sci. Fd Agric.* **22**, 335–7.

Wilson, J.R. (1982). In "Nutritional Limits to Animal Production from Pastures." (Ed. J.B. Hacker.) pp. 111–31. (Commw. Agric. Bureaux: Farnham Royal, UK.)

Wilson, J.R., and Ford, C.W. (1971). Aust. *J. Agric. Res.* **22**, 563–71.

Wilson, J.R., and Hattersley, P.W. (1983). *Grass Forage Sci.* **38**, 219–23.

Wilson, J.R., and Minson, D.J. (1980). *Trop. Grasslds.* **14**, 253–9.

Wilson, J.R., Brown, R.H., and Windham, W.R. (1983). *Crop Sci.* **23**, 141–6.

5

SECONDARY COMPOUNDS OF FORAGES

by T.N. Barry and B.J. Blaney

Secondary compounds have been so named because they were originally thought not to be involved in the biochemical processes which form the basis of growth and reproduction in plants. Subsequently, it has been found that secondary compounds are produced by some plants as defence mechanisms against insect and fungal attack, and in others as a mechanism for restricting grazing by herbivores, thus ensuring survival of the plant. In this review the term secondary compound will also include compounds produced by fungi growing on forage plants that are eaten by herbivores.

Rather than attempt to cover the vast array of secondary compounds produced by plants and fungi, the review will be divided into sections each dealing with representatives of a separate group of compounds. Within each section, sub-sections will describe the chemistry of the compound(s), factors affecting their concentration in plants, and the effect of the secondary compound(s) upon animals and upon forage nutritive value. Secondary compounds may be either toxic to animals (e.g. cyanide, nitrate and fluoroacetate), cause sub-clinical losses in productivity, (e.g. some mycotoxins and high concentrations of condensed tannins), enhance nutritive value (e.g. low concentrations of condensed tannins), reduce mineral availability (e.g. oxalate) or improve plant persistency (e.g. peramine). The important groups of secondary compounds are listed in Table 1, together with appropriate references.

TANNINS

Tannins and lignins are phenolic compounds that are found in all classes of vascular plants and are derived from the shikimic acid pathway. Tannins help to inhibit attack on lignified tissues by fungi and bacteria. This not only defends against parasitism, but also reduces saprophytic attack on plant detritus in the soil, ensuring a better supply of soil nutrients such as nitrogen and phosphorus for subsequent plant regeneration. Tannins are also of great importance as a defence mechanism against over-grazing by vertebrate herbivores and insect attack (Swain 1979). Recent research, reviewed here, has been aimed at examining animal responses as the concentration of tannin in the diet is increased, and in the case of condensed tannins has identified detrimental effects at high concentrations but beneficial effects at low concentrations.

NUTRITION OF HERBIVORES
ISBN 0 12 750052 9

Table 1. Common secondary compounds affecting the feeding value of forages.

Class	Examples	References
Alkaloids	Pyrrolizidine alkaloids	Bull *et al.* (1968)
	Fescue alkaloids	Cheeke and Shull (1985)
Glycosides	Cyanogenic glycosides	Conn (1979)
	Coumarin	Cheeke and Shull (1985)
	Isoflavones and coumestans	Reviewed here
Steroids and terpenes	Saponins	Applebaum and Birk (1979)
Proteins and Amino acids	Bloat-producing protein	Majak *et al.* (1983)
	Mimosine	Hegarty (1982)
	Indospicine	Hegarty and Pound (1970)
	Brassica anaemic factor	Smith (1980) Barry *et al.* (1985)
Simple acids and their salts	Oxalates	Reviewed here
	Nitrates	Buck *et al.* (1976)
	Fluoroacetate	Reviewed here
Polyphenols	Tannins	Reviewed here McLeod (1974)
Mycotoxins	Zearalenone	Reviewed here
	Trichothecenes	Reviewed here
	Sporidesmin	Reviewed here
	Phomopsin	Reviewed here
	Lolitrems	Reviewed here
	Swainsonine	Broquist *et al.* (1985)
	Slaframine	Broquist *et al.* (1985)
	Ergot alkaloids	Cheeke and Shull (1985)

Chemical properties

Tannins comprise both condensed and hydrolysable tannins. Their structures have been described by McLeod (1974) and Reid *et al.* (1974). Condensed tannins (CT) occur in the leaves and stems of sainfoin (*Onobrychis viciifolia*; MW 17,000–28,000) and of *Lotus* species (MW 6,000–7,000), in the flower petals only of white clover (*Trifolium repens*; MW 8,500–9,100); they have not been detected in lucerne or in the temperate grass species (Jones *et al.* 1976). During disintegration of plant material, such as results from chewing by animals, CT reacts with plant proteins by hydrogen bonding to form a complex. The complex is stable and insoluble at pH 3.5–7.0, but unstable and releases protein at pH < 3.0 and at pH 8.0 (Jones and Mangan 1977). Hydrolysable tannins are mainly found in fruit pods and plant galls (McLeod 1974), and have not been detected in temperate grasses and legumes grown in New Zealand. Hydrolysable tannins bind strongly to proteins at pH 3–4, but to a continuously decreasing extent above pH 5 (Loomis and Battaile 1966).

Because CT and hydrolysable tannins are likely to occur in different plants and exert different actions upon grazing herbivores, it is important that chemical methods used to measure the concentration of the tannin should be specific to the particular tannin under investigation. Analytical methods that measure total phenolic compounds in plants give no indication of their likely effects upon herbivores. In New Zealand, CT in temperate legumes are determined by the vanillin–HCl procedure (Broadhurst and Jones 1978). Barry and Forss (1983) defined free CT as that not protein-bound and precipitated by high speed centrifugation of plant mascerates, and thus available to react with other proteins such as those of rumen micro-organisms and the gut wall. Because of the paucity of data on hydrolysable tannins, only CT will henceforth be considered in this section.

Table 2. Effect of three years annual single superphosphate application upon the condensed tannin concentration in primary growth *Lotus pedunculatus* grown under low soil fertility conditions.

	Fertiliser application			
kg P/ha	0	9	18	27
kg S/ha	0	11	22	33
	(3rd year of annual fertiliser application)			
Available soil P (µg/ml)	4	5	5	7
Available soil SO_4-S (mg/kg)	5	7	10	21
DM yield (kg/ha)	150	829	1124	1347
Condensed tannin (g/kg DM)	80	54	41	36

From Barry and Forss (1983). Soil pH 4.7. CT = 85.3 − 0.038 DMY

Factors affecting CT concentration in plants

Concentrations of CT in forages are affected by plant genotype and level of soil fertility. Lowther *et al.* (1987) found that when grown under low soil fertility (pH 5.2) conditions, concentrations of CT in *Lotus corniculatus* (1.3 – 3.9 g /kg DM) were lower than in *Lotus pedunculatus* (cv. Grasslands Maku; 58 – 98 g /kg DM). Within *Lotus corniculatus*, CT concentrations were lower for semi-erect (1.3 – 8.4 g /kg DM) than for erect types (11.6 – 39.0 g /kg DM). Application of P+S fertiliser stimulates DM yield and reduces forage CT concentration in *Lotus pedunculatus* (cv. Grasslands Maku) grown under cold conditions in acid low fertility soils (Table 2). When grown under warm conditions in high fertility soils, *Lotus pedunculatus* contains 20 g CT /kg DM (John and Lancashire 1981), whilst Fakir and Renumex cultivars of sainfoin contained respectively 30 and 28 g CT /kg DM (Barry and Manley 1986).

Polyethylene glycol (PEG; MW 3,350) binds with CT, and can thus prevent CT reacting with protein or displace CT from a CT:protein complex (Jones and Mangan 1977; Barry and Manley 1986). Spraying legumes containing high concentrations of CT with varying rates of PEG can thus be used as a means of reducing the "reactive" CT concentration.

Table 3. Effect of reducing dietary-reactive condensed tannin concentration in *Lotus pedunculatus* grown under low soil fertility conditions upon cell wall digestion, voluntary intake and growth of young sheep. Unconditioned lambs were transferred direct from grazing ryegrass/clover pasture to lotus, whereas conditioned lambs from the same flock grazed non-experimental high CT lotus for 8 weeks before being introduced to the experimental areas.

	Control	PEG supplemented	SE
Apparent digestibility:†			
Cellulose	0.53	0.58	0.013
Hemicellulose	0.51	0.59	0.010
Voluntary intake:†			
MJ ME/kg $W^{0.75}$ /d	0.48	0.69	0.049
Reactive condensed tannin			
Total†	62.9 (5.4)*	6.8 (0.5)	
g/kg DM‡	90.0 (12.4)		
Liveweight gain (g/day)			
Unconditioned‡	45	106	
Conditioned	68	83	10.9

† From Barry and Duncan (1984) ‡ From Barry (1985)
* Free condensed tannin (g/kg DM)

Effects of CT upon forage nutritive value

Several authors have observed poor performance in animals fed forages high in CT, and PEG supplementation experiments (Table 3) specifically show that high concentrations of *Lotus* CT (60–90 g/kg DM) depress

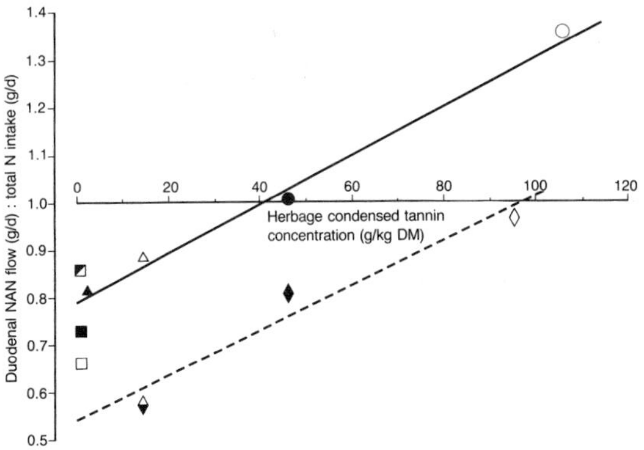

Figure 1. Duodenal non-ammonia nitrogen (NAN) flow per unit of total N intake in sheep fed fresh *Lotus* species. ——primary growth (equation 1); ---- secondary growth (equation 2). o, ● high and low tannin *Lotus pedunculatus*. △ , ▲ high and low tannin *Lotus corniculatus*. Non-tannin containing species perennial ryegrass (□), white clover (■) and short rotation ryegrass (◪). ◊ , ♦ , ◊ *Lotus pedunculatus* treated with zero, low and high rates of PEG (MW 3,350) to lower reactive tannin concentration. Tannin refers to condensed tannin only. From Barry *et al.* (1986).

cell wall digestion, voluntary intake and liveweight gain (LWG). These data also indicate that sheep can at least partially adapt to high dietary CT concentrations.

Due to their reactivity with proteins, increasing dietary reactive CT concentration in the range 0–110 g/kg DM linearly increases the duodenal flow of non-ammonia nitrogen (NAN) per unit of total N consumed (D) in sheep fed fresh lotus diets (Fig. 1). The equations for primary (1) and secondary (2) growth plant material are:

$$D = 0.79 + 0.0052\ CT \qquad r = 0.99 \qquad (1)$$

$$D = 0.54 + 0.0047\ CT \qquad r = 0.92 \qquad (2)$$

Thus, CT can be used to prevent the large loss of N across the rumen which occurs in sheep and cattle fed fresh diets of ryegrass and white clover, generally equivalent to about 30% of the N eaten (Fig. 1; Beever and Siddons 1984). The same protein-precipitating action also prevents bloat (Jones *et al.* 1973), and legumes containing > 20 g CT/kg DM are non-bloating.

Although apparent post-ruminal NAN absorption (proportion duodenal flow) in sheep fed lotus diets containing 46 and 106 g CT /kg DM (0.71 and 0.67 respectively; Barry and Manley 1984) was less than found for ryegrass and white clover (0.76; MacRae and Ulyatt 1974), the overall values obtained are nevertheless quite high. This indicates that the bonds between CT and protein must be reversible under the low pH conditions of the abomasum and the high pH of the small intestine

(Jones and Mangan 1977). Using PEG-treated lotus as a reference, Waghorn *et al.* (1986) found that the presence of 22 g CT /kg DM in *Lotus corniculatus* increased abomasal flow and apparent absorption from the small intestine of essential amino acids (EAA) more than that of non-essential amino acids (NEAA; Table 4). Possible reasons are selective protection of EAA in the forage protein, which seems unlikely, or modification of the rumen microbial population by free CT, which seems more probable.

Table 4. The effect of condensed tannins (22 g/kg DM) upon the digestion of amino acids in sheep fed fresh *Lotus corniculatus*.

	Essential*		Non-essential	
	Control	PEG	Control	PEG
Intake (g/d)	98.9	98.9	97.9	97.9
Abomasal flow				
grams per day	84.7	55.5	68.6	59.1
as proportion of intake	0.86	0.56	0.70	0.60
Apparent absorption from small intestine				
grams per day	58.8	36.2	37.4	41.3
as proportion of abomasal flow	0.67	0.67	0.54	0.67
as proportion of intake	0.59	0.37	0.38	0.42

From Waghorn *et al.* (1986). Groups (n=3) were continuously infused into the rumen with water (control) or polyethylene glycol (PEG; MW 3,350), with the latter designed to bind all the condensed tannin.
* excluding arginine, because of co-elution of another compound in some samples.

Lotus CT up to 25 g /kg DM appear to have no effect upon rumen carbohydrate (CHO) digestion (John and Lancashire 1981; Waghorn *et al.* 1986), but concentrations between 25 and 100 g /kg DM cause depressions in a dose-dependent manner (Barry and Manley 1984; Barry *et al.* 1986a). Hemicellulose digestion is depressed most, followed by water-soluble CHO and pectin, with cellulose digestion being least affected. Barry and Manley (1986) suggested that these effects are caused by free CT precipitating CHO-degrading enzymes excreted by rumen micro-organisms. However, these depressions in the rumen of sheep fed high CT lotus were compensated by increased post-ruminal digestion of water-soluble CHO, pectin and hemicellulose, with apparent digestibility in the entire gastro-intestinal tract being little affected.

Purchas and Keogh (1984) found the carcasses of lambs grazing *Lotus pedunculatus* containing 10–30 g CT /kg DM to contain less fat than the carcasses of lambs grazing white clover. Barry *et al.* (1986b) attributed this to both lower rates of lipogenesis and faster rates of lipolysis in lambs grazing lotus compared with those grazing white clover, with the effects being attributable to CT.

In contrast to high concentrations of lotus CT (90–100 g /kg DM), it therefore appears that low concentrations of lotus CT (20–30 g /kg DM) improve nutrient utilisation by ruminants, principally by reducing forage protein degradation in the rumen, thus simultaneously eliminating bloat and increasing amino acid supply to the animal. The selective increase in EAA absorption (Waghorn *et al.* 1986) and increase in growth hormone secretion (Barry *et al.* 1986b) caused by CT could be contributing reasons why N retention in sheep fed legumes containing low CT concentrations is greater than might be expected from duodenal NAN flows (Egan and Ulyatt 1980; John and Lancashire 1981).

MYCOTOXINS

Mycotoxins are defined as fungal metabolites that are toxic to animals. Vertebrate foragers are exposed to mycotoxins from fungi growing saprophytically on dead leaf litter in pasture, in endophytes of pasture grasses and in infected grains and stubbles of forage crops. Some important representatives of mycotoxigenic fungi affecting foraging animals are discussed below.

Acremonium loliae

Acremonium loliae is a fungal endophyte that infects perennial but not annual or short rotation ryegrasses. Hyphae of the endophyte grow in between plant cells throughout all parts of infected plants. Recent studies show an association between the amount of endophyte in perennial ryegrass (*Lolium perenne*, PRG) and the incidence and severity of the neurotoxic disorder ryegrass staggers (Fletcher and Harvey 1981; Fletcher 1982, 1986). Also, presence of the endophyte in PRG has been associated with increased resistance of the plant to attack by Argentine stem weevil (*Listronotus bonariensis*). Concurrently, the compounds responsible for causing ryegrass staggers (Gallagher *et al.* 1981) and conferring insect resistance (Rowan *et al.* 1986) have been isolated and their chemical structures determined.

Chemistry. Ryegrass staggers is caused by the production of Lolitrem B by the endophyte in infected plants. This complex compound (Fig. 2a) is a substituted indole of MW 685 (Gallagher *et al.* 1984), is strongly lipotrophic and can be routinely determined by HPLC with fluorescence detection (Gallagher *et al.* 1985). Insect resistance is believed to be caused by the production of peramine, which has a structure as shown in Fig. 2b, is of MW 247, and is strongly polar (Rowan *et al.* 1986).

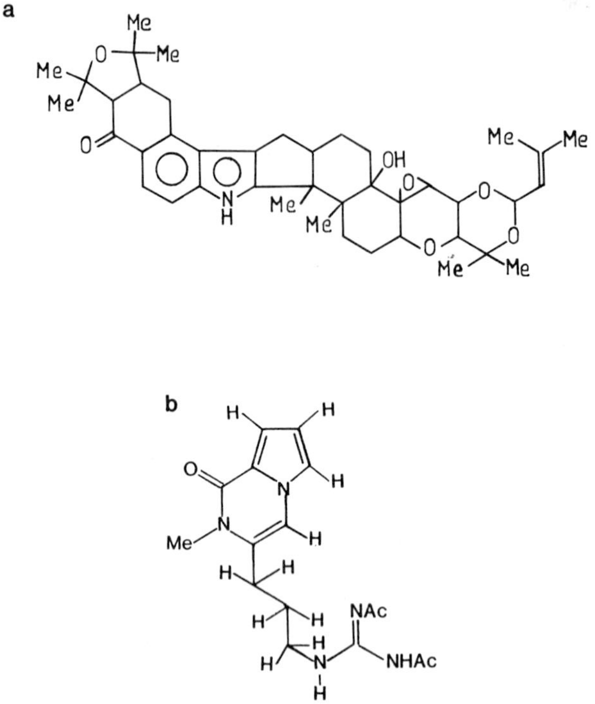

Figure 2. Chemical structure of compounds produced by the fungal endophyte *Acremonium loliae* in infected perennial ryegrass plants (a) Lolitrem B, MW 685, non-polar, responsible for the neurotoxic disorder ryegrass staggers in grazing herbivores (Gallagher *et al.* 1984, 1985). (b) Peramine, MW 247, strongly polar, acts as a feeding deterrent to adult Argentine stem weevils (Rowan *et al.* 1986).

Concentration in plants. The amount of total endophyte in infected PRG follows a strong seasonal trend, with maximum values in mid-summer and minumum values in mid-winter (di Menna and Waller 1986). Within PRG, some varieties or cultivars contain higher concentrations of endophyte than others. Old pasture ecotypes tend to have high concentrations, as probably these are the plants that have survived insect attack for a long period (i.e. 20 years). Endophyte is spread due to its incorporation into seed, and treatment of high endophyte seed with a fungicide, notably prochloraz, is a good experimental method of producing low and high endophyte plants of the same cultivar (Latch 1983).

There is little information to date on the actual concentrations of lolitrem B and peramine in infected PRG. Mean concentration of lolitrem B in the whole plant of low and high toxin samples of PRG were respectively 0.2 and 1.7 μg /g DM (Gallagher *et al.* 1985). However, high toxin varieties of PRG sampled in the spring show higher concentrations in the basal leaf sheath/stem (2.7 μg /g DM) than in the upper leaf sheath/stem (1.4 μg /g DM) or in the leaf blade (0.2 μg /g DM; Gallagher *et al.* 1985). Ryegrass seed cleanings which produced symptoms of ryegrass staggers when fed to horses contained 5.3 μg lolitrem B /g DM (Munday *et al.* 1985). In PRG seedlings grown from endophyte-infected seed, maximum concentrations of lolitrem B (25 μg /g DM) occurred in the lower leaf sheath after 5 weeks, followed by a slow decline to 15–20 μg /g DM. High values (20 μg /g DM) have also been recorded in inflorescence tissue (R.T. Gallagher, personal communication). In preliminary investigations, the concentration of peramine in endophyte-infected PRG was in the range 5.0–7.5 μg /g DM, which was the minimum effective concentration needed to deter feeding of adult Argentine stem weevils (Gaynor and Rowan 1986).

Effects upon plants and animals. Ryegrass staggers, caused by the ingestion of lolitrem B, is a nervous disorder affecting sheep, cattle, deer and horses grazing PRG-dominant pastures. In laboratory studies with mice, lolitrem B has been shown to be a very potent tremorgen, with single injections inducing tremors over a longer period than single injections of the well-known tremorgenic neurotoxin aflatrem, produced by the fungus *Aspergillus flavus* (Gallagher and Hawkes 1986). In grazing herbivores the disease is characterised by severe inco-ordination, with the front limbs being worst affected, and normally occurs in mid-summer when endophyte concentrations in PRG are highest. Severely affected animals fall over at the slightest excitement, such as during stock shifts, and hence the disorder affects animal management procedures. The disorder itself does not cause death, but stock losses can occur from effects of the inco-ordination, such as falling into dams and water troughs and drowning. An unusual feature of the disorder is lack of any specific lesions, and the eventual complete recovery of affected animals, generally within 6–8 weeks of cessation of ingestion. In addition to the above clinical symptoms of ryegrass staggers, preceding sub-clinical effects are a reduction in liveweight gain in sheep and a reduction in plasma prolactin concentration (Fletcher and Barrell 1984; Fletcher 1986; Table 5).

In contrast to the detrimental effects of high endophyte levels upon animal performance, presence of high levels of endophyte increase DM yields and tiller density of PRG cultivars grown in a dry summer location (Table 5). This effect is probably due to the presence of peramine in the high endophyte plots, as presence of the endophyte has been associated with reduced adult feeding of the Argentine stem weevil, and consequently reduced egg laying and larval development (Gaynor and Hunt 1983; Barker *et al.* 1984 a, b).

The endophyte story therefore presents a dilemma for New Zealand grazing systems at this time, which are dependent upon a 75:25 mixture of PRG with white clover. High endophyte strains of PRG will have greater plant DM yields and persistency, but will cause ryegrass staggers and depressed animal growth. One possibility, currently being followed by New Zealand researchers, is to screen populations of endophyte with a view to establishing if strains exist that produce peramine but not lolitrem B. Future research should also establish the physiological and biochemical effects of lolitrem B and peramine upon animals; for instance, as peramine at 5 µg /g DM acts as a feeding deterrent to adult Argentine stem weevils it is also possible that at the same or higher concentration it may deter grazing by herbivores and so reduce their voluntary intake. As symptoms of ryegrass staggers have also been reported in Australia, South Africa and the UK, the implications of endophyte association with PRG is probably widespread.

Table 5. Effect of endophyte status upon plant vigour and yield, together with animal growth, in young sheep grazing plots of perennial ryegrass during the spring–early summer period*

Cultivar	Ariki		Nui		
Endophyte status	low	high	low	high	SE
Tillers containing endophyte (%)	0	82	1.7	90	
Tiller density (1000/m^2)	2.1	12.6	3.9	9.2	0.93
DM yield (t/ha)					
23 Oct – 24 Dec	2.9	3.6	2.1	2.5	
25 Dec – 27 Feb	0.9	3.2	1.4	3.1	
Liveweight gain (g/d)*	133	115	113	102	

From Fletcher (1986) * 21 Sept–4 Jan

* Plots located near Christchurch, New Zealand, with low lines produced by storing seed at room temperature for 3 years until endophyte died out. All plots grazed at equal pasture allowances with yearling sheep (initial liveweight 40 kg).

Pithomyces chartarum

Pithomyces chartarum is a fungal contaminant of pasture that causes animal health problems in the North Island of New Zealand and also occurs sporadically in Victoria and Western Australia. Spore numbers build up in herbage as the summer progresses, and very large increases

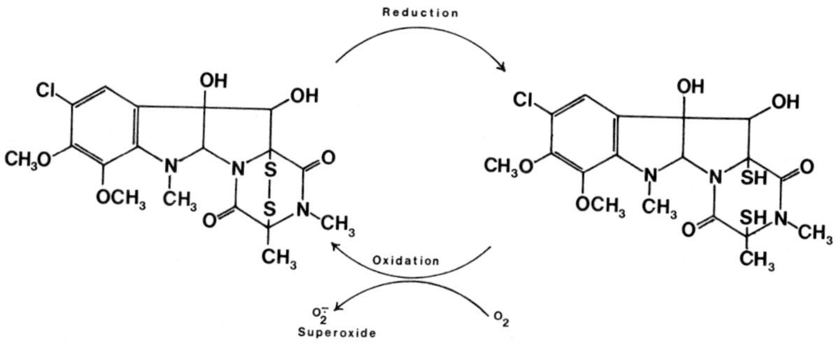

Figure 3. Chemical structure of sporidesmin, and its proposed cyclic mechanism of autoxidation/reduction to generate the superoxide (0_2^-) anion, now suggested as the agent causing damage to bile ducts in the liver which lead to symptoms of the disease "facial eczema". (From Munday 1982, 1984, 1985).

in numbers occur when humidity is high (close to 100%) and temperatures at soil level are above 12°C for a period of three or more days. In particular, spore numbers build up under these conditions if the pasture is very dry, or contains substantial amounts of dead plant material.

Chemistry. Spores produced by *Pithomyces chartarum* contain the compound sporidesmin (Fig. 3), which is known to cause the disease facial eczema in animals ingesting herbage with a high spore count.

Effects upon animals. Symptoms of facial eczema (FE) have been described previously, and are caused by inflammation and damage to bile ducts in the liver by sporidesmin. Consequently, degradation products of chlorophyll (phylloerythrin) and bile are released into the general blood circulation, causing jaundice, photo-sensitisation of the skin, loss of production and death due to liver failure in extreme cases. Sub-clinical FE causes losses in production, such as reduced reproductive performance, which are not immediately obvious to the farmer (Towers 1986). Recent research (Munday 1982, 1985) has shown the disulphide group of sporidesmin to undergo auto-oxidation with molecular oxygen (O_2), yielding reduced (dithiol) sporidesmin and the superoxide anion (O_2^-; Fig. 3). The latter is now believed to be the causal agent of bile duct inflammation and subsequent liver degeneration, and the auto-oxidation reaction producing it is strongly catalised by copper ions. Serum concentration of the enzyme γ-glutamyl-transferase (GGT; EC 2.3.2.2.) has been shown to be an accurate predictor of liver damage (Towers and Stratton 1978; Towers and Smith 1978), and is now routinely used as an index of liver damage caused by facial eczema.

Protective measures. Following prediction of spore build up in pasture from weather forecasting, treatment of pasture with fungicides and use of "safe" crops such as brassicas have been effective (Towers 1986). More recent findings are the protective effect of drenching with zinc salts (Smith *et al.* 1977; Towers and Smith 1978). Zinc is believed to exert this effect through forming a mercaptide with sporidesmin, and through reducing the intestinal absorption and other reactions of copper, thus leading to a reduction in superoxide anion formation (Munday 1984, 1985). Another successful development has been screening of rams for tolerance to sporidesmin, using serum GGT concentrations as an index of liver damage, with FE tolerance being passed on to the progeny (Towers *et al.* 1983; Towers 1986).

Phomopsis leptostromiformis

Phomopsis leptostromiformis is a pathogen and saprophyte largely confined to *Lupinus* spp. It produces several mycotoxins which cause the disease lupinosis in sheep and cattle grazing lupin stubble (Gardiner and Parr 1967; Gardiner 1967). The structure of the principal mycotoxin, phomopsin, has only recently been clarified (Fig. 4a; Mackay *et al.* 1986). Lupinosis was first reported in Germany and subsequently Australia and South Africa. Lupinosis is regarded as the most economically damaging poisoning disease of animals in Australia (Culvenor 1985), and lupin plantings have been severely reduced for fear of lupinosis. The disease in sheep presents itself as depression, inappettance, loss of weight and jaundice.

The mechanism of phomopsin toxicity has been reported by Peterson (1986). Phomopsin binds to tubulin and thus by inhibiting microtubule assembly and function, it causes arrest of mitosis and interference with microtubule-dependant intracellular transport mechanisms (e.g. movement of lipids and some bile components). Phomopsin also affects some transport associated cell activities. Acute toxicity can be largely related to these effects but chronic toxicity involves additional mechanisms that are not shared by other microtubule poisons. The liver is the principal organ in both acute and chronic poisoning and failure of liver function is the common cause of death.

In Western Australia, increased infection levels of lupin crops seem to be associated with higher rainfall and with susceptibility of some cultivars (Wood and Petterson 1985). In lupin stubbles, phomopsin A levels are normally low at plant senescence and remain so unless seasonal and environmental conditions favour toxin production. The majority of field samples contain less than 20 mg phomopsin A /kg (Culvenor 1986), but the hazard to sheep may be seen from the cumulative nature of the toxicity; the oral LD_{50} of phomopsin A in sheep is *ca* 1 - 2 mg /kg liveweight (Peterson 1986). Some success in reducing the level of infection of lupin stubbles and consequent toxicity

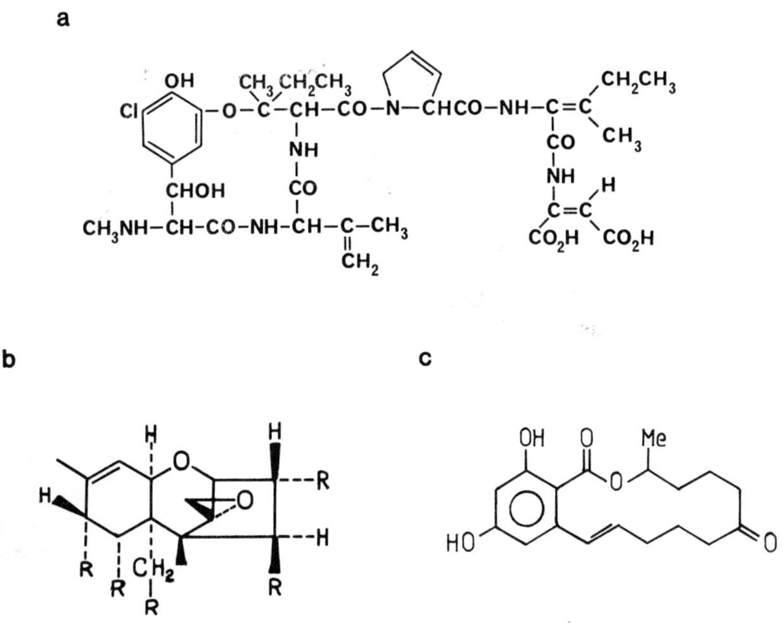

Figure 4. Chemical structure of (a) Phomopsin A, produced by *Phomopsis leptostromiformis*, (b) Basic nucleus of the trichothecenes, produced by *Fusarium* species, and (c) Zearalenone, produced by *Fusarium* species in ryegrass/clover pasture.

has been achieved by preparing hay from the stubbles (Allen *et al.* 1978).

Fusarium species

Fusarium species are a very important group of fungi which includes many plant pathogens. Grains infected with *Fusarium* spp. also cause a number of diseases of humans and domestic animals. The most important group of mycotoxins produced by *Fusarium* spp. are the trichothecenes (Fig. 4b), which are also produced by *Myrothecium* spp., *Trichoderma* spp., *Cephalosporium* spp. and *Stachybotrys* spp. (Cole and Cox 1981, pp. 152–263; Ichinoe and Kurata 1983). Biochemically, trichothecenes are potent inhibitors of DNA and protein synthesis in mammalian cells. Other mycotoxins produced by *Fusarium* species include butenolide, moniliformin and zearalenone (Yoshizawa 1983).

Toxigenic *Fusarium* spp. include *F. graminearum*, *F. sporotrichiodes*, *F. equiseti*, *F. poae*, *F. culmorum* and *F. moniliforme*.

There are several instances on record of trichothecenes causing disease characterised by inappetence, oral lesions, skin necrosis, haemorrhage, scouring and decreased production in domestic animals such as cattle, sheep, horses and water buffalo (Ribelin 1978; Kurmanov 1978, Szathmary 1983). Most proven cases have involved infected grains but *Fusarium* species also attack the stalks of grain crops used as forages and some pasture grasses, and so trichothecenes are likely to be found there, once improved methods are available for their detection.

In Australia, disease of cattle or sheep occurs sporadically from grazing forage crops and crop stubbles. In one report (Fisher *et al.* 1967), dairy cattle grazing a maize crop in Victoria developed inappetence, scouring, ataxia and decreased milk production. Several *Fusarium* species were isolated, including *F. culmorum* and *F. graminearum*, which are now known to be capable of producing trichothecenes. The trichothecene deoxynivalenol and zearalenone have recently been detected at high concentrations in wheat straw infected with *F. graminearum* in Queensland (B.J. Blaney and G.B. Wildermuth unpublished). *F. graminearum* has also been reported to infect Kikuyu grass *Pennisetum clandestinum* pastures on the Atherton tableland of north Queensland (Burgess 1985). *Fusarium* species are a common component of pasture microflora in the North Island of New Zealand, and are present in greatest concentration from February to April (10^5 microconidia /g wet leaves; di Menna and Parle 1970). Strains have been isolated that produce zearalenone (Gallagher 1985; Fig. 4c), and this compound has been isolated from pasture leaf samples collected in February and March (di Menna *et al.* 1985). Recent research on the effect of zearalenone on sheep fertility is reviewed below in the section on oestrogens.

Research into vaccines

A new line of research concerns production of vaccines against FE, ryegrass staggers and zearalenone, using new ways of coupling sporidesmin, lolitrem B and zearalenone to proteins that elicit immunological responses. Antibodies raised by herbivores against these antigens may bind and inactivate sporidesmin, lolitrem B and zearalenone before they reach their normal target sites, thus protecting the animal against detrimental effects produced by these compounds (R.T. Gallagher, pers. comm.). Field studies are currently being carried out with these vaccines in New Zealand. If successful, the technique could potentially be used to counteract effects of other mycotoxins in grazing herbivores.

OESTROGENIC SUBSTANCES

Zearalenone

This compound has a steroid-like structure (Fig. 4c) and has been shown to have oestrogenic properties and to depress reproductive performance in cattle (Mirocha *et al.* 1968, 1978). When zearalenone alone (25 mg /day) or a culture of *Fusarium* spp. supplying an equivalent amount of zearalenone, was administered to ewes (55 kg) for 10 days pre-mating, all aspects of reproductive performance were severely depressed (Table 6; Smith *et al.* 1986). This corresponds to a dietary concentration of 16 mg /kg DM, which is higher than the levels found in pasture samples to date (0.2 - 2.6 mg /kg DM), but nevertheless indicates the potential of zearalenone produced by *Fusarium* spp. to depress reproductive performance of grazing animals. The depression in reproductive performance is greater than produced by equivalent dietary concentrations of zearalenone fed to pigs, and may be explained through conversion of zearalenone to zearalenol by rumen protozoa (Kiessling *et al.* 1984). Zearalenol has a greater absorption rate and is more oestrogenic than zearalenone (Mirocha *et al.* 1978).

Table 6. Effect of daily administration of zearalenone (25mg), and of a *Fusarium* culture supplying an equivalent amount of zearalenone, for 10 days upon reproduction in the ewe.

Variate	Treatment Groups (n=33)			
	Zearalenone	Fusarium	Control	SE
Incidence of oestrus (%)	87.8	72.7	93.9	5.9
Duration of oestrus (h)	39.7	25.5	16.0	3.7
Incidence of ovulation (%)	54.4	53.1	87.9	7.8
Ovulation rate (1)	1.55	1.50	1.86	0.12
Ewes yielding eggs of those ovulating (%)	61.1	52.9	65.5	11.1
Egg recovery rate (%) (2)	57.1	53.8	48.1	8.8
Ewes with fertilized eggs of those yielding eggs (%)	27.3	55.5	100.0	15.8
Fertilization rate (%) (3)	28.8	42.9	100.0	11.9
Ewes with fertilized eggs of those treated (%)	9.1	15.6	57.6	6.8

From Smith *et al.* (1986)

(1) Number of ovulations/number of ewes ovulating

(2) Number of eggs recovered/number of ovulations as a percentage

(3) Number of fertilized eggs/number of eggs recovered as a percentage.

Isoflavones and coumestans

These compounds occur in some legumes, are oestrogenic and can cause reproductive disorders and infertility in grazing animals, more commonly in sheep than in cattle (Schutt 1976). Isoflavones occur typically in the clovers; their concentration is low in white clover but high in *Trifolium pratense* (red clover) and can reach 50 g /kg DM in *T. subterraneum* (subterranean clover). Formononetin is the most important isoflavone; it is not oestrogenic itself, but is converted in the rumino-reticulum to the weakly oestrogenic equol, which is the biologically active agent. In contrast, the other isoflavones, intrinsically active, are converted to inactive forms.

Coumestans, the second class of compounds, are present in only small amounts in healthy plants but can increase very markedly in white clover and lucerne (*Medicago sativa*) in response to attack by fungi (Loper and Hanson 1964; Wong and Latch 1971) or insects (Loper 1968; Kain and Biggs 1980). Smith *et al.* (1979) have reported a negative, linear relationship between coumestan (CM; mg /kg DM) concentration in lucerne and ovulation rate (OR) in ewes:

$$OR = 1.38 - 0.004\ CM; \quad r = -0.97^{***} \quad (3).$$

In New Zealand significant infertility in sheep due to phyto-oestrogens occurs only in association with grazing lucerne or red clover. Reproductive inefficiency of ewes grazing lucerne has been known for 25 years (Coop and Clark 1960) and has been associated with fungal infection of the plant (Smith *et al.* 1979). The effects of grazing red clover on the reproductive performance of ewes have been reported by Kelly *et al.* (1979, 1980), Kelly and Shackell (1982), Shackell and Kelly (1984 and unpublished) and can be summarised as follows:

Formononetin concentration in the cultivars "Pawera", "Hamua" and "Turoa" (6 - 14 g /kg DM) is above the "safe" level (3 g /kg DM). Grazing of Pawera containing 10 and 7 g /kg DM as formononetin for 25 and 10 days, respectively, during mating depressed ovulation rate by 50% and 20%, increased returns to service, and reduced lambing percentage. Grazing of Pawera for 4-month periods during summer, terminating one month prior to mating, had no effect upon ovulation rate, but increased both returns to service and barrenness in both the following and in future matings. Ewe lambs grazing Pawera from 3-7 months suffered no depression in fertility when mated at 1.5 years. Making the legume into field-cured hay reduced formononetin concentration to safe levels.

ORGANO-FLUORINE COMPOUNDS

Fluoroacetate is the most common of the range of organo-fluorine compounds that occurs in plants (McEwan 1978). It was first identified in the southern African shrub *Dichapetalum cymosum* (Marais 1944) and also occurs in other African *Dichapetalum* species (Vickery and Vickery

1973). In east Africa, it was more recently identified in *Spondianthus preussii* (Sere *et al.* 1982), a tree growing from 5–25 m in height. The toxicity of these plants has been known for hundreds of years and the seeds, fruit and leaves have been used to poison dogs and rodents. In Brazil, *Palicourea marcgravii* contains fluoroacetate (de Oliveira 1963). In Australia, fluoroacetate was first identified as the toxic principle of *Acacia georginae* (a small tree growing to 6 m), which is localized near the border of Queensland and the Northern Territory (Oelrichs and McEwan 1961; Pedley 1978). It also occurs in 33 Australian species of *Gastrolobium* and *Oxylobium*, shrubs of the family Leguminosae (McEwan 1964; Aplin 1971a, b). Despite the toxicity of fluoroacetate-containing plants, some of these species represent an important food resource for foraging animals in certain areas.

Conditions of poisoning of animals

Fluoroacetic acid is toxic to animals by means of a 'lethal synthesis' (Peters *et al.* 1953), in which it is converted to fluorocitrate which blocks the Kreb's cycle. Other naturally occurring organo-fluorides are thought to be converted to fluoroacetate following ingestion by animals. Poisoning of domestic animals is influenced by the availability of other food sources and the growth cycles of the plants. Leaf of *A. georginae* usually contains low concentrations of fluoroacetate ($<$ 0.001–0.01 g /kg DM) and is an important feed for cattle, particularly during droughts. Poisoning is cumulative in short-term dosing (McEwan 1978), and generally occurs only when other food sources are depleted and large amounts of the more toxic pods (containing up to 0.4 g /kg DM) are available (Bell *et al.* 1955). However, concentrations of fluoroacetate tend to be higher in new growth and leaves from suckering plants, so that the emergence of larger amounts of such material, which occurs irregularly, also increases the hazard. *D. cymosum* is toxic at all times, but cattle deaths are greater in spring and autumn when the plant produces highly toxic new leaves (Vickery and Vickery 1973). Poisoning of cattle and goats by *S. preussii* occurs when dry conditions deplete other food stores and animals are attracted by the green foliage of the plant (Sere *et al.* 1982). Australian species of *Gastrolobium* and *Oxylobium* are toxic throughout the year, but they are considered to be most toxic when new shoots appear, or when in the flowering and fruiting stage. Fluoroacetate concentrations up to 12.5 g /kg DM have been recorded in leaf, so that 30 g of leaf may be sufficient to kill a sheep (McEwan 1978). Cattle, sheep, goats, horses, pigs, dogs and cats have all been poisoned, and *Oxylobium graniticum* has also been reported as poisoning camels (Aplin 1971a, b). However, native marsupials and birds were observed to consume the seeds without ill effect (Aplin 1967a, b).

The distribution and role of organo-fluorine compounds in plants

There are indications that organo-fluorine compounds confer an advantage on the plants producing them in terms of evolutionary selection pressure. The distributions of inorganic and organic fluoride in *A. georginae*, *P. marcgravii*, several *Dichapetalum* species, *Oxylobium parviflorum* and several *Gastrolobium* species are higher in seeds than in leaf, stem or root (Hall 1972; McEwan 1978). There could be several reasons for accumulation of a secondary compound in seed, but one of the more likely is as a defence against predation by insects, birds or small animals. Fluoracetate and its derivatives are known to be effective insecticides against aphids and flies (David and Gardiner 1959; Tahori 1963). Apart from seed, the parts accumulating highest concentrations of organo-fluoride in some plants are young leaves, roots and flowers (Hall 1972). High concentrations in flowers might be a defence against non-pollinating insects. Toxic constituents are common in flowers, and possibly serve to maximize flower constancy in pollinators (Rhoades 1979).

Further evidence to suggest that the organo-fluorides in *A. georginae* are primarily a protection against insects, is the much greater concentration of organo-fluorine in a waxy coating on its leaves than is present within the leaves. Hall (1972) noted this and interpreted it as an excretory mechanism. In some plant species, such a waxy coating is a protection against water loss. However, how much more effective would be a waxy layer containing organo-fluorides as a defence against leaf-sucking insects!

In comparison to a tree such as *A. georginae*, there could have been more need for protection against mammals in the case of low-growing shrubs. This might be expected to exert pressure to broaden the emphasis from seed and sucker protection as in *A. georginae* towards leaf, flower and fruit protection, and perhaps make very high fluoroacetate concentrations throughout the plant more essential to survival. This appears to be the case with *Gastrolobium* and *Oxylobium* species. These latter species are palatable and nutritious legumes, which form a major proportion of the available forage in parts of the south-west of Western Australia, and may be dominant following bush fires. Oliver *et al.* (1977) and King *et al.* (1978) have uncovered an interesting story of co-evolution between these fluoroacetate-containing plants and several native mammals which eat them. They found that the tolerance to fluoroacetate of the western grey kangaroo (*Macropus fuliginosus*), the brush-tailed possum (*Trichosurus vulpecula*), and the bush rat (*Rattus fuscipes*) from the south-west of Western Australia were unusually high. In contrast, *T. vulpecula* and *R. fuscipes* from eastern Australia were much more susceptible. They suggested that this different tolerance, which is about 100-fold, was an adaptation to the presence of fluoroacetate in plants growing within the range of these mammals. They also pointed out that the fluoroacetate contents of a number of *Gastrolobium* and *Oxylobium* species appeared to be unnecessarily high in view of its toxicity to most mammals and insects, unless some herbivores had evolved a degree of resistance to it. Subsequent investigations (Oliver *et al.* 1979; King *et al.* 1981) have supported the theory, and fluoroacetate tolerance has been used to trace the recent evolutionary history of the western grey kangaroo, the bush rat, the tammar wallaby (*Macropus eugenii*) and the rat kangaroo (*Bettongia* species). The mechanism of fluoroacetate tolerance in these species is under study (Mead *et al.* 1979; Twigg *et al.* 1986) and may provide a lead to methods for protecting domestic animals. It appears to involve several mechanisms, including a more rapid rate of defluorination in erythrocytes and liver. Destruction by intestinal microorganisms does not seem to be important. However, one possible approach to protecting domestic ruminants may be to transfer the ability to defluorinate fluoroacetate possessed by some soil microorganisms into a ruminal microorganism by gene cloning. Use of an intra-ruminal device containing an ion-exchange resin to absorb the strongly acidic fluoroacetate ion has also been suggested (B.J. Blaney unpublished).

OXALATES

Chemistry and occurrence of oxalates in plants

Oxalic acid and its salts are widely distributed throughout the plant kingdom. Oxalate is simply a waste product in some plants, but in others it may function in structural support, calcium storage and elimination, maintenance of ionic balance, or protection against predators (Hodgkinson 1977).

In most plants that produce oxalate in other than trace amounts, much of the calcium is present as calcium oxalate crystals. These crystals are held within cell vacuoles (Arnott and Pautard 1970), and their formation is related to the calcium supply. This might be a strategy aimed at keeping the concentration of ionic calcium low, since high levels of ionic calcium are deleterious to growth (Franceschi and Horner 1980).

Some groups of plants accumulate high concentrations of soluble oxalate, which may also be stored in cell vacuoles. Soluble oxalate-accumulators can be divided into those with a sap pH of about 2 and those with a sap pH around 5-6. Potassium hydrogen oxalate is common in those with acid sap, while sodium, calcium and magnesium oxalates are more common in the others (Michael 1959). However, plants with an acid sap also contain calcium oxalate crystals, due to compartmentation in the plant (Hodgkinson 1977). Apparently, cations freely substitute for each other in oxalate salts depending upon their relative availability to the plant. In saltbushes (*Atriplex* spp.), and in the tropical grass *Setaria sphacelata*, the oxalate ion is produced to maintain ionic balance between cations (potassium, sodium, ammonium, calcium and magnesium) and anions (nitrate, chloride, phosphate and sulphate) (Osmond 1963; Osmond and Avadhani 1968; Smith 1972). Oxalate may serve a similar purpose in most accumulators of soluble oxalate. Representatives of several important groups of oxalate-containing plants are discussed below.

Shrubs and herbs. This includes plants with total oxalate concentrations that can exceed 100 g /kg DM. They are widely consumed by foragers. Mathams and Sutherland (1952) found that 46 out of 80 species of plants grazed by sheep in Queensland contained more than 5 g /kg DM total oxalate, and 17 species more than 4 g /kg DM. Only one (*Panicum antidotale*) was a grass, the others were forbs. Poisoning of cattle and sheep had been recorded on species of *Portulaca*, *Salsola*, *Trianthema* and *Threlkeldia*. Other species containing 40 - 160 g /kg DM were from the genera *Amaranthus*, *Atriplex*, *Chenopodium*, *Enchylaena*, *Oxalis* and *Tetragonia*. Continued consumption of these plants may also lead to urolithiasis in sheep (Sutherland 1958). In South Australia, *Oxalis pes-caprae* containing up to 140 g oxalate /kg DM is widespread and has caused heavy sheep losses (Dodson 1959). In the USA, *Halogeton glomeratus* is a major cause of sheep poisoning (James 1978).

Tropical grasses. Many tropical grasses native to Africa and the Americas accumulate oxalates. Highest concentrations have been found in *Setaria sphacelata* (60 - 100 g /kg DM), and up to 42g /kg DM has been recorded in buffel grass (*Cenchrus ciliaris*). Kikuyu (*Pennisetum clandestinum*), green panic (*Panicum maximum*), Para (*Brachiaria mutica*), pangola (*Digitaria decumbens*) and other related grasses usually contain 5 - 20 g /kg DM. In grasses, oxalate accumulates during rapid growth, particularly following fertilization with potassium and nitrogen (Jones and Ford 1972). In Australia, acute oxalate poisoning has been recorded in cattle and horses grazing *S. sphacelata* (Jones *et al.* 1970; McKenzie 1985) and in sheep grazing *C. ciliaris* (R.A. McKenzie pers. comm.). The availability of calcium in these grasses is low, since most is present as calcium oxalate crystals (McKenzie and Schultz 1983).

Paddy plant. Paddy straw (*Oryza sativa*) which is widely used as feed for ruminants in the tropics, can contain up to 40 g oxalates /kg DM, and causes negative calcium balances in cattle and water buffalo (Talapatra *et al.* 1948).

Herbaceous legumes. There are no reports of soluble oxalate accumulation in legumes. However, about 33% of the calcium in lucerne (*Medicago sativa*) is unavailable to cattle, since it is present as calcium oxalate crystals (Ward *et al.* 1979).

Forage trees. Many tree leaves contain calcium oxalate crystals (Hodgkinson 1977). Mulga (*Acacia aneura*), which is used widely as fodder for ruminants in Australia, contains sufficient insoluble oxalate to account for all of the calcium present. Consequently, the availability of the calcium has been questioned (Gartner and Hurwood 1976).

Desert plants. Some desert plants such as prickly pear (*Opuntia*) and *Yucca* contain very high concentrations (up to 85% by dry weight) of calcium oxalate, but are used as forage by rodents (Hodgkinson 1977).

Utilization of oxalates by foragers

Most plants containing high concentrations of oxalates are palatable and can form the major portion of the diet. The digestive system of an animal is presented with a mixture of oxalates contained within plant vacuoles. Once these vacuoles are breached by intestinal secretions or micro-organisms, soluble oxalates and insoluble calcium oxalate crystals are released. Depending upon the pH of the intestine at any particular point, soluble oxalate may associate with free calcium ions present, probably to form a colloid, since crystallization of calcium oxalate is unlikely in the complex, viscous environment of intestinal contents, regardless of the pH. Such a colloid might hinder absorption of calcium and oxalate from the intestine, but not degradation by enzymes. In acid regions of the intestinal tract, the colloid could rapidly dissociate into free calcium and oxalate ions. The free oxalate ions will be either absorbed into the vascular system, or degraded by intestinal secretions and microorganisms. The insoluble calcium oxalate crystals in many plants are located in the sheath cells surrounding vascular bundles in the leaves (Ward *et al.* 1979; McKenzie and Shultz 1983, Hans *et al.* 1984). These structures tend to be highly lignified and resist digestion (Akin and Burdick 1975). If calcium oxalate crystals are released from these cells degradation will be much slower than for either soluble oxalates or colloidal calcium oxalate, because digestive processes can only work at the crystal surfaces. The degree of dissolution of calcium oxalate crystals will therefore depend on the rate of release of crystals from cells, the rate of dissolution of crystals by intestinal secretions or microorganisms, and their rate of passage through the digestive tract.

Ruminants. Consumption of oxalates can lead to poisoning, mineral deficiences or urolithiasis, depending upon the amount and rate of oxalate ingestion and upon the animal species. Oxalate poisoning is complex. Although deaths have been mainly attributed to hypocalcaemia or uraemia resulting from kidney damage, a number of other factors are involved. In cattle and sheep, haemorrhagic rumenitis, shock and an interference with energy metabolism may contribute to death (James 1978). When ruminants are fed diets containing gradually increasing amounts of oxalates, they become tolerant to concentrations that would be lethal to non-adapted animals. This adaptation is paralleled by an increase in the rate of oxalate degradation by ruminal microorganisms, some of which produce stoichiometric amounts of carbon dioxide and formate from oxalate as the sole carbon source (Dawson *et al.* 1980). The bacteria can utilize calcium oxalate as well as soluble oxalates (Allison 1985), albeit more slowly. Consequently, cattle may utilize up to 50% of the calcium in tropical grasses containing high levels of soluble oxalate, even though more than 90% of the calcium in these grasses may be present as calcium oxalate (Blaney *et al.* 1982). By comparison, Ward *et al.* (1979) found that very few of the calcium oxalate crystals in lucerne were degraded in the rumen. The differences between these results might be explained either by the greater release of crystals from tropical grasses in the rumen or by greater activity of oxalate-degrading rumen bacteria in the case of tropical grasses, since the total oxalate content of lucerne is very low. The reasons for negative calcium retention in cattle and water buffalo fed paddy plant are believed to be reduced true digestibility of calcium in the intestinal tract due to alkalosis from breakdown of soluble oxalates to bicarbonate (Negi 1971; Hans *et al.* 1983), combined with poor utilization of calcium oxalate crystals. Hans *et al.* (1984) reported that one fraction of the calcium oxalate was more resistant to degradation by rumen liquor than the remaining calcium oxalate and soluble oxalate in paddy straw.

Non-ruminants. The adapted rumen provides a barrier against absorption of large amounts of soluble oxalates, but this protection is unavailable for monogastrics and hind-gut fermenters such as pigs, horses and rabbits. Oxalate-degrading bacteria have been isolated from caecal contents of pigs, guinea-pigs and wild rats (Allison 1985) and from horse faeces (McKenzie 1985). The rates of oxalate degradation by mixed microbial populations from rectal contents of a horse and caecal contents from pigs also increase markedly when oxalate concentrations in the diet are increased, showing a similar adaptive response to the rumen (Allison 1985). While this allows use of oxalate as an energy source, it provides less protection against absorption of soluble oxalates which occurs in preceding segments of the intestinal tract. It might be expected, therefore, that these species would be much more susceptible to oxalate poisoning than ruminants. At least with horses, this is not necessarily the case. Although oxalate poisoning has been recorded in horses grazing the tropical grass *S. sphacelata* cv. Kazungula, concentrations of soluble oxalate up to at least 39 g /kg DM in the diet have been tolerated for 3 weeks without apparent ill effects (McKenzie *et al.* 1981a; McKenzie 1985). Additional mechanisms for detoxification of oxalate may have evolved in non-ruminants, such as either microbiological destruction of oxalates in the stomach and small intestine, decreased absorption of oxalate, increased urinary excretion of oxalate or rapid metabolism by mammalian enzymes. Oxalate metabolism has been studied in North American pack rats (*Neotoma albigula*) and Egyptian sand rats (*Psammomys obesus*) which consume large amounts of soluble and insoluble oxalates in their staple diet (Shirley and Schmidt-Nielsen 1967). Both species metabolised dietary sodium and calcium oxalate, although the pack rats metabolised more calcium oxalate. The results indicated that most degradation occurred in the intestinal tract. In horses fed large amounts of soluble oxalates, urinary excretion is significantly greater than in horses receiving little oxalate, but accounts for only about 2% of the ingested soluble oxalate (McKenzie *et al.* 1981a; Blaney *et al.* 1981a).

Although most animal species can degrade calcium oxalate, efficient utilization of the calcium depends upon the site of calcium oxalate degradation in relation to the site of calcium absorption. Pigs are able to utilize some calcium from calcium oxalate (Brune and Bredehorn 1961). Perhaps there is substantial dissociation of calcium oxalate in the acid stomach, which might also be the case with pack rats. Horses cannot efficiently utilize calcium present as calcium oxalate crystals, since it is released after the site of calcium absorption, even following adaptation to a high-oxalate diet (Blaney *et al.* 1981b). In comparable circumstances, cattle absorbed 27% of a dose of calcium oxalate while horses absorbed < 6% (Blaney *et al.* 1982). This leads to calcium deficiency in horses given limited feeding options. The risk to horses is increased relative to cattle by the horse's more selective grazing of young leaf containing a higher oxalate:calcium ratio than mature leaf and stalk (McKenzie *et al.* 1981b). In hind-gut fermenting rabbits, coprophagia would be a useful method of utilizing the calcium in calcium oxalate.

Oxalate-induced calcium deficiency in horses can be overcome by providing calcium and phosphorus supplements, mixed in molasses (Gartner *et al.* 1981; McKenzie *et al.* 1981b). Lactating ruminants may also require calcium supplementation when fed oxalate-containing fodders. The risk of acute oxalate poisoning can be reduced by allowing for a period of adaptation to oxalate metabolism by the intestinal microorganisms.

CONCLUSIONS AND FUTURE WORK

Traditionally, most secondary compounds produced by plants and mycotoxins produced by fungi have come to the attention of animal nutritionists because of deleterious effects produced in grazing herbivores. This chapter presents an up-date of such effects produced by tannins, fluoroacetate, oxalate, oestrogenic compounds, and a number of mycotoxins.

Whilst high concentrations (90-100 g /kg DM) of condensed tannin (CT) have been shown to be detrimental, low concentrations (20-30 g /kg DM) have been shown to be beneficial in fresh forage diets of ruminants, due to their reducing ruminal protein degradation and increasing the absorption of essential amino acids. Much future work is likely to centre on transferring genes coded for leaf production of condensed tannin from sainfoin and lotus into legumes that have greater agronomic potential, such as lucerne and white clover. This is currently under way in Australian and New Zealand laboratories, using somatic fusion of protoplasts.

A further recent advance has been the identification of

lolitrem B and peramine produced by the endophyte infecting perennial ryegrass as the respective causes of ryegrass staggers and resistance to Argentine Stem Weevil. Production of vaccines against mycotoxins through coupling of the low molecular weight mycotoxins to proteins, is likely to be a very important future area of research. Other approaches to reducing deleterious effects of secondary compounds upon domestic animals include breeding of plant varieties with low concentrations of these compounds, genetic manipulation of intestinal microorganisms to destroy secondary compounds and the use of intra-ruminal devices to specifically absorb secondary compounds. All are likely to be avenues of future research.

Akin, D.E., and Burdick, D. (1975). *Crop Sci.* **15**, 661-8.

Allen, J.G., Wood, P. McR., and O'Donnell, F.M. (1978). *Aust. Vet. J.* **54**, 19-22.

Allison, M.J. (1985). In "Plant Toxicology." Proceedings of the Australia-USA Poisonous Plants Symposium. (Eds A.A. Seawright, M.P. Hegarty, L.F. James and R.F. Keeler.) pp. 120-6. (Qld. Dept. Prim. Ind.: Yeerongpilly, Brisbane.)

Aplin, T.E.H. (1967a). *J. Agric. West. Aust.* **8**, 42-52.

Aplin, T.E.H. (1967b). *J. Agric. West. Aust.* **8**, 200-6.

Aplin, T.E.H. (1969). *J. Agric. West. Aust.* **10**, 248-57.

Aplin, T.E.H. (1971a). *J. Agric. West. Aust.* **12**, 12-8.

Aplin, T.E.H. (1971b). *J. Agric. West. Aust.* **12**, 154-9.

Applebaum, S.B. and Birk, Y. (1979). In "Herbivores; Their Interaction with Secondary Plant Metabolites." (Eds G.A. Rosenthal and D.H. Janzen.) pp. 539-66. (Academic: NY.)

Arnott, H.J., and Pautard, F.G.E. (1970). In "Biological Calcification; Cellular and Molecular Aspects." (Ed. H. Schraer.) pp. 375-446. (North Holland: Amsterdam.)

Barker, G.M., Pottinger, R.P., Addison, P.J., and Prestidge, R.A. (1984a). *N.Z. J. Agric. Res.* **27**, 271-7.

Barker, G.M., Pottinger, R.P., and Addison, P.J. (1984b). *N.Z. J. Agric. Res.* **27**, 279-81.

Barry, T.N. (1985). *Br. J. Nutr.* **54**, 211-7.

Barry, T.N., and Forss, D.A. (1983). *J. Sci. Food Agric.* **34**, 1047-56.

Barry, T.N., and Duncan, S.J. (1984). *Br. J. Nutr.* **51**, 485-91.

Barry, T.N., and Manley, T.R. (1984). *Br. J. Nutr.* **51**, 493-504.

Barry, T.N., and Manley, T.R. (1986). *J. Sci. Food Agric.* **37**, 248-54.

Barry, T.N., Manley, T.R., and Duncan, S.J. (1986). *Br. J. Nutr.* **55**, 123-37.

Barry, T.N., Allsop, T.F., and Redekopp, C. (1986). *Br. J. Nutr.* **56**, *607-14*.

Barry, T.N., Manley, T.R., Redekopp, C., and Allsop, T.F. (1985). *Br. J. Nutr.* **54**, 165-73.

Beever, D.E., and Siddons, R.C. (1984). In "Control of Digestion and Metabolism in Ruminants." (Eds L.P. Milligan, W.L. Grovum and A. Dobson.) pp. 479-97. (Prentice-Hall: New Jersey, USA.)

Bell, A.T., Newton, L.G., Everist, S.L., and Legg, J. (1955). *Aust. Vet. J.* **31**, 249–57.

Bell, E.A. (1978). In "Biochemical Aspects of Plant and Animal Coevolution." (Ed. J.B. Harborne.) pp. 143–61. (Academic: London.)

Blaney, B.J., Gartner, R.J.W., and McKenzie, R.A. (1981a). *J. Agric. Sci.* **97**, 507–14.

Blaney, B.J., Gartner, R.J.W., and McKenzie, R.A. (1981b). *J. Agric. Sci.* **97**, 639–41.

Blaney, B.J., Gartner, R.J.W, and Head, T.A. (1982). *J. Agric. Sci.* **99**, 533–9.

Broadhurst, R.B., and Jones, W.T. (1978). *J. Sci. Food Agric.* **29**, 788–94.

Broquist, H.P., Mason, P.S., Wickwire, B., Homann, R., Schneider, M.J., and Harris, T.M. (1985). In "Plant Toxicology." Proceedings of the Australia-USA Poisonous Plants Symposium. (Eds A.A. Seawright, M.P. Hegarty, L.F. James and R.F. Keeler.) pp. 301–8. (Qld. Dept. Prim. Ind.: Yeerongpilly, Brisbane.)

Brune, H., and Bredehorn, H.L. (1961). *Z. Tierphysiol. Tierernaehr. Futtermittelkd.* **16**, 214–36.

Buck, W.B., Osweiler, G.D., and Van Gelder, G.A. (1976). In "Clinical and Diagnostic Veterinary Toxicology." 2nd Ed. (Ed. G.A. Van Gelder.) pp. 109–16. (Kendall Hunt: Dubuque, Iowa.)

Bull, L.B., Culvenor, C.C.J., and Dick, A.T. (1968). "The Pyrrolizidine Alkaloids: their Chemistry, Pathogenicity and other Biological Properties." (North Holland: Amsterdam.)

Burgess, L.W. (1985). In "Trichothecenes and Other Mycotoxins." (Ed. J. Lacey.) pp. 15–9. (Wiley: Chichester, UK.)

Cheeke, P.R., and Shull, L.R. (1985). "Natural Toxicants in Feeds and Poisonous Plants." (AVI: Westport, Connecticut.)

Cole, R.J., and Cox, R.H. (1981). "Handbook of Toxic Fungal Metabolites." (Academic: London.)

Conn, E.E. (1979). In "Herbivores. Their Interaction with Secondary Plant Metabolites." (Eds G.A. Rosenthal and D.H. Janzen.) pp. 387–412. (Academic: New York.)

Coop, I.E., and Clark, V.R. (1960). *N.Z. J. Agric. Res.* **3**, 922–33.

Culvenor, C.C.J. (1985). In "Plant Toxicology." Proceedings of the Australia-USA Poisonous Plants Symposium. (Eds A.A. Seawright, M.P. Hegarty, L.F. James and R.F. Keeler.) pp. 3–13. (Qld. Dept. Prim. Ind.: Yeerongpilly, Brisbane.)

Culvenor, C.C.J. (1986). *Proc. Int. Lupin Conf.* **4** (in press.)

David, W.A.L., and Gardiner, B.O.C. (1959). *Bull. Ent. Res.*. **50**, 25–38.

Dawson, K.A., Allison, M.J., and Hartman, P.A. (1980). *Appl. Environ. Microbiol.* **40**, 833–9.

di Menna, M.E., and Parle, J.N. (1970). *N.Z. J. Agric. Res.* **13**, 57–8.

di Menna, M.E., Lauren, D.R., and Holland, P.T. (1985). *N.Z. Vet. J.* **33**, 193.

di Menna, M.E., and Waller, J.E. (1986). *N.Z. J. Agric. Res.* **29**, 111–6.

Dodson, M.E. (1959). *Aust. Vet. J.* **35**, 225–33.

Edgar, J.A., Culvenor, C.C.J., Frahn, J.L., Jones, A.J., Gorst-Allman, C.P., Marasas, W.F.O., Steyn, P.S., Vleggaar, R., and Wessels, P.L. (1985). In "Trichothecenes and Other Mycotoxins." (Ed. J. Lacey.) pp. 317-24. (Wiley : Chichester, UK.)

Egan, A.R., and Ulyatt, M.J. (1980). *J. Agric. Sci.*, **94**, 47-56.

Fisher, E.E., Kellock, A.W., and Wellington, N.A.M. (1967). *Nature* (London) **215**, 322.

Fletcher, L.R. (1982). *N.Z. J. Exp. Agric.* **10**, 203-7.

Fletcher, L.R. (1986). *Proc. N.Z. Grassld Assoc.* **47**, 49-105.

Fletcher, L.R., and Harvey, I.C. (1981). *N.Z. Vet. J.* **29**, 185-6.

Fletcher, L.R., and Barrell, G.K. (1984). *N.Z. Vet. J.* **32**, 139-40.

Franceschi, V.R., and Horner, H.T. (1980). *Bot. Rev.* **46**, 361-427.

Gallagher, R.T. (1985). *N.Z. Vet. J.* **33**, 37-8.

Gallagher, R.T., White, E.P., and Mortimer, P.H. (1981). *N.Z. Vet. J.* **29**, 189-90.

Gallagher, R.T., Hawkes, A.D., Steyn, P.S., and Vleggaar, R. (1984). *J. Chem. Soc. Commun.*, 614-6.

Gallagher, R.T., Hawkes, A.D., and Stewart, J.M. (1985). *J. Chromatography* **321**, 217-26.

Gallagher, R.T., and Hawkes, A.D. (1986). *Experientia* **42**, 823-5.

Gardiner, M.R. (1967). *J. Comp. Path.* **77**, 63-9.

Gardiner, M.R., and Parr, W.H. (1967). *J. Comp. Path.* **77**, 51-62.

Gartner, R.J.W., and Hurwood, I.S. (1979). *Aust. Vet. J.* **52**, 194-5.

Gartner, R.J.W., Blaney, B.J., and McKenzie, R.A. (1981). *J. Agric. Sci.* **97**, 581-9.

Gaynor, D.L., and Rowan, D.D. (1986). *Proc. N.Z. Grassld Assoc.* **47**, 115-20.

Hall, R.J. (1972). *New Phytol.* **71**, 855-71.

Hans, M., Singh, S., and Sareen, V.K. (1983). *Z. Tierphysiol. Tierernaehr. Futtermittelkd.* **49**, 57-62.

Hans, M., Singh, S., and Sareen, V.K. (1984). *Z. Tierphysiol. Tierernaehr. Futtermittelkd.* **52**, 1-7.

Hegarty, M.P. (1982). In "Nutritional Limits to Animal Production from Pastures." (Ed. J.B. Hacker.) pp. 130-55. (Comm. Agric. Bur.: Farnham Royal, UK.)

Hegarty, M.P., and Pound, A.W. (1970). *Aust. J. Biol. Sci.* **23**, 831-42.

Hodgkinson, A. (1977). "Oxalic Acid in Biology and Medicine." (Academic: London.)

Ichinoe, M., and Kurata, H. (1983). In "Trichothecenes; Chemical, Biological and Toxicological Aspects." (Ed. Y. Ueno.) pp. 73-82. (Elsevier: Amsterdam.)

James, L.F. (1978). In "Effects of Poisonous Plants on Livestock." (Eds R.F. Keeler, K.R. Van Kampen and L.F. James.) pp. 135-45. (Academic: New York.)

John, A., and Lancashire, J.A. (1981). *Proc. N.Z. Grassld Assoc.* **42**, 152-9.

Jones, R.J., and Ford, C.W. (1972). *Aust. J. Exp. Agric. Anim. Husb.* **12**, 400-6.

Jones, R.J., Seawright, A.A., and Little, D.A. (1970). *J. Aust. Inst. Agric. Sci.* **36**, 41-3.

Jones, W.T., Anderson, L.B., and Ross, M.D. (1973). *N.Z. J. Agric. Res.* **16**, 441-6.

Jones, W.T., Broadhurst, R.B., and Lyttleton, J.W. (1976). *Phytochemistry* **15**, 1407-9.

Jones, W.T., and Mangan, J.L. (1977). *J. Sci. Food Agric.* **28**, 126-36.

Kain, W.M., and Biggs, D.R. (1980). *N.Z. J. Agric. Res.* **23**, 563-8.

Keeler, R.F., Van Kampen, K.R. and James, L.F. (1978). "Effects of Poisonous Plants on Livestock." (Academic: New York.)

Kelly, R.W., Hay, R.J.M., and Shackell, G.H. (1979). *N.Z. J. Exp. Agric.* **7**, 131-4.

Kelly, R.W., and Shackell, G.H. (1982). *Proc. N.Z. Soc. Anim. Prod.* **42**, 29-31.

Kelly, R.W., Shackell, G.H., and Allison, A.J. (1980). *N.Z. J. Exp. Agric.* **8**, 87-91.

Kiessling, K.H., Pettersson, H., Sandholm, K., and Olsen, M. (1984). *App. Env. Microbiol.* **47**, 1070-3.

King, D.R., Oliver, A.J., and Mead, R.J. (1978). *Aust. J. Zool.* **26**, 699-712.

King, D.R., Oliver, A.J., and Mead, R.J. (1981). *Aust. Wildl. Res.* **8**, 529-36.

Kurmanov, I.A. (1978). (Translated by A. Novacky.) In "Mycotoxic Fungi, Mycotoxins and Mycotoxicoses - an Encyclopedic Handbook." Vol. 2 (Eds T.D. Wyllie and L.G. Morehouse.) pp. 85-110. (Marcel Dekker: New York.)

Latch, G.C.M. (1983). *Proc. N.Z. Grassld Assoc.* **44**, 251-3.

Loomis, W.D., and Battoile, J. (1966). *Phytochem.* **5**, 423.

Loper, G.M. (1968). *Crop Sci.* **8**, 104-6.

Loper, G.M., and Hanson, C.H. (1964). *Crop Sci.* **4**, 480-2.

Lowther, W.L., Manley, T.R., and Barry, T.N. (1987). *N.Z. J. Agric. Res.* (in press.)

McEwan, T. (1964). *Nature* (London) **201**, 827.

McEwan, T. (1978). In "Effects of Poisonous Plants on Livestock." (Eds R.F. Keeler, K.R. Van Kampen and L.F. James.) pp. 147-58. (Academic: New York.)

McKenzie, R.A. (1985). In "Plant Toxicology." Proceedings of the Australia-USA Poisonous Plants Symposium. (Eds A.A. Seawright, M.P. Hegarty, L.F. James and R.F. Keeler.) pp. 150-4. (Qld. Dept. Prim. Ind.: Yeerongpilly, Brisbane.)

McKenzie, R.A., Blaney, B.J., and Gartner, R.J.W. (1981a). *J. Agric. Sci.* **97**, 69-74.

McKenzie, R.A., Gartner, R.J.W., Blaney, B.J., and Glanville, R.J. (1981b). *Aust. Vet. J.* **57**, 554-7.

McKenzie, R.A., and Schultz, K. (1983). *J. Agric. Sci.* **100**, 249-50.

McLeod, N.M. (1974). *Nutr. Abstr. Rev.* **44**, 803-15.

MacRae, J.C., and Ulyatt, M.J. (1974). *J. Agric. Sci.* **82**, 309-19.

Mackay, M.F., Van Donkelaar, A., and Culvenor, C.C.J. (1986). *J. Chem. Soc. Chem. Comm.*, 1219–21.
Majak, W., Howarth, R.E., Cheng, K.J., and Hall, J.W. (1983). *J. Dairy Sci.* **66**, 1683–8.
Marais, T.S.C. (1944). *Onderstepoort J. Vet. Sci. Anim. Ind.* **20**, 67–73.
Mathams, R.H., and Sutherland, A.K. (1952). *Qld. Agric. J.* **9**, 317–34.
Mead, R.J., Oliver, A.J., and King, D.R. (1979). *Aust. J. Biol. Sci.* **32**, 15–26.
Michael, P.W. (1959). *Aust. Vet. J.* **35**, 431–2.
Mirocha, C.J., Harrison, J., Nichols, A.A., and McClintock, M. (1968). *Appl. Microbiol.* **16**, 797–8.
Mirocha, C.J., Pathre, S.V., Behrens, J., and Schavenhamer, B. (1978). *Appl. Env. Microbiol.* **35**, 986–7.
Munday, B.L., Monkhouse, I.M., and Gallagher, R.T. (1985). *Aust. Vet. J.* **62**, 207.
Munday, R. (1982). *Chem. Biol. Interactions* **41**, 361–74.
Munday, R. (1984). *J. Appl. Toxicol.* **4**, 182–6.
Munday, R. (1985). *J. Appl. Toxicol.***5**, 69–73.
Negi, S.S. (1971). *Indian J. Anim. Sci.* **41**, 913–21.
Oelrichs, P.B., and McEwan, T. (1961). *Nature* (London) **190**, 808–9.
Oliver, A.J., King, D.R., and Mead, R.J. (1977). *Search* (Sydney) **8**, 130–2.
Oliver, A.J., King, D.R., and Mead, R.J. (1979). *Aust. J. Zool.* **27**, 363–72.
Osmond, C.B. (1963). *Nature* (London) **198**, 503–4.
Osmond, C.B., and Avadhani, P.N. (1968). *Aust. J. Biol. Sci.* **21**, 917–27.
Pedley, L. (1978). A revision of *Acacia* Mill. in Queensland. *Austrobaileya* **1**, 75–234.
Peters, R.A., Buffa, R., Wakelin, R.W., and Thomas, L.C. (1953). *Proc. R. Soc. B* **140**, 497–507.
Peterson, J.E. (1986). *Proc. Int. Lupin Conf.* **4**, (in press.)
Pierce, E.C., and Appleman, C.D. (1943). *Plant. Physiol.* **18**, 224–38.
Purchas, R.W., and Keogh, R.G. (1984). *Proc. N.Z. Soc. Anim. Prod.* **44**, 219–21.
Reid, C.S.W., Ulyatt, M.J., and Wilson, J.M. (1974). *Proc. N.Z. Soc. Anim. Prod.* **34**, 82–93.
Rhoades, D.F. (1979). In "Herbivores. Their Interaction with Secondary Plant Metabolites." (Eds G.A. Rosenthal and D.H. Janzen.) pp. 4–52. (Academic: New York.)
Ribelin, W.E. (1978). In "Mycotoxic Fungi, Mycotoxins and Mycotoxicosis – an Encyclopedic Handbook." Vol. 2 (Eds T.D. Wyllie and L.G. Morehouse.) pp. 36–45. (Marcel Dekker: New York.)
Rowan, D.D., Hunt, M.B., and Gaynor, D.L. (1986). *J. Chem. Soc. Commun.*, 935–6.
Schutt, D.A. (1976). *Endeavour* 35, 110–4.

Seawright, A.A., Hegarty, M.P., James, L.F. and Keeler, R.F. (1985). (Eds). "Plant Toxicology." Proceedings of the Australia-USA Poisonous Plants Symposium. (Qld. Dept. Prim. Ind.: Yeerongpilly, Brisbane.)

Sere, A., Tayou Kamgue, R., Ake Assi, L., and Ba, A.C. (1982). *Rev. Elev. Med. Vet. Pays Trop.* **35**, 73-82.

Shackell, G.H., and Kelly, R.W. (1984). *N.Z. J. Exp. Agric.* **12,** 113-7.

Shirley, E.K., and Schmidt-Nielsen, K. (1967). *J. Nutr.* **91**, 496-502.

Smith, B.L, Embling, P.P., Towers, N.R., Wright, D.E., and Payne, E. (1977). *N.Z. Vet. J.* **25**, 124-7.

Smith, F.W. (1972). *Aust. J. Agric. Res.* **23**, 969-80.

Smith, J.F., Jagusch, K.T., Brunswich, L.F.C., and Kelly, R.W. (1979). *N.Z. J. Agric. Res.* **22**, 411-6.

Smith, J.F., di Menna, M.E., and McGowan, L.T. (1986). *Proc. N.Z. Soc. Anim. Prod.* **46** (in press.)

Smith, R.H. (1980). *Vet. Rec.* **107**, 12-5.

Sutherland, A.K. (1958). *Aust. Vet. J.* **34**, 44-7.

Swain, T. (1979). In "Herbivores; Their Interaction with Secondary Plant Metabolites." (Eds G.A. Rosenthal and D.H. Janzen.) pp. 657-82. (Academic: New York.)

Szathmary, C.I. (1983). In "Trichothecenes; Chemical, Biological and Toxicological Aspects." (Ed. Y. Ueno.) pp. 229-50. (Elsevier: Amsterdam.)

Tahori, A.S. (1963). *J. Econ. Entomol.* **56**, 67-9.

Talapatra, S.K., Ray, S.C., Kehar, N.D., and Sen, K.C. (1948). *Science and Culture* **8**, 209-13.

Towers, N.R. (1986). *Proc. N.Z. Grassld Assoc.* (in press.)

Towers, N.R., Meyer, H.H., Campbell, A.G., and Southey, C.A. (1983). *Proc. N.Z. Soc. Anim. Prod.* **43**, 221-3.

Towers, N.R., and Smith, B.L. (1978). *N.Z. Vet. J.* **26**, 119-203.

Towers, N.R., and Stratton, G.C. (1978). *N.Z. Vet. J.***26**, 109-12.

Twigg, L.E., Mead, R.J., and King, D.R. (1986). *Aust. J. Biol. Sci.* **39**, 1-15.

Vickery, B., and Vickery, M.L. (1973). *Vet. Bull.* **43**, 537-42.

Waghorn, G.C., Ulyatt, M.J., John, A., and Fisher, M.T. (1987). *Br. J. Nutr.* (in press.)

Ward, G., Harbers, L.H., and Blaha, J.L. (1979). *J. Dairy Sci.* **62**, 712-22.

Wong, E., and Latch, C.G.M. (1971). *N.Z. J. Agric. Res.* **14**, 633-8.

Wood, P. McR., and Petterson, D.S. (1985). *Aust. J. Exp. Agric.* **25**, 164-8.

Yoshizawa, T. (1983). In "Trichothecenes; Chemical, Biological and Toxicological Aspects." (Ed. Y. Ueno.) pp. 60-9. (Elsevier: Amsterdam.)

6

DIET SELECTION BY GRAZING AND BROWSING LIVESTOCK

by John C. Malechek and David F. Balph

Diet selection is fundamentally a behavioural process that constitutes the interface between the free-ranging animal and the plant community upon which it feeds. While the existence of linkages between diet selection and nutritional status of the animal is obvious, the components and dynamics of these linkages are exceedingly complex and only incompletely understood. A few research workers (McClymont 1967; Arnold and Dudzinski 1978) stand prominently in the literature for their pioneering efforts to bring organization and insight to this relationship, but much remains to be learned before we are able to predict the botanical and nutritive composition of diets for animals under a given set of circumstances. The purpose of this chapter is not a comprehensive review of the literature on animal diet composition, for that would be massive and not particularly illuminating. Rather, our purpose is an attempt to frame diet selection in its proper behavioural context and to place it in the perspective of some new theoretical concepts pertaining to free-foraging herbivores and, especially, plants and their defence mechanisms against herbivores. These concepts hold promise for rapidly advancing the progress of our research and management. For the most part, we have limited our discussion and our thinking to domestic ruminants, and we have not pursued the comparative aspects of diet selection by different species of herbivores.

BASIS FOR DIET SELECTION

We believe that the best understanding of diet selection can be obtained through an examination of its underlying function and causal mechanisms. The question of function and cause can be approached through optimal foraging theory (Pyke 1984) and through the principles of reinforcement (Reynolds 1975). In this section we present an overview of these two concepts and see how they might be applied to diet choice by livestock.

Optimal foraging

Behavioural ecologists believe that variation in behavioural performance has, to some degree, a genetic base. They believe that natural selection operates on genetic variation to promote the survival of genes that underlie behavioural characteristics that are advantageous. This logic is the basis for the concept of optimality, a major tenet of behavioural ecology.

NUTRITION OF HERBIVORES
ISBN 0 12 750052 9

Optimal foraging theorists have applied the concept to foraging behaviour, on the assumption that the ability to forage efficiently (optimally) is an advantageous trait. Discrimination in the choice of foods is one means of foraging optimally.

Optimal foraging theory as originally conceived (MacArthur and Pianka 1966), postulates that animals should feed in such a way as to gain the most calories per unit time spent feeding. Further developments of the theory have included such considerations as when an animal should move to a new feeding area (Charnov 1976) and how central-place foragers should select dietary items at varying distances from where they live (Orians and Pearson 1979).

More recently, researchers have realized that there may be constraints on animals that prevent them from foraging optimally. Models have been developed to take into account foraging while taking care of young, defending a territory, or other special needs (Krebs *et al.* 1983). The question often asked is whether or not an animal is feeding optimally in the broader context of other activities necessary for its survival.

Tests of hypotheses based on optimal foraging theory have had mixed results. More often than not, however, predictions derived from foraging theory have been qualitatively correct. Perhaps the greatest value of the theory is that it has stimulated much research of a type that has increased our understanding of foraging behaviour.

There is some difficulty in using optimal foraging as a theoretical concept for exploring dietary choice in livestock (Stephens and Krebs 1986). First, the theory stresses optimality while the array of plant defence mechanisms that confront the animals calls for an emphasis on constraints. Rather than ask whether or not the animals are behaving optimally, we believe it is more appropriate to ask to what extent, under specified conditions, animals can optimize, and to what extent they do optimize.

This relativistic approach to foraging performance allows for differences in foraging efficiency between species and circumstances to be explored. We suspect that selection for optimal foraging would be greater for an animal with a high metabolic rate under stress than a large-body animal in an ideal foraging environment. It is not surprising that some of the best examples of optimal foraging are birds with young to feed. Livestock on pasture may not be pressed to forage very efficiently, even if they could.

Another difficulty with applying the theory to livestock is that the concept has been developed primarily for predators where the quality of prey tissue is of little importance relative to the rate at which prey can be obtained. In contrast, plants differ markedly in their nutritional value (Crawley 1983). This suggests that an optimal foraging model for herbivores should be nutrient rather than energy based with plant defence systems as constraints. The rate of harvest still must be considered, and indeed Belovsky (1978) has shown that the moose (*Alces malchis*) appears to be a rate-maximizer within the constraint of obtaining salt.

If the currency to be optimized is primarily nutrients, then an herbivore must be able to discriminate between the nutrient values of different types of forage. Althought rats have been shown to have such "nutritional wisdom" (Richter 1943) and the diets of livestock are more nutritionally dense that the average forage on offer (Van Dyne *et al.* 1980), it seems clear that vertebrates do not have specific nutrient hungers other than for salt (Booth 1985).

This brings up a point that J. R. Krebs emphasized in his address to the Animal Behavior Society in 1986. If an animal is to be an efficient forager, if must have information about its foraging environment, and discovering how animals gather such information should be the next thrust in foraging research. We believe that the best chance of discovering the informational mechanism for diet selection lies in understanding the consequences of eating specific foods.

Plants as reinforcers

Experimental psychologists believe that one of the more powerful causal mechanisms in behaviour is B. F. Skinner's dictum that the consequences of past actions determine subsequent behavioural responses. This concept, applied to foraging behaviour, predicts that diet selection is driven by the consequence of sampling or eating specific plants, with the plants acting as positive or negative reinforcers for the foraging herbivore.

The consequences of eating a specific plant can be both immediate and delayed. The immediate consequences are associated with the smell, taste, and texture of the plant. After sampling the plant, the animal may choose to increase or decrease its consumption of that type of plant based on its sensory experience (Arnold and Hill 1972).

Those factors that set the immediate reinforcing value of potential foods must include the match between a food and an animal's mechanisms for harvesting and digesting the food. Of more interest to us, however, is the develoment of these values in livestock through early experience (Provenza and Balph 1986).

Exposing livestock during weaning to palatable food they will encounter later in life enhances the acceptance of the food (Arnold and Maller 1977). There may even be an imprinting process associated with this learning (Lobato *et al.* 1980). In continuing experiments at Utah State University, R. Squibb has identified the critical period in lambs as about four weeks of age.

The social environment may also be important in setting the value of dietary items (Galef 1985). The presence of the mother as a foraging partner for the lamb seems to influence its acceptance of foods (Green *et al.* 1984). As predicted by social modelling theory (Bandura 1977), the mother has more influence in this regard than a ewe that is not the mother, or a peer; the basis for the learning seems to be interaction while foraging rather than observational learning (A. G. Thorhallsdottier unpublished).

Of special interest is how herbivores might selectively avoid less nutritious diets is the delayed consequences of foraging. This has been studied extensively by pychologists as conditional food aversion (Braveman and Bronstein 1985). Their interest stems from the fact that the learning that occurs is unique in two ways. The animals associate gastronomic distress with what they have eaten rather than with something else (cue-consequence specificity) and they can make the association when the two events are separate in time (long-delay learning).

Rats cautiously sample all new food or food in new locations. If they suffer no adverse consequences, they increase their consumption of the food. Should they become ill or experience adverse affects (including nutritional affects), they tend to reject novel foods (or current diet) and only eat what has been safe in the past (Rozin 1976). Through a system of sampling foods and the conditioning of aversions by nutritional effects, it has been shown conclusively that rats are able to select a nutritious diet (Booth 1985).

Livestock may possess the rudiments of the same system. Sheep are known to sample periodically available foods, to reject foods that make them ill following a delay, and to remember what they have learned (Thorhallsdottier *et al*. 1986). Cattle are also able to reject, through delayed learning, a single food type that is paired with a poison (Olsen and Ralphs 1986). Zahorik and Houpt (1981), however, have shown that some livestock are unable to link, over the short term, the gastronomic consequences of eating a particular food when other foods are eaten at the same time; and it is common knowledge that some livestock die from eating poisonous plants. Clearly, research should be done on the response of livestock to a variety of foods over the long-term to determine how close the animals come to a nutritionally optimal diet. We believe that livestock may not come very close, but then they may not have a continuous need for a nutritionally balanced diet (Crawley 1983; Booth 1985), and they face antiherbivory constraints that may play a major role in their diet selection.

PLANT DEFENCES

Over the past decade, a new sub-discipline of ecology has emerged that deals with plant-herbivore interactions in general, and more specifically, with the nature and role of plant defences in these interactions. While most of the theory and experimentation has been concerned mainly with plant-insect relations (Rhoades 1985; Rosenthal 1986), many of the concepts have potential for generalization to vertebrate herbivores, including free-ranging livestock. The notion that plants are passive participants in the plant-herbivore interaction is no longer tenable. To understand more fully diet selection, increased effort is needed in discovering how plants defend themselves and what the responses of herbivores are to these defences. Our working hypothesis is that these defences, whether they be chemical or physical, constitute the constraint to efficient foraging mentioned earlier.

Theory

Rhoades (1979) originally outlined the basic assumptions of plant defence theory. He considered that plants allocate defences (secondary biochemicals, structural features) in ways which maximize their "fitness" or the ability to transmit their genes to the succeeding generation. These defences are expensive in terms of energy and nutrients diverted from primary plant functions. They have evolved in direct proportion to the risk from predators and in inverse proportion to the cost of the defence, with costly defences being more effective than less costly ones. Within plants, defences are allocated to particular parts and organs in relation to the value of these parts in terms of loss of fitness if eaten. Considering the relatively high metabolic cost of producing and maintaining most chemical defences, they are decreased when risk of herbivory is low and increased when attacked. Defences are a function of the total energy and nutrient budget of the plant.

Resource limitation with respect to browse plants

Recently these concepts have been re-defined and broadened, particularly as they apply to woody plants, and emphasis has been placed on how the availability of resources (light, water, nutrients) in particular environments affects types and quantities of plant defences (Bryant *et al.* 1983; Coley *et al.* 1985; Bryant *et al.* 1987). In resource-poor environments, plants with *inherently* slow growth rates have been favoured by natural selection. These plants have low potential to replace tissue lost to browsing and are characterized by high investments in chemical defences, generally of the "quantitative" type, that are relatively immobile in the plant (e.g. tannins). In resource-rich environments, *inherently* fast growing plants are favoured. They can readily replace lost or damaged tissue and commit relatively little energy and nutrients to defences. When they do, defences are generally of the "qualitative" type that are relatively mobile (e.g. alkaloids). The general outcome of these relationships is that fast-growing species are relatively more palatable and nutritious than are slow-growing or evergreen species.

Following this line of theory, browsing by herbivores may elicit several complex alternative responses by woody plants, depending upon the particular plant species, the part of the plant that is browsed, the age of the plant, the extent of tissue removed, the season in which browsing occurs, and the prevailing environmental circumstances with respect to availability of resources. These relations are discussed in detail and in a management context by Bryant *et al.* (1987), but several of the major points are summarized here because of the obvious ramifications on diet selection and nutrition.

One response of major consequence is termed "long-term defensive induction" (Haukioja 1980; Bryant *et al.* 1987) where severe defoliation of deciduous woody plants reduces subsequent palatability and nutritive value for 1-3 years. Under these conditions, both leaf growth and nitrogen concentration in leaf tissue are reduced, suggesting a nitrogen-limited environment. Additionally, photosynthetic rates are increased, favouring production of carbon-rich secondary metabolites and fibre.

Other woody species exhibit a different set of responses to browsing. The current year's twigs of many woody plants found in the deserts of North America (for example *Artemisia* spp., *Coleogyne ramosissima*, *Larrea tridentata*) have considerably higher concentrations of aversive chemicals and are less palatable to herbivores than are older twigs (Bryant *et al.* 1987). Any browsing that does occur tends to stimulate production of twigs from dormant axillary and adventitious buds, leading to an overall increase in the proportion of current year twigs and a decreased palatability of the plant. Provenza and Malechek (1983) showed that when the browse *Coleogyne ramosissima* is eaten by goats during the winter dormant period, this increased the current year's twig production with a corresponding 240% increase in tannin concentrations. Even though crude protein content and *in vitro* digestibility were also higher in the new twigs, goats preferred to eat the older, less nutritious twigs from previously unbrowsed plants (Provenza and Malechek 1984). Goats forced to eat the new high-tannin twigs lost less weight over the browsing season than did those eating older twigs.

Cutting or other physical removal of the aerial parts of adult trees and shrubs of particular species induces regeneration of shoots (coppice) from buds in the stump. This juvenile growth differs in major physiological and morphological ways from the adult tissue. Coppice shoots often have accelerated growth rates, larger leaves that are retained longer into the dormant period (Hardesty 1984) and elevated concentrations of secondary chemicals in relation to adult-phase twigs (Bryant *et al.* 1987).

The juvenile form of even fast-growing plants on resource-rich sites may be strongly defended, presumably because of the vulnerability of juvenile tissue to browsing herbivores (Bryant *et al.* 1983). However, with repeated browsing or removal of the juvenile coppice shoots, the plant is forced to reallocate carbon and nutrients to compensatory growth rather than to defensive chemicals (Bryant *et al.* 1987). At this stage, newly growing shoots would presumably become more palatable.

Grasses

Grazing ecosystems based on graminoid species appear not to be so heavily engaged in chemical defences as are those composed of broad-leafed herbs and, particularly, woody species. Rather, the strategy seems to be one of tolerance to grazing, at least at moderate levels (Ellison 1960) or reductions in the likelihood of being grazed through development of prostrate growth forms that render the individual cryptic or less accessible (Detling and Painter 1983; Carman and Briske 1985).

The possibility has been raised for an "inducible defence" based on leaf blade silica content of three East African grasses (McNaughton and Tarrants 1983). In a controlled laboratory study, plants collected from heavily grazed areas accumulated more leaf silica than did con-specific individuals collected from less heavily grazed areas. Silica levels were also increased by defoliation treatments applied in the laboratory. North American researchers (Brizuela *et al.* 1986), attempting to validate this hypothesis in respect to the two grass species *Agropyron smithii* and *Schizachyrium scoparium*, also found higher levels of elemental silicon in plants from areas heavily grazed by native prairie dogs (*Cynomys ludovicianus*) than in less-grazed areas. However, these researchers were unable to demonstrate elevated silicon levels in response to controlled defoliation studies. The existence of a general chemical defence system in grasses still remains to be demonstrated.

Toxic substances which may be interpreted as constraining mechanisms under the optimal foraging paradigm are well documented in a few economically important grass genera (e.g. *Phalaris*, *Lolium*, *Festuca*, *Setaria*). However, this involvement seems small in light of the large number of genera and species where no apparent problems with toxicity exist. Overall, the chemical relationships between grasses and grazers appear markedly more subtle than those found existing between woody plants and browsers. In individual grass plants and grass communities, a much stronger argument can be made for the importance of structural features (cell wall composition, physical arrangement of leaf versus stem structures, sward architecture) that alter forage intake than for secondary chemicals that directly affect dietary selection.

Community-level considerations

While the main theme of this paper is diet selection at the individual plant level, a host of environmental and animal-related factors interact to determine the physical location of the herbivore in time and space and the particular behaviour (for example, feeding, ruminating, resting) displayed. However, several aspects of plant community organization and structure can be considered in the context of plant defences against mobile herbivores.

Variability in space and time. Increased variability in the forage resource in terms of the location and distribution of palatable plants would presumably lead to greater outlays of time and energy expended in searching by the foraging animal, with associated risks of separation from herd mates and, in the case of wild herbivores, potential death by predation. Arnold (1970) observed that even in relatively simple forage grass stands, as the overall availability of food diminished, sheep spent a disproportionately large amount of time searching for the preferred species, and daily forage intake rates declined.

In this context, a singularly impressive feature of our preliminary studies on chemical defences of shrubs is variation of these chemicals among con-specific individuals within a stand. For example, Yabann (1984) found that in a 0.06 ha plot containing approximately 650 individuals of the shrub *Artemisia tridentata* subsp. *tridentata*, total monoterpenoid concentrations in leaves of individual plants varied from 0.06% to 1.8% (dry-weight basis) on a given day. This large plant-to-plant variation, coupled with major seasonal (Yabann 1984) and, probably, shorter-term changes complicates the foraging process by making the presence of acceptable forage less predictable, even on long-lived evergreen shrubs. Yabann (1984) and Yabann *et al.* (1985) reported that sheep refused to eat *Artemisia* that contained on average more than 0.3% total monoterpenoids, resulting in a utilization pattern that ranged from total defoliation of some plants to virtual refusal of others in the same small plot. Because there may be no stable cues to this variation, animals are probably obligated to sample each potential dietary item by one or more sensory modalities. Rather than the conventional view of this variability as a statistical sampling problem, we should be attempting to better understand how such variability affects the way animals learn to select a diet. Logically, this variability could promote sampling and trial-and-error learning rather than imprinting or social learning.

In a like manner, variability among individual animals in a flock or herd with respect to their individual taste and odour preferences or tolerances is apparently large (Arnold *et al.* 1980). This phenomenon has received surprisingly little attention in view of the potentially large ecological consequences. As suggested by the limited data considered by Arnold and Dudzinski (1978), even though selectivity by individuals is probably large, selectivity by a flock of sheep as a whole may be relatively small when the range of individual preferences is considered.

Browse vs. grass. Many rangelands of the world support plant communities that are mixtures of shrubs, grasses, and forbs. Questions of chemical defences, palatability, and nutritional factors aside, the physical presentation of leaf and stem tissue on shrubs probably modifies feeding efficiency and diet selection to an equal or greater extent than in grasses. In our own work in Utah we have observed that sheep use at least three feeding behaviours when browsing on shrubs: plucking of individual leaves; stripping of leaves, twig tips, and bark from current-season's twigs; and biting and ingestion of entire twigs of current-season's growth. The particular tactic employed depends on the species of plant being browsed and the time of year. Although we have not attempted to quantify the dry matter harvested per unit of feeding time by each of these tactics, rates probably differ widely. Preliminary results from trials where 4-month-old lambs having prior browsing experience were presented with entire plants of the palatable shrub *Amelanchier alnifolia* indicate ingestion rates of 4.6 g DM /min of feeding time (E. Flores and F. D. Provenza, personal communication). The primary feeding mode involved plucking individual leaves. In contrast, lambs with no prior browsing experience consumed only 3.7 g DM /min. E. Flores (pers. comm.) observed that experienced lambs appeared to have better developed motor skills for manipulating the plants and maintaining a relatively rapid rate of biting (33 bites /min.) compared to the inexperienced group (24 bites /min.). The ingestion rates of 4.6 g DM /min for the experienced lambs compares favourably with the average maximum rate of about 5 g DM /min that Penning (1986) reported for adult sheep on uniform grass pastures in the United Kingdom. However, on native pastures where shrubs are distributed in patches and there is often large variability in their palatability (see above), one would expect to see overall lower ingestion rates due to the large amount of search time required.

Ingestion rates of graminoids appear to be particularly sensitive to the physical arrangement of the forage sward (see recent review by Hodgson 1985). How such sward characteristics as height, tiller density, proportions of leaf *vs.* stem and green *vs.* dead, and standing biomass of various plant species affect diet selection is unclear. Until now, work in this area has focussed principally on food intake rates. However, considering how changes in these sward variables can elicit major changes in ingestive behaviours (biting rate, bite size, daily grazing time), we should expect that they would greatly influence diet selection as well. All other factors being equal, the grazing animal would predictably select the plant that is the easiest to prehend. On the other hand, there logically exists some lower threshold of availability and accessibility for preferred species beyond which the grazing animal is unwilling to continue to seek them. Rather, it will turn its attention to less preferred but more available species, or it will simply reduce its intake if the alternative choices are not palatable (Arnold 1970). The challenge facing researchers is to better define the characteristics of forage plants that facilitate (or hinder) their harvest by the grazing animal. These findings will have far-reaching implications to plant breeders and to grazing managers alike.

CONCLUSION AND RESEARCH PERSPECTIVES

While many aspects of diet selection remain unclear, much progress has been made since McClymont's (1967) benchmark paper. A prime contributor to this progress has been the interchange of ideas among scientists from diverse disciplines. This is apparent when one considers the literature sources of the major points made in this paper. These points are first, that the function of diet selection is to meet, with at least some level of satisfaction, the nutrient and energy needs of herbivores. The concept of optimal foraging has utility as a standard against which foraging performance can be compared. Second, animals may be able to select a nutritious diet through conditioned preferences, and particularly, conditioned avoidances. Third, learning early in life plays a role in the initial establishment of preferences or tolerances. Last, plant defences are major constraints to selecting dietary items, and in the case of browsers may be of overriding importance.

Several areas of research appear to us to be particularly important, given the present state of knowledge on diet selection. High in priority is the need for a better understanding of conditioned taste aversions in ruminants. Just how sensitive is the system and how fast are animals able to adjust their diets based on nutritional consequences?

There is also a need for more information about the perceptional world of ruminants. This includes not only their sensory capabilities but also their ability to sense and respond to plant communities and landscapes beyond their immediate feeding station (R.L. Senft *et al.* unpublished). Are livestock attentive to landscape features that are associated with the plant communities they prefer?

Better research approaches and techniques are also needed. Given the complex nature of diet selection, field trials with limited control on variables are certain to produce only more descriptive reports that contribute little to the understanding of determinants of diet selection.

One promising tool in foraging research is operant analysis, employed so successfully in experimental psychology. The few applications that have been made to livestock (Matthews and Kilgour 1980) are highly encouraging. This approach would allow, for example, the independent assessment of gustatory and nutritional properties of foods.

In order better to understand the plant side of the forage-animal equation, much more sophisticated approaches to plant chemistry will be required. Relating animal feeding responses to generic mixes of chemicals (for example "tannins", "phenolics" and "volatile oils") may suffice for demonstrating correlations (Provenza and Malechek 1984; Cooper and Owen-Smith 1985) but are likely to be inadequate for determining causation.

Finally, selective breeding programs for both plants and animals would appear to profit from greater attention to principles of plant-animal relationships. The apparent wide range of genetic variability among individual plants with respect to production of secondary phytochemicals and the similarly wide range of individual dietary preferences and tolerances of animals would offer wide latitude for artificial selection by the breeder. However, the large initial step - adequately describing and quantifying this variability - still remains to be done.

This work was supported in-part by the United States Agency for International Development Title XII Small Ruminants Collaborative Research Support Program under Grant No. AID/DSAN/XII-G-0049, and by the Utah Agricultural Experiment Station.

Arnold, G.W. (1970). In "Digestion and Metabolism in the Ruminant." (Ed. A.T. Phillipson.) pp. 264-76. (Oriel: Cambridge.)

Arnold, G.W., de Boer E.S., and Boundy, C.A.P. (1980). *Aust. J. Agric. Res.* **31**, 571-87.

Arnold, G.W., and Dudzinski, M.L. (1978). "Ethology of Freeranging Domestic Animals." (Elsevier: New York.)

Arnold. G. W., and Hill. J.L. (1972). In "Phytochemical Ecology." (Ed. J.B. Harboone.) pp. 71-101. (Academic: New York.)

Arnold, G.W., and Maller, R.A. (1977). *Appl. Anim. Ethol.* **3**, 5-26.

Bandura, A. (1977). "Social Learning Theory." (Prentice-Hall: Englewood Cliffs, New Jersey.)

Belovsky, G.E. (1978). *Theor. Popul. Biol.* **14**, 105-34.

Booth, D.A. (1985). *Ann. N.Y. Acad. Sci.*, **443**, 21-41.

Braveman, N.S., and Bronstein, P. (Eds) (1985). "Experimental assessments and clinical applications of conditioned food aversions." *Ann. N. Y. Acad. Sci.*, Vol. 443. (N. Y. Acad. Sci.: New York.)

Brizuela, M.A., Detling, J.K., and Cid, M.S. (1986). *Ecology* **67**, 1098-101.

Bryant, J.P., Chapin, F.D., and Klein, D.R. (1983). *Oikos* **40**, 357-68.

Bryant, J.P., Provenza, F.D., and Gobena. A. (1987). "Environmental Controls over Woody Plant Chemical Defenses: Implications for Goat Management." Proc. 4th Int. Conf. on Goats, Brasilia, Brazil. (in press.)

Carman, J.G., and Briske, D.D. (1985). *Oecologia* (Berlin) **66**, 332-7.

Charnov, E.L. (1976). *Theor. Popul. Biol.* **9**, 129-36.

Coley, P.D., Bryant, J.P., and Chapin, F.S. 1985. *Science* **230**, 895-9.

Cooper, S.M., and Owen-Smith, N. (1985). *Oecologia* (Berlin) **67**, 142-6.

Crawley, M.J. (1983). "Herbivory." (Blackwell: Oxford.)

Delting, J.K., and Painter, D.L. (1983). *Oecologia* (Berlin) **57**, 65-71.

Ellison, L. (1960). *Bot. Rev.* **26**, 1-78.

Galef, B.G. (1985). Ann. N.Y. Acad. Sci. **443**. 203-15.

Green, G.C., Elwin, R.L., Mottershead, B.E., Keogh, R.G., and Lynch, J.J. (1984). *Proc. Aust. Soc. Anim. Prod.* **15**, 373-5.

Hardesty, L. (1984). Coppicing to improve dry season forage production in the semi-arid tropics. Working Papers of the Second Int. Rangeland Cong., Adelaide, Aust., sec. 6-C.

Haukioja, E. (1980). *Oikos* **35**, 202-3.

Hodgson, J. (1985). *Proc. Nutr. Soc.* **44**, 339-46.

Krebs, J.R., Stephens D.W., and Sutherland, W.J. (1983). In "Perspectives in Ornithology." (Eds A.H. Brush and G.A.Clark.) pp. 165-216. (Cambridge Univ.: New York.)

Lobato, J.F.P., Pearce G.R., and Beilharz R.G. (1980). *Appl. Anim. Ethol.* **6**, 149-61.

MacAuthor, R.A., and Pianka, E.R. (1966). *Amer. Nat.* **100**, 603–9.
Matthews, L.R., and Kilgour, R. (1980). In "Digestive Physiology and Metabolism in Ruminants." (Eds Y. Ruckebush and P. Thivend.) pp. 123–44. (AVI: Westport, Connecticut.)
McClymont, G.L. (1967). In "Handbook of Physiology." Section 6. Alimentary Canal. (Ed. C.F. Cole.) pp. 129–37. (Am. Physiol. Soc.: Washington, DC.)
McNaughton, S.J., and Tarrants, J.L. (1983). *Proc. Nat. Acad. Sci.* **80**, 790–1.
Olsen, J.D., and Ralphs, M.H. (1986). *Am. J. Vet. Res.* **74**, 1829–33.
Orians, G.H. and N.E. Pearson. (1979). In "Analysis of Ecological Systems." (Eds D.J. Horn, R.D. Mitchell and G.R. Stairs.) pp. 154–77. (Ohio State Univ.: Columbus, Ohio.)
Penning, P.D. (1986). In "Grazing Research at Northern Latitudes." (Ed. O. Gudmundsson.) NATO ASI Series A, Vol. 108, pp. 219–26. (Plenum: New York.)
Provenza, F.D., and Balph, D.F. (1986). *J. Appl. Anim. Behav. Sci.* (in press.)
Provenza, F.D., and Malechek, J.C. (1983). *Biochem. Syst. Ecol.* **3**, 223–38.
Provenza, F.D., and Malechek, J.C. (1984). *J. Appl. Ecol.* **21**, 831–41.
Pyke, G.H. (1984). *Ann. Rev. Ecol. Syst.* **15**, 523–75.
Reynolds, G.S. (1975). "A Primer of Operant Conditioning." (Scott, Forseman: Glenview, Illinois, USA.)
Rhoades, D.F. (1979). In "Herbivores, their Interaction with Secondary Plant Metabolites." (Eds D.A. Rosenthal and D.H.Janzen.) pp. 3–54. (Academic: New York.)
Rhoades, D.F. (1985). *Amer. Natur.* **125**, 205–38.
Richter, C.P. (1943). *Harvey Lect.* **38**, 63–103.
Rosenthal, G.A. (1986). *Sci. Amer.* **254**, 94–9.
Rozin, P. (1976). In "Advances in the Study of Behavior." (Eds J.S. Rosenblatt, R.A. Hinde, E. Shaw and C. Beer.) pp. 21–76. (Academic: New York.)
Stephens, D.W., and Krebs, J.R. (1986). "Foraging Theory." (Princeton University: Princeton, NJ.) (in press.)
Thorhallsdottier, A.G., Provenza, F.D., and Balph, D.F. (1986). *J. Appl. Behav. Sci.* (in press.)
Van Dyne, G.M., Brockington, N.R., Szocs, Z., Duek, J., and Ribic, C.A. (1980). In "Grasslands, Systems Analysis, and Man." (Eds A.J. Brockington and G.M. Van Dyne.) pp. 269–537. (Cambridge Univ.:London.)
Yabann, W.K. (1984). MS Thesis, Utah State Univ., USA.
Yabann, W.K., Burritt, E.A., and Malechek, J.C. (1985.) In "Plant-Herbivore Interactions." (Eds F.D. Provenza and J.T. Flinders.) Proc. 4th Ann. Wildland Shrub Symp. USDA Forest Serv. Gen. Tech. Rep.
Zahorik, D.M., and Houpt, K.A. (1981). "Foraging behavior." (Eds A.C. Kamil and T.D. Sargent.) pp. 289–310. (Garland: New York.)

7

COMPARITIVE ASPECTS OF FOOD INTAKE

R.H. Weston and D.P. Poppi

The very survival of any mammalian species indicates that the species has a robust system for regulating its feed intake, a system that in addition to facilitating growth and development, also ensures the partitioning of nutrients, as required, to the utero-foetal complex and the mammary gland(s). Basic similarities might be expected in the regulation of feed intake of all mammalian herbivores as the species have the same basic biology, are similar with respect to intermediary metabolism, produce tissues comparable in chemical composition, generally possess the same assemblage of internal organs and need to secrete milk for their young. However, some variation in the detail of the regulation might also be expected because of herbivore diversity. Thus the species vary in body size, gastro-intestinal tract (GIT) anatomy and function, reproduction strategies, selected diet composition and evolutionary background.

In contrast to many omnivorous and carnivorous mammals, herbivores may have particular problems to overcome in relation to the availability and the physical and chemical characteristics of their diets. For example, appropriate feed may be sparsely spread over large areas, difficult to prehend, highly resistant to structural degradation during ingestion, refractory to digestion, high in content of deleterious secondary compounds, electrolytes and water, and in content of fibre components available as energy sources only after fermentation in the GIT.

Research into feed intake regulation in the herbivore has ranged from controlled studies with single cells in the central nervous system (CNS) to diet selection observations in freely-ranging wild animals. As might be expected the major part of the data relevant to this review has been obtained with commercially important species - sheep and cattle in the case of foregut fermenters, and horses and rabbits in the case of hindgut fermenters. Unfortunately few adequate comparisons of feed intake between various species are available. Many comparisons provide only equivocal data because of lack of parity in various animal, diet and environmental factors. In this context the extreme practical and technical difficulties encountered by many research workers, particularly those studying non-domesticated herbivores under freely-ranging conditions, have to be recognized and emphasized.

A multiplicity of factors is involved in feed intake regulation and in the light of this complexity only a general outline can be provided here. The subject is considered in this review basically in two parts, *viz.* regulation under optimal conditions of diet and

NUTRITION OF HERBIVORES
ISBN 0 12 750052 9

environment, and regulation where constraints to feed intake exist. In the former situation, but not the latter, the animal has the capacity to express its genetic potential to use energy. The optimal situation is considered to prevail when environmental stresses are minimal and the diet is palatable, readily accessible, adequate with respect to essential nutrients, and sufficiently digestible such that its undigested residue is easily handled by the GIT. Many factors may act as constraints to feed intake. In this review, discussion is mainly confined to constraints relating to the feed acquisition, feed ingestion and the flow of undigested material through the GIT. Various other factors such as diet palatability, between animal competition, climate, water availability, toxic secondary compounds and parasitic disease are discussed elsewhere in this Symposium.

Clear distinction is drawn here between meal eating regulation and the longer term regulation (> 2–3 days) which is referred to here as feed intake regulation. This difference is emphasized as many determinants of meal eating behaviour are not determinants of feed intake. Again, in this review for the sake of conciseness, the abbreviations DE, ME and NE are used to denote digestible energy, metabolizable energy and net energy respectively, the abbreviations DM, OM and CWC are used to designate dry matter, organic matter and cell wall constituents respectively, and BMR, BW and MW are used to designate basal metabolic rate, body weight and metabolic weight respectively, the latter being equivalent to BW $kg^{0.75}$. Further, in calculations it has been assumed that 1 g digestible DM provides 19 kJ of DE.

FEED INTAKE UNDER OPTIMAL CONDITIONS

Growth and development phase

Clear evidence exists that under optimal conditions a range of herbivores control feed intake in relation to their energy needs for sustaining growth or energy accretion at a set rate. The evidence is shown in feed intake responses to alteration of the DE content of the diet and to the direct introduction of nutrients into the GIT via gastric fistulae. For example, with growing rabbits decrease in the diet DE content in the range 14.7 MJ/kg DM – 8.7 MJ/kg DM was accompanied by increases in feed intake, such that DE intake and rate of BW gain remained constant (Spreadbury and Davidson 1978). Similar feed intake responses were shown in equines when the dietary DE content was decreased from 14.2 MJ/kg DM to 10.9 MJ/kg DM (Laut *et al.* 1985). Compensatory changes in feed intake to maintain DE intake and BW gain constant are also well documented in ruminants. Such data, as summarized by Baumgardt (1970), show feed intake compensation with sheep and cattle over a diet DE concentration range of 14.1 MJ/kg DM – 10.0 MJ/kg DM. Decreases in feed intake accompanied the direct introduction of nutrient into the abomasum of sheep (Weston 1971;

Seoane *et al.* 1976), these changes being such that both DE intake and rate of BW gain remained constant.

The foregoing indicates that under optimal conditions the feed intake of herbivores during growth and development is probably regulated in relation to the energy required to maintain a particular growth rate. It would seem reasonable to assume that this growth rate is determined according to a particular genetic programming. Accordingly the genetic programme for growth, together with the body maintenance need, would set an upper limit to feed intake, a limit that would reflect the genetic potential of the animal to consume feed.

Differences in the potential to use energy would be expected at different stages of growth and development. Pertinent data in this context have been obtained with the growing crossbred sheep. As shown in Fig. 1 the rates of body growth, energy deposition and protein deposition varied at different stages of maturity. In general terms the demand for energy in pathways of synthesis, per unit of body weight, decreased progressively with age, and this change was accompanied by a decrease in BMR. In accordance with these changes ME intake per unit of body size continuously decreased. It is noteworthy that ME intake was not directly proportional to BMR although the two were directly related.

Reproduction cycle

Few adequately controlled studies have been made of feed intake throughout pregnancy and lactation in herbivores. Practical problems of some magnitude exist with studies in species having long gestation periods and in several studies the animals used have been both pregnant and lactating.

Feeding to meet a particular energy demand during pregnancy is shown in the rabbit. Thus in the last trimester does fed a diet of low DE content increased feed intake in a compensatory manner relative to DE intake (Butcher *et al.* 1983). No comparable studies appear to have been done with other species. The studies of Butcher *et al.* (1983) also show feed intake to increase with decrease in diet DE content in lactating does. However, the compensation was incomplete, with intake increasing by only 0.15 for a 0.3 reduction in DE concentration. In dairy cows, feed intake increase tends to compensate for decrease in DE concentration down to about 11.3 MJ/kg DM (Baumgardt 1970).

Feed intake data for an entire reproductive cycle under optimal conditions appear to be available only for the rabbit, although data for various parts of the cycle are available for a few species. With pregnancy the energy needs for the utero-foetal complex and the mammary gland(s) are relatively low during the first two trimesters but then progressively increase until parturition. No convincing evidence exists for a consistent enhancement of feed intake at any stage of pregnancy although some such change would be expected. With does and ewes no increase was observed in the first two trimesters (Reid

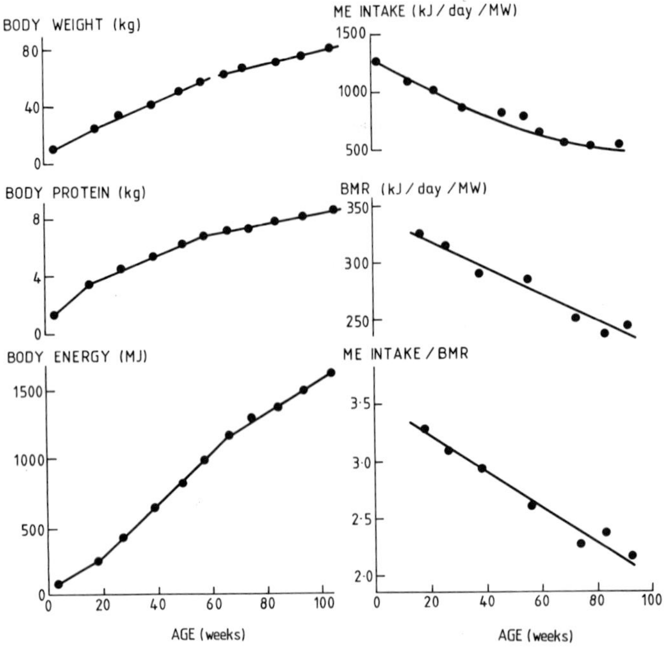

Figure 1. Changes in body weight, body protein, body energy, metabolizable energy (ME) intake and basal metabolic rate (BMR), with increasing age in crossbred sheep, as obtained by Graham and Searle (1972) and Graham *et al.* (1974).

and Hinks 1962; Lebas 1975). Again, in the third trimester, intake does not rise to match the increasing demands for reproduction at least in the rabbit, sheep, and cow (Forbes 1970; Anderson *et al.* 1979; Butcher *et al.* 1983). Rather the intake is more likely to decline at least with rabbits, cattle and sheep (Lebas 1975; Weston 1982). Further, in does, ewes and cows a marked decrease in intake is consistently observed in the last few days of pregnancy (Reid and Hinks 1962; Reyne *et al.* 1977; Erb *et al.* 1982). Such declines are unlikely to be related to physical limitations to GIT function because they appear independent of the number of foetuses and diet DE content (Forbes 1970; Butcher *et al.* 1983). The failure of feed intake to increase in relation to the increased energy demands for pregnancy indicates that during late pregnancy the rate of accretion of maternal body energy reserves must diminish.

Increase in feed consumption during lactation is probably universal in herbivores. Enhanced intake prevails with increase in litter size at least in sheep and rabbits, this presumably being a reflection of a higher energy need (Prud'Hon 1968; Weston 1982). During early lactation in the rabbit, dairy cow and the sheep, intake may not increase sufficiently to match the enhanced energy demand (Bauman and Currie 1980; Cowan *et al.* 1980). This situation can be reversed later in lactation when intake declines at a slower rate than the energy demand for milk (Bauman and Currie 1980). In rabbits, sheep and cattle, those animals with high levels of body energy stores at parturition consume less feed in early lactation than their leaner counterparts (Lebas 1975; Lodge *et al.* 1975; Foot and Russel 1979). In the dairy cow at least, the failure of feed consumption increase to match the additional energy needs in lactation is not necessarily due to inadequate rumen capacity (Mowat 1963). It would appear that in early lactation, a tendency for preferential use of some maternal body energy reserves may sometimes occur.

MEAL EATING REGULATION UNDER OPTIMAL CONDITIONS

Many herbivores under free ranging conditions have a small number of comparatively long grazing periods each 24 h (Arnold and Dudzinski 1978). However, under optimal conditions, horses, cattle, sheep, goats, rabbits and guinea-pigs consume small meals at frequent intervals (Baile 1975; Hirsch 1973; Reyne *et al.* 1977; Laut *et al.* 1985), a meal eating pattern that also prevails with many omnivores. It would appear that the pattern of long and infrequent meals under free ranging conditions is not a general herbivore trait but relates to the environment.

The environment, diet characteristics, the GIT, the liver and various metabolic transactions are among the many factors that may contribute to the regulation of meal eating. It is to be noted that the regulation besides being concerned with nutrient intake, must also serve to protect the internal milieu, with only small deviations from constancy being possible in osmotic, acid/base, and thermal balances; some protection of the GIT from overload is also required.

Meal eating regulation is still poorly understood even in the laboratory rat, the species most studied. The major components of the regulation and some interactions for optimal conditions are shown schematically in Fig. 2 and the components are discussed separately below. The transactions for the brain as shown in Fig. 2 are essentially those in the scheme of Morley *et al.* (1984) for the laboratory rat; in the absence of significant amounts of data for herbivores this scheme for the rat provides a basis for appropriate discussion.

Gastro-intestinal tract and pancreas

Entry of feed into the GIT initiates the secretion of a range of gastro-intestinal tract and pancreatic hormones. In the laboratory rat cholecystokinin peptides, particularly the octapeptide, somatostatin,

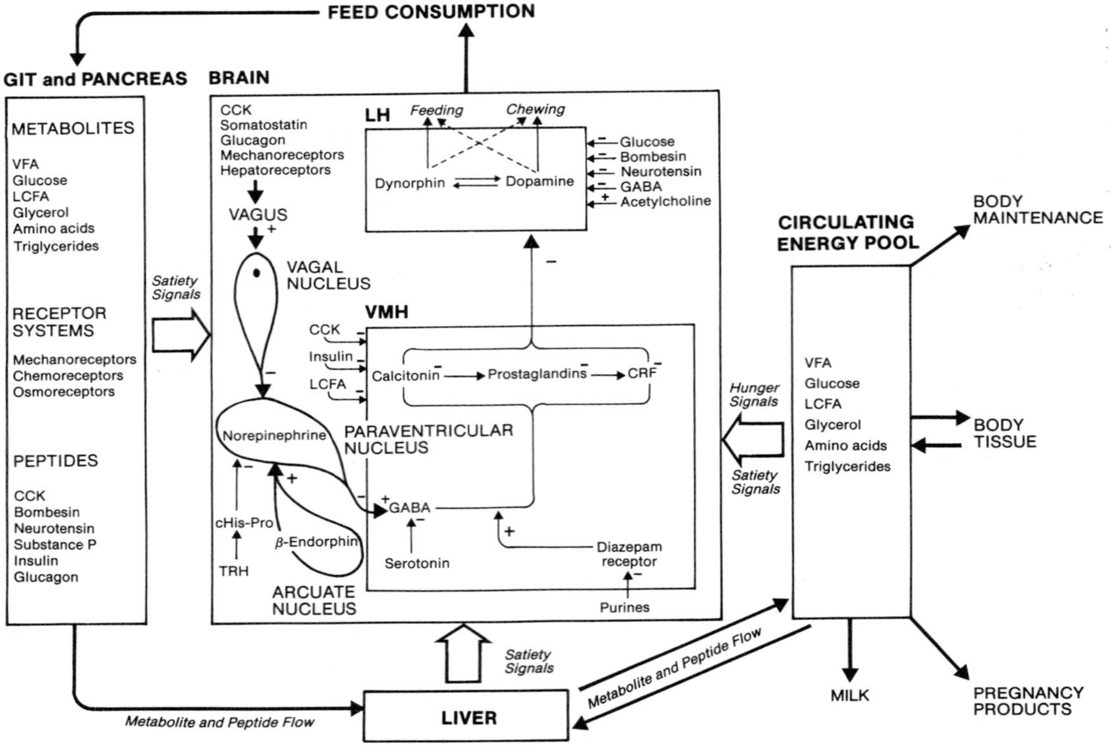

Figure 2. Diagrammatic scheme of the regulation of feed intake; the components of the system within the brain are largely as postulated by Morley *et al.* (1984) for the rat.

neurotensin and substance P, have been shown to contribute to satiety and accordingly to meal termination.

Few data on GIT peptides are available for herbivores. Cholecystokinin appears to promote satiety in the rabbit as in the rat (Houpt *et al.* 1978). However in the sheep, GIT cholecystokinin has little effect on satiety although brain cholecystokinin has a marked effect (Baile *et al.* 1986). Bombesin has been shown to be a potent satiety-inducing peptide in the sheep but data for other herbivores seem lacking. No data appear to be available with herbivores for somatostatin, neurotensin and substance P.

Spontaneous feeding in the sheep induces rapid increases in the venous concentrations of insulin and glucagon. In the sheep, insulin given via the jugular or portal vein depresses meal eating, and intravenous injection of glucose reduces feed intake. In reviewing research on the peptides, de Jong (1986) concluded that in ruminants, whereas insulin was unlikely to be implicated in meal eating regulation, glucagon could well be involved as a satiety agent.

During the course of feed digestion, various digestion end products increase in concentration in the GIT and may generate cues influencing meal eating. The volatile fatty acids (VFA) are of particular significance for herbivores. With ruminants, de Jong (1986) has concluded that rapid and large changes in the concentration of VFA in rumen liquor could influence meal eating behaviour in schedule-fed animals via chemoreceptors in the rumen wall. However, he concluded that this was unlikely in freely-fed animals as neither the onset nor termination of meal eating was related to ruminal VFA level. The same situation could apply in the equine with the caecum being the significant GIT compartment. Thus intra-caecal infusion of VFA at appropriate levels has been shown to decrease feeding activity in schedule-fed equines (Ralston *et al.* 1983). Few data are available for meal eating in relation to GIT concentrations of glucose and long-chain fatty acids. Immediate meal eating activity in schedule-fed equines is reduced following intragastric administration of glucose (Ralston and

---

(see Figure 2 opposite)
Abbreviations:

TRH	thyrotropin-releasing factor
cHis-Pro	cyclohistidyl proline diketopiperazine
CRF	corticotrophin-releasing factor
CCK	cholecystokinin
GABA	gamma-amino butyric acid
GIT	gastro-intestinal tract
VFA	volatile fatty acids
LCFA	long chain fatty acids
LH	lateral hypothalamus
VMH	ventromedial hypothalamus

Baile 1982) and normal meal eating in freely-fed rabbits is reduced by duodenal glucose administration (Novin *et al.* 1974). No immediate effects of corn oil given orally were found with the equine (Ralston and Baile 1983). No role is postulated for amino acid levels in the GIT as the peaks of concentration occur well after meals are completed.

Osmoreceptors may be active in promoting satiety in herbivores. Thus meal size in the schedule-fed sheep was found to be inversely related to the osmolality of rumen liquor (Ternouth and Beattie 1971; Phillip *et al.* 1981) and the onset of meal eating was delayed in schedule-fed equines following intragastric administration of electrolytes (Ralston and Baile 1983). Mechanoreceptors responsive to stretch and tactile stimulation are present in the rumen and presumably generate cues relating to meal eating behaviour. In schedule-fed sheep and cattle, meal eating has decreased linearly with increase in the volume of water-filled balloons inserted in the rumen to occupy space (Campling and Balch 1961; Adams and Forbes 1981). Corresponding studies on the possible role of GIT mechanoreceptors in other types of herbivore do not appear to have been conducted. Although osmoreceptors and mechanoreceptors may affect meal eating in schedule-fed herbivores, the nature of the role they could play in the freely-fed animal has yet to be clarified.

Liver

Controversy still exists as to the role the liver could play in meal eating regulation in both monogastric animals and herbivores (Forbes 1982; Novin *et al.* 1985; de Jong 1986). As most absorbed substances have to pass through the liver, this organ is strategically placed to monitor uptake from the GIT and contribute to meal eating regulation. The following findings, all relating to glucose metabolism (Forbes 1982; de Jong 1986), suggest that the liver could play a role in meal eating regulation. First, meal eating following an overnight fast is depressed in rabbits following intraportal administration of glucose, and meal eating in freely-fed sheep, but not goats, is depressed by similar administration of propionate. Second, in both the rabbit and sheep, section of vagal afferents in the liver overcomes the meal eating depression prevailing above. Third, the activity of the hepatic branch of the vagus in perfused guinea-pig liver is related to glucose concentration in the infusate.

Circulating energy pool

A mixture of components collected from the gastro-intestinal tract, liver, adipose tissue, muscle mass and kidney comprise the bulk of the pool of energy substrates available to the body's metabolizing cells.

No clear picture has emerged as to the role in meal eating regulation in herbivores of the levels of individual metabolites in the circulating energy pool. The intravenous infusion of various VFAs generally depresses meal eating in ruminants (Baile and Forbes 1974) and in the rabbit (Tournier 1972). However as discussed by Baile and

Forbes (1974) no evidence exists for VFA receptors in the CNS and there is no apparent relation between blood VFA levels and eating patterns (de Jong 1986). In the ruminant meal patterning is not related to blood glucose level (de Jong 1986) and in the equine hyperglycaemia has no effect on meal eating (Ralston and Baile 1982). In the sheep intravenous infusion of long-chain fatty acids reduces feeding (Vandermeerschen-Doize and Paquay 1984); however, in schedule-fed ruminants venous levels of long-chain fatty acids are at their highest just before feeding when hunger is at its peak. In reviewing the role of blood metabolites in feed intake regulation of ruminants, de Jong (1986) concluded that long-chain fatty acids in the blood were unlikely to play a major role in meal patterning. He also pointed out that clear evidence was lacking for any involvement of glycerol, ketone bodies or amino acids.

Nicolaidis and Even (1985) have shown that the onset of meal eating in the laboratory rat is related to a relatively sharp decline in non-locomotor heat production. They consider the decline to result from a reduction in the rate of nutrient metabolism. With feed intake in the rat being regulated in relation to the animal's energy need, as in herbivores, this mechanism could be of particular significance. It provides a necessary link between meal eating regulation and energy metabolism. Moreover, it would be consistent with the existence of a regulatory scheme that need not rely on the blood levels of any specific metabolite.

Spontaneous meals in ruminants are usually preceded by a peak in plasma growth hormone (Forbes 1980) and meal eating in schedule-fed ruminants causes plasma growth hormone to decrease. However exogenous growth hormone does not directly affect feed intake (Forbes 1980) so circulating growth hormone levels presumably do not influence meal patterning.

The rate of energy utilization varies with the metabolic state of the herbivore and such variation must be accompanied by differences in the rate of substrate depletion in the circulating energy pool. It is of some interest that in lactation, when the rate of energy utilization is high, rabbits and sheep respond to the increase in rate of substrate depletion by increasing both meal size and meal frequency (Reyne *et al.* 1977; Forbes 1980).

Finally, Baile and Forbes (1974) in their review consider that the heat produced during meal eating in ruminants is unlikely to be involved in meal eating regulation.

Brain

The hypothalamus is generally considered to be the region of the brain most involved with meal eating regulation. Early studies led to the concept of excitatory and inhibitory centres in the hypothalamus with a feeding centre in the lateral region (LH) and a satiety centre in the ventro-medial region (VMH). This simple concept has been modified in recent years to permit involvement of various nerve fibre tracts and the operation of various neuropeptides and monoamines.

The central control of meal eating has been studied much more extensively in the laboratory rat than in herbivores and, as indicated earlier, the scheme presented in Figure 2 is essentially one proposed for the rat by Morley *et al.* (1984). The regulation involves at least 4 major nerve fibre tracts, endogenous opiates and other neuropeptides and is subject to modulation by various hormones and nutrients . The nerve fibre tracts include (1) a dopaminergic tract that passes through the VMH, (2) a serotoninergic tract that passes through the LH, (3) the ventral non-adrenergic tract that passes near the VMH and projects to the LH, and (4) vagal afferents with a series of projections into the paraventricular nucleus and the hypothalamus, with these afferents being involved in conveying peripheral signals.

Morley *et al.* (1984) postulate that feeding is facilitated at least in part by an interaction between dynorphin and dopamine. This facilitation is considered to be inhibited by a series of primary inhibitors including corticotropin releasing factor, prostaglandins, bombesin, neurotensin and calcitonin. The primary inhibiting factors are considered to be subject to a nor-adrenergic/gamma-amino butyric acid (GABA) system that enhances feeding, with this system in turn being influenced by serotonin, neurotensin and thyroid releasing hormone. The modulating effects of various nutrients, GIT receptors and hormones, are superimposed on this complex.

Some, but not all, of the data obtained with herbivores are consistent with this model. For example, data consistent with the operation of an opiate peptide system in initiating feeding have been obtained for the rabbit, guinea-pig, sheep and deer (Plotka *et al.* 1985; Baile *et al.* 1986; Sanger and McCarthy, unpublished data). Again data have been obtained for the sheep indicating that dopaminergic, nor-adrenergic and GABA systems are involved in facilitating feeding, whereas bombesin and various prostaglandins inhibit feeding (Baile 1975; Seoane *et al.* 1984; Girard *et al.* 1985). However, unlike the rat, the sheep shows no responsiveness to glucose, and α-adrenoreceptor agonists elicit feeding in sheep much more readily than in cattle (Baile and Forbes 1974; Baile 1975).

The foregoing discussion indicates that a wide range of signals may be concerned with meal eating regulation in herbivores under optimal conditions. However, no clear picture of the regulation has emerged, with no single metabolite or receptor system being predominantly implicated. The concerted action of many factors is probably involved. The data of Adams and Forbes (1981) are consistent with this situation, at least with respect to satiety; they show the effects of three different satiety systems to be generally additive. With regard to the onset of meal eating it seems probable that some as yet undefined signal(s) relating to substrate energy availability (Nicolaidis and Even 1985) could be directly involved.

FEED INTAKE REGULATION UNDER OPTIMAL CONDITIONS

Feed intake in the herbivore under optimal conditions is largely determined by energy need as discussed above. Accordingly, feed intake regulation has to be centred about energy transactions.

Most of the peripheral factors concerned with meal eating regulation have no obvious or postulated quantitative link with energy transactions. The possibility that they are not essential in feed intake regulation under optimal conditions has to be considered. In this context, findings in studies of many days' duration raise doubts about an essential role in feed intake regulation, under optimal conditions for rumen wall receptors responsive to VFA, hepatoreceptors responsive to propionate, GIT mechanoreceptors and GIT osmoreceptors. The relevant findings are

1. intraruminal administration of VFA, including propionate at 137 g/day, decreased feed intake by sheep by no more than would be expected from their ME content (Weston 1985),
2. with post-ruminal infusion of nutrients, sheep decreased feed intake in a manner compensatory for DE as discussed earlier, even though rumen VFA production would have been reduced and the assemblage of digestion end-products altered appreciably,
3. liver denervation as performed by Anil and Forbes (1980) profoundly affected meal eating pattern but did not affect feed intake,
4. intraruminal administration of sawdust to sheep fed an optimal diet had no effect on feed intake or DE intake but appreciably enhanced rumen digesta load and the quantity of indigestible DM passing through the GIT (Egan 1972) and
5. the continuous intraruminal infusion of large quantities of electrolytes had little effect on feed intake in sheep (Weston 1966 and unpublished data). No quantitative link between the GIT peptides and energy transactions is apparent or proposed so their direct involvement in feed intake regulation may also be in question. In this connection, the studies of West *et al.* (1984) suggest that cholecystokinin may not be involved in feed intake regulation in the rat although there is evidence for its possible involvement in meal eating. In these studies the administration of cholecystokinin with each meal for several days appreciably reduced meal size but meal frequency gradually increased so that food intake was essentially unaltered.

The ease with which a well established pattern of meal eating can be altered drastically without change in feed intake further raises doubts about an essential role in feed intake regulation for various components of the meal eating regulation system. Thus, when freely-fed sheep, established on an optimal diet and consuming some 18 meals daily at relatively regular intervals, were changed to a feeding regime permitting access to feed for only one hour at intervals of 8 hours, their feed intakes initially declined; however, it took them only 4-5 days to adjust their meal eating pattern such that feed intake was restored to normal (Weston and Davis unpublished data). It seems possible that many components of meal eating regulation may form part of a network of signals designed mainly to protect the internal milieu from disturbance and the GIT from overload; in this context they need not be necessary for feed intake regulation.

The essential mechanisms in feed intake regulation remain unknown. In the growth phase, intake appears to be regulated in relation to a set growth rate, with this rate being possibly determined on the basis of some type of comparison of the prevailing body tissue mass (or its components) with the body tissue mass required, according to the genetic programme. Mechanisms involved in the growth rate assessment and the transmission of this information to the tissues have not been described. However, once the tissues are programmed for a particular rate of synthesis, the mechanisms relating onset of meal eating to a decline in the rate of tissue energy metabolism (Nicolaidis and Even 1985) could facilitate intake regulation. Accordingly, nutrients resulting from meal eating, as initiated by energy metabolism decline, would replenish the circulating energy pool and thus sustain tissue synthesis. It could be postulated that such a precise mechanism for initiating feeding, together with a satiety system that need not be precise in controlling meal size, would permit adequate regulation of feed intake; within this framework errors in meal size control would be readily corrected by appropriate increase or decrease in inter-meal interval. This type of regulatory system, although generally adequate, would not by itself, accommodate the delays in adjusting energy intake to an enhanced expenditure, such as occur with the onset of cold exposure or of work (Wodzicka-Tomaszewska 1963). No direct involvement of known hormones in feed intake regulation in herbivores has been demonstrated although growth hormone, thyroid hormone and others are required for the expression of maximum intake. There is considerable evidence for a role for insulin in feed regulation in monogastric animals (de Jong 1986) but this possibility has been little explored in herbivores.

The failure of feed intake in late pregnancy to respond to the energy needs of the conceptus and the apparent mis-match between energy intake and energy expenditure in early lactation do not preclude feed intake being regulated by energy demand. Major alterations in maternal adipose tissue metabolism occur in late pregnancy and early lactation due to hormonal changes (Bauman and Currie 1980; Bauman 1984). Thus marked decreases occur in the capacity of adipose tissue to synthesize fat, and fat mobilization increases. Decline in the rate of fatty acid synthesis must reduce the rate of depletion of the circulating energy pool and increased lipolysis must add substrate to it. Accordingly, less dietary energy is needed by the adipose tissue and feed intake can be appropriately adjusted. Should the amount of fat mobilized at these times vary in proportion to the size of the adipose tissue stores, the lower feed intake response by fatter animals in early lactation, as discussed earlier, would be expected, as would the tendency for fat ewes to be susceptible to feed intake decline in late pregnancy (Foot and Russel 1979).

Deficiencies of essential nutrients in the diet are consistently associated with sub-optimal feed intake (Minson 1982). These deficiencies, which include insufficient glucogenic substrate, generally reduce the rate at which body tissues can use energy. Thus various pathways of tissue synthesis, oxidation and excretion are blocked or impaired either directly or indirectly by lack of necessary substrates (for example, amino acids) or by inadequate amounts of co-factors such as minerals, vitamins or vitamin derivatives. The position is the same in the herbivore as in the omnivore with the added complexity of varying rates of synthesis in the GIT. The rates of synthesis may well vary with herbivore type, and the availability of the products of synthesis could be greater in foregut fermenters, because of greater opportunity for absorption. Within the hindgut fermenters, availability would be greater in coprophagous species.

FEED INTAKE REGULATION WHERE CONSTRAINTS PREVAIL

As discussed above, herbivore consumption of optimal diets given in ground form increases with decrease in the DE content of the diet such that DE intake remains relatively constant. However, the consumption of unprocessed forages (long or coarsely chopped) generally shows the reverse situation. With such feeds, decrease in DE content, generally a reflection of increasing fibre, tends to cause a decrease in feed intake and accordingly is associated with a reduced intake of DE. This trend is well documented for ruminants (Minson 1982) and is shown in limited data for foregut fermenter marsupials (McIntosh 1966; Forbes and Tribe 1970). The studies of Darlington and Hershberger (1968) show that it also applies with the equine. It may be calculated from their data that daily DE intake, besides declining with increasing forage fibre content ($r = -0.86$), also declined with decreasing DE content as follows:

DE intake (MJ/MW) = 180 x DE content (MJ/kg DM) − 1140 ($r = -0.88$).

The estimated rate of decrease in DE intake is appreciable, being some 55% for decline in DE content from 13.5 MJ/kg DM (DM digestibility *ca* 0.7) to 9.6 MJ/kg DM (DM digestibility *ca* 0.5). Limited data show this trend could also occur in the capybara (*Hydrochoerus hydrochaeris*), rock hyrax (*Procavia capensis*) and cape porcupine (*Hystrix africaeaustralis*) (Baldizan *et al.* 1983; Eloff *et al.* 1983; Jaarsveld 1983).

Both chemical and physical constraints could cause or contribute to the decreasing DE intake that accompanies declining herbage DE content, the latter usually being associated with increasing fibre content and advancing plant maturity. The chemical constraints include change in palatability (taste, odour) increase in toxic secondary compounds and reduction in capacity to provide essential nutrients for tissue metabolism. The physical constraints, as considered below, include the resistance of the forage to structural degradation during ingestive mastication (which is a physical component of palatability) and the rate at which the undigested feed can flow through the GIT. The increase in forage intake by herbivores that accompanies forage grinding (Weston and Kennedy 1984) has generally been assumed to provide presumptive evidence that physical constraints apply to the intake of unprocessed forage.

Herbage resistance to structural degradation during eating

Herbages vary widely in their resistance to structural degradation as measured by the amount of energy required for their breakdown by mechanical grinding. Thus comminution energy values of about 30 kJ/kg DM may be obtained with temperate leafy pasture material (Weston 1985) compared to up to 550 kJ/kg DM with tropical grass stem (Laredo and Minson 1973). The work involved by the herbivore in structurally degrading forage in the course of forming deglutition boluses could be expected to be generally related to the herbage comminution energy. However, this relation remains to be studied. In this connection, it could be pertinent that, over a narrow range of diets, the time required by sheep to consume unit weight of herbage is directly related to the herbage comminution energy (Weston 1985).

The extent to which a high resistance to structural degradation in a herbage could act as a constraint to its intake by herbivores needs elucidation. Recent studies indicate that it was probably significant in weaner lambs fed diets based on cereal straw (Weston and Davis 1986). Herbivores would be expected to vary in the extent to which a high resistance to structural degradation in the herbage would affect intake. Less effect on intake would be expected in those species that normally comminute their feed poorly, and in those having strong jaw musculature and appropriately favourable dentition. Tessema (cited by van Soest 1982) reported that coarse low quality Guinea grass (*Panicum maximum*) was eaten by cattle but not by sheep. As sheep comminute coarse forage to a greater degree than cattle (Poppi *et al*. 1981b) this difference between the two in forage acceptability could possibly relate to the proportionately greater chewing effort that may be needed by sheep during coarse forage ingestion. Uden and van Soest (1982) reported in studies involving chopped timothy hay (670 g/kg CWC), that, whereas feed intakes by sheep, goats, equines and cattle were at least adequate for body maintenance, the intake by rabbits was insufficient to prevent body weight loss. This may point to a physical constraint to the ingestion of coarse herbage by the rabbit which extensively comminutes its feed (Cheeke *et al*. 1985). As feed energy requirement per unit BW generally decreases with increasing BW, greater body size must confer an advantage with respect to the chewing effort needed in satisfying body energy requirements.

As noted above, mechanical reduction of particle size generally increases herbage intake. These increases have usually been attributed to a reduction in the constraint to particle flow through the GIT. However, part of the response with herbages of high resistance to structural degradation may be related to a reduction in the effort required by the herbivore to ingest the forage.

Flow through the GIT

It is well accepted that constraints to the flow of undigested feed through a rate limiting compartment (RLC) of the GIT may influence forage intake by the herbivore (Van Soest 1982). In ruminants the RLC of the GIT appears to be the rumen (Campling 1970); in other foregut fermenters and in hindgut fermenters the compartment containing the most digesta *viz.* the fermentation compartment, has been considered to be the RLC, but there is a lack of data to show that this is indeed the case.

The rate of flow of material through the RLC is related to the amount of material therein (digesta load) and the mean residence time as follows:

rate of flow = digesta load /mean residence time

Rate of digesta flow: The upper physiological limits to digesta propulsion through RLCs of herbivores remain undefined. It may be deduced with various species, however, that propulsive capacity is unlikely to be a constraint to the intake of unprocessed forage. This deduction can be made by comparing faecal DM outputs and feed DM intakes, as obtained with ground forages, with those obtained with unprocessed forages as fed *ad libitum*. As shown in Figure 3, daily faecal DM outputs (g/kg BW) of the order of 60, 30, 20 and 20 have been obtained with rabbits, weaner lambs, equines and dairy cows respectively. In terms of MW these output values (g/MW) are 63, 70, 78 and 97 respectively, and they correspond to DM intakes of 130, 130, 120 and 110 (g/MW). These values together provide some minimal estimates of the propulsive capacity of the GIT with respect to DM. For example, the RLC of the equine can handle digesta DM at least equal to that prevailing when a diet consumed at 120 g DM/MW provides faeces at 78 g DM/MW. Our examination of available feed intake and digestion data for equines fed unprocessed herbage shows that these levels of intake and output have not been achieved. Similar considerations apply for data from rabbits, sheep and dairy cows. Hence the propulsive capacity of the RLC *per se* may not generally limit the consumption of unprocessed herbages over a range of herbivores.

For any particular forage, hindgut fermenters have to handle more undigested feed CWC in the GIT proximal to the caecum than do foregut fermenters, in the GIT proximal to the caecum. Thus with

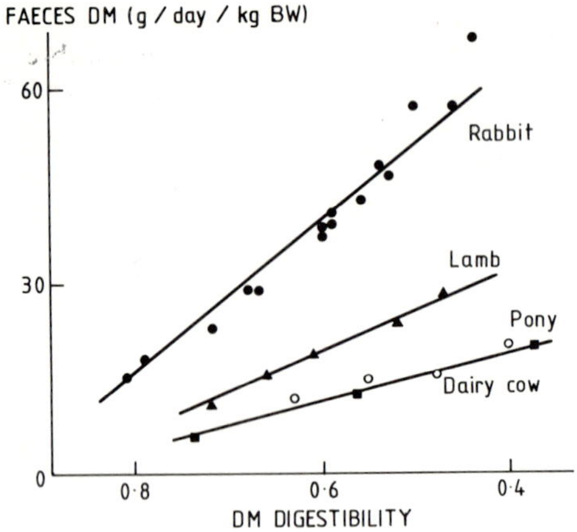

Figure 3. The relationship between faeces dry matter (DM) output and DM digestibility in rabbits (●) (Spreadbury and Davidson 1978), lambs (▲) (Owen *et al.* 1967), ponies (■) (Laut *et al.* 1985) and dairy cows (o) (Nelson *et al.* 1968, as calculated by Baumgardt 1970).

sheep fed phalaris forage (750 g CWC/kg DM), 460 g CWC left the stomach per kg DM consumed (Hogan *et al.* 1969) and assuming negligible CWC digestion in the small intestine, this quantity would have entered the caecum. By contrast the quantity passing from the mouth to the caecum would be 750 g/kg DM in the hindgut fermenter, assuming no pre-caecum digestion.

Digesta load: The herbivore characteristically has a relatively large amount of digesta in its GIT. From data on some 23 species of herbivores, as tabulated by Parra (1978), mean digesta loads of 13.7% of BW (C of V 28%) and 11.6% of BW (C of V 31%) may be calculated for the entire GIT and the major fermentation compartment respectively. A general increase in digesta load with increasing herbivore body size was shown by Van Soest (1982) in his examination of data from 59 herbivore species; over the range 0.1 kg - 1000 kg, the load in the fermentation compartment increased progressively, on average, from 10.8% of BW to 14.5% of BW.

The digesta load that herbivores are prepared to maintain in the RLC could affect forage intake in two ways. First, the existence of an upper physiological limit to load would rigidly set intake in direct relation to the rate of digesta clearance. Second, where flow through the RLC is determined by its load, the magnitude of the load can indirectly affect intake. For example, increase in rumen DM load will enhance rumination and the microbial population, which in turn will indirectly tend to increase DM flow rate, by permitting an enhanced rate of particle degradation.

Few adequate comparative data are available on the digesta load various herbivores are prepared to maintain in the RLC. Poppi *et al.* (1981a) found rumen load to be similar in cattle and sheep over a range of forages. When fed lucerne hay, Hume (1977) found digesta load in the rumen of the sheep (15.3% of BW) to be appreciably higher than that in the stomach of the red-necked wallaby (*Macropus rufogriseus*) (9.5% of BW) or the pademelon (*Thylogale thetis*) (9.5% of BW). Further the tammar wallaby (*Macropus eugenii*) maintains a lower digesta load in the stomach than does the pademelon (Dellow 1982; Dellow and Hume 1982b). The stomach digesta of pademelons and tammar wallabies were found to contain more DM than the rumen digesta of the sheep (16.3% and 18.5% vs 14.7% Hume 1977); hence the comparative digesta DM loads between these animals differ less than the digesta loads.

The load the herbivore is willing to maintain in the RLC varies widely under conditions of *ad libitum* feeding in ruminants, but data are not available to show if equivalent variation exists in other herbivores. Perusal of the literature shows about a three-fold variation in rumen digesta and rumen digesta DM loads in both sheep and cattle (Weston 1985). Factors such as low forage palatability, low content of essential nutrients in forages, and the age and metabolic state of the animal all affect rumen digesta load mainly through their effects on forage intake. Over a range of diets, the load maintained in the rumen has been shown to be related to DE intake, with higher DE intake being associated with lower load (Weston 1985). Similar trends have been shown in freely-grazing cattle, sheep and kangaroos, with load tending to be lower with better quality pasture than with poorer pasture (Ellis *et al.* 1984; Weyreter and Engelhardt 1984; Goodchild 1985; McCollum and Galyean 1985; Taylor 1985; Cruickshank 1986). However, this relationship is not consistently shown, and various studies show similarity in load between forages despite differences in DE intake (Ulyatt *et al.* 1967; Poppi *et al.* 1981a). Rumen digesta load has been found to increase progressively during the course of the day (Ulyatt 1971; Thomson *et al.* 1985) in sheep grazing good quality pasture. Similar trends may be expected in other herbivores with increasing duration of the grazing period.

Mean retention time: The time undigested feed particles spend in the RLC can obviously play a critical role in the GIT constraint to forage intake. Feed DM is removed from the RLC by propulsion and by absorption and eructation with the latter being mainly dependent on digestion. Accordingly, delay in feed DM removal can result from inadequate propulsion, slow digestion rate or the existence of mechanisms to selectively retain fractions of digesta.

Many herbivores selectively retain DM in sections of the GIT. Selective retention of DM in large particle form occurs in the rumen (e.g. Poppi *et al.* 1981c), in the stomach of foregut fermenter marsupials including pademelon, tammar wallaby and kangaroo (Warner 1981a; Dellow 1982), and in the caecum and post caecum sections of the GIT in the equine (Uden *et al.* 1982). Selective retention of fine particulate DM is shown in the hindgut of coprophagous herbivores including the rabbit and ringtail possum (*Pseudocheirus peregrinus*) (Warner 1981b; Chilcott and Hume 1985).

The retention time of forage DM in the RLC varies with forage type and herbivore species. Poppi *et al.* (1981b) found that sheep fed tropical grass stem diets had on average 1.38 kg OM in the rumen for each kg of forage OM eaten. The hourly clearance rate of this forage OM from the rumen may be calculated to be 31 g/kg; with tropical grass leaf diets, on the other hand, the mean clearance rate was 39 g/kg. More rapid rates prevail with better quality forages, values of 77 g/kg being recorded for immature clover (Doyle 1985) and of 80 g/kg for high quality forage oats (Weston 1984). Clearance rate is faster in sheep than in cattle, an average advantage of some 30% being shown for sheep over a range of herbages (Poppi *et al.* 1981e). With lucerne hay diets at least, DM clearance rates from the stomach of the red-necked wallaby, the tammar wallaby and the pademelon (80 – 100 g/kg) exceed that from the sheep's rumen (*ca* 50 g/kg) (Hume 1977). Lack of data preclude the estimation of clearance rates for hindgut fermenters. With small hindgut fermenters the rates must be high. The rate of clearance of a particulate marker from the caecum and post caecum section of the rabbit GIT may be calculated to be 19 g/kg from the studies of Uden *et al.* (1982), and the DM clearance rate would have to exceed this.

The preceding discussion indicates that the extent to which mean retention time in the RLC constrains the intake of unprocessed forage would depend on the species of forage and the herbivore. Nutrient recycling to the RLC increases digestion rate and thus may facilitate DM removal with appropriately deficient diets. In this connection, it is significant that the recycling of nitrogen has been shown to be greater in water buffalo than in cattle (P.M. Kennedy, pers. comm.) and appears greater in the goat than in the sheep (Alam *et al.* 1985), as is evident from rumen ammonia concentrations.

Herbage acquisition constraints

Trudell and White (1981) observed that reindeer (*Rangifer tarandus tarandus*) took *ca* 1.7 h to obtain 1.0 kg DM when grazing lichens, compared with only 0.7 h when given harvested lichens in pens. Such data exemplify the obvious constraints to herbage acquisition that apply with the freely grazing herbivore. Two types of constraint seem involved, limitation to foraging rate (time spent foraging per unit feed intake) and restriction of the time available for foraging. Clearly the size and characteristics of the herbage biomass would be expected to influence foraging rate whereas the need to pursue vital activities other than foraging could place limits on foraging time.

Rate of foraging: Within any plant community, the level of herbage availability will depend on the species of herbivore concerned, due to species diversity in both herbage preferences and physical attributes pertinent to foraging (Hoogesteijn Reul 1979; Squires 1982). Care needs to be exercised in the interpretation and use of data relating feed intake to biomass as obtained in various experimental situations. Much of the data has been obtained from studies using models involving measurements with:

1. animals continuously maintained at different herbage biomass levels for prolonged periods (> 21 days),
2. animals continuously maintained for short periods (7d - 15d) in a situation of rapid and progressive defoliation, and
3. animals allowed access to herbage for short periods (3 min - 60 min) following a period of fast (3 h - 18 h).

In the first situation, selection opportunity increases with biomass level, hence diet composition and biomass are confounded. In the second and third situations adequate behavioural adaptation cannot take place, and in the third situation, eating rate and diet selection pattern can be altered because the animal is hungry (Langlands 1965; Jung and Koong 1985).

The herbage intake response to increasing levels of biomass generally comprises an ascending section in which prehension constraints prevail, followed by a plateau region where nutritional constraints of the type discussed earlier apply. More than four-fold variation exists in estimates of the level of biomass below which prehension constraints apply for cattle and sheep grazing in the long term experimental situation, with much of this variation being attributable to differences between grasslands in scope for diet selection (Hodgson 1977; Allison 1985). In short term studies with sheep, Allden and Whittaker (1970) found foraging rate to be constant at 2.8 h/kg DM for biomass values in excess of 1500 kg DM/ha. However, at levels below this threshold constraints progressively increased such that the foraging rate was at 11 h/kg DM at *ca* 400 kg DM/ha biomass. Responses of a similar nature have been obtained with elk (*Alces alces*) and mule deer (*Cariacus macrotis*) (Wickstrom *et al.* 1984). Various data indicate that the larger herbivore may be at a disadvantage relative to its smaller counterpart in feed acquisition at low biomass levels. For example, Wickstrom *et al.* (1984) found threshold values for foraging constraint with grassland at levels of 1500 kg DM/ha with the elk (160 kg BW) compared with only 200 kg/ha with the mule deer (35 kg BW); similarly, Allden and Whittaker (1970) found a higher biomass threshold for yearling sheep (43 kg BW) than for lambs (25 kg BW).

Physical factors such as snow cover can obviously affect available biomass; a snow depth of 20 cm was estimated by T.A. Hanley and J.D. McKendreck (Wickstrom *et al.* 1984) to reduce herbage availability for deer by some 60% in Alaskan conifer forests.

Herbage biomass has been shown to affect various activities indirectly concerned with feed ingestion. Thus low biomass level is associated with a reduction in the proportion of the foraging time spent by reindeer in ingestion (Trudell and White 1981), and with an increase in the velocity of walking in the elk and mule deer (Wickstrom *et al.* 1984).

Constraints to rate of foraging have been related to various characteristics of plant communities as reviewed by Hodgson (1982, 1985). Plant height or tiller length appear to be important in this context for temperate grasslands. Allden and Whittaker (1970) found that whereas foraging rate with sheep was relatively constant at 2.5 h /kg DM with tiller lengths in the range 7 cm - 35 cm, marked constraints prevailed with shorter herbage. Thus foraging rate was 6.7 h /kg DM with tiller length at 3.7 cm. Low bulk density of herbage has been shown to be associated with reduced intake per bite (Stobbs 1975) and hence should affect foraging rate but definitive data are not yet available. Increase in the ratio of leaf to stem in tropical legume pasture was found to be associated with less constraint to herbage intake by cattle (Hendricksen and Minson 1980). Again, leaf area is important with browse. Wickstrom *et al.* (1984) found rate of foraging with elk to be 2.2 h /kg DM with leaf area of 6 cm^2 compared to 0.8 h/kg DM with leaf area of 30 cm^2. The resistance of herbage to defoliation could play a role in the modulation of rate of foraging. Higher resistance to defoliation is associated with greater constraint to foraging in the few studies available. Thus Hendricksen and Minson (1980) found, for a tropical legume, the shear load needed for cattle to prehend leaves was less than that required with stems, and also observed that rate of eating with the legume was faster as the ratio of leaf to stem advanced. Again Evans (1967) reported breaking strength to be greater for perennial ryegrass than for short rotation ryegrass and sheep have been observed to spend more time in grazing with the former (Lancashire and Keogh 1966).

The studies of Curll and Davidson (1983) suggest that appreciable time may be required for herbivores to adapt behaviourally to a change in herbage biomass, which is very relevant to data interpretation and to their use in computer models. Thus when sheep were changed from a pasture of high biomass to one of low biomass, foraging rate changed from 5.6 h /kg DM to some 36 h /kg DM; however, over the course of 10 days the rate increased progressively to about 15 h/kg DM.

Time constraints: The time available for foraging will vary according to the herbivore's time commitment to other essential activities including water acquisition, resting, general surveillance, predator avoidance, rumination, migratory travel and the need to nuture. The extent to which such activities could compete for time with foraging must vary with the species and circumstances, and is difficult to estimate.

Many herbivores have a marked capacity for foraging. On a daily basis, reindeer have been observed to forage for 16 h - 20 h (Trudell and White 1981), African elephants for 16 h (Owen-Smith 1982), horses and cattle for 14 h - 15 h (Arnold and Dudzinski 1978) and black rhinoceros (*Diceros bicornis*), kudu (*Tragelaphus* sp.), zebra and sheep for 13 h - 14 h (Bircham 1981; Owen-Smith 1982). There is no reason to believe that these values are necessarily maxima. Nevertheless, the majority of grazing time observations with sheep, cattle and horses fall well below the respective values for these species as given above. For example, grazing times in 92% of some 400 studies with sheep, as collated by Arnold and Dudzinski (1978), fall below 11 h. Hence it seems likely that foraging time capability may not act as a constraint to herbage intake in many situations.

Behavioural adaptation with respect to time spent grazing has been shown in the studies of Curll and Davidson (1983) discussed above. When the sheep were changed from high biomass to low biomass pastures mean daily grazing time increased from some 5.5 h in the week before the change to *ca* 6.4 h in the week that followed, and to *ca* 7.5 h in the succeeding week.

In general, a wide range of studies has indicated that various properties of the herbage biomass may apply constraints to herbage intake. The constraints are not absolute, however, as an increase in the animal's capacity to use energy, such as occurs due to lactation or to thermal considerations following shearing in sheep, also increases rate of eating and/or time spent grazing (Wheeler *et al.* 1963; Arnold and Dudzinski 1967). There appear to be no convincing data to show the physiological basis for feed intake limitation when acquisition constraint prevails. Thus circumstances remain to be defined in which the limitation can be specifically attributed to muscular fatigue, to an untoward response to the high work load or difficulty involved in selection and prehension, or to the time available for foraging.

SOME GENOTYPE CONSIDERATIONS

Few meaningful comparisons can be made of the herbage DM or DE intakes between different herbivore genotypes apart from those of sheep and cattle. Lack of parity in animal, herbage and environmental factors has generally militated against the acquisition of adequate comparative data. Further, no precise scaler is available with respect to body size. Hence the method used to compare herbivores has to be chosen according to the use to which the comparative intake data are to be put.

Inasmuch as BMR could reflect the metabolic capacity of the animal to use energy, higher BMR should be associated with higher intake. This is certainly so, at least in general terms, for many comparisons. Thus cattle eat more than do sheep per unit of MW but also have a higher BMR per unit of BW. The two groups have generally comparable BMRs and intakes per unit of $BW^{0.9}$. In the pen and field comparisons of sheep, oryx (*Oryx beisa*), and cattle by Stanley Price (1985a, 1985b), the intake differences, apparent with use of MW as a scaler, largely disappear when the $BW^{0.9}$ scaler is used. Again, the intakes of breed genotypes of cattle and sheep are more closely related to BMR than to MW (Blaxter *et al.* 1966b; Frisch and Vercoe 1984). Further, seasonal variations in BMR in sheep are accompanied by corresponding variation in feed intake (Blaxter and Boyne 1982) and the deep winter trough in feed intake in species such as white-tailed deer (*Odocoileus virginianus*) accompanies a major drop in BMR (McMillin *et al.* 1980). Foregut-fermenter marsupials eat less DE than do sheep (McIntosh 1966; Hume 1974; Dellow and Hume 1982b) and have lower BMRs (Hume 1982), and species such as the rock hyrax (*Procavia capensis*) and the three-toed sloth (*Bradypus variegatus*) with very low BMRs have low energy intakes (Nagy and Montgomery 1980; Eloff 1983; Hayssen and Lacy 1985).

Although BMR and DE intake are broadly related, a consistent proportionality between the two does not prevail. For example, intake per unit of BMR decreases appreciably with age (Fig. 1), the tammar wallaby has a BMR similar to that of various other marsupials but consistently eats less (Dellow 1982; Dellow and Hume 1982a), and the DE intake differences between various marsupials and the sheep (McIntosh 1966; Foot and Tribe 1970; Hume 1974) appreciably exceed the corresponding BMR differences (Blaxter *et al.* 1966a; Hume 1982).

Attention has been drawn to the possible disadvantage in the use of high fibre herbage, of the ruminant type of GIT compared with the more tubular type of GIT, such as prevails with non-ruminant foregut fermenters and hindgut fermenters (Janis 1976; Hume 1978, 1982). A lower intake of high fibre herbage, and accordingly a nutritional disadvantage, is postulated for the ruminants, as greater constraint would apply to the passage of coarse particles through the GIT with this group. This interesting theory merits further attention. At present data are not available adequately to support or refute it. The postulate of Janis (1976) of a superiority of equines over ruminants was mainly based on findings of Darlington and Hershberger (1968) that equines showed little intake decrease over a range of forages, varying in crude fibre content from 280 to 380 g/kg DM, and in DE content from 13.1 MJ/kg DM to 10.4 MJ/kg DM; appreciable intake decrease would be expected by ruminants over this range. However, in the nutritional context, DE intake rather than DM intake needs to be considered, and as indicated earlier in reference to these data, an appreciable decline in DE intake did occur with the equines. Hume (1978, 1982) noted that when Foot and Romberg (1965) fed young kangaroos a poor quality hay instead of medium quality hay their intakes of DE decreased less than that observed with adult sheep given the same herbages. Accordingly he considered that the more tubular type of GIT could confer an advantage. However, this type of difference between kangaroo and sheep in DE intake is not shown in comparisons of adults (McIntosh 1966; Foot and Tribe 1970; Hume 1974). Need for more detailed study in this area is clearly indicated.

Feed requirement per kg of BW increases with decreasing BW, some three times as much feed being needed per kg of BW with BW of 5 kg than with BW of 450 kg, assuming the feed requirement to be proportional to MW. This situation, together with obvious limitations to rumen capacity and function, has led to proposals of a minimal possible size for ruminants (Van Soest 1982; Moir 1984). The recent acquisition of data on herbage DM clearance rates from the rumen (Weston 1984; Doyle 1985) permits the question of size to be re-examined, and the data as discussed below show a much lower body size limit than was previously considered possible. Thus the data of Weston (1984) show the capacity of the sheep's rumen to be at least 32 g OM/kg BW. The data also show the hourly clearance rate of herbage OM from the rumen with sheep to be 80 g/kg for oats herbage containing 11.8 MJ of ME per kg OM, about half of which would be derived from digested CWC (Hogan and Weston 1969). It may be calculated that a daily intake of 61 g OM/kg BW from the oats could be achieved without exceeding the rumen load of 32 g OM/kg BW. This intake of OM being equivalent to *ca* 720 kJ of ME/kg BW would be ample for growth and reproduction in a ruminant weighing 1.0 kg (MW 1.0); moreover, ruminal digestion of CWC could contribute appreciably to the ME.

CONCEPTUAL SCHEME FOR REGULATION

The complexity of feed intake regulation is increased when constraints arising from essential nutrient deficiency, GIT limitations, feed acquisition limitations etc. are superimposed on a control system that is trying to match energy intake with energy demand. No attempt can be made here to discuss regulation under these conditions in any detail. However, the possible interplay of the major contributing processes is indicated in a conceptual scheme suggested for herbage intake regulation in the ruminant (Weston 1982, 1985). The scheme envisages major roles for the balance between energy supply and demand on the one hand, and for rumen digesta load on the other. The assumptions on which the scheme is based are described briefly below.

1. the animal's capacity to use energy is the vital energy regulation component - this equals the genetic potential of the animal to use energy unless essential nutrient deficiency prevails and in which case it is appropriately lower.

2. constraints with herbage feeding prevent the animal from consuming enough feed to meet its capacity to use energy and accordingly an energy deficit prevails which is of variable magnitude depending on circumstances.
3. the energy deficit generates a hunger signal whose magnitude is proportional to the size of the deficit.
4. the digesta load provides a satiety or inhibitory signal the intensity of which is directly related to load size.
5. low palatability provides an inhibitory signal.
6. environmental factors such as the untoward effects of the high work load necessary in locating and prehending feed, muscular fatigue etc. provide inhibitory signals.
7. basic hunger drive of the animal is overcome when the inhibitory signals from load, palatability and the environment more than balance the hunger signal.
8. the clearance rate of herbage OM from the rumen is directly related to the rumen load.

It is necessary to superimpose on this scheme possible constraints relating to the time the freely-ranging animal could have available for foraging each day after other essential activities are allowed for.

This conceptual scheme accommodates variation in digesta load with change in energy demand, and variation in herbage quality and accessibility. For example, higher quality herbages are consumed in greater quantity than their lower quality counterparts and provide more DE. Accordingly, lower energy deficits and lower hunger signal intensities prevail when they are fed, so lesser digesta loads are needed to promote satiety. The proposed scheme is necessarily speculative in that the relevant quantitative data are lacking and necessary physiological mechanisms remain to be delineated. Nevertheless, it provides a simple framework within which many herbage intake responses can be accommodated, permits linking of energy metabolism with GIT function and is broadly applicable in principle to herbivores generally.

CONCLUDING REMARKS

Experimentation in the area of herbage intake regulation in herbivores is most demanding, as the apparent interplay between the animal's energy transactions and a wide range of physical, chemical and behavioural constraints necessitates the measurement or control of many variables. The need for adequate assessment of the physical properties of herbage is now well established, as is the desirability of determining the capacity of the herbivore to use energy in the prevailing study situation. Again it is clear that data are needed on upper physiological limits with respect to foraging capability, mastication capability, GIT capacity etc. At present insufficient data are available to permit the construction of computer models that could facilitate prediction with any useful precision. Accordingly, much more data from adequately controlled studies will be needed before the subject of herbage intake regulation can be confidently treated both experimentally and theoretically in quantitative terms.

Adams, G.B., and Forbes, J.M. (1981). *Proc. Nutr. Soc.* **40**, 44A.

Alam, M.R., Poppi, D.P., and Sykes, A.R. (1985). *Proc. N.Z. Soc. Anim. Prod.* **45**, 107–11.

Allden, W.G., and Whittaker, I.A.McD. (1970). *Aust. J. Agric. Res.* **21**, 755–66.

Allison, C.D. (1985). *J. Range Manage.* **38**, 305–11.

Anderson, G.B., Koong, L.J., and Garrett, W.N. (1979). *J. Anim. Sci. Suppl.* **49**, 354–5.

Anil, M.H., and Forbes, J.M. (1980). *J. Physiol.*, London **298**, 407–14.

Arnold, G.W., and Dudzinski, M.L. (1967). *Aust. J. Agric. Res.* **18**, 349–59.

Arnold, G.W., and Dudzinski, M.L. (1978). In "Ethology of Free-ranging Domestic Animals." (Elsevier: Amsterdam.)

Baile, C.A. (1975). In "Digestion and Metabolism in the Ruminant." (Eds I.W. McDonald and A.C.I. Warner.) pp. 333–50. (University of New England: Armidale, Australia.)

Baile, C.A., and Forbes, J.M. (1974). *Physiol. Rev.* **54**, 160–214.

Baile, C.A., McLaughlin, C.L., and Della-Fera, M.A. (1986). *Physiol. Rev.* **66**, 172–234.

Baldizan, A., Dixon, R.M., and Parra, R. (1983). *S. Afr. J. Anim. Sci.* **13**, 27–8.

Bauman, D.E. (1984). In "Herbivore Nutrition in the Subtropics and Tropics." (Eds F.M.C. Gilchrist and R.I. Mackie.) pp. 505–24. (The Science Press: Craighall, South Africa.)

Bauman, D.E., and Currie, W.B. (1980). *J. Dairy Sci.* **63**, 1514–29.

Baumgardt, B.R. (1970). In "Physiology of Digestion and Metabolism in the Ruminant." (Ed. A.T. Phillipson.) pp. 235–53. (Oriel: Newcastle upon Tyne, UK.)

Bircham, J.S. (1981). Ph.D. Thesis, University of Edinburgh.

Blaxter, K.L., and Boyne, A.W. (1982). *J. Agric. Sci.* **99**, 611–20.

Blaxter, K.L., Clapperton, J.L., and Wainman, F.W. (1966a). *Br. J. Nutr.* **20**, 283–94.

Blaxter, K.L., Wainman, F.W., and Davidson, J.L. (1966b). *Anim. Prod.* **8**, 75–83.

Butcher, C., Bryant, M.J., Owen, E., Leach, I., and Machin, D.H. (1983). *Anim. Prod.* **36**, 229–36.

Campling, R.C. (1970). In "Physiology of Digestion and Metabolism in the Ruminant." (Ed. A.T. Phillipson.) pp. 226–34. (Oriel: Newcastle upon Tyne, UK.)

Campling, R.C., and Balch, C.C. (1961). *Br. J. Nutr.* **15**, 523–30.

Cheeke, P.R., Akin, D.E., Robinson, K.L., and Patton, N.M. (1985). *J. Appl. Rabbit Res.* **8**, 26–30.
Chilcott, M.J., and Hume, I.D. (1985). *Aust. J. Zool.* **33**, 1–15.
Cowan, R.T., Robinson, J.J., McDonald, I., and Smart, R. (1980). *J. Agric. Sci.* **95**, 497–514.
Cruickshank, G.J. (1986). Ph.D. Thesis, University of Canterbury, New Zealand.
Curll, M.L., and Davidson, J.L. (1983). *Grass and Forage Sci.* **38**, 159–67.
Darlington, J.M., and Hershberger, T.V. (1968). *J. Anim. Sci.* **27**, 1572–6.
de Jong, A. (1986). In "Control of Digestion and Metabolism in Ruminants." (Eds L.P. Milligan, W.L. Grovum and A. Dobson.) pp. 459–78. (Prentice-Hall: Englewood Cliffs, NJ.)
Dellow, D.W. (1982). *Aust. J. Zool.* **30**, 751–65.
Dellow, D.W., and Hume, I.D. (1982a). *Aust. J. Zool.* **30**, 391–8.
Dellow, D.W., and Hume, I.D. (1982b). *Aust. J. Zool.* **30**, 767–77.
Doyle, P.T. (1985). In "Ruminant Physiology, Concepts and Consequences." (Eds S.K. Baker, J.M. Gawthorne, J.B. Mackintosh and D.B. Purser.) pp. 105–16. (University of Western Australia: Perth.)
Egan, A.R. (1972). *Aust. J. Agric. Res.* **23**, 347–61.
Ellis, W.C., Matis, J.H., Pond, K.R., Lascano, C.E., and Telford, J.P. (1984). In "Herbivore Nutrition in the Subtropics and Tropics." (Eds F.M.C. Gilchrist and R.I. Mackie.) pp. 269–93. (The Science Press: Craighall, South Africa.)
Eloff, A.K. (1983). *S. Afr. J. Anim. Sci.* **13**, 28–30.
Erb, R.E., Malven, P.V., Stewart, T.S., Zamet, C.N., and Chew, B.P. (1982). *J. Dairy Sci.* **65**, 937–43.
Evans, P.S. (1967). *J. Agric. Sci.* **69**, 175–181.
Foot, J.Z., and Romberg, B. (1965). *Aust. J. Agric. Res.* **16**, 429–35.
Foot, J.Z., and Russel, A.J.F. (1979). *Anim. Prod.* **28**, 25–39.
Forbes, D.K., and Tribe, D.E. (1970). *Aust. J. Zool.* **18**, 247–56.
Forbes, J.M. (1970). *J. Anim. Sci.* **31**, 1222–7.
Forbes, J.M. (1980). In "Digestive Physiology and Metabolism in Ruminants." (Eds Y. Ruckebusch and P. Thivend.) pp. 145–60. (MTP: Lancaster, UK.)
Forbes, J.M. (1982). *Proc. Nutr. Soc.* **41**, 123.
Frisch, J.E., and Vercoe, J.E. (1984). *J. Agric. Sci.* **103**, 137–53.
Girard, C.L., Seoane, J.R., and Matte, J.J. (1985). *Can. J. Physiol. Pharm.* **63**, 1297–1301.
Goodchild, A.V. (1985). *Anim. Prod.* **40**, 455–63.
Graham, N.McC., and Searle, T.W. (1972). *J. Agric. Sci.* **79**, 383–9.
Graham, N.McC., Searle, T.W., and Griffiths, D.A. (1974). *Aust. J. Agric. Res.* **25**, 957–71.
Hayssen, V., and Lacy, R.C. (1985). *Comp. Biochem. Physiol.* **81A**, 741–54.
Hendricksen, R., and Minson, D.J. (1980). *J. Agric. Sci.* **95**, 547–54.
Hirsch, E. (1973). *Physiol. Behav.* **11**, 687–704.

Hodgson, J. (1977). In "International Meeting on Animal Production from Temperate Grasslands, Proceedings." (Ed. B. Gilsenan.) pp. 70–5. (Irish Grassland and Animal Production Association: Dublin.)

Hodgson, J. (1982). In "Nutritional Limits to Animal Production from Pastures." (Ed. J.B. Hacker.) pp. 153–66. (Commonwealth Agricultural Bureaux: Farnham Royal, UK.)

Hodgson, J. (1985). *Proc. Nutr. Soc.* **44**, 339–46.

Hogan, J.P., and Weston, R.H. (1969). *Aust. J. Agric. Res.* **20**, 347–63.

Hogan, J.P., Weston, R.H., and Lindsay, J.R. (1969). *Aust. J. Agric. Res.* **20**, 925–40.

Hoogesteijn Reul, R. (1979). *World Anim. Rev.* **32**, 18–24.

Houpt, T.R., Anika, S.M., and Wolff, N.C. (1978). *Am. J. Physiol.* **235**, R23–8.

Hume, I.D. (1974). *Aust. J. Zool.* **22**, 13–23.

Hume, I.D. (1977). *Comp. Biochem. Physiol.* **56A**, 299–304.

Hume, I.D. (1978). *Aust. Mammalogy.* **2**, 37–42.

Hume, I.D. (1982). "Digestive Physiology and Nutrition of Marsupials." (University of Cambridge: Cambridge.)

Jaarsveld, A.S. van (1983). *S. Afr. J. Anim. Sci.* **13**, 31–3.

Janis, C. (1976). *Evolution.* **30**, 757–74.

Jung, H.G., and Koong, L.J. (1985). *J. Range Manage.* **38**, 302–5.

Lancashire, J.A., and Keogh, R.G. (1966). *Proc. N.Z. Assoc. Anim. Prod.* **26**, 22–34.

Langlands, J.P. (1965). *Nature*, London. **207**, 666–7.

Laredo, M.A., and Minson, D.J. (1973). *Aust. J. Agric. Res.* **24**, 875–88.

Laut, J.E., Houpt, K.A., Hintz, H.F., and Houpt, T.R. (1985). *Physiol. Behav.* **35**, 549–54.

Lebas, F. (1975). *Ann. Zootech.* **24**, 267–79.

Lodge, G.A., Fisher, L.J., and Lessard, J.R. (1975). *J. Dairy Sci.* **58**, 696–702.

McCollum, F.T., and Galyean, M.L. (1985). *J. Range Manage.* **38**, 543–6.

McIntosh, D.L. (1966). *C.S.I.R.O. Wildl. Res.* **11**, 125–35.

McMillin, J.M., Seal, U.S., and Karns, P.D. (1980). *Fed. Proc.* **39**, 2964–8.

Minson, D.J. (1982). In "Nutritional Limits to Animal Production from Pastures." (Ed. J.B. Hacker.) pp. 167–82. (Commonwealth Agricultural Bureaux: Farnham Royal, UK.)

Moir, R.J. (1984). In "Herbivore Nutrition in the Subtropics and Tropics." (Eds F.M.C. Gilchrist and R.I. Mackie.) pp. 734–43. (The Science Press: Craighall, South Africa.)

Morley, J.E., Levine, A.S., Gosnell, B.A., and Billington, C.J. (1984). *Fed. Proc.* **43**, 2903–7.

Mowat, D.N. (1963). Ph.D. Thesis, Cornell University.

Nagy, K.A., and Montgomery, G.G. (1980). *J. Mamm.* **61**, 465–72.

Nelson, B.D., Ellzey, H.D., Morgan, E.B., and Allen, M. (1968). *J. Dairy. Sci.* **51**, 1796–1800.

Nicolaidis, S., and Even, P. (1985). *Am. J. Clin. Nutr.* **42**, 1083–92.

Novin, D., Sanderson, J.D., and Vanderweele, D.A. (1974). *Physiol. Behav.* **13**, 3–7.

Novin, D., Robinson, K., Culbreth, L.A., and Tordoff, M.G. (1985). *Am. J. Clin. Nutr.* **42**, 1050–62.

Owen, J.B., Davies, D.A.R., Miller, E.L., and Ridgman, W.J. (1967). *Anim. Prod.* **9**, 509–20.

Owen-Smith, N. (1982). In "Ecology of Tropical Savannas." (Eds B.J. Huntley and B.H. Walker.) pp. 359–404. (Springer- Verlag: New York.)

Parra, R. (1978). In "The Ecology of Arboreal Folivores." (Ed. G. Gene Montgomery.) pp. 205–29. (Smithsonian Institution: Washington, DC.)

Phillip, L.E., Buchanan-Smith, J.G., and Grovum, W.L. (1981). *J. Agric. Sci.* **96**, 439–45.

Plotka, E.D., Morley, J.E., Levine, A.S., and Seal, U.S. (1985). *Physiol. Behav.* **35**, 965–9.

Poppi, D.P., Minson, D.J., and Ternouth, J.H. (1981a) *Aust. J. Agric. Res.* **32**, 99–108.

Poppi, D.P., Minson, D.J., and Ternouth, J.H. (1981b). *Aust. J. Agric. Res.* **32**, 109–21.

Poppi, D.P., Minson, D.J., and Ternouth, J.H. (1981c). *Aust. J. Agric. Res.* **32**, 123–37.

Prud'Hon. M. (1968). *Coniglicoltura* **5**, 13–32.

Ralston, S.L., and Baile, C.A. (1982). *J. Anim. Sci.* **54**, 1132–37.

Ralston, S.L., and Baile, C.A. (1983). *J. Anim. Sci.* **56**, 302–8.

Ralston, S.L., Freeman, D.E., and Baile, C.A. (1983). *J. Anim. Sci.* **57**, 815–25.

Reid, R.L., and Hinks, N.T. (1962). *Aust. J. Agric. Res.* **13**, 1092.

Reyne, G., Prud'Hon, M., Debicki, A.M., and Goussopoulous, J. (1977). *Ann. Żootech.* **27**, 211–23.

Sanger, D.J., and McCarthy, P.S. Unpublished manuscript cited in: Sanger D.J. (1981). *Appetite* **2**, 193–208.

Seoane, R.J., Dumont, F., Girard, C.L., Bedard, L., and Matte, J.J. (1984). *Can. J. Physiol. Pharm.* **62**, 1296–9.

Seoane, J.R., Gorrill, A.D.L., and Nicholson, J.W.G. (1976). *Can. J. Anim. Sci.* **56**, 305–15.

Spreadbury, D., and Davidson, J. (1978). *J. Sci. Fd Agric.* **29**, 640–8.

Squires, V.R. (1982). *J. Range Manage.* **35**, 116–9.

Stanley Price, M.R. (1985a). *J. Agric. Sci.* **104**, 367–74.

Stanley Price, M.R. (1985b). *J. Agric. Sci.* **104**, 375–82.

Stobbs, T.H. (1975). *Aust. J. Agric. Res.* **26**, 997–1007.

Taylor, R.J. (1985). *J. Appl. Ecol.* **22**, 717–25.

Ternouth, J.H., and Beattie, A.W. (1971). *Br. J. Nutr.* **25**, 153–64.

Thomson, B.C., Cruickshank, G.J., Poppi, D.P., and Sykes, A.R. (1985). *Proc. N.Z. Soc. Anim. Prod.* **45**, 117–20.

Tournier, F. (1972). Thesis, Ecole Nationale Veterinaire d'Alfort.

Trudell, J., and White, R.G. (1981). *J. Appl. Ecol.* **18**, 63–81.

Uden, P., Rounsaville, T.R., Wiggans, G.R., and Van Soest, P.J. (1982). *Br. J. Nutr.* **48**, 329-39.
Uden, P., and Van Soest, P.J. (1982). *Br. J. Nutr.* **47**, 267-72.
Ulyatt, M.J. (1971). *N.Z. J. Agric. Res.* **14**, 352-67.
Ulyatt, M.J., Blaxter, K.L., and McDonald, I. (1967). *Anim. Prod.* **9**, 463-70.
Vandermeerschen-Doize, F., and Paquay, R. (1984). *Appetite* **5**, 137-46.
Van Soest, P.J. (1982). "Nutritional Ecology of the Ruminant." pp. 325-44. (O. and B. Books: Oregon.)
Warner, A.C.I. (1981a). *Aust. J. Zool.* **29**, 759-71.
Warner, A.C.I. (1981b). *Nutr. Abstr. Rev. B.* **51**, 789-820.
West, D.B., Fey, D., and Woods, S.C. (1984). *Am. J. Physiol.* **246**, R776-R787.
Weston, R.H. (1966). *Aust. J. Agric. Res.* **17**, 939-54.
Weston, R.H. (1971). *Aust. J. Agric. Res.* **22**, 469-79.
Weston, R.H. (1982). In "Nutritional Limits to Animal Production from Pastures." (Ed. J.B. Hacker.) pp. 183-98. (Commonwealth Agricultural Bureaux: Farnham Royal, UK.)
Weston, R.H. (1984). *Can. J. Anim. Sci.* **64** (Suppl.), 324-5.
Weston, R.H. (1985). *Proc. Nutr. Soc. Aust.* **10**, 55.
Weston, R.H., and Davis, P. (1986). *Proc. Nutr. Soc. Aust.* **11**, 172-5.
Weston, R.H., and Kennedy, P.M. (1984). In "Techniques in Particle Size Analysis of Feed and Digesta in Ruminants." (Ed. P.M. Kennedy.) (Can. Soc. Anim. Sci.: Edmonton, Canada.)
Weyreter, H., and Engelhardt, W.V. (1984). *Can. J. Anim. Sci.*, **64**, Suppl., 152-3.
Wheeler, J.L., Reardon, T.F., and Lambourne, L.J. (1963). *Aust. J. Agric. Res.* **14**, 364-72.
Wickstrom, M.L., Robbins, C.T., Hanley, T.A., Spalinger, D.E., and Parish, S.M. (1984). *J. Wildl. Manage.* **48**, 1285-1301.
Wodzicka-Tomaszewska, M. (1963). *N.Z. J. Agric. Res.* **6**, 440.

8
THE EFFECT OF CLIMATE UPON INTAKE

by B.A. Young

Appetite is a complex process influenced by factors both internal and external to the animal. The desire for food and the amount ingested is determined by the animal's need for energy, but is also modulated by the animal's past and present dietary and environmental experiences. The idea of climate affecting food intake is entrenched, but presently there is inadequate information to understand the physiological mechanisms linking climatic cues to the control of food intake by herbivores.

This review commences with the presentation of a simplified model of energy balance regulation. Evidence is then summarized to identify the possible effects of season and climatic components on the food intake of a number of larger mammalian herbivores (Table 1). The effects on food intake of photoperiod (hours of light and dark each day) and of the complexity of factors comprising the thermal environment are given most attention. Despite the risks in the translation of laboratory observations and data from a few animals to the wider spectrum of herbivores in naturally occurring multi-variable environments, speculations are put forth in this review to stimulate further research in an area of herbivore nutrition which has been neglected.

ENERGY BALANCE REGULATION AND FOOD INTAKE

The ingestion of food is not haphazard but part of a regulated system involving the interplay of long-term, presumably genetically-programmed, goals upon which there are short-term temporary diversions. There is general agreement on the central importance of long-term energy balance in determining the appetite and thus food intake (Balch and Campling 1969; Baumgardt 1970; Baile and Forbes 1974; Weston 1982; Kissileff and Ban Italle 1982). While uncertainty exists in adapting the general models to herbivores, a simple negative feedback scheme based on the goal of accumulated body energy balance aids in understanding control of food intake (Fig. 1). Modifiers to the basic scheme may be only temporary and easily overcome by minor behavioural or physiological adjustments. Other modifiers may persist and result in a substantial deviation from energy sufficiency.

Negative feedback processes involving the central nervous system (CNS) play essential roles in the regulation of autonomic functions in mammals. There are recognized mechanisms for sensing internal and environmental cues, compilation and integration of afferent

NUTRITION OF HERBIVORES
ISBN 0 12 750052 9

neural and humoral signals, and the generation of appropriate efferent autonomic and behavioural responses. Involvement of specific areas of the CNS in feeding activity is well established (Andersson 1962; Balch and Campling 1969; Baile and Forbes 1974; Morley *et al.* (1984). While the exact humoral or neural mechanisms are not well understood, the regulation of accumulated body energy is probably similar in principle to the regulation of other autonomic functions in mammals.

Table 1. Scientific and corresponding common names of animals in this review.

Scientific name	Common name
Bos taurus	Domestic cattle, temperate breeds Holstein, Brown Swiss, Hereford, etc.
Bos indicus	Brahman cattle, tropical breeds.
Ovis aries	Domestic sheep eg, Suffolk breed. The Soay sheep is a semi-domestic breed from islands off Scotland.
Capra hircus	Goat.
Odocoileus virginianus	White-tailed deer.
Cervus elaphus	Red deer, European origin.
Rangifer tarandus	Reindeer, Euro-Asian origin. Caribou, American origin.
Megaleia rufa	Red kangaroo.
Macropus robustus erubescens	Euro.
Oryctolagus cuniculus	Rabbit.

The proposed scheme recognizes the accumulated energy balance of the animal as the source of primary afferent inputs to the CNS and other input factors are viewed as modifiers to the primary drive. Extensive studies (Frisch and Vercoe 1977; ARC 1980, Blaxter 1982), have shown that the appetite of ruminants is proportional to their rate of energy metabolism, implying a balance between inputs and outputs. This idea is also supported by much evidence indicative of long-term energy balance. Individuals of a species tend towards a remarkably similar mature body weight when subjected to wide variations in nutritional and climatic circumstances. Once mature, an individual fluctuates around a rather consistent body weight over many years, despite seasonal and short-term energy imbalances. This is clearly illustrated in the classical studies on the white-tailed deer by Wood *et al.* (1962). The above suggest genetic programming to achieve a target mature size and body composition that an individual attempts to attain and defend. Mechanisms for sensing the "sufficiency" or "deficiency" signals remain to be elucidated. In the proposed scheme, growth and gains in body energy are viewed as processes in attaining the desired status and, in principle, cease upon reaching the accumulated energy goal. Accumulated energy balance is defined as the difference between the metabolizable energy obtained from dietary sources and the total energy output from the animal.

The hub of the proposed scheme is the negative feedback process. Its function is illustrated as follows. An increase in the rate of energy output by an animal will cause a transient energy deficit. This deficit is sensed and appropriate afferent signals are sent to the CNS and a shift towards "hunger" occurs in the "satiety-hunger" balance. If the opportunity exists there is increased feeding behaviour, the ingestion and digestion of additional food and subsequent absorption of metabolizable energy to compensate for the original deficit. Only one of two physiological states may arise in the CNS: "satiety" or "hunger". Satiety or nutritional well-being is manifested as a lack of drive for feeding activity. In contrast, hunger (or appetite for food) is a state resulting in the drive for food-seeking. The dominant state is probably derived from the weighted inputs of the various afferent signals and reciprocal inhibition between efferent pathways similar to those proposed by Sherrington (1911) for muscle motor control, and Bligh (1973) for thermoregulation.

The regulation of body energy content is complicated by the immediate "on/off" activity of meal feeding behaviour and by transient factors influencing diurnal and seasonal levels of energy expenditure and intake. Input modifiers of dietary, digestive, environmental, disease or other origin may be viewed as warning signals which impinge on the primary negative feedback process and usually result in reduced food intake. In Fig. 1 boxes 1 to 7 are possible modifiers of the primary feedback system. Often of greatest importance to herbivores are the digestive constraints (Box 6). The specialization of the gastrointestinal tract and symbiosis with gut-dwelling microbiota are essential to the herbivore's ability to utilize bulky, fibrous diets. However, a ceiling in food intake is often reached because of the predominance of digesta load (Conrad *et al.* 1964; Forbes 1970; Weston 1979a, 1980, 1984; Grovum 1984). Similarly, temporal or spatial displacement of behavioural activity (Box 3) or accessibility to food or water (Box 4) may deny an animal feeding opportunities. These constraints to intake do not necessarily reflect satiety but rather that the animal is prevented from achieving optimal food intake and a state of energy deficit may persist.

The food intake may be suppressed by various humoral or neural signals arising peripherally or centrally resulting from conditions which tax, or threaten to tax, some physiological process. Conrad *et al.* (1964) showed that when fibrous feeds were offered to dairy cows there was a direct relationship between digestible energy density of a diet and level of intake. However, when the constraint of digesta bulk was removed by offering a grain-based diet, the level of energy intake was immediately increased. Similarly, Hidari (1984) has shown that removing food residues via a ruminal fistula stimulated subsequent

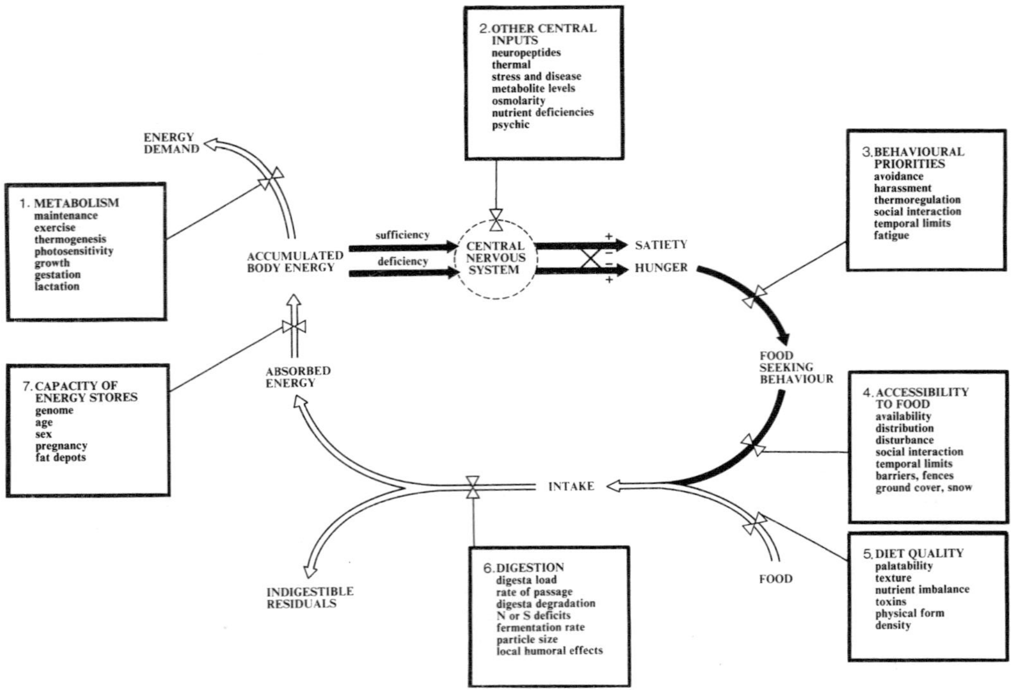

Figure 1. Regulation of body energy balance illustrating interrelationships with food intake. Material (open arrows) and information (solid arrows) flows and modifying factors (boxes) are linked in a negative feedback system (Young 1987).

feeding activity in domestic sheep. The importance of indigestible fibre and digesta load in limiting food intake by herbivores has been confirmed by several investigators (e.g. Weston 1980, 1984). These aspects are considered further by other authors at this symposium. The remainder of this review will focus on seasonal and climatic components and their possible influences on the food intake by individual mammalian herbivores.

SEASON AND APPETITE

Under natural conditions, herbivores encounter major seasonal cyclic changes in climatic conditions and in the amount and quality of available feed. These cycles are either the wet/dry seasons common in tropical and semi-tropical areas, or the marked summer-winter annual cycles which are particularly pronounced in the cool temperate zones and the more polar latitudes. The latter will be the basis of the present discussion.

In the temperate and polar regions of the world, many domestic and wild herbivores show pronounced seasonal cycles in reproductive activity, body weight, hair growth and food intake (McEwan and Whitehead 1970; Wood *et al.* 1962; Moen 1978; Milne *et al.* 1978). Such reports indicate that appetite rises to a peak in summer, falls during autumn and is low in the midwinter period. The reduced appetite often causes a marked check in the growth of wild ungulates and intake may even fall below maintenance requirements despite access to feed (Wood *et al.* 1962; Kay 1979; Milne *et al.* 1978). At first glance, the basic concept of stability of accumulated body energy, as shown in Fig. 1, seems incompatible with the observed seasonal fluctuations. The apparent inconsistency arises because the natural environment is continually changing nutritionally and climatically, so the animal makes progressive responses to environmental cues to minimize stress and enhance survival probability (Wood *et al.* 1962; Blaxter 1982; Kay 1985). Exact synchrony cannot be expected because of the time delay in physiological adaptation.

With natural seasonal variation, it is difficult to identify from the available data the relative importance of individual nutritional and climatic cues in determining food intake.

There is probably significant interdependence of factors, some acting more directly than others. Such conflicting drives may be responsible for some of the variations in food intake at different times of the year (Fig. 2; Ozoga and Verme 1970; Chao and Brown 1974; Gilbert and Bateman 1983; Adams *et al.* 1986). For example, Ozoga and Verme (1970) found that the normal food intake pattern of white-tailed deer in covered outside pens and given free access to fresh browse, consisted of high levels of intake in early winter and in spring, but a markedly reduced intake in midwinter. Adult does entering winter in an underweight condition showed more fluctuation in intake than does entering winter in fat condition. Arnold (1984) summarized data which showed that for some deer species there was considerable variation in the time of peak food intake during the summer period. Peak intake occurred earlier in the summer when animals had access to feed of high quality rather than when only poor quality feed was available. Early peaking of summer intakes tended to be associated with rapid attainment of maximum body weights due to the better feed quality. Rapid weight gain together with a decline in appetite as animals fattened led to declines in food intake earlier than expected from the natural photoperiod cue (see below).

A number of studies have clearly established the importance of photoperiod and thermal cues as factors influencing the food intake of herbivores. Evidence for possible effects of changes in atmospheric pressure will be considered briefly after discussing effects of photoperiod and thermal environment on food intake.

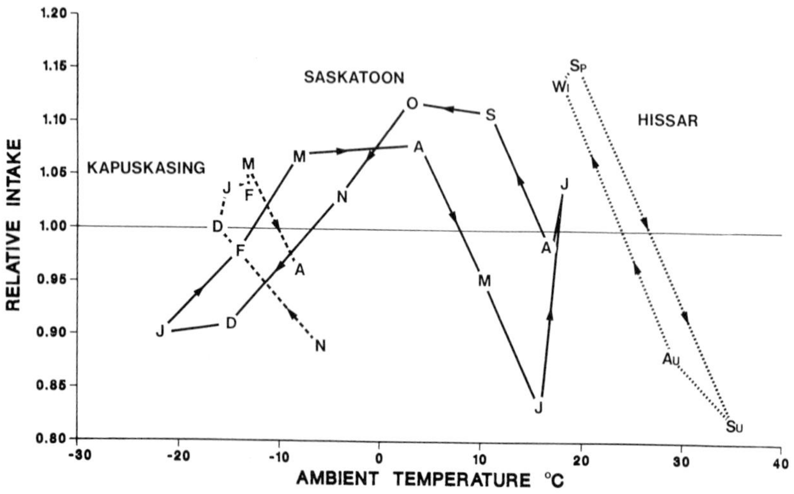

Figure 2. Relative food intake of cattle exposed to natural seasonal variations in temperature and other climatic variables. Symbols represent month of the year for data from Kapuskasing, Canada (Latitude 49° north; Hidiroglou and Lessard 1971) and from Saskatoon, Canada (Latitude 52° north; Milligan and Christison 1974), and seasons for data from Hissar, India (Latitude 29° north; Sharma and Kehar 1961).

PHOTOPERIOD AND APPETITE

Much of the pronounced seasonal fluctuation in reproduction and body weight observed, particularly in wild ungulates, appears to be associated with endocrine activities apparently entrained by photoperiod (Ringberg *et al.* 1978; Ryg and Jacobsen 1982a; Tucker *et al.* 1984; Kay 1985). Recent experiments have established that photoperiod also affects the food intake of domestic sheep and wild ungulates fed to appetite in controlled situations where other environmental cues, such as ambient temperature and diet quality, were held constant. In the natural situation, daily and seasonal light patterns may impinge on voluntary food intake in additional ways. For example, the amount of incident light affects plant growth and therefore the quality and quantity of available feed; grazing behaviour may be altered because of lack of light at night or because of the high intensity of daytime solar radiation (Dulphy *et al.* 1980). The effects of seasonal and diurnal light:dark cycles where photosensitivity apparently arises from subtle changes in light patterns will be considered in this section.

Food intake values are expressed in this review in relative terms to overcome the problem of different terminology and units used by various authors. Some of the data were derived from graphic presentations and precise values are not available.

Natural photoperiod

Table 2 lists relative food intake values from domestic sheep, red deer, white-tailed deer, reindeer and caribou given free access to food and exposed to natural seasonal photoperiod and temperatures. While these data do not exclude confounding from numerous factors they show consistently lower food intakes during the winter period. Sexually active adult males showed greatest fluctuations in intake. Castration reduced but did not eliminate the seasonal cycle in appetite. Wood *et al.* (1962) and Moen (1973) reported that the appetite of deer stags declined suddenly in the fall to virtually zero during the rut.

Artificial photoperiod

When sheep (Forbes *et al.* 1979; Schanbacher and Crouse 1980) and dairy cows (Tucker *et al.* 1984) were exposed to a constant daily photoperiod of 8 hours light and 16 hours dark (8L:16D), their intakes were .94 to .88 of those for similar animals kept under 16L:8D photoperiod. In the studies of Forbes *et al.* (1979) on growing sheep and Peters *et al.* (1980) on dairy cows, those animals which received the greater amount of light each day not only ate more but also grew faster or produced more milk than animals on lesser light regimes. Similarly, Peters *et al.* (1981) reported that the dry matter intake of heifers kept in 16 hours of light daily was greater than when they were kept in shorter natural daily light periods. However, a decrease in intake was observed when similar heifers were exposed to continuous lighting (Peters *et al.* 1980). The effectiveness of supplementary light on levels of production and the commercial implications thereof has been reviewed by Tucker *et al.* (1984) and Tucker (1985).

Experiments in which sheep and deer were exposed in illuminated rooms to programmed lighting designed to mimic natural cycles at an accelerated rate of two cycles per year, produced clear cyclic effects on voluntary food intake (Figure 3; Table 3). However, in young growing Suffolk cross rams (Simpson *et al.* 1984), the cyclic effect on food intake was not as distinct as with more mature sheep or red deer.

Table 2. Effect of natural photoperiod on the relative food intake of large herbivores. Animals were penned, exposed to the naturally occurring environment and given free access to a prepared diet. Responses were considered by the reporting authors to be due to photoperiod effects.

Animal	Diet	Location (latitude)	Relative intake*	Reference
Sheep, castrates	grass pellets	Edinburgh, UK. (56° north)	.81	Milne *et al.* (1978)
White-tailed deer male	concentrate mix	Univ. Park, PA, USA (41° north)	.47	French *et al.* (1960)
White-tailed deer fawns	concentrate pellets	(Durham, NH, USA (43° north)	.62	Holter *et al.* (1977)
Red deer castrates	grass pellets	Edinburgh, UK. (56 north)	.69	Milne *et al.* (1978)
Reindeer female	hay pellets	Vancouver, Canada (49° north)	.75	McEwan & Whitehead (1970)
Reindeer male	hay pellet	Vancouver, Canada (49° north)	.68	McEwan & Whitehead
Reindeer male	concentrate pellets	Tromso, Norway (70° north)	.43	Nilssen & Ringberg (1980)
Caribou female	hay pellets	Vancouver, Canada (49° north)	.69	McEwan & Whitehead (1970)
Caribou male	hay pellets	Vancouver, Canada (49° north)	.55	McEwan & Whitehead (1970)

*Ratio of food intake during winter (short photoperiod) relative to summer intake.

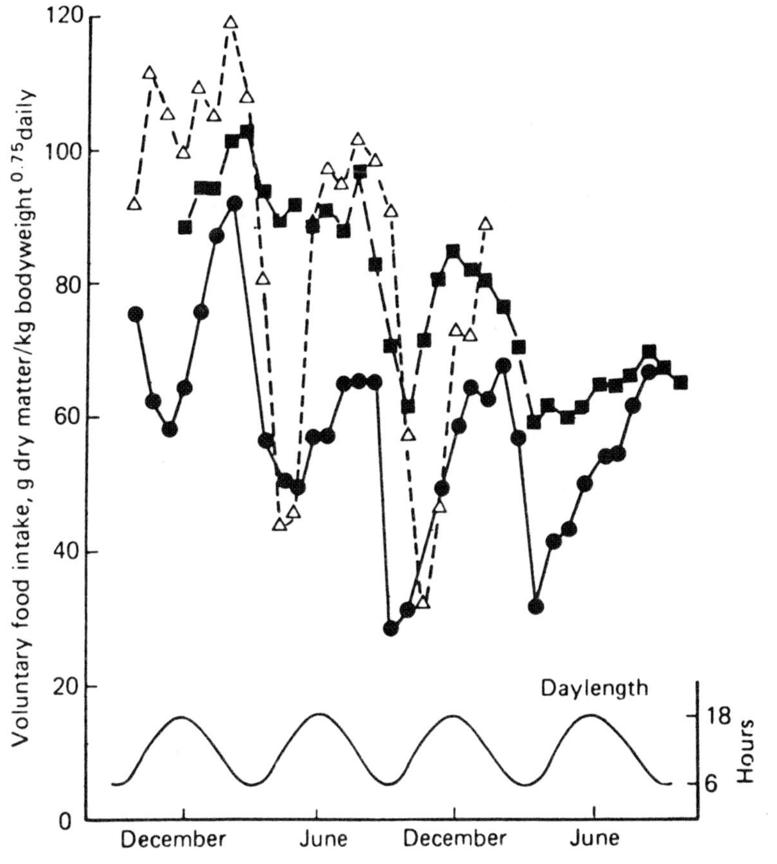

Figure 3. Voluntary food intake per unit metabolic body size by red deer (●), Suffolk cross sheep (■) and Soay sheep (△) subjected to 6-month photoperiod cycles (Kay 1979).

Minimal information is available on the effect of the source of artificial light, illumination intensity, or diurnal exposure patterns.

Fluorescent, incandescent and mercury vapour lamps with intensities over 100 lux at animal height seem equally effective in initiating photosensitive responses and subsequent changes in food intake (Tucker *et al.* 1984; Tucker 1985). Longer diurnal light generally promoted food intake but continuous light proved to be no better, and at times even poorer, than 18L:6D (Peters *et al.* 1980; Tanida *et al.* 1984). Sheep given a diurnal pattern of 7L:9D:1L:7D consumed more feed than similar animals exposed to 8L:16D each day, but consumed about the same as sheep exposed to 16L:8D (Schanbacher and Crouse 1980). These observations imply that the length of individual intervals of light or dark rather than the total number of hours of each, may be the important cue. The importance of the specific amount of light or dark each day relative to the subtle diurnal changes in photoperiod that occurs naturally is as yet unresolved. A cursory examination of available data indicates apparent greater response to natural or artificial cyclic changes in photoperiod than to fixed light regimes. Some caution should be exercised in interpreting these findings, as the lower values observed with the fixed light regimes were obtained exclusively from domestic species which may be less photosensitive than wild species (Milne *et al.* 1978; Kay 1985). Furthermore, evidence from the studies of Kay and associates (Suttie *et al.* 1984; Simpson *et al.* 1984; Kay 1985), showed there was a lag of 10 to 16 weeks between a change in the phases of a photoperiod cycle and the expected appetite response. The lag phase for a shift from one constant light regime to another is not known.

Table 3. Estimates of cyclic peak and trough food intakes by red deer, semidomestic Soay sheep and Suffolk cross sheep exposed to artificial cyclic illumination mimicking the day length amplitudes of Aberdeen, Scotland (approximately 18L:6D to 6L:18D) but with two cycles of day lengths each calendar year.

	Diet	Daily intake kg DM /day Peak	Trough	Relative intake*	Reference
RED DEER					
stags	pelleted conc.	2.7	1.3	.48	Kay (1979)
castrates	pelleted conc.	1.7	1.3	.76	Kay (1979)
stags	pelleted conc.	2.6	1.3	.50	Kay (1979)
young males	barley and fishmeal	2.8	1.4	.50	Simpson *et al.*(1984)
SOAY SHEEP					
rams	pelleted conc.	1.30	.35	.26	Kay (1979)
rams	barley straw	1.35	.30	.24	Suttie *et al.*(1984)
SUFFOLK CROSS SHEEP					
rams	pelleted conc.	2.40	1.75	.75	Kay (1979)
ram lambs	barley straw plus conc.	no clear cyclic pattern			Simpson *et al.*(1984)

*Ratio of trough to peak food intake.

Mechanism of photoperiod effect

The physiological basis of photosensitive variations in appetite and of specific endocrine involvements, is still rather speculative. Recently, Kay (1985) suggested that the mechanism of the photo-induced appetite cycle could be similar to that for the breeding cycle. He proposed that during the diurnal light phase, impulses from the retina stimulate a neural centre in the hypothalamus, which functions as an "internal clock". This centre sends neural signals to the pineal gland, stimulating the formation of melatonin, which in turn influences secretion of gonodotrophic hormones. Denervation of the pineal gland by removal of the superior cervical sympathetic ganglion reduced the amplitude of the photo-determined appetite cycle in Soay sheep (Suttie *et al.* 1984). Furthermore, when Kay and co-workers (Kay 1985) placed Soay sheep and red deer in an artificial photoperiod such that two cycles occurred during one calendar year, the gonadal activity, growth and appetite cycle adjusted accordingly. Male deer grew a set of antlers during each six month cycle. This demonstrated that the photosensitive breeding and appetite cycles were activated by similar, or at least parallel, mechanisms.

Other studies at the Rowett Research Institute on Soay sheep and red deer (Simpson 1976 and Argo 1985, as cited by Kay 1985) indicated that the high energy demand of breeding activities and gestation may override any direct photoperiodic effect on appetite. Normally, the appetite of hinds and ewes increased in summer, paralleling the increased metabolic demand from lactation and prewinter fat accumulation, while appetite decreased in autumn and remained low during winter when metabolic demands were lower. However, when the breeding cycle of ewes was artificially shifted so they lambed in January (midwinter) instead of March, the appetite cycle was also changed. In this modified situation, appetite was low during the breeding and gestation periods, but increased after parturition with the onset of lactation and the associated increased energy demand. Thus, the feed intake cycle was three months out of phase from that expected with the natural photoperiodic cycle, but in phase with energy demand.

Are seasonal photoperiod-induced changes in appetite primarily coupled to changes in energy demand? The proposed model (Fig. 1) indicates that increased metabolism increases energy demand and thus appetite drive. Is there evidence for photosensitivity of metabolic rate? If such an effect exists, it is possible that this variation in energy demand may be responsible for the influences of photoperiod on appetite. Silver *et al.* (1969) showed that the fasting metabolic rate of white-tailed deer exposed to natural seasonal cycles decreased during the winter and increased in summer, and Blaxter and Boyne (1982) found that the rate of metabolic heat production of sheep fed at near-maintenance level cycled in accord with the natural seasons. The 10 to 12 week lag between the peak metabolic response and seasonal cycle observed by Blaxter and Boyne (1982) is consistent with the lag observed by Simpson *et al.* (1984) for the photosensitive response in the appetite of red deer and sheep.

The specific physiological events that mediate the photoperiod-induced changes in appetite have not been satisfactorily identified. Lincoln (1979), Peters *et al.* (1980), Ryg and Jacobsen (1982a, b) and Tucker *et al.* (1984) have documented seasonal photosensitive variations in circulating levels of growth hormone, insulin, thyroid hormones, prolactin, as well as several reproductive hormones. As level of food intake and some non-photoperiod components of the environment were not always controlled in the above studies, their interpretation is somewhat difficult. Levels of thyroid hormones, for example, seem to be determined more by the amount of food eaten rather than being determinants of appetite (Ryg and Jacobsen 1982a). In a recent review, Tucker (1985) focussed on prolactin as a hormone likely to be sensitive to changes in photoperiod since there was a direct relationship between serum prolactin level and hours of light each day. Furthermore, pinealectomy blocked photo-induced changes in the circulating level of prolactin in sheep (Brown and Forbes 1980). Beyond this, the link between prolactin, or any other hormone, and photosensitive changes in appetite is still an enigma.

Why have photosensitive appetite?

Does the long photoperiod of summer stimulate voluntary food intake or does the short photoperiod of fall and winter depress appetite? Using intake values adjusted for metabolic size, Kay (1979) indicated that peak food intake of red deer in summer was less than both summer and winter food intake by domestic sheep (Fig. 3). Similarly, Milne *et al.* (1978) found that when red deer and domestic sheep were offered the same diet of chopped grass in late fall and summer, the organic matter intake by the deer was 57.3 and 82.3 g $/kg^{0.75}$ respectively, compared with 71.3 and 88.2 g $/kg^{0.75}$ for the sheep. However, in a second experiment the same authors reported that deer consumed twice as much roughage of low digestibility (0.5) in winter and spring as did sheep.

The observed pattern of photosensitive variation in appetite is present in domestic sheep and cattle but is not as marked as in the wild ungulates (Tucker 1975; Milne *et al.* 1978; Kay 1979; Simpson *et al.* 1984). Blaxter (1982) and Kay (1985) suggested that genetic selection in the processes of domestication may have played a role in the observed photosensitive differences between domestic and wild ungulates. Furthermore, Kay (1985) showed that the Soay sheep had amore pronounced seasonal cycle in appetite than did the domesticated Suffolk sheep. Kay suggested that the provision of shelter and winter feed to domestic species may have reduced the natural selective pressures that maintain the seasonal cycle in food intake.

The photosensitive annual cycle in appetite was considered by French *et al.* (1985) and Kay (1979, 1985) to be an adaption to take advantage of the season cycle in quality and quantity of available forage. Forage of high quality is usually abundant in summer when the photoperiod cue allows animals to eat well above their immediate maintenance needs and lay down substantial body energy reserves. In contrast, forage is often scarce and of poor quality in winter and appetite declines, reducing an animal's expenditure of energy in foraging activity and allowing more time to shelter from inclement weather. Body fat reserves accumulated in summer are used during the winter to meet maintenance needs.

A decrease in appetite during fall and winter may enhance the opportunity for animals to congregate for breeding and to survive in a natural environment with inadequate winter feed. As indicated above, photoperiod may be the factor that links seasonal variations in energy demands to seasonal availability of feed resources, making it important to the nutrition, reproduction and survival of wild herbivores. However, reduced winter production is considered undesirable in commercial agriculture, and intensive genetic selection of domestic livestock for high growth rate and year-round productivity has considerably increased their appetite ceiling and reduced the effects arising from the photoperiod cue.

THERMAL ENVIRONMENT AND APPETITE

Wild and domestic herbivores thrive in a wide range of thermal environments, from the warm equatorial wet tropics which have relatively little variation in ambient temperature, to the deserts and the cooler temperate and polar regions which have marked seasonal and diurnal variations in ambient temperature. Does the thermal environment significantly affect the appetite of herbivores? If so, by what mechanisms?

The many reviews on food intake conclude that animals increase intake following exposure to cold and decrease intake in hot conditions (Bianca 1965; Balch and Campling 1969; Thompson 1973; Baile and Forbes 1974; Weston 1979, 1982; Arnold 1984). Research data is lacking for the wider spectrum of herbivore species as is critical examination of the physiological mechanisms underlying the thermal influence on food intake. The following section identifies the former and discusses aspects of the latter.

Attempts to relate natural seasonal or diurnal changes in ambient temperature to appetite are easily confounded by other variables (Fig. 2). The cooler winter temperatures experienced in temperate regions increase metabolic energy demand for thermo-regulation and thus would be expected to promote increased food intake. However, as discussed above, the short photoperiod during autumn and winter has the opposite effect.

These conflicting drives may explain some of the variability in food intake observed in animals exposed to natural environments. After

Table 4. Effect of ambient temperature on the relative food intake. Animals were confined in controlled environmental chambers.

Ambient temp. °C	Relative intake* 0-10	10-20	20-25	25-30	30-35	35-40	Ref.[#]
CATTLE							
Growing, 1 to 12 mo.		1.00		.92			1
B. taurus, 5 to 7 mo.		1.00		.86			2
B. indicus, 5 to 7 mo.		1.00		.89			2
Lact. cows, wind <0.2 m/s		1.00		.88		.56	3
Lact. cows, wind 3.5-4 m/s		.97		.93		.76	3
Dry cows, wind <0.2 m/s		1.00		.92		.79	3
Dry cows, wind 3.5-4 m/s		.97		.96		.83	3
Lact. cows, rel. humid. <50%		1.00		.88	.87	.74	4
Lact. cows, rel. humid. >50%		1.00	.97	.87	.79		4
B. taurus, conc. diet	1.30	1.00			.83		5
B. taurus, rough. diet	1.10	1.00			.60		5
B. indicus, conc. diet	1.38	1.00			.88		5
B. indicus, rough. diet	1.07	1.00			.87		5
SHEEP							
Growing, 0.65 rough. diet	1.00			.76			6
Growing, 0.35 rough. diet	1.00			.83			6
Growing, 0.65 rough. diet		1.00		.86			6
Growing, 0.30 rough. diet		1.00		.99			6
Growing, conc. diet			1.00		.50		7
Growing, rough. diet			1.00		.86		7
Growing, conc. diet			1.00		.85		8
Growing, 0.25 rough. diet			1.00		.82		9
Growing, 0.50 rough. diet			1.00		.87		9
Growing, 0.75 rough. diet			1.00		.97		9
GOATS							
Growing		1.00				.75	10
RABBITS							
Acclim. to 10°C	1.00		.65	.21	.11		11
Acclim. to 28°C		1.00	.90	.63	.21	.08	11

*Ratio of food intake to intake at thermoneutrality.
[#]References: 1. Johnson *et al.* (1960); 2. Colditz and Kellaway (1972); 3. Brody *et al.* (1954); 4. Johnson *et al.* (1963); 5. Olbrich *et al.* (1973); 6. Moose *et al.* (1969); 7. Bhattachary and Hussain (1974); 8. Valtora *et al.* (1982); 9. Bhattachary and Uwayjan (1975); 10. Appleman and Delouche (1958); 11. Johnson *et al.* (1958).

an extensive study of white-tailed deer over-wintered in Maine, USA, Gilbert and Bateman (1983) concluded that feeding activity was dependent upon the interaction of several factors. Solar radiation accounted for about 28 percent of the observed variation. Maximum food intake was recorded at a mean daily temperature of -15°C, and minimum intake occurred at -4°C. Intakes decreased markedly at temperatures below -15°C. Furthermore, in this study feeding activity increased when skies were overcast and atmospheric pressure was falling.

Measurements of the effects of the thermal environment on voluntary food intake of herbivores has come mainly from studies on domestic sheep and cattle confined to controlled climate laboratories. Results from selected studies are summarized in Table 4. Food intake values have been expressed relative to estimates made at air temperatures of 10 to 20°C, or another temperature which was considered not to be thermally stressful. While in principle there is an inverse relationship between ambient temperature and food intake, the response was variable and dependent on the animal's thermal susceptibility, acclimation and diet. In other trials, goats (Appleman and Delouche 1958) and cattle (Ragsdale *et al.* 1950; Worstell and Brody 1953) exposed to temperatures of 40°C and above, developed hyperthermia and refused food. Under conditions of less severe continuous mild heat stress cattle had a negative energy balance and lost substantial body weight due to reduced intake (Ragsdale *et al.* 1950). Food intake of cattle has been reported to increase under conditions of continuous cold, to meet the apparent increased metabolic demand (Ragsdale *et al.* 1949, 1950; Worstell and Brody 1953; Moose *et al.* 1969; Olbrich *et al.* 1973). However, with acute severe cold, feeding behaviour was abruptly curtailed in adult cattle (Malechek and Smith 1976; Adams *et al.* 1986) and suckling lambs (Alexander and Williams 1966). These sudden reductions in appetite with acute cold were associated respectively with shelter-seeking behaviour and the development of hypothermia.

A shift in voluntary food intake due to thermal influences apparently arises from two basic sources: i) in response to a short-term thermal stress which challenges the homeothermy of susceptible animals, and ii) as a consequence of metabolic and digestive functional changes arising from thermal acclimation.

Thermal susceptibility

An animal's thermal susceptibility is dependent on a complex array of individual and environmental factors affecting the rate of gain or dissipation of body heat. One can conclude from the many reviews on environmental influences on the thermal balance of animals that the large mammalian herbivores are quite thermally resistant due to their substantial body mass, effective thermal insulation and efficient dissipation of excess body heat by respiratory evaporation. Within their normal environment, well adapted, healthy herbivores rarely suffer from overt thermal stress. Thermal stress more often occurs as a consequence of an abrupt change in the animals or the environment. If there is a failure to maintain homeothermy, food intake is affected.

In cold weather, the animal's pelage provides substantial thermal protection by reducing convective and conductive heat exchange. Shearing of sheep has been associated with increases in food intake. When recently shorn sheep were exposed to temperatures below their lower critical temperature food intake increased (Wheeler *et al.* 1963) but no increase was observed in sheep exposed to warmer conditions (Moose *et al.* 1969; Minson and Ternouth 1971).

In tropical climates, high ambient temperatures and solar radiation add heat loads to animals and reduce the thermal gradient between the body core and the surface, thus inhibiting dissipation of endogenous body heat. Failure to compensate for the added heat load results in hyperthermia and reduced food intake (Worstell and Brody 1953; Johnson *et al.* 1963). High humidity and low air movement intensify the effects of high ambient temperature due to the reduced ability of animals to dissipate body heat by evaporative cooling. In contrast, natural rain or artificial sprinkling systems and wind or fans enhance body cooling and reduce the heat load on animals in hot environments (Bond *et al.* 1957; McDowell 1972; Morrison *et al.* 1974) but may add to the thermal drain on animals in cold stress situations (Young 1983). In a cold environment, melting snow, mud and dampness also add to the drain on body heat, thus increasing the risk of hypothermia in susceptible animals and of possible effects on food intake (NRC 1981).

Endogenous heat production

Elevated endogenous metabolic heat production increases the risk of hyperthermia in hot climates. Effects on appetite are particularly marked in high-producing dairy cows which have endogenous metabolic heat-production greater than twice that of their non-lactating counterparts (Moe and Tyrrell 1979). Consequently, depressed food intake is a major commercial problem for high-producing dairy herds in hot climates (McDowell 1972). Increased metabolic rate during exercise, feeding activity and ingestion of food also results in increased endogenous heat production, which assists in maintaining body homeothermy in cold climates but contributes to the risk of hyperthermia in hot climates.

The thermogenic effect of food is substantial in ruminants, and presumably other herbivores, and the effect is greatest with high fibrous feeds (Balch and Campling 1969). An immediate .3 to .7 increase in metabolic heat production occurs during grazing and feeding activity by sheep (Young and Corbett 1972; Osuji 1974) and cattle (Adam *et al.* 1984). This increased metabolism persists throughout the feeding period, and is proportional to the time spent feeding rather than to the amount of food actually consumed. As most herbivores spend a substantial portion of each day in grazing activities, increased endogenous heat production may substantially add to an animal's thermal load in a hot environment. Greater endogenous heat production would be expected in the acquisition of a diet of low quality under sparse grazing conditions. Parallel consequences have been identified in commercial feedlot situations, where Moose *et al.* (1969) reported that the high endogenous heat production seriously impaired the intake and efficiency in utilizing high roughage diets by growing sheep. They reported it was advantageous for animals exposed to high ambient temperatures to be switched to concentrate diets because of the lower diet-induced thermogenesis.

The thermoneutral metabolic intensity of sheep and cattle is also increased by .2 to .4 as a consequence of acclimation to cold and conversely is decreased with acclimation to heat, as discussed in the following section on thermal acclimation. These adaptive shifts in metabolism take several weeks to develop fully and they persist for a similar time after a change in the thermal environment. With thermal acclimation metabolic demand changes, giving rise to possible influences on appetite (Fig. 1).

Thermal imbalance

Thermal imbalance and a shift in body temperature occur when the rates of heat gain and heat loss are unequal. Mammals have highly developed regulatory mechanisms and respond behaviourally and physiologically to counter unacceptable variations in body temperature. Peripheral vasomotor tone, endogenous heat production, evaporative cooling and shelter-seeking behaviour are closely linked to the attainment of thermal balance, and may impinge on feeding behaviour and food intake.

Worstell and Brody (1953) showed that in dairy cattle heat stressed in climate chambers, the beginning of the marked depression in food intake coincided with a rise in rectal temperature (Fig. 4). However, the actual ambient temperature at which the increase in body temperature occurred was dependent upon the breed of cattle and level of milk production. High-producing dairy breeds responded at lower ambient temperatures than did dry cows or *Bos indicus* breeds. Similarly, Johnson *et al.* (1963) showed that heat stress induced by combinations of high temperature and humidity, depressed food intake as body temperature rose above about 39°C.

Andersson and associates (Andersson 1962; Andersson and Larsson 1961), using implanted thermodes, showed that the heating or cooling of the hypothalamic region of the brain could modify the feeding behaviour of goats. Heating the thermodes decreased feeding activity while cooling of specific areas promoted food intake. The unphysiological level of heating or cooling necessary to elicit a response in these experiments throws some doubt on the physiological role of central neural temperature in regulation of appetite. However, the sensitivity of the central nervous system to temperature is indisputable.

Gengler *et al.* (1970) found that direct heating or cooling of the rumen reduced the food intake by dairy cows. However,

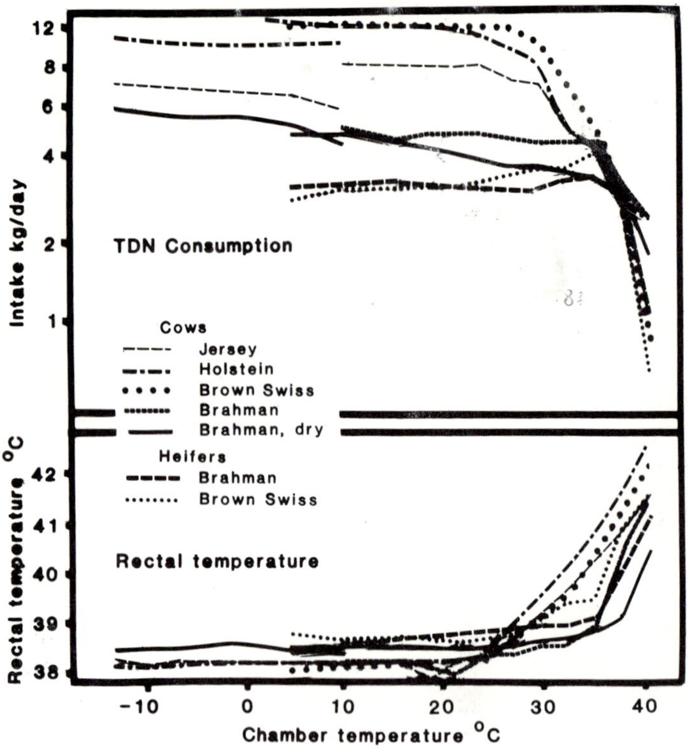

Figure 4. Food intake (total digestible nutrients; TDN) and rectal temperature of lactating cows in relation to decreasing and increasing ambient temperature (Worstell and Brody 1953).

Bhattacharya and Warner (1968) were able to stimulate the intake of a pelleted ration by adding cold water to the rumen of sheep but not the intake of long hay. Cooled drinking water is sometimes provided to commercial feedlot cattle in hot climates to promote food intake (Ittner *et al.* 1951; Lofgreen *et al.* 1975; Fuquay 1981).

As indicated earlier, high ambient temperatures and solar radiation affect the thermal balance of herbivores in hot climates. The actual heat gain is, however, dependent upon intensity of solar radiation, accessibility to shade and the animal's pelage. In high solar radiative conditions, heat input to exposed sheep (Porter and Gates 1969), cattle (Finch 1984) and kangaroos (Dawson 1972a) has been estimated to be up to 160 watts /m^2. These levels of solar radiation have major effects on thermal balance and reduce the time spent grazing (Finch 1984; Bennett *et al.* 1985). Climate laboratory studies by Brody and associates (Brody *et al.* 1954b) showed that the voluntary intake by dairy cows was decreased by .2 where radiation levels were high and when ambient temperature was above 20°C but by only .1 when the temperature was 10°C.

As discussed above, there is a substantial diet-induced thermogenic response with consequent transient increases in the body heat content (McLean *et al.* 1984) and of central (Forbes 1970; Baile and Forbes 1974) and surface temperatures (Ingram and Whittow 1962). Brobeck's (1948) thermostatic theory: "animals eat to keep warm and stop eating to prevent hyperthermia", has been largely rejected by recent evidence (Balch and Campling 1969; Baumgardt 1970; Forbes 1970; Baile and Della-Fera 1981). However, while central temperature is no longer considered to be the regulator of appetite, homeothermy seems to be necessary for normal food intake. Severe hypothermia or hyperthermia constrains feeding behaviour.

Behavioural priorities

Animals adjust their behaviour to avoid unpleasant situations. For example, during the heat of the day or during inclement weather there may be temporal displacement of feeding behaviour and reduction of food intake. Dawson (1972b) reported that red kangaroos and euros rarely ventured out during periods of high solar radiation and restricted their foraging times to early morning and late evening. Similarly, during periods of high temperature and solar radiation, cattle actively seek shade (Seath and Miller 1946; Payne *et al.* 1951; Frisch and Vercoe 1977; Bennett *et al.* 1985). Shade, whether natural or artificial, tends to lower effective temperature and reduce the amount of ground radiation (Ittner and Kelly 1951), thus providing some thermal relief. Pontif *et al.* (1974) reported that cattle under artificial shade constantly had higher food intakes relative to animals without access to shade. Furthermore, under natural conditions, cows were observed to spend more time grazing under shade than in areas without shade (McDaniel and Roark 1956). While reduced daytime grazing is often observed in hot conditions, Hutchinson and McRae (1969) have reported reduced night-time grazing in newly shorn sheep during cold weather. This cold weather response was presumably to minimize night radiation efflux of body heat.

In climates with high solar radiation but with substantial night-time cooling, animals behaviourally adapt by moving feeding activity with its associated endogenous heat production to the night period (McDowell 1972; Bhattacharya and Uwayjan 1975; Finch 1984; Bennett *et al.* 1985). Without the advantage of night cooling, compensation in food intake does not occur (Seath and Miller 1946; McDowell 1972; Faquay 1981).

Feeding activity also may be constrained when animals are

disturbed or harassed by predators, insects or even the well-meaning tourist. Such disturbances not only reduce feeding opportunity but escape or avoidance behaviour and sudden activity increase the risk of hyperthermia. These problems are compounded in hot climates when combined with sparse grazing conditions as in such conditions herbivores must forage for long periods, often during the heat of the day. In their reviews, Hafez (1968) and Moen (1973) concluded that thermoregulation is generally a higher behavioural priority of herbivores than is feeding activity. They, along with McDowell (1972), Arnold (1984) and Bennett *et al.* (1985), reported that during the hottest part of the day, animals sought shade rather than graze in regions of high solar radiation. Similarly, in winter beef cattle will curtail grazing and seek shelter to avoid the unpleasantness of inclement winter conditions (Malenchek and Smith 1976; Adams *et al.* 1986). Ozoga and Verme (1970) and Moen (1978) reported that when white-tailed deer were wintered under conditions of food shortage, they gave behavioural priority in cold weather to reduced physical activity and conservation of body energy rather than to foraging and possible exposure to cold stress. Such behavioural strategies have obvious advantages for large herbivores with substantial body energy reserves and resistance to a thermal challenge. Similar strategies may not be appropriate for small animals which are more dependent on increased metabolic heat production for maintenance of homeothermy.

Thermal acclimation

Individuals have capacities to adapt to the environment to which they are exposed and display appropriate morphological and physiological adjustments to become better suited to their environment. Adaptive changes occur in thermal insulation of pelage, metabolic intensity and digestive functions as a consequence of a change in the thermal environment (Robertshaw 1974; Christopherson 1985; Young 1985). These changes have potential consequences to the voluntary food intake by herbivores, and may provide insight into possible mechanisms for the inverse relationship between ambient temperature and food intake (Table 4).

With acclimation to cold there is a 0.3 to 0.4 elevation in resting metabolic rate in sheep and cattle (Young 1975, 1985; Christopherson 1976), which is independent of level of feeding but linked with increased thyroid hormone activity (Westra and Christopherson 1976; Christopherson 1985). The increased metabolic energy demand due to elevated thyroid levels potentially increases appetite drive. In contrast, prolonged exposure to a hot environment has been reported to decrease thyroid activity (Thompson 1973; Valtorta *et al.* 1982) and reduce metabolic intensity (Kamal *et al.* 1962).

Senft and Rittenhouse (1985) recently proposed a descriptive model which incorporates the immediate response, as well as the effects of thermal acclimation, on the food intake of cattle exposed to outdoor environments (Fig. 5). They suggested that with physiological acclimation

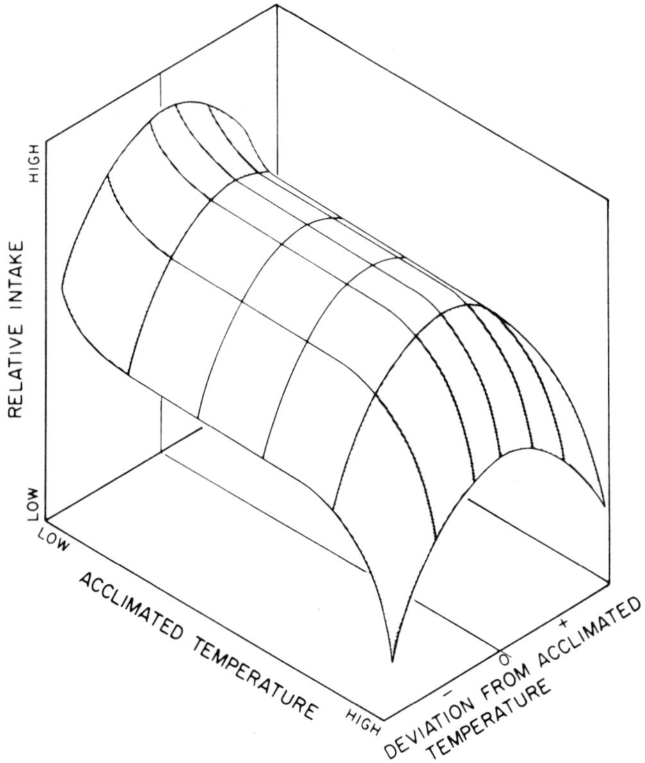

Figure 5. Model of relative food intake by cattle in relation to acclimation and current ambient temperatures (Senft and Rittenhouse 1985).

there is an integration of the animal's thermal experiences over several days. Senft and Rittenhouse used a calculated running average of mean daily temperatures to derive estimates of acclimation temperature. Such calculations take into account natural day-to-day fluctuations in temperature and the shifting adaptive state of the animal which is unlikely to be in synchrony with short-term thermal fluctuations. It was proposed that an immediate behavioural or physiological response is made to the difference between the current thermal environment and the acclimated state. While the Senft and Rittenhouse model requires further validation, it does assist in describing some available field observations of variation in winter foraging behaviour of cattle (Malechek and Smith 1976). Furthermore, it recognizes the idea that cold-acclimatised animals are more sensitive to a mild heat stress

(Webster 1970; Young 1975; Gonyou *et al.* 1979) and heat acclimated animals are more sensitive to a mild cold stress (Kamal *et al.* 1962).

Results from our laboratory have shown that when shorn sheep or cattle were acclimatised to cold there was increased rumination (Gonyou *et al.* 1979) and reticulo-ruminal motility (Westra and Christopherson 1976), with a reduction in reticulo-ruminal volume and an increase in rate of passage of digesta (Kennedy *et al.* 1986). Faster clearance of fibrous residues from the reticulo-rumen permits ruminants with free access to food to increase their intake (Christopherson 1985: Kennedy *et al.* 1986). Furthermore, with the faster clearance from the forestomachs there is increased protein flow to the intestines, which has been shown by Egan (1977) to increase appetite. The increased intestinal protein flow in cold acclimatised animals arises from three sources. With the increased flow of digesta from the reticulo-rumen, more dietary protein escapes fermentation, with a second concurrent increase in efficiency of microbial protein synthesis in the rumen (Kennedy and Milligan 1978; Kennedy *et al.* 1982). The third possible source is from the increased flow of pancreatic secretions. Kato and Young (1985) found an increase in pancreatic secretion in sheep during cold exposure compared with similarly fed sheep in warm conditions.

Summary of thermally induced mechanisms

Overall, there are several possible mechanisms for the effect of the thermal environment on the food intake of herbivores. The role of each appears to be dependent upon the duration and degree of thermal strain imposed on the animal as well as the ability of the animal to maintain homeothermy while under challenge. The thermal environment influences metabolic energy demand in two ways. During immediate cold stress, metabolic heat production is increased to compensate for the faster rate of heat loss to the environment and, provided the cold is not severe enough to cause hypothermia, the animal usually compensates by increasing its food intake to meet the increased metabolic demand. Thermal acclimation also affects food intake through the induced shifts in metabolic intensity and digestive functions. With acclimation to hot conditions, resting metabolic rate decreases, while acclimation to cold results in increased resting metabolism and metabolic demand. Concomitant adjustments in digestive function and rate of digesta flow have direct consequences by reducing constraints imposed by digesta load with bulky foods. The more rapid clearance of digesta in cold acclimated animals may reduce the constraint of digesta load thus allowing for an increase in food intake.

In summary, the thermal environment may affect voluntary food intake of herbivores in several ways. Both extreme heat and cold stress cause thermal imbalance along with a severely depressed food intake. In less severe situations, cold tends to stimulate food intake, while mild heat reduces food intake. The inverse relationship between acclimation temperature and metabolic demand is probably the basis for much of the observed variation in food intake during exposure to mild thermal environments. The digestive mechanisms allowing increases in food intake with cooler temperatures, are probably associated with the increase in rate of passage of digesta residues and the greater amounts of protein reaching the intestines.

ATMOSPHERIC PRESSURE AND APPETITE

From a field study on the overwintering of beef cows in northern Utah, Malechek and Smith (1976) reported that the grazing and ruminating times of the cows were increased in response to both increasing and decreasing atmospheric pressures. Similarly, Gilbert and Bateman (1983) observed the feeding activity of white-tailed deer fawns was greater when atmospheric pressure was falling than when the pressure was stable or rising. Although the correlations between feeding activities and changes in atmospheric pressure were significant, the authors did not attempt to provide an explanation for their observations. The possible influences of atmospheric pressure on feed intake of herbivores should be viewed with caution until more data is available.

CONCLUDING REMARKS

Seasonal and climatic factors influence the food intake of herbivores, but the physiological mechanisms involved are still poorly understood. Clearly, subtle diurnal changes in photoperiod and various factors contributing to an animal's thermal environment are important. The influences on food intake are probably primarily through modified metabolic intensity and metabolic energy demand or, with more severe climatic deviations, by impinging on the behavioural pattern or thermoregulatory ability of the animal. Distortions occur in complex natural environments and consistency of response to a climatic variable is not often as evident as in controlled laboratory situations.

Adam, I., Young, B.A., Nicol, A.M., and Degen, A.A. (1984). *Anim. Prod.* **38**, 53-6.

Agricultural Research Council. (1980). "The Nutrient Requirements of Ruminant Livestock." (Comm. Agric. Bur.: Farnham Royal, UK.)

Adams, D.C., Nelsen, T.C., Reynolds, W.L., and Knapp, B.W. (1986). *J. Anim. Sci.* **62**, 1240-6.

Alexander, G., and Williams, D. (1966). *J. Agric. Sci.* **67**, 181-9.

Andersson, B. (1962). pp. 117-185. In "Brain and Behavior", Vol. 11. (Ed. M.A.B. Brazer.) pp. 117-85. (Am. Inst. Biol.: Washington, DC.)

Andersson, B., and Larsson, B. (1961). *Acta. Physiol. Scand.* **52**, 175-89.

Appleman, R.D., and Delouche, J.C. (1958). *J. Anim. Sci.* **17**, 326-35.

Argo, C.M. (1985). Ph.D. Thesis, University of Aberdeen.

Arnold, G.W. 1984. "Bioenergetics of Wild Herbivores." (Eds R.J. Hudson and R.G. White.) (CRC Press, Inc.: Florida.)

Baile, C.A., and Della-Fera, M.A. (1981). *J. Dairy Sci.* **64**, 1140-52.

Baile, C.A., and Forbes, J.M. (1974). *Physiol. Rev.* **54**, 160-214.

Balch, C.C., and Campling, R.C. (1969). In "Handbuch der Tierernahrung." (Eds W. Lenkeit, K. Breinem and E. Crasemann.) pp. 554-79. (Verlag Paul Parcy: Hamburg.)

Baumgardt, B.R. (1970). In "Physiology of Digestion and Metabolism in the Ruminant." (Ed. A.T. Phillipson). pp. 235-253. (Oriel: Newcastle-upon-Tyne, UK.)

Bennett, I.L., Finch, Virginia A., and Holmes, C.R. (1985). *Applied Anim. Behavior Sci.* **13**, 227-36.

Bhattacharya, A.N., and Hussain, F. (1974). *J. Anim. Sci.* **38**, 877-86.

Bhattacharya, A.N., and Warner, R.G. (1968). *J. Dairy Sci.* **51**, 1481-89.

Bhattacharya, A.N., and Uwayjan, M. (1975). *J. Anim. Sci.* **40**, 320-28.

Bianca, W. (1965). *J. Dairy Res.* **32**, 291-345.

Blaxter, K.L., and Boyne, A.W. (1982). *J. Agric. Sci.* **99**, 611-20.

Blaxter, K.L. (1982). In "Food, Nutrition, and Climate." (Ed. K.L. Blaxter and L. Fowden.) pp. 247-265. (Butterworths: London.

Bligh, J. (1973). "Temperatures Regulation in Mammals and Other Vertebrates." (North-Holland Publ.: Amsterdam.)

Bond, T.E., Kelly, C.F., and Ittner, N.R. (1957). *Agr. Eng.* **38**, 308.

Brobeck, J.R. (1948). *Yale J. Biol. Med.* **20**, 545-52.

Brody, S., Ragsdale, A.C., Johnson, H.J., and Worstell, D.M. (1954a). Mo. Agr. Exp. Res. Sta. Bull. No. 545.

Brody, S., Ragsdale, A.C., Thompson, H.J., and Worstell, D.M. (1954b). Mo. Agr. Exp. Res. Sta. Bull. No. 556.

Brown, W.B., and Forbes, J.M. (1980). *J. Endocrinol.* **84**, 91-9.

Chao, C.C., and Brown, R.D. (1984). *Comp. Biochem. Physiol.* **A 77**, 299-306.

Christopherson, R.J. (1976). *Can. J. Anim. Sci.* **56**, 201.

Christopherson, R.J. (1985). In "Stress Physiology in Livestock, Volume 1: Basic Principles." (Ed. M.K. Yousef.) pp.163-80. (CRC Inc: Florida.)

Colditz, P.J., and Kellaway, R.C. (1972). *Aust. J. Agric. Res.* **23**, 717-25.

Conrad, H.R., Pratt, A.D., and Hobbs, J.W. (1964). *J. Dairy Sci.* **47**, 54-62.

Dawson, T.J. (1972a). *Aust. J. Zool.* **20**, 17-22.

Dawson, T.J. (1972b). *Symp. Zool. Soc. Lond.* **31**, 133-46.

Dulphy, J.P., Remond, B. and Theriez, M. (1980). In "Digestive Physiology and Metabolism in Ruminants." (Eds Y. Ruckebusch and P. Thivend.) pp.103-22. (MTP: Lancaster, UK.)

Egan, A.R. (1977). *Aust. J. Agric. Res.* **28**, 907-15.

Finch, V.A. (1984). In "Herbivore Nutrition in the Subtropics and Tropics." (Eds F.M.C. Gilchrist and R.I. Mackie.) pp. 89-105. (Science Press: South Africa).

Forbes, J.M. (1970). *Br. Vet. J.* **126**, 1-11.

Forbes, J.M., El Shahat A.A., Jones, R., Duncan, J.G.S., and Boaz, T.G. (1979). *Anim. Prod.* **29**, 33–42.

French, C.E., McEwen, L.C., Ingram, R.H., and Swift, R.W. (1955). Pennsylvania Agr. Exp. Stat. Bull. No. 600.

French, C.E., McEwen, L.C., Magruder, N.D., Rader, T., Long, T.A., and Swift, R.W. (1960). *J. Wildl. Manage.* **41**, 23–9.

Frisch, J.E. and Vercoe, J.E. (1977). *Anim. Prod.* **25**, 343–58.

Fuquay, J.W. (1981). *J. Anim. Sci.* **52**, 164–74.

Gengler, W.R., Martz, F.A., Johnson, H.D., Krause, G.F., and Hahn, L. (1970). *J. Dairy Sci.* **53**, 434–37.

Gilbert, F.F., and Bateman, M.C. (1983). *Can. Field Naturalist* **97**, 391–400.

Gonyou, H.W., Christopherson, R.J., and Young, B.A. (1979). *Appl. Anim. Ethol.* **5**, 113–24.

Grovum, W.L. (1984). In "Herbivore Nutrition." (Eds F.M.C. Gilchrist and R.I. Mackie.) pp. 224–68. (Science Press: South Africa.)

Hafez, E.S.E. (1968). In "Adaptation of Domestic Animals." (Ed. E.S.E. Hafez.) pp. 202–14. (Lea and Febiger: Philadelphia.)

Hidari, H. (1984). *Can. J. Anim. Sci.* **64** (Suppl.) 310–21.

Hidiroglou, M., and Lessard, J.R. (1971). *Can. J. Anim. Sci.* **51**, 111–20.

Holter, J.B., Urban, W.E. Jr., and Hayes, H.H. (1977). *J. Anim. Sci.* **45**, 365–76.

Hutchinson, K.J., and McRae, B.H. (1969). *Aust. J. Agric. Res.* **20**, 513–21.

Ingram, D.L., and Whittow, G.C. (1962). *J. Physiol.* **163**, 211–17.

Ittner, N.R., and Kelly, C.F. (1951). *J. Anim. Sci.* **10**, 184–94.

Ittner, N.R., Kelly, C.F., and Guilbert, H.R. (1951). *J. Anim. Sci.* **10**, 742–51.

Johnson, H.D., Ragsdale, A.C., Yeck, R.G., and Jones, Joan F. (1960). Mo. Agr. Exp. Sta. Res. Bull. No. 786.

Johnson, H.D., Ragsdale, A.C., Berry, I.L., and Shanklin, M.D. (1963). Mo. Agric. Exp. Sta. Res. Bull. No. 846.

Johnson, H.D, Cheng, C.S., and Ragsdale, S.C. (1958). Mo. Agr. Exp. Sta. Res. Bull. No. 648.

Kamal, T.H., Johnson, H.D., and Ragsdale, A.C. (1962). Mo. Agr. Exp. Res. Bull. No. 785.

Kato, S. and Young, B.A. (1984). *Can. J. Anim. Sci.* **64**, 263–64.

Kay, R.N.B. (1979). ARC Res. Rev. 5, 13–15.

Kay, R.N.B. (1985). In "Recent Advances in Animal Nutrition" (Eds W. Haresign and D.J.A. Cole.) pp. 199–210. (Butterworths: London.)

Kennedy, P.M., and Milligan, L.P. (1978). *Brit. J. Nutr.* **39**, 105–17.

Kennedy, P.M., Christopherson, R.J., and Milligan, L.P. (1982). *Brit. J. Nutr.* **47**, 521–5.

Kennedy, P.M., Christopherson, R.J., and Milligan, L.P. (1986). In "Control of Digestion and Metabolism in Ruminants." (Eds L.P. Milligan, W.L. Grovum, and A. Dobson.) pp. 285–306. (Prentice-Hall: New Jersey.)

Kissileff, H.R., and Ban Italle, T.B. (1985). *Ann. Rev. Nutr.* **2**, 371–418.

Lincoln, G.A. (1979). *J. Endocrinol.* **82**, 135–43.

Lofgreen, G.P., Givens, R.L., Morrison, S.R., and Bond, T.E. (1975). *J. Anim. Sci.* **40**, 223–29.

Malechek, J.C., and Smith, B.M. (1976). *J. Range Manage.* **29**, 9–12

McDaniel, A.H., and Roark, C.B. (1956). *J. Anim. Sci.* **15**, 59–63.

McDowell, R.E. 1972. "Improvement of Livestock Production in Warm Climates." (W.H. Freeman: San Francisco.)

McEwan, E.H., and Whitehead, P.E. (1970). *Can. J. Zool.* **48**, 905–13.

McLean, J., Whitmore, W.T., Young, B.A., and Weingardt, R. (1984). *Can. J. Anim. Sci.* **64**, 641–53.

Milligan, J.D., and Christison, G.I. (1974). *Can. J. Anim. Sci.* **54**, 605–10.

Milne, J.A., MacRae, J.C., Spence, Angela M., and Wilson, S. (1978). *Br. J. Nutr.* **40**, 347–57.

Minson, D.J., and Ternouth, J.H. (1971). *Br. J. Nutr.* **26**, 31–9.

Moe, P.W., and Tyrrell, H.F. (1979). *J. Dairy Sci.* **62**, 447–54.

Moen, A.N. (1973). "Wildlife Ecology." (W.H. Freeman: San Francisco.)

Moen, A.N. (1978). *J. Wildl. Manage.* **42**, 715–38.

Moose, M.G., Ross, C.V., and Pfander, W.H. (1969). *J. Anim. Sci.* **29**, 619.

Morley, J.E., Levine, A.S., Gosnell, B.A., and Billington, C.J. (1984). *Fed. Proc.* **43**, 2903–7.

Morrison, S.R., Lofgreen, G.P., and Givens, R.L. (1974). In "Livestock Environment." Proc. International Livestock Environment Symposium. ASAE, St. Joseph, Michigan.

Morrison, S.R., and Lofgreen, G.P. (1979). *Trans. ASAE* **22**, 861–2.

Nilssen, K.J. and Ringberg, T. (1980). In Proc. 2nd Intl. Reindeer/Caribou Symp. (Eds E. Reimers and S. Skjenneberg.) pp. 329–31. (Roros: Norway.)

NRC. (1981). "Effect of Environment on Nutrient Requirements of Domestic Animals." (National Academy: Washington, DC.)

Olbrich, S.E., Martz, F.A., and Hildebrand, E.S. (1973). *J. Anim. Sci.* **37**, 574–80.

Osuji, P.O. (1974). *J. Range Manage.* **27**, 437–44.

Ozoga, J.J. and Verme, L.J. (1970). *J. Wildl. Manage.* **34**, 431–39.

Payne, W.J.A., Laing, W.I., and Raivoka, E.N. (1951). *Nature*, (London) **167**, 610–1.

Peters, R.R., Chapin, L.T., Emery, R.S., and Tucker, H.A. (1981). *J. Dairy Sci.* **64**, 1671–8.

Peters, R.R., Chapin, L.T., Emery, R.S., and Tucker, H.A. (1980). *J. Anim. Sci.* **51**, 1148–53.

Pontif, J.E., Nipper, W.A., Loycano, A.F., and Braud, H.J. (1974). In "Livestock Environment." Proc. International Livestock Environment Symposium. ASAE, St. Joseph, Michigan.

Porter, W.P., and Gates, D.M. (1969). *Ecol. Monogr.* **39**, 227–44.

Ragsdale, A.C., Thompson, H.J., Worstell, D.M., and Brody, S. (1950). Mo. Agr. Exp. Sta. Res. Bull. No. 460.

Ragsdale, A.C., Worstell, D.M., Thompson, H.J., and Brody, S. (1949). Mo. Agr. Exp. Sta. Res. Bull. No. 449.

Ringberg, T., Jacobsen, E., Ryg, M. and Krog, J. (1978). *Comp. Biochem. Physiol.* **60A**, 123–6.

Robertshaw, D. (Ed.). (1974). "Environmental Physiology." (University Park: Baltimore.)

Ryg, M., and Jacobsen, E. (1982a). *Can. J. Zool.* **60**, 15–23.

Ryg, M., and Jacobsen, E. (1982b). *Can. J. Zool.* **60**, 1562–7.

Schanbacher, B.D., and Crouse, J.D. (1980). *J. Anim. Sci.* **51**, 943–8.

Seath, D.M., and Miller, G.D. (1946). *J. Dairy Sci.* **29**, 199–206.

Senft, R.L., and Rittenhouse, L.R. (1985). *J. Anim. Sci.* **61**, 297–306.

Sharma, D.C., and Kehar, N.D. (1961). *J. Appl. Physiol.* **16**, 611–6.

Sherrington, D.S. (1911). "The Integrative Action of the Nervous System." (Yale Univ.: New Haven.)

Silver, H., Colovos, N.F., Holter, J.B., and Hayes, H.H. (1969). *J. Wildl. Manage.* **33**, 490–8.

Simpson, A.M. (1976). Ph.D. Thesis, University of Aberdeen.

Simpson, A.M., Suttie, J.M., and Kay, R.N.B. (1984). *Anim. Reproduction Sci.* **6**, 291–9.

Suttie, J.M., Kay, R.N.B., and Goodall, E.D. (1984). *Livestock Prod. Sci.* **11**, 529–34.

Tanida, H.L., Swanson, L.V., and Hohenboken, W.D. (1984). *J. Dairy. Sci.* **67**, 585.

Thompson, G.E. (1973). *J. Dairy Res.* **40**, 441–73.

Tucker, H.A. (1985). In "Recent Advances in Animal Nutrition." (Eds W. Haresign and D.J.A. Cole.) pp. 211–21. (Butterworths: London.)

Tucker, H.A., Petitclerc, D., and Zinn, S.A. (1984). *J. Anim. Sci.* 59, 1610–20.

Valtorta, S., Hahn, L., and Johnson, H.D. (1982). *Proc. Soc. Expt. Biol. Med.* **169**, 260–5.

Webster, A.J.F., Chlumecky, J., and Young, B.A. (1970). Can. J. Anim. Sci. **50**, 89–100.

Weston, R.H. (1979a). *Aust. J. Agric. Res.* **30**, 533–41.

Weston, R.H. (1979b). In "Physiological and Environmental Limitations to Wool Growth." (Eds J.L. Black and P.J. Reis.) pp. 163–77. (Univ. of New England: Armidale, Aust.)

Weston, R.H. (1980). *Proc. Nutr. Soc. Aust.* 5, 191.

Weston, R.H. (1982). In "Nutritional Limits to Animal Production from Pastures." (Ed. J.B. Hacker). pp. 183–98 (Comm. Agric. Bur.: Farnham Royal, UK.)

Weston, R.H. (1984). *Can. J. Anim. Sci.* **64** (Suppl.), 324–5.

Westra, R., and Christopherson, R.J. (1976). *Can. J. Anim. Sci.* **56**, 699–708.

Wheeler, J.L., Reardon, T.F., and Lambourne, J.L. (1963). *Aust. J. Agric. Res.* **14**, 364–72.

Wood, A.J., Cowan, I.McT., and Nordan, H.C. (1962). *Can. J. Zool.* **40**, 593–603.

Worstell, D.M., and Brody, S. (1953). Mo. Agr. Exp. Res. Bull. No. 515.
Young, B.A. (1975). *Can. J. Physiol. Pharmacol.* **53**, 947–53.
Young, B.A. (1983). *J. Anim. Sci.* **57**, 1601–7.
Young, B.A. (1985). In "Stress Physiology in Livestock, Vol. II. Ungulates." (Ed. M.K. Yousef.) (CRC Press: Boca Raton, USA.)
Young, B.A. (1987). "Diagrams in Animal Bioenergetics." (Univ. Alberta: Edmonton, Canada.)
Young, B.A. and Corbett, J.L. (1972). *Aust. J. Agric. Res.* **23**, 57–76.

9
ECTOPARASITES AND HERBIVORE NUTRITION

by R.W. Sutherst

There are numerous and diverse examples of arthropod parasites affecting the nutrition of their herbivore hosts. The most significant arthropod parasites of herbivores are ixodid ticks, biting flies such as *Haematobia* (horn flies, buffalo flies), *Glossina* (tsetse flies), *Similium* (black flies), Tabanidae (March flies), Culicidae (mosquitoes), Ceratopogonidae (biting midges), myiasis flies such as Calliphoridae, Oestridae and Gastrophilidae (screw-worm flies, sheep blowflies, bot and warble flies) and lice and mites of various genera. Many of these arthropods not only affect their host directly, but also act as vectors for protozoal, rikettsial, viral or helminth infections. They share the host's habitat and so may rely on the plants it eats or its pelage for shelter or its tissues or dung for breeding. The parasites and their hosts are both subject to the influence of climate which may have effects that favour the domination of one over the other or help maintain a balance between them. Management practices aimed at improving host nutrition can concurrently affect the parasites' survival and conversely, some parasite control measures may affect their host's nutritional status. The outcome of these interactions will determine the overall performance of herbivores under domestication and place constraints on some otherwise attractive management options for improving nutrition or controlling parasites.

The extent to which arthropod parasites influence the nutritional status of their hosts depends partly on their ability to multiply under the conditions that prevail in both their free-living and parasitic phases. The response of different host individuals to parasite challenge also varies and determines whether that host is more or less affected by the parasites than are other members of the herd or flock. This paper reviews the nutrition-related factors that affect the abundance of arthropod parasites and the extent and mechanisms by which these parasites affect the productivity of herbivores. The effects of the nutrition of animals on their arthropod parasites has recently been reviewed by Nelson (1984).

FACTORS THAT AFFECT THE ABUNDANCE OF ARTHROPOD PARASITES

Climate

The effects of climate on parasite populations are complex, as they involve combined responses of host and parasite to local regimens of temperature, moisture and day-length. These responses influence the

NUTRITION OF HERBIVORES
ISBN 0 12 750052 9

size of the parasite populations and the extent to which parasite feeding coincides with seasons that are either favourable or stressful to their hosts. Depending on species, peak arthropod parasite populations may occur at any time of the year.

In temperate zones synchronised cohorts of the damaging adult stages of the economically important species of ticks (*Ixodes ricinus* in Europe, *Haemaphysalis longicornis* in Asia, *Rhipicephalus appendiculatus* in Africa and *Dermacentor andersoni* in N. America) attach to their hosts in spring and summer when the host is growing rapidly. Biting flies are also active in warm wet weather when their hosts are usually growing most rapidly. In the dry tropics most arthropod parasites are active in the wet season but their populations are less synchronised because these parasites have short life cycles. This pattern of feeding subjects the host to a brief but severe attack anytime during the year.

Table 1. Mean number of females of *Boophilus microplus* engorging on *Bos indicus* x *B. taurus* cattle at 3 latitudes in Australia. (From Sutherst *et al.* 1979; Johnston *et al.* 1981 and Mahoney *et al.* 1981).

Latitude	Quarter of Calendar Year			
	1	2	3	4
19°S	12	30	30	35
23°S	37	58	15	49
27°S	13	44	7	1
Comments	Hot and wet	Warm: pasture senescent	Cool and dry	Warm and dry with storms

The case of parasites with multiple generations is well illustrated by cattle ticks of the genus *Boophilus*, for which data on incidence are available for a range of habitats (Table 1). Seasonal incidence in the dry tropics may vary little with a peak at the end of the dry season caused by severe nutritional stress of the host and a smaller one at the end of the rains following the period of successful propagation of the ticks and lactation-stress of female hosts (Table 1). There is a trend away from spring to autumn periods of peak incidence as we move away from the equator. In subtropical and temperate areas the spring population is the smallest of the 3 or 4 annual generations and the vast majority of ticks occur in autumn (Table 1). Near the Tropic of Capricorn in Australia the intermediate pattern is seen with moderate spring and larger autumn peaks. In wet tropical habitats no seasonal activity patterns are discernible.

Lice and a few tick species, such as *Margaropus winthemi* and

Dermacentor albipictus, feed in winter and cause problems in late winter or early spring. Poor nutritional condition of hosts at this time delays the shedding of the winter coat with its load of lice and so exacerbates the lice problem (Murray 1957). The tick *D. albipictus* causes premature hair loss by stimulating intense grooming by its moose (*Alces alces*) host in Canada, leading to the phenomenon of "ghost moose" (Samuel and Barker 1979). Premature loss of hair leaves the animals vulnerable to the cold and often leads to their death through exposure or predation.

These examples from the tropics and temperate areas illustrate how parasites may occur in any season provided there is a reasonably even summer rainfall in the latter zone. The arid and mediterranean environments are less conducive to arthropod parasite problems.

Vegetation types

The freeliving phases of many parasites depend on vegetation, to create favourable microclimates for their development or survival whilst awaiting a host. As most domesticated herbivores are grazers, it is usually pastures that create these microclimates. Pasture plants affect ticks both by providing shelter from the heat or by creating a moist matt at the soil surface and by providing stems of varying favourability for the ticks to ascend in search of a host. In addition, the amount of tree cover is an important factor determining the incidence of some major pests such as screw-worm fly (*Chrysomya bezziana* and *Cochliomyia hominivorax*) and "bot" fly (*Dermatobia hominis*). Widespread planting of forage trees such as *Leucaena* spp. in tropical pastures may therefore exacerbate these very severe problems. They will also moderate soil surface temperatures in grass pastures in hot tropical areas to the advantage of parasites such as ticks.

Improved pastures increase dung nitrogen levels which are advantageous to dung-breeding arthropod pests such as *Haematobia* and *Musca* spp. (Macqueen *et al.* 1986). There is also increasing evidence that even the type of pasture grass can determine whether a tick problem is mild or severe.

The temperate species of ixodid ticks *I. ricinus* in Britain and *H. longicornis* in Asia both rely heavily on 'rough' pastures with high levels of moisture retained in the matt (Milne 1944, Mutch 1966). Traditional methods for controlling these species have relied on drainage and removal of local areas of unpalatable plants such as *Juncus* spp. However, Milne (1948) concluded that improvement of hill pasture was not economic and that treatment of sheep with acaricides was more likely to be acceptable. Recent neglect of bracken control in pastures in the north of England and in Scotland has apparently led to a resurgence of *I. ricinus* (Dobson and Hudson 1986). This tick successfully multiplies on sheep but it also attacks grouse and transmits louping ill virus. Recent declines in the population of these game birds in some areas have been attributed to a high incidence of louping ill virus (Phillips and Duncan 1978).

In the tropics native pastures tend to provide sparse ground cover which is replaced by a denser sward when improved grasses such as green panic (*Panicum maximum*), buffel (*Cenchrus ciliaris*), signal grass (*Brachiaria decumbens*) and legumes are introduced. Emphasis on understanding the effects of different pasture types on tick propagation has characterised the ecological research carried out on ticks in South America. This work arose from the reported rarity of ticks on cattle grazing molasses grass (*Melinis minutiflora*) in the Philippines (de Jesus 1934). Thompson *et al.* (1978) showed that larvae of *Boophilus microplus* survived for shorter periods on *M. minutiflora*, *Pennisteum clandestinum* or *Cynodon dactylon* than on *B. decumbens* or *Hyparrhenia rufa* or *Andropogon gayanus*. Aycardi *et al.* (1984) later showed that cattle grazing *M. minutiflora* or native Colombian pastures carried fewer ticks than animals on *B. decumbens* or *A. gayanus*. As there were only four animals on each pasture, these results need confirmation because variability in host resistance to ticks and can readily exceed that arising from pasture differences when such small experimental units are involved. The relatively low productivity and stress tolerance of *M. minutiflora* has discouraged its widespread use in tick control.

An alternative group of pasture plants that show promise for exploitation, both as a source of soil nitrogen and dry season nutrients for cattle and as a deterrent to tick attack, are the perennial tropical legumes *Stylosanthes* spp. (stylos) (Sutherst *et al.* 1982). Glandular trichomes and hairs on the stems of *S. viscosa* and *S. scabra* trap the tick larvae (Zimmerman *et al.* 1984). A survey of the tick-trapping ability of other members of the genus revealed several other promising species and cultivars (Sutherst and Wilson 1986). Laboratory studies have shown that tick larvae are repelled from ascending the stems of *Stylosanthes* spp. (Norval *et al.* 1983), but in the field this effect is only found in very young plants (Wilson and Sutherst 1986). Nevertheless, there has not yet been a demonstration that these legumes actually reduce tick numbers on cattle. Sutherst and Wilson (1986) were cautious about the prospects of success, mainly because the thin main stems of the plants provided much less opportunity for larvae to come into contact with them than with the broader tufts of grass in mixed grass-legume pastures. Exudates from *Stylosanthes* spp. are toxic to ticks, containing the volatile, insecticidal and acaricidal compounds α- and β- pinene, and are likely to have a fumigant effect. Pure *Stylosanthes* pastures of any variety may greatly suppress tick propagation because they provide no humid refuge at the soil surface for the development of tick eggs. In fact, cattle grazing *S. humilis* and *S. hamata* pastures in northern Queensland were observed to be virtually tick free in autumn (Sutherst unpublished), when tick numbers normally are expected to be high in that environment (Table 1).

Any pasture improvement that enhances the nutritional status of herbivores, particularly in the dry or cold seasons, will assist the animals to withstand parasite attack. However, as the observations on the pasture types above show, the concurrent improvement in the microclimate for tick development or dung nitrogen for fly development may more than offset such an advantage.

In summary, the competing needs for improving host nutrition and for parasite control rarely allow simple solutions to either problem, based on pasture plants *per se*. The greater potential return from pasture plants as sources of nutrients means that parasite control will be a secondary consideration. In general, any change of vegetation, that increases the density of a pasture sward or provides shade, will assist propagation of ticks and myiasis flies, although this may be offset by the enhancement of host resistance that results from improved levels of nutrition (Sutherst *et al.* 1983a). Benefits of improved nutrition will only be expressed in reduced parasitism when there is adequate genetic potential in the host population to express a useful level of host resistance.

Grazing management

Management of pastures, to conserve fodder for the dry or cold seasons or to control parasites, often involves changes in host density or movement. In addition, the traditional and widespread practice of burning of pastures in the tropics and subtropics has consequences both for host nutrition and for parasite control.

Traditional systems of pasture rotation have involved both short and long distance animal movements. In southern Africa, low lying areas and river flats ('dambos') are often grazed in the dry season as the standing water recedes and cattle graze the savanna the rest of the year (Pegram *et al.* 1986). Alternatively, these 'dambos' may be exploited in the wet season when higher ground is committed to growing of crops. In the dry season the livestock survive on crop wastes whilst the 'dambos' are left bare and inhospitable to any ticks remaining there.

Interest in rotational grazing as a means of controlling ticks (Wilkinson 1964; Rudder *et al.* 1976) and helminths (Morley and Donald 1980) has declined in recent years. These practices have usually found little favour because of the paucity of data demonstrating clear increases in productivity in the face of major management demands. A contributing reason for such negative views, I believe, has been the lack of ecological inputs into the design of rotation programs. The phenology of pasture plants and the dynamics of parasite populations must be matched if there are to be overall benefits. For example, whilst regular rotations have been rejected as impractical for control of the cattle tick in Australia (Wharton and Norris 1980), there appears to be scope for single spelling in summer or autumn in specific situations. When integrated with the management of pasture legumes such as siratro (*Macroptilium atropurpureum*), autumn spelling can both control ticks (*B. microplus*) prior to the period of heaviest infestation (Table 1) and increase seeding to conserve the legume (Jones *et al.* 1983). Intermittent grazing of *Leucaena leucocephala*, as advocated by R.J. Jones (Sutherst 1983) is also likely to reduce tick populations as a secondary benefit.

In some situations pasture spelling can be very dangerous to livestock. Some losses of cattle due to excessive infestation with the tick *H. longicornis* in Australia result from spelling of pastures over winter. Adults of this species can survive for at least a year in the pasture and cattle introduced into spelled pasture in spring stimulate the hungry ticks to attach in unison. Cattle have been exsanguinated in 4–6 days and occasionally herds have been decimated in northern New South Wales (Dicker 1978).

Long distance migrations of wild herbivores are legendary, as those animals follow seasonal supplies of feed. Animal migrations of hundreds of kilometres are made by caribou (*Rangifer tarandus*) in the Arctic, and by wildebeest (*Connochaetes taurinus*) in tropical Africa (e.g. Cloudsley-Thompson 1978). The migrations of moose (*Alces alces*) up and down mountainous areas in Canada and of antelope such as eland (*Taurotragus oryx*) in Africa appear to be in search of supplies of specialised plant foods, but some of these movements may occur in response to heavy attack by ticks as discussed below. These natural migrations probably result in reduced but highly synchronised parasite attack.

Nomadic tribes have emulated these natural migrations with annual transhumance movements which can also cover many hundreds of kilometres. In both the Sudan (Asher 1985) and in Nigeria (Strickland 1961) these migrations are long distance and involve large numbers of livestock. Attempts to avoid the need to migrate, by providing dams and wells, have resulted in severe degradation of the African landscape both from wild animals such as elephant (Parker and Amin 1983) and from livestock. One consequence of such long distance movements of livestock in search of food has been the dissemination of some very severe parasites such as the old world screw-worm fly (*C. bezziana*) in South Africa (Baker *et al.* 1968).

One of the best examples of a stable, integrated program of pasture rotation to achieve the dual objectives of continuous feed supplies and the control of parasites was that of the Masai in Tanzania (Branagan 1974). Cattle grazed the Ngorongoro crater in the dry season and moved down onto the surrounding arid plains at the onset of the wet season. Movements were carefully timed to avoid contact between cattle and both ticks and helminths in the crater, which are aroused from dormancy by rain. The wild herbivores were temporarily deterred from grazing the plains by fencing-off of the waterholes with thorn bushes prior to reintroduction of the cattle. The return to the crater also involved a complex ritual to reduce the numbers of the tick *Rhipicephalus appendiculatus*, the vector of *Theileria parva* which causes the severe disease known as East Coast Fever. The *Themeda triandra* pastures in the crater were first burnt to kill ticks. Grazing by sheep and goats then harvested any surviving ticks. The burning killed ticks and also controlled the unpalatable grass *Pennisetum schimperi* which formed dense tussocks favourable for tick survival. Only after these procedures were completed were cattle reintroduced. Unfortunately, that elegant system was lost when game conservation measures were imposed and the disruption of the system led to devastating outbreaks of East Coast Fever.

Both animal density and animal movements can be exploited to control parasites. Firstly, the deliberate overgrazing of rangelands to reduce pasture cover has been carried out, again by the Masai in East Africa. The lack of cover almost eliminates those species of ticks with a 3-host type of life cycle from traditionally grazed areas of Africa (Norval 1977), whilst neighbouring commercial ranches are heavily infested. This effect occurs because the host-seeking nymphs and adults of 3-host ticks need more shelter than do larvae. The more subtle effects of small changes in stocking rates are complicated by concurrent changes in pasture cover, host finding rates of the ticks and host resistance resulting from different nutritional status of the host population. In Australia, bullocks are finished for sale on different country which may affect stocking rates, host resistance and tick survival which together probably explain the lower infestations of the tick *B. microplus* found on them compared with other classes of cattle. Such observations cannot be generalised to ticks with a 3-host type of life cycle, which have long-lived nymphal and adult free-living stages that benefit from the greater pasture cover that prevails under reduced stocking pressure.

Secondly, an alternative to rotational grazing or changes in stocking densities of livestock is to graze the pasture with alternating host species. This practice has been explored with sheep and cattle or horse rotations to control helminths (Morley and Donald 1980) with mixed success because the helminths are not all host specific. Similar alternation of sheep and cattle has been reported to virtually eliminate the tick *B. microplus* from pastures in Brazil (D.E. Evans unpublished).

Burning of pastures late in the dry season is still a regular practice throughout the tropics. The practice is aimed at removing old material and stimulating new growth but it has the potential both to burn ticks on the grass and to denude the ground and so remove the shelter that the ticks require. All the published evidence suggests that the benefits of fire are marginal because ticks shelter beneath stones and other debris (Hoch *et al.* 1972). However, the increase in infestations of Canadian populations of the tick *D. albipictus* on moose during the present century may have arisen from the strict control of fires (W.M. Samuel unpublished). The consequences of burning in large areas are sometimes complicated by the subsequent movement of large populations of herbivores onto the burnt area in search of new plant growth. In Africa, the congregation of game animals, which drop their ticks in these burnt areas, negates much of the benefit that accrues from killing ticks on the pasture (Minshull and Norval 1982).

Pasture fouling or contamination

Pastures intensively grazed at stocking densities of 25 cattle /ha become fouled with animal dung which requires mechanical dispersal. The need for such dispersal has been obviated in parts of California following the introduction of African dung beetles for control of horn flies (Legner 1985). Similar claims have been made by stockowners in areas of Australia with a mediterranean type of climate, following the release of African beetles in an attempt to control breeding of bush flies (*Musca vetustissima*). In northern Australia, almost all the burial of cattle dung is done by introduced African beetles. Many biting flies breed in cattle dung and their activities can accelerate the release of nutrients such as ammonia into the atmosphere (Harris *et al.* 1980). Dung beetles, which are used in the biological control of dung-breeding flies, reduce the rate of loss of ammonia and also incorporate dung nutrients into their tissues (Macqueen and Beirne 1975; Harris *et al.* 1980). The benefits for animal nutrition of dung burial under normal grazing conditions were questioned by McKinney and Morley (1975), on the grounds that it would have negligible impact because of the small proportion of the surface area affected. These authors appear to have overlooked two points in their argument. Firstly, a substantial proportion (two fifths in the case reported by Robbins (1984)) of nitrogen turnover in grazed pastures is recycled in the form of animal dung. Secondly, the importance of dung recycling in the soil nutrient balance will depend more on the proportion of the available herbage that is consumed than on the surface area affected. A more comprehensive understanding of energy and mineral balances in pasture systems is needed before the impact of arthropods on the recycling of dung nutrients can be evaluated.

A different problem arises from contamination of pasture by parasites. Cattle have been shown to be able to detect and avoid areas of pasture that carry high densities of larvae of the cattle tick, *B. microplus* in Australia (Sutherst *et al.* 1986). They are so adept at this behaviour that it limits the rate of growth of tick populations. As engorged female ticks detach from their host at dawn, they are concentrated in the vicinity of campsites. Such concentrations may contribute to the regular re-establishment of campsites by cattle, but as most of the tick larvae only live for a few weeks, the proportion of the pasture affected may be too small to affect feed intake. Sheep display a similar pattern of selective grazing, avoiding concentrations of helminths (Southcott *et al.* 1970; Cabaret *et al.* 1986).

In summary, there is almost total overlap in the effects of pasture improvement and pasture management on both the productivity of herbivores and on their parasites. Compatible management systems are rare and rely on detailed ecological knowledge of the whole system. Further, parasite control and pasture management are usually competing interests and the 'opportunity costs' of diverting funds into either activity need to be calculated. If maximum benefits are to be obtained and adverse consequences avoided, herbivore nutritionists, agronomists and parasitologists need to communicate more closely with each other.

Host nutrition
Herbivores mount an immune response to parasite attack that stimulates defensive behaviour against biting flies (Haufe 1986) and prevents ticks from feeding to maturity (Wagland 1978). The immune response to arthropod parasites is often affected by the nutritional status of the host. Milne (1947) reported that lean sheep carried 50% more *I. ricinus* ticks than did fat sheep in England, while starvation reduced resistance of cattle to *Boophilus annulatus* (Gladney et al. 1973), *B. microplus* (O'Kelly and Seifert 1969) and *Damalinia bovis* lice (Utech *et al.* 1969).

Resistance to tick feeding has since been shown to be very sensitive to the nutritional status of cattle (Sutherst *et al.* 1983a), independent of their breed. Those authors found that the tick resistance of cattle increased when a high grade supplement of lucerne was fed, even when they were still gaining weight in autumn in southern Queensland. They attributed the response to the special physiological demands being made on the cattle to adapt to the oncoming winter. A second finding was that cattle on native pasture recovered much of their resistance in late winter at a time when the resistance of supplemented cattle was exhibiting its maximum response to the lucerne. It was concluded that recovery of resistance occurred spontaneously when adaptation to winter was complete and that the supplement accelerated and enhanced the recovery. These seasonal changes in resistance of cattle need further study. A third finding of the study was the very prolonged period during which the supplemented group maintained its advantage in terms of host resistance after the supplement was stopped in the spring. They allowed only half as many ticks to survive on them for the following 6 months and 25% fewer in the following autumn and winter (Table 2). These results have obvious relevance to the field, where seasons vary from year to year, as particularly stressful dry seasons are likely to leave the cattle prone to tick attack throughout the following wet season. They will be especially severe in weaners which tend to have lower tick resistance levels than calves or older cattle (Sutherst *et al.* 1979). In the dry tropics the loss of resistance during the dry season can be very severe (Sutherst unpublished) and is followed by a slow recovery. Even in the wet tropics at Darwin, responses to nutritional supplements have been observed during short dry periods (Sutherst *et al.* unpublished).

Quantitative data on the effect of nutritional stress on resistance to biting flies have not been published, but anecdotal reports claim that fly numbers are higher on the less well nourished animals in a herd. On the other hand, Vale (1981) showed that tsetse fly (*Glossina* spp) were more attracted to cattle on a high plane of nutrition.

The obvious sensitivity of tick resistance of cattle to nutritional stress raises the question of the geographical incidence of such stress. McCown (1981) showed that the degree of nutritional stress of cattle in the tropics was related to the length of the dry season. Using the climate matching program, CLIMEX (Sutherst and

Table 2. The proportion of larvae of *B. microplus* maturing to adults on *Bos indicus* x *B. taurus* steers on different nutritional regimens in the subtropics. The lucerne treatment group was transferred to native pasture in October 1978.

Treatment	'Winter' 1978	'Spring' 1978	'Summer-Autumn' 1979	'Autumn-Winter' 1979
(1) Native pasture	0.17	0.05	0.03	0.11
(2) Native pasture and lucerne	0.05	0.02	0.02	0.03
(3) Lucerne/native	0.07	0.03	–	–
pasture	–	–	0.02	0.06

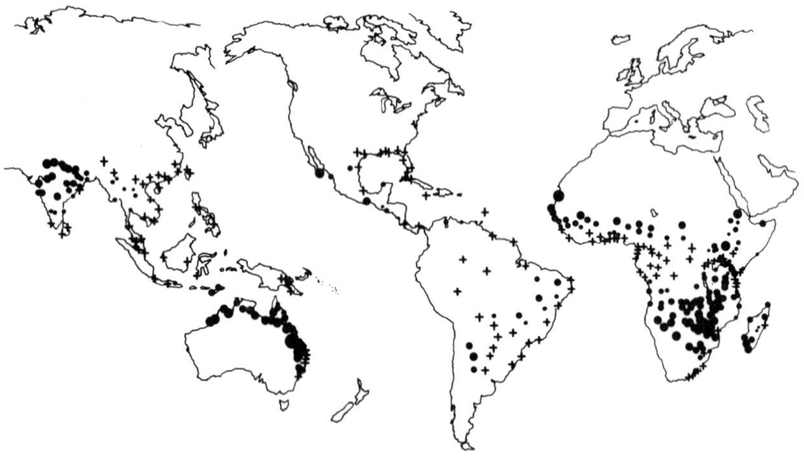

Figure 1. The predicted relative severity of nutritional stress in environments in which tick infestations are likely to occur on domestic herbivores. Predicted stress is proportional to the areas of the circles; locations for which meteorological data was used but for which the stress was zero are marked with a cross.

Maywald 1985), it is possible to compare the extent of climate induced nutritional stress in different cattle raising areas around the world, where ticks are a problem (Fig. 1). These figures need modification in areas where good quality standing hay provides adequate nutrition in the dry season. Knowledge of the role of nutritional stress in tick parasitism has enabled interpretation of the seasonal severity of tick infestations on cattle in countries such as Colombia by using the CLIMEX model.

Attempts to measure responses to a wide variety of supplements or feed additives have had different results. Utech and Wharton (1972) found no response of tick resistance to a supplement of sulphur, perhaps because the control group was fed lucerne which itself has a high sulphur content. Sutherst *et al.* (1983b) found that neither a multi-vitamin regimen nor a maintenance ration of molasses and urea prevented a loss of resistance. O'Kelly (1984) observed that host resistance to *B. microplus* was reduced in cattle supplemented with whole cottonseed which contained 20% fat and produced hyperlipidaemia in the cattle. Nelson *et al.* (1968) detected an increase in the resistance to the tick *Dermacentor andersoni* in vitamin A deficient sheep when given vitamin A supplement. When such a supplement was fed to deficient sheep it also increased their resistance to sheep ked (*Melophagus ovinus*) (Nelson and Hironaka 1966) and to warble fly (*Hypoderma lineatum*) (Gingrich and Barrett 1975).

These results highlight the interaction between the nutrition of herbivores and their ability both to limit parasite population growth and to withstand parasite attack.

THE EFFECTS OF ARTHROPOD PARASITES ON THE NUTRITIONAL STATUS OF HERBIVORES

Arthropod parasites can affect their hosts in three quite different ways;

1. they can so irritate the host that its behaviour is modified, leading to reduced grazing and resting, and so influence feed intake or conversion efficiency; they do this by stimulating defensive immune responses in the host which may limit the parasites' reproduction rate, but at the same time cause the host to divert energy away from its own growth and reproduction;
2. the arthropod parasites remove nutrients from the blood or other epidermal tissues on which they feed; such losses are more serious in hosts that have an immune response that is inadequate to prevent massive numbers of large parasites from feeding on them;
3. they secrete or excrete toxins that disrupt their hosts' metabolism; and
4. the mere presence of arthropod parasites, or of the diseases they transmit, can force livestock owners to protect their animals from exposure to attack by confining them in various ways. These practices deprive the animals of grazing time and so reduce feed intake.

Examples will be taken mainly from a cattle tick (*B.*

microplus), two species of biting fly (*Haematobia irritans irritans*, *Stomoxys calcitrans*) and a tick-borne protozoan (*Theileria parva*) that causes both chronic and acute effects on its host, under the syndrome known as East Coast Fever.

Cattle tick (*B. microplus*)

Extensive studies have been carried out into both the effect of ticks on the growth of cattle (Sutherst and Utech 1981) and on the mechanisms by which the ticks affect the metabolism of their host (Seebeck *et al.* 1971; O'Kelly *et al.* 1971; Springell *et al.* 1971). Ticks remove blood during feeding, at the rate of 0.3–0.5 ml per engorged female tick (Seifert *et al.* 1968), and also secrete toxins which interfere with their host's feed intake and utilisation.

Table 3. Mean daily feed intake and liveweight gain (LWG; kg /day) of tick infested and control groups of steers during a 77-day period of tick infestation, and their mean daily LWG during variable post infestation periods until slaughter (Seebeck *et al.* 1971).

Experimental group	Feed intake	LWG during infestation	LWG post infestation
Tick-free *ad lib.* feed	5.66	0.97	0.53
Tick infested *ad lib.* feed	4.73	0.44	0.63
Tick-free pair-fed to match intake of infested group	4.68	0.64	0.89

Estimates of reductions in liveweight gain of cattle caused by each engorging tick average about 0.6–1.5 g (Sutherst and Utech 1981; Sutherst *et al.* 1983a), of which two thirds is caused by reduced feed intake ('anorectic effect') and one third by the 'specific effect' of the ticks on host metabolism and nutrient loss (Seebeck *et al.* 1971). Tick feeding depresses the levels of many blood components including serum albumin, cholesterol, haematocrit, haemoglobin and several enzymes. The amount of albumin, haemoglobin and cholesterol lost by the steers roughly corresponds to the amounts withdrawn by ticks (Springell *et al.* 1971). The parasite also elevates serum globulins, probably by stimulating the animal's immune system. The effects of tick infestation on the feed intake and growth of steers are shown in Table 3, together with the relative growth rates of infested and tick-free groups during a tick-free period following the infestations. Following the infestation period, the tick-free group, which was pair-fed with a ration that matched the intake of the infested group, grew much more rapidly than either the previously tick-free or infested groups that had been fed *ad libitum*. There was limited compensatory growth of the tick-infested animals after the infestations were stopped and much of the increase in weight was attributed to water accretion (Springell *et al*. 1971). The results indicate the severe immediate and delayed effects of ticks on the metabolism of cattle.

Biting flies

The horn fly (*H. irritans irritans*) is a small but prolific biting fly in the Northern Hemisphere that spends its whole adult life on cattle. It is rated as the most economically important arthropod parasite of cattle in North America (Drummond *et al*. 1981). Its close relative, the 'buffalo fly', *H. irritans exigua*, infests cattle across northern Australia and south-east Asia. The stable fly (*Stomoxys calcitrans*) is a cosmopolitan pest of livestock, particularly when held under intensive conditions such as in stables or feedlots. Populations are usually much lower than are those of *Haematobia* spp. but the fly has a painful bite.

Technical difficulties have led to numerous inconclusive attempts to measure production losses caused by both *Haematobia* and *Stomoxys*, but some recent studies have shed light on such losses and on the mechanisms involved. The most impressive growth responses of cattle to season-long protection against horn fly with insecticide were reported by Haufe (1979, 1986). In a longterm study annual losses of at least 16% in growth were reported. Responses of 35-45% loss of potential gain were recorded in years of heavy fly infestation (500 /animal). It is not clear from these studies whether the insecticide protected the cattle against any other biting insects that may have contributed to these losses. Horn fly infestations often exceed 1000 /animal in climatically suitable environments and they cause intense irritation by feeding about every half hour. Their activities affect the host's vital functions and feed utilisation efficiency (Tables 4 and 5). In addition, they disrupt normal patterns of social and grazing behaviour (Harvey and Launchbaugh 1982) and so probably reduce feed intake under grazing conditions, although intake of concentrates was unaffected in the above trials. Heavily infested cattle walk and rest more than cattle protected with insecticide and also masticate each bolus fewer times and for shorter periods. In the Middle East, infestations of a related species, *Lyperosia minuta* (khareef fly), are claimed to be so severe that cattle are unable to graze during the day in the wet season (Mellor 1978). These flies do, however, also feed at night so that nocturnal grazing will also be disturbed.

Other biting flies, such as mosquitoes, have no effect on feed intake of animals in fly proof cages, but reduce growth rates of cattle fed on low-energy rations based on rice straw (Steelman *et al*. 1972).

The positive correlation between cattle growth rates and burdens of *Haematobia* spp. on individual animals (Haufe 1974; Holroyd *et al*. 1984) supports the view that blood losses are relatively unimportant and that it is the energetically expensive and stressful defensive reactions of the less tolerant animals that reduce their

Table 4. The effect of horn fly (*Haematobia irritans irritans*) and stablefly (*Stomoxys calcitrans*) on the daily feed intake, daily liveweight gain (LWG) and feed-to-gain conversion ratio of young cattle infested in fly-proof enclosures (from Campbell *et al.* 1977; Kinzer *et al.* 1977).

Flies/animal	Daily feed intake (kg)	Daily LWG (kg)	Conversion ratio (kg feed/kg LWG)
Haematobia irritans irritans			
0	6.09	0.75	8.13
1300	6.28	0.59	10.72
Stomoxys calcitians			
Growing ration			
0	6.87	0.68a	10.11
50	6.84	0.59b	11.60
Finishing ration			
0	7.70	1.10a	7.00
100	6.90	0.88b	7.84

Within pairs values not followed by a different letter do not differ significantly ($P > 0.05$).

growth rates. Unfortunately, the thriving tolerant animals enable the fly populations to grow more rapidly than they could on more defensive and irritable animals and so are responsible for the large challenges on the defensive individuals.

The lack of an anorectic effect of biting flies on their host contrasts with the large anorectic response of cattle to cattle tick feeding. It appears that the response to ticks is caused by metabolic disturbance induced by toxins whilst the fly effect is caused by the animals' energies being diverted from growth to defensive behaviour.

Vector-borne protozoan infections

There is a surprising lack of data on the effect of tick-borne diseases on the nutritional status of their hosts. Pathological studies have concentrated on blood composition and no studies comparable to those on the arthropod vectors could be found. The usual effect of *Babesia* spp., *Anaplasma* spp. and *Theileria* spp. is anaemia with the associated excretion of nutrients and in acute cases death may occur. Whilst *Babesia* causes acute fever and, in extreme cases 'Redwater', no data on weight losses and subsequent recovery are available. Loss of haemoglobin and albumin occur in the form of proteinuria during acute infection (Rogers 1971; Wright and Goodger 1979). In the terminal stages of babesiosis extensive skeletal muscle degeneration, with its associated loss of nutrients, occurs (Wright *et al.* 1981). *Anaplasma*

Table 5. The effect of horn fly on the physiology and daily nutrient status of yearling steers in fly proof enclosures from Schwinghammer *et al*. (1986).

Flies/ animal	Feed intake (kg)	Nitrogen retained (%)	Water intake (l)	Urine output (kg)	Urine nitrogen (g)	Heart rate beats/min	Respiration rate breaths/min	Rectal temperature (°C)
0	5.61a	52.3a	16.7a	3.83a	24.6a	76.6a	44.6a	38.8a
100	5.56a	41.3b	16.3a	4.25a	31.1ab	89.1b	52.7b	39.0b
500	5.61a	40.1b	24.9b	12.16b	34.7b	101.0c	62.1c	39.1c

Means in the same columns differ significantly if followed by a different letter ($p < 0.05$).

marginale has been shown to cause chronic anaemia (pcv = 20%) which is associated with maximal weight losses of 5–6% 4 weeks after infection (Otim *et al.* 1980). In that experiment compensatory growth returned the weight to normal 10 weeks after infection.

Theileriosis, caused by *Theileria parva* (East Coast Fever) and *T. mutans*, is claimed to be one of the most important factors in the stunting of calf development in Kenya (Moll *et al.* 1986). Infections occur early in life and lead to loss of appetite with reduced or negative weight gains, increased body temperature, chronic anaemia and, in a small proportion of cases, diarrhoea. As the two parasites occur successively under natural conditions in the Mara district, because *T. mutans* has a much higher rate of transmission, they lead to separate episodes of illness and all calves show signs of morbidity at 6 months of age (Moll *et al.* 1986). Clearly the productivity of such animals is severely constrained by these parasites.

CONCLUSIONS

Interactions between herbivore nutrition, arthropod parasites and vector-borne diseases pervade the whole animal production system. The data on the various aspects are uneven in quality and quantity. They nevertheless indicate the importance of pasture type and management and of host nutritional stress in affecting the survival of arthropod parasites. The apparent sensitivity of arthropod parasites to the nutritional status of the host contrasts with many reports on the relative insensitivity of helminths to host nutritional stress. This suggests that host grooming behaviour and local reactions in the skin are more important than systemic factors in regulating arthropod parasite numbers.

The reciprocal effects of parasites on the nutritional status of their hosts are also unevenly documented. Comparison of the limited data indicates that different groups of parasites have different effects. Ticks depress appetite and reduce feed conversion efficiency, biting flies appear mainly to divert energy from growth to defensive behaviour while tick-borne protozoan parasites cause losses of nutrients by altering the host's biochemical and physiological functions.

As always, I am indebted to my colleagues G.F. Maywald and J.D. Kerr for assistance with computing and data analyses. L. Sexton and D.D. Dunn typed the manuscript expeditiously and carefully and several colleagues made helpful comments on the manuscript.

Asher, M. (1985). *New Sci.* **1450**, 9–11.

Aycardi, E., Benavides, E., Garcia, O., Mateus, G., Henao, F., and Zuluaga, F.N. (1984). *Trop. Anim. Health. Prod.* **16**, 78–84.

Baker, J.A.F., McHardy, W.M., Thorburn, J.A., and Thompson, G.E. (1968). *J. S. Afr. Med. Assoc.* **39**, 3-11.

Branagan, D. (1974). In "Tourism in Africa and the Management Related Resources." pp. 67-74. Proc. Sem. Centre African Studies, Univ. Edinburgh, May, 1974.

Cabaret, J., Mangeon, N., and Gruner, L. (1986). *Vet. Parasitol.* **19**, 315-20.

Campbell, J.B., White, R.G., Wright, J.E., Crookshank, R., and Clanton, D.C. (1977). *J. Econ. Entomol.* **70**, 592-4.

Cloudsley-Thompson, J. (1978). "Animal Migration." (G.P.Putnam and Sons: New York.)

de Jesus, Z. (1934). *Philipp. J. Anim. Ind.* **1**, 193-207.

Dicker, R.W. (1978). *Agric. Gaz. NSW.* **89**(4), 10-2.

Dobson, H.P., and Hudson, P.J. (1986). *Tree* **1**, 11-5.

Drummond, R.O., Lambert, G., Smalley, H.E., and Terrill, C.E. (1981). In "CRC Handbook of Pest Management in Agriculture." (Ed. D. Pimentel.) Vol. 1. pp.111-27. (CRC Press: Boca Raton, Fla.)

Gingrich, R.E., and Barrett, C.C. (1975). *J. Med. Entomol.* **12**, 13-5.

Gladney, W.J., Graham, O.H., Trevino, J.L., and Ernst, S.E. (1973). J. *Med. Entomol.* **10**, 123-30.

Harris, R.L., Ilcken, E.H., Blume, R.R., and Oehler, D.D. (1980). *Southwest Entomol.* **5**, 104-6.

Harvey, T.L., and Launchbaugh, J.L. (1982). *J. Econ. Entomol.* **75**, 25-7.

Haufe, W.O. (1974). Proc. Int. Livestock Environment Symp. pp. 203-8. (Amer. Soc. Agric. Eng.)

Haufe, W.O. (1979). In "Research Highlights 1978." (Eds G.C.R. Croome and N.B. Holmes.) pp. 61-3. (Agriculture Canada Research Station: Lethbridge, Alberta.)

Haufe, W.O. (1986). In "Parasitology - Quo Vadit?" Proc. 6th Int. Congr. Parasitology. (Ed. M.J. Howell.) pp. 607-14. (Aust. Acad. Sci.: Canberra.)

Hoch, A.L., Semtner, P.J., Barker, R.W., and Hair, J.A. (1972). *J. Med. Entomol.* **9**, 446-51.

Holroyd, R.G., Hirst, D.J., Merrifield, A.W., and Toleman, M.A. (1984). *Aust. J. Agric. Res.* **3**, 595-608.

Johnston, L.A.Y., Haydock, K.P., and Leatch, G. (1981). *Aust. J. Exp. Agric. Anim. Husb.* **21**, 256-67.

Jones, R.M., Tothill, J.C., and Mannetje, L. 't (1983). In "Siratro in south-east Queensland." (Ed. R.F.Brown.) pp. 45-57. (Qld. Dept. Prim. Ind., Brisbane.)

Kinzer, H.G., Houghton, W.E., Reeves, J.M., Kunz, S.E., Wallace, J.D., and Urquhart, N.S. (1984). *Southwest. Entomol.* **9**, 212-7.

Legner, E.F. (1985). In "Biological Control of Muscoid Flies." (Eds R.S. Patterson and D.A. Rutz.) Misc. Publ. Entomol. Soc. America **61**, pp. 120-31.

Macqueen, A., and Beirne, B.P. (1975). *Can. Entomol.* **107**, 1255-64.

Macqueen, A., Wallace M.M.H., and Doube, B.M. (1986). *J. Aust. Entomol. Soc.* **25**, 23–9.

Mahoney, D.F., Wright, I.G., Goodger, B.V., Mirre, G.B., Sutherst, R.W., and Utech, K.B.W. (1981). *Aust. Vet. J.* **57**, 461–9.

McCown, R.L. (1981). *Agric. Syst.* **6**, 303–17.

McKinney, G.T., and Morley, F.H.W. (1975). *J. Appl. Ecol.* **12**, 831–7.

Mellor, P.S. (1978). *Trop. Anim. Health Prod.* **10**, 167–9.

Milne, A. (1944). *Parasitol.* **35**, 186–96.

Milne, A. (1947). *Parasitol.* **38**, 34–50.

Milne, A. (1948). *Ann. Appl. Biol.* **35**, 369–78.

Minshull, J.I., and Norval, R.A.I. (1982). *S. Afr. J. Wildl. Res.* **12**, 118–23.

Moll, G., Lohding, A., Young, A.S., and Leitch, B.L. (1986). *Vet. Parasitol.* **19**, 255–73.

Morley, F.H.W., and Donald, A.D. (1980). *Vet. Parasitol.* **6**, 105–34.

Murray, M.D. (1957). *Aust. J. Zool.* **5**, 183–7.

Mutch, G.V. (1966). *N.Z. J. Agric.* **112**, 62–3.

Nelson, W.A. (1984). *J. Med. Entomol.* **21**: 621–35.

Nelson, W.A., and Hironaka, R. (1966). *Exp. Parasitol.* **18**, 274–80.

Nelson, W.A., Weintraub, J., and Hironaka, R. (1968). *Exp. Parasitol.* **22**, 240–4.

Norval, R.A.I. (1977). *Rhod. Vet. J.* **8**, 33–8.

Norval, R.A.I., Tebele, N., Short, N.J., and Clatworthy, J.N. (1983). *Zimbabwe Vet. J.* **14**, 26–9.

O'Kelly, J.C. (1984). *Vet. Parasitol.* **15**, 151–7.

O'Kelly, J.C., Seebeck, R.M., and Springell, P.H. (1971). *Aust. J. Biol. Sci.* **24**, 381–9

O'Kelly, J.C., and Seifert, G.W. (1969). *Aust. J. Biol. Sci.* **22**, 1497–506.

Otim, C., Wilson, A.J., and Campbell, R.S.F. (1980). *Aust. Vet. J.* **56**, 262–6.

Parker, I., and Amin, M. (1983). "Ivory Crisis." (Chatto and Windus: London.)

Pegram, R.G., Perry, B.D., Musisi, F.L., and Mwanaumo, B. (1986). *Exp. Appl. Acarol.* **2**, 25–45.

Phillips, J., and Duncan,J.S. (1978). *Ann. Appl. Biol.* **88**, 346–7.

Robbins, G.B. (1984). Ph.D. Thesis, Univ. Queensland.

Rogers, R.J. (1971). *Aust. Vet. J.* **47**, 242–7.

Rudder, T.H., Hale, K.B., and Wildin, J.H. (1976). *Brahman News.* No. **33**, 3–16.

Samuel, W.M., and Barker, M.T. (1979). *Proc. N. Am. Moose Conf. Workshop* **15**, 303–48.

Schwinghammer, K.A., Knapp, F.W., Boling, J.A., and Schillop, K.K. (1986). *J. Econ. Entomol.* **79**, 1010–5.

Seebeck, R.M., Springell, P.H., and O'Kelly, J.C. (1971). *Aust. J. Biol. Sci.* **24**, 373–80.

Seifert, G.W., Springell, P.H., and Tatchell, R.J. (1968). *Parasitol.* **58**, 415–30.

Southcott, W.H., Langlands, J.P., and Heath, D.D. (1970). Proc. XI Int. Grassland Congress, Surfers Paradise, Australia, pp. 888-90.

Springell, P.H., O'Kelly, J.C., and Seebeck, R.M. (1971). *Aust. J. Biol. Sci.* **24**, 1033-45.

Steelman, C.D., White, T.W., and Schilling, P.E. (1972). *J. Econ. Entomol.* **65**, 462-6.

Strickland, K.L. (1961). M.Sc. Thesis, Univ. Dublin.

Sutherst, R.W. (1983). In "Tropical Parasitoses and Parasitic Zoonoses." (Ed. J.D. Dunsmore.) pp. 41-56. Proc. 10th Int. Conf. World Assoc. Advance Vet. Parasitol., Perth.

Sutherst, R.W., and Maywald, G.F. (1985). *Agric. Ecosyst. Environ.* **13**, 281-99.

Sutherst, R.W., and Utech, K.B.W. (1981). In "CRC Handbook of Pest Management in Agriculture." (Ed. D. Pimentel.) Vol. II. pp. 385-407. (CRC Press: Boca Raton, Fla.)

Sutherst, R.W., and Wilson, L.J. (1986). In "Insects and the Plant Surface." (Eds B.E. Juniper and T.R.E. Southwood.) pp. 185-194. (Edward Arnold: London.)

Sutherst, R.W., Floyd, R.B., Bourne, A.S., and Dallwitz, M.J. (1986) *Experientia* **42**, 194-6.

Sutherst, R.W., Jones, R.J., and Schnitzerling, H.J. (1982). *Nature* (Lond.) **295**, 320-1.

Sutherst, R.W., Kerr, J.D., Maywald, G.F., and Stegeman, D.A. (1983a). *Aust. J. Agric. Res.* **34**, 329-39.

Sutherst, R.W., Maywald, G.F., Kerr, J.C., and Stegeman, D.A. (1983b). *Aust. J. Agric. Res.* **34**, 317-27.

Sutherst, R.W., Wharton, R.H., Cook, I.M., Sutherland, I.D., and Bourne, A.S. (1979). *Aust. J. Agric. Res.* **30**, 353-8.

Thompson, K.C., Roa, E.J., and Romero, N.T. (1978). *Trop. Anim. Health Prod.* **10**, 179-82.

Utech, K.B.W., and Wharton, R.H. (1972). *Aust. Vet. J.* **48**, 73-4.

Utech, K.B.W., Wharton, R.H., and Wooderson, L.A. (1969). *Aust. Vet. J.* **45**, 414-6.

Vale, G.A. (1981). *Bull. Entomol. Res.* **71**, 259-65.

Wagland, B.M. (1978). *Aust. J. Agric. Res.* **29**, 395-400.

Wharton, R.H., and Norris, K.R. (1980). *Vet. Parasitol.* **6**, 135-64.

Wilkinson, P.R. (1964). *Aust. J. Agric. Res.* **15**, 822-40.

Wilson, L.J., and Sutherst, R.W. (1986). *Aust. J. Entomol. Soc.* **25**, 353-8.

Wright, I.G., and Goodger, B.V. (1979). *Z. Parasitenkd.* **59**, 115-9.

Wright, I.G., McKenna, R.V., and Goodger, B.V. (1981). *Z. Parasitenkd.* **64**, 297-302.

Zimmerman, R.H., Garris, G.I., and Beaver, J.S. (1984). In "Impact of Diseases on Livestock Production in the Tropics." (Eds H.P. Riemann and M.J. Burridge.) pp. 579-88. (Elsevier: Amsterdam.)

10

ENDOPARASITES AND HERBIVORE NUTRITION

by A.R. Sykes

The grazing herbivore is in a state of continuous immunological response to the challenge to the alimentary tract presented by helminth larvae ingested with herbage. This response is weak in the young inexperienced animal and strongest in the mature, experienced and non-productive animal, although the ability to respond may be lost by animals in certain physiological states, such as immediately *post-partum*. The impact of endoparasites on the nutrition of the host is determined by the extent to which the immunological response limits invasion of host tissues. The situation is, however, complicated by recent, and as yet incomplete, evidence that the nutritional status of the host may also affect its ability to mount an immunological response to larval challenge.

This subject was very thoroughly reviewed by Miller (1984). Briefly, a host which has recently experienced larval intake and has developed resistance to and eliminated such infection, may reject ingested larvae as soon as they enter the alimentary tract with herbage. Elimination may also occur after they have established in the gut, but before they reach adulthood (Dineen and Wagland 1966) when their presence would be recognised in egg production in faeces. On the other hand, expulsion of adult worms may occur during the course of several months. In other cases the host response may simply cause stunting of the parasite, a reduction in its fecundity or inhibition of development of infective larvae after they have reached their particular niche in the intestinal mucosa.

Rapid expulsion of larvae by the very immune host is probably effected by combined secretion of mucus glycoprotein and serum-derived γ-globulins (IgG) and by an intestinal mucosa replete with inflammatory cells; the mucosa is probably bathed in parasite-specific IgA and mast cells coated with parasite-specific IgE. This local hypersensitivity reaction is probably followed by a boosting of local production of IgA immunoblasts and of IgA in lymph (Smith *et al.* 1983).

In the inexperienced host, in which parasites develop to maturity, expulsion of the adult worms probably involves the thymus and T-cells and is associated with intestinal epithelial villous atrophy, crypt hyperplasia, changes in brush border biochemistry and cell hyperplasia and, in the lamina propria, increased levels of eosinophils, basophils, mast cells and lymphocytes.

The resistant state, judged by low worm burdens or small numbers of eggs in faeces, can be attained in well nourished sheep by 12 months of age (Gibson and Parfitt 1972) and within 3-4 months of

NUTRITION OF HERBIVORES
ISBN 0 12 750052 9

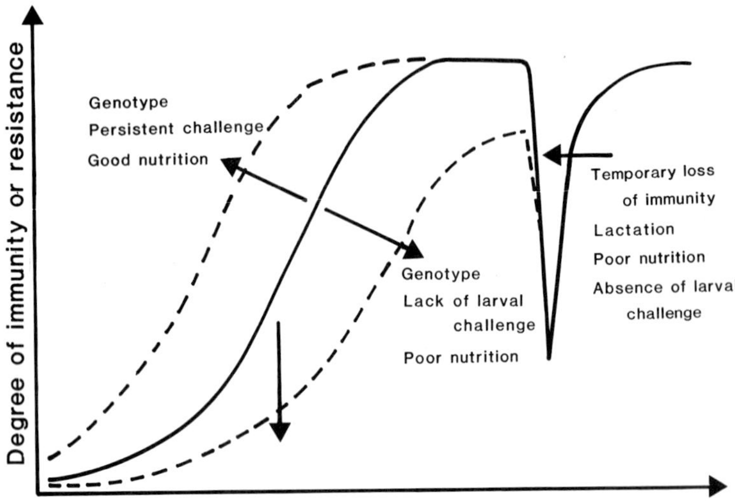

Figure 1. The general factors responsible for variation in host resistance to parasitism.

persistent challenge, although field studies suggest full resistance may not necessarily be reached until 2.5 years of age (Familton, McAnulty and Nicol pers. comm.). Resistance is inevitably lost during the 6 to 8 weeks immediately post-partum (Leyva *et al.* 1982), but is usually subsequently rapidly regained. Continuous stimulation by incoming larvae seems to be important in maintaining the resistant state (Dineen and Wagland 1966; Leyva et al. 1982) which limits damaging contact between the host and parasite. A generalized summary of the change in susceptibility of the host during its lifetime is given in Fig. 1.

Variation between breeds of sheep occurs in worm establishment and in host metabolic response in controlled infections with *Haemonchus contortus* (Altaif and Dargie 1978; Preston and Allonby 1979a, b) which appears to be associated with haemoglobin type. Recognition of such variation, reviewed by Dargie (1982), is extremely important in extrapolating evidence for parasite-nutrition interactions from one environment to another.

This review will be concerned, predominantly, with helminth endoparasite-nutrition interactions in sheep. Nutritional studies have been numerous in this species both for convenience and because sheep appear to be more vulnerable to endo-parasitism than other farmed species such as cattle or deer (Johnston et al. 1984). The recent increase in the breeding of goats has exposed its susceptibility to parasitism but its previous rejection by western societies has led to a paucity of systematic data.

THE NATURE OF NUTRITION-PARASITE INTERACTIONS

The subject of nutrition-parasite interactions is potentially complex and studies must recognise the following possibilities:

1. a direct effect of infection on the absorption, endogenous loss and metabolism of specific nutrients as a direct result of damage to tissues at the site of predilection in the body and perturbation of the endocrine system of the host;
2. direct effects of specific nutrient intake (digesta concentration) on establishment and maintenance of nematode populations within the tract, and
3. effect of current and previous nutrient status on the ability of the host to mount its immune response to limit parasite establishment or effect rejection of an established infection.

A different and specific approach to assessment might be required for each nutrient. The major part of the review will be concerned with section 1 since there are few data on sections 2 and 3.

THE NATURE OF DAMAGE TO THE HOST CAUSED BY ENDO-PARASITES

Pathophysiological studies have concentrated on the primary infection of the young host because it suffers the most severe clinical symptoms of disease.

Damage to the abomasum

Developing larvae invade the gastric glands before emerging to inhabit the mucosal surface as adults. Differentiation of parietal and chief cells is lost and, probably as a consequence, acid and pepsinogen secretion are diminished. Abomasal pH may rise to 5 or 6. Cell proliferation increases (Rowe *et al.* 1982) as do numbers of mucus-producing goblet cells (Murray *et al.* 1970). Blood loss may occur with some species (Dargie 1980).

Damage to the intestine

The proximal 0.2 (3-5m length) from the pylorus is the major site of infection of the small intestine of sheep. Some species such as *Nematodirus battus* invade submucosal tissues before emerging as adults to inhabit the lumen of the tract. Others, for example, *Trichostrongylus colubriformis* and *T.vitrinus* (Barker 1973) adopt a more superficial intraepithelial association and cause extensive villous atrophy and thickening of submucosal layers of the tract. Infections do occur in more distal parts of the tract, for example, *Oesophagostomum* and *Cooperia* spp., but these have received less attention (Bawden 1969a). Proliferation in mucus-producing goblet cells has been observed at sites of infection in the proximal (Coop et al. 1979) and distal (Dobson 1967) intestine.

Impairment of function of epithelial cells has not been established. These cells appear to migrate more quickly towards the apices of villi and become immature and crypt like, with poorly developed brush borders (Symons 1965). The brush-border enzymes alkaline phosphatase, leucine amino peptidase, maltase and glycyl-l-leucine dipeptidase are reduced as are the mucosal activities of the pancreatic enzymes trypsin and chromotrypsin (Jones 1982). The impact on nutrient digestion and metabolism is not known. Knowledge that peptidase activity is present in the mucosa throughout the whole intestine and shows maximal activity at 0.6 of the length (Richardson and Jouan 1986), well beyond the major site of parasite infection, must influence our assessment of the significance of the changes.

Changes in gastrointestinal motility and in the endocrine environment
Stable subclinical infections are generally associated with reduced rate of digesta flow (Gregory 1985; Bown 1986) but there is contrary evidence (Roseby 1977).

The effect on the complex of hormones concerned with local regulation of the gastrointestinal tract is not well understood. Hyposecretion of acid by poorly differentiated parietal cells in the abomasum during infection of that organ occurs in an environment of increased stimulation by gastrin (Anderson *et al.* 1981). There may, however, be a dominant negative stimulus on acid secretion from other gastrointestinal hormones, as yet unidentified, in concurrent infections in the intestine (Barker and Titchen 1982).

Hormones which regulate general metabolism have received little attention. Circulating levels of corticosteroids may increase and of insulin decrease (Pritchard *et al.* 1974), which may reflect either a direct effect of the infection or an adaptation of the metabolism of the host to change in nutrient supply.

NUTRIENT DIGESTION AND TRANSACTIONS IN THE ALIMENTARY TRACT

Protein digestion

General considerations. The previous discussion raises issues about the competence of the alimentary tract for protein digestion and absorption at particular sites of infection. Assessment of the impact of damage on protein digestion has generally been achieved from the net disappearance of N over the whole tract. Such methods are unreliable because unabsorbed protein can be degraded in the caecum and N absorbed as NH_3, thus leading to over-estimation of true protein digestion. Any change in endogenous protein losses into the tract could be masked by the same process.There are now sufficient reliable data on nutrient absorption from specific sites in the digestive tract to allow rejection of data from conventional digestibility trials. The techniques involved and their application to studies of digestion in gastrointestinal nematode infections were discussed in detail by Poppi (1986). A problem for such studies has been the establishment of sufficiently stable infections for the application of nutritional techniques. Only data where this has been achieved by chronic infection will be considered here.

Abomasal infections. There is no quantitative evidence that abomasal protein digestion is impaired despite an increase in abomasal pH. For example, a 50% increase in flow of non-ammonia nitrogen (NAN) from the abomasum was reduced to 15-16% at the ileum, indicating that 1.8 times more NAN disappeared along the intestine in infected than in pair-fed control sheep (Steel 1978). Similarly, Rowe et al. (1982) found that an additional flow of 6 g N/day at the duodenum, as a consequence of infection with *H. contortus*, had disappeared at the ileum. This raises the interesting question as to the importance of acidification of digesta in the abomasum for protein digestion.

Intestinal infections. Increased flow of NAN has been observed at the ileum implying reduced net protein digestion (Steel and Hogan 1972; Poppi et al. 1981). The latter workers, however, showed no difference in the disappearance of ^{35}S-labelled bacteria introduced into the abomasum, confirming the conclusions of Symons and Jones (1970), based on disappearance of casein and ^{14}C-labelled *Chlorella* protein across the infected intestine of both sheep and rodents, that true digestibility and absorption of protein is not affected by infection.

Simultaneous infection of the abomasum and small intestine. Animals in the field are often exposed to simultaneous infections in the abomasum and small intestine which could have more serious consequences than simple infections. Bown *et al.* (1984) recently investigated this with mixed infection with *Ostertagia circumcincta* and *T. colubriformis* using novel cannulation techniques at sites just distal to the major sites of infection (0.2 length of small intestine). They showed, firstly, that 85% of protein digestion and absorption in normal animals occurs distal to the sites of infection, as might have been expected from knowledge of the enzyme capability of the mucosa, *vide supra*. Secondly, infected animals actually absorbed 25-40% more protein in the distal small intestine than did pair-fed controls and, overall, protein digestion and absorption was not impaired.

The conclusion that infection does not affect protein digestion and absorption seems, at this stage, to be sound.

Endogenous losses of protein

The greater rates of digestion of protein across the small intestine in infected sheep reflect the additional endogenous protein losses presented for absorption. Leakage of circulating proteins, particularly of albumin, into the tract may increase 5-fold in infections of the abomasum or small intestine (Holmes 1986). Whether the losses are greatest during the initial phase of infection or during development of resistance is difficult to judge (Steel 1978; Steel *et al.* 1982) and may well depend on size and pattern of infection and the initial immune status of the host. Technical difficulties have prevented direct quantification of the protein losses associated with the increase in sloughing of epithelial cells and production of mucus. Such changes may, in fact, extend to regions distal to the site of infection since Symons (1978) demonstrated increased mitotic rate of crypt cells in the distal ileum and colon of infected rats and Symons and Jones (1983) showed increased fractional synthetic rates of protein in the distal small intestine and in the large intestine of guinea pigs infected only in the proximal intestine.

Two indirect estimates of the additional endogenous protein loss from combined mucus secretion and cell sloughing exist. Poppi *et al.* (1981) assumed that the increase in NAN appearing at the ileum represented a similar unabsorbed proportion of the increase in true endogenous secretions as did ileal ^{35}S-labelled bacterial protein to total ^{35}S-bacteria entering the duodenum of sheep infected in the proximal intestine. More recently Bown (1986) assumed that the increase in ileal flow of NAN reflected the same indigestible fraction of total endogenous protein losses as did appearance of ^{125}I-labelled albumin at the ileum to ^{125}I-labelled albumin introduced in the proximal small intestine. The two methods suggested that mucus and cell sloughing contributed between 4- and 10- fold more endogenous protein than did plasma leakage measured in the same studies.

Energy digestion

The evidence for reduction in digestibility and metabolizability of feed energy is conflicting. Thus in chronic abomasal parasitism induced with *O.circumcincta* larvae, Sykes and Coop (1977) observed a 4% reduction in energy digestibility; this was confirmed by Coop *et al.* (1982). In the latter study there was some evidence for a greater effect around the time that adult worms were first establishing (4 weeks). Rate of infection, ranging from 1000–5000 larvae /d, did not seem to be important.

In infections of the small intestine, using *T.colubriformis* or *T.vitrinus* larvae, Sykes and Coop (1976) and Sykes *et al.* (1979) found no significant effect on energy digestibility. On the other hand MacRae *et al.* (1982) observed reductions in feed metabolizability of 7–10% due to change in digestibility. Very recently, however, Sykes, Poppi and Elliot (unpublished) found no consistent effect on DM digestibility in a large mixed-infection experiment involving 4 levels of *O.circumcincta* (0–3000 larvae /d) and of *T.colubriformis* (0–2500 larvae /d) applied factorially. The question of effect on food energy digestibility is not, therefore, fully resolved.

UTILIZATION OF ABSORBED PROTEIN AND ENERGY

Despite lack of firm evidence for reduced digestion of protein and energy the fact remains that infected animals grow less well and deposit less fat and protein than do uninfected contemporaries consuming the same amount of food (Sykes 1982, 1983). Studies of nutrient utilization have been conducted at three levels:

(a) physiological studies of protein synthetic rates of tissues,

(b) studies of the gross efficiency with which nutrients are deposited in the body using comparative slaughter techniques, and

(c) indirect estimates of efficiency based on calorimetric studies of heat production.

Physiological studies have shown that infected animals may synthesise 50 g protein /d (20%) more than pair-fed controls (Jones and Symons 1982), probably due to increased synthesis of fixed liver and plasma proteins and to support gastrointestinal cell turnover. This protein synthesis, concerned with the integrity of the body, represents an additional maintenance cost which, since these tissues are normally considered to utilize 20% of whole body O_2 consumption (Webster 1983), implies substantial change in whole body metabolism. Quantitative estimates of its magnitude, based on comparative slaughter studies, suggest reductions of up to 50% in gross efficiency of use of digested energy in both abomasal (Sykes and Coop 1977; Coop *et al.* 1982) and particularly intestinal infection (Sykes and Coop 1976; Sykes *et al.* 1979). Sykes (1983) calculated, albeit crudely because of lack of precise information on energy costs of protein synthesis, that the observed increases in endogenous protein secretions, which have been calculated to average about 50 g /d *vide supra*, could explain the additional heat production implied in the reduced gross efficiency of use of digested energy for body energy gain. On the other hand MacRae *et al.* (1982) concluded from indirect calorimetry that efficiency of use of metabolizable energy was unchanged. This latter finding seems plausible only if additional energy expenditure incurred in endogenous protein synthesis was balanced by corresponding reductions in heat production in other tissues.

MINERAL ABSORPTION AND UTILIZATION

There are comparatively few data on the effect of parasitism on the capacity of the alimentary tract for mineral absorption. Progress is severely impeded, moreover, by the lack of a clear understanding of the significance of specific sites in the tract for mineral absorption.

There are no reliable data for the effect on metabolism of trace elements. Very recently Hucker and Yong (1986) observed greater loss of liver Cu in sheep infected with *Trichostrongylus axei* and *T.colubriformis* which they attributed to impaired Cu absorption. It is difficult to judge the significance of these data, however, since pair-fed animals were not used nor were data on feed intake given. Similarly, the possibility of greater endogenous loss cannot be excluded.

There are more data for the major minerals, Ca and P. The nature of control of homeostasis, however, makes it difficult to judge the extent to which impaired capability for mineral absorption *per se* is responsible. Net Ca and P absorption (retention) in ruminants is heavily dependent on Ca demand, especially for skeletal growth or for lactation. Calcium absorption is regulated by effect of Ca demand on plasma ionic Ca concentration (Care *et al.* 1980). Secretion of endogenous P into the tract, particularly in saliva, is involved in P homeostasis (Care et al. 1980). Accretion of Ca and P in the body is clearly diminished in both abomasal and intestinal infections (Sykes and Coop 1976, 1977). Bone histology and chemical pathology, however, have clearly implicated reduced deposition of bone collagen matrix (Sykes 1982) a consequence, presumably, of the diversion of amino acids towards homeostasis in the alimentary tract. Such a reduction in bone matrix deposition will, itself, reduce demand for net absorption of Ca and for net retention of P for bone mineralization. There is, however, evidence for specific reduction in the ability of intestinal tissue to transport P but not Ca in infections of the proximal small intestine induced by *T.colubriformis* (Wilson and Field 1983; Poppi *et al.* 1985). Wilson and Field (1983) also showed increased endogenous losses of Ca and P across the small intestine. Abomasal infection with *O.cirumcincta*, on the other hand, was without effect on Ca or P absorption. These findings confirm the general conclusions about the primary cause of impaired bone growth derived from changes in bone histological and chemical analyses and from changes in plasma mineral concentration, summarized by Sykes (1982). These latter have shown mineral and matrix osteoporosis of bone coupled with hypophosphataemia and normocalcaemia in intestinal infections, suggestive of combined protein and phosphorus deficiency. In contrast, the findings of osteoporosis of bone with normocalcaemia and normophosphataemia in abomasal infections have suggested that, in this case, a simple protein deficiency is involved in impairment of growth and mineralization.

There are few quantitative studies of the metabolism of Na and K. The increased glandular secretions in abomasal parasitism have caused a 3-fold increase in flow of Na from the abomasum but this was completely reabsorbed by the ileum (Wilson and Field 1983). This may not necessarily be the case in acutely affected animals and may contribute to the diarrhoea often seen in such cases.

FOOD INTAKE

Anorexia has been a universal finding in endoparasitism and may be virtually complete in immunologically naive animals exposed to large acute infections. In similar naive animals (Sykes and Coop 1976; Coop *et al.* 1982), or even susceptible adult ewes (Leyva *et al.* 1982, Thomas and Ali 1983) exposed to chronic subclinical infections the reduction has generally been less than 20%. There are few reliable data on the effect of infection on intake of immunologically competent animals although Leyva et al. (1982) found no effect in sheep exposed to *O.circumcincta* during late pregnancy and in which helminth larval development was suppressed.

This effect on feed intake is extremely important and probably has a greater overall effect on animal productivity than change in nutrient digestion or in efficiency of nutrient utilization, simply by its effect in reducing the amount of energy available above maintenance for productive processes. The relative importance of diminished feed intake and reduced efficiency of nutrient utilization is illustrated diagramatically in Fig. 2, which is drawn from unpublished

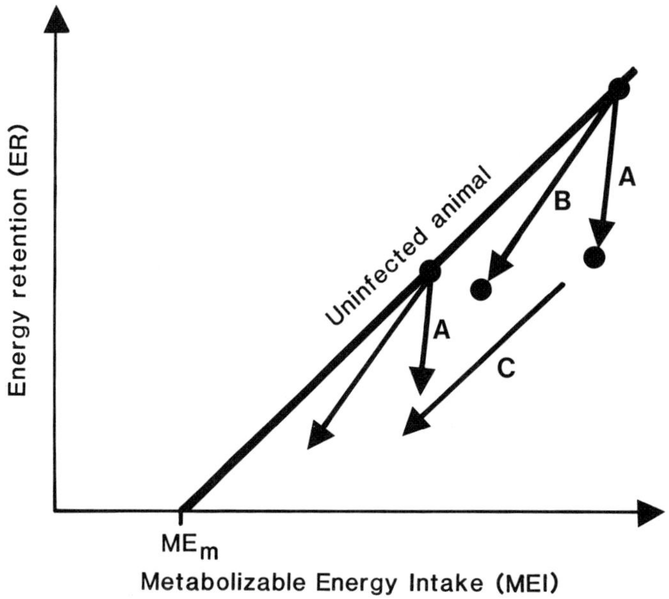

Figure 2. Relationship between metabolizable energy intake (MEI) and energy retention (ER) in uninfected and parasitised sheep. Reduction in ER due to parasitism may occur; as a result, predominantly, of reduction in efficiency of use of ME (A) – typical of mild intestinal infections; or due, predominantly, to reduction in food intake (B) – typical of mild abomasal infection. With increasing level of infection of either site or in mixed infection reduction in feed intake (C) increases in importance. (From Sykes, Poppi and Elliot (unpublished) and other data).

data of Sykes, Poppi and Elliot for sheep exposed to mixed infections and from the data available in the literature (Sykes 1983). Whereas chronic subclinical infection with *T.colubriformis* has tended to reduce energy retention (ER) through reduced efficiency of use of ME, abomasal infections appear to have exerted greater effect on ER through reduction in feed intake. Mixed infections and increasing levels of infection with either species appear to have major additional effects on ER through further depression of intake.

The precise mechanisms through which the anorexia resulting from parasitism occurs is not understood; they are undoubtedly complex. Symons (1985) reviewed the sparse information on this subject very effectively. The virtual complete anorexia of acute infection in naive animals may very well reflect the rapid response of the host to pain at the site of infection (Miller 1979) or changes in gastrointestinal hormone secretions such as cholecystokinin with local or systemic effects. The importance of such phenomena may be less in more chronic 'subclinical' infections and in more resistant hosts. In view of the evidence for change in metabolism of protein, energy and minerals as a consequence of infection, the possible involvement of metabolic factors in appetite depression operating in comparable manner to a simple nutrient deficiency might be anticipated. The situation is probably, however, extremely complex. In infected animals greater flux of amino acid through the liver must be an inevitable consequence of the cycling of protein between the small intestine and the body. The importance of metabolite flow through the liver in the regulation of intake is still, however, in dispute (Forbes 1986; de Jong 1986). Moreover, a reduction in nutrient demand for carcase protein synthesis, consequent upon reduced skeletal growth, may not necessarily be highly correlated with increase in tissue synthesis in the alimentary tract. These changes and their net effect on total energy demand may fluctuate during infection but could be important in determining response of feed intake (Sykes 1982).

THE MATURE ANIMAL

There are few quantitative data on the metabolic costs of infection in the mature, resistant animal. Large, 15–20%, reductions in wool growth in grazing sheep, judged to be resistant to infection on the basis of failure of larval challenge to develop to patency (Barger and Southcott 1975), is evidence for this, although it may simply have reflected reduced intake. Leyva *et al.* (1982), however, found no evidence for any effect on feed intake, weight change or wool production of daily infection with 4000 *O.circumcincta* larvae in mature sheep during late pregnancy. After parturition the same sheep exposed to the same rate of infection suffered a 25% reduction in milk production, a 100% increase in weight loss and a 20% reduction in wool production. Evidence for abomasal damage, in elevated plasma pepsinogen concentrations, was the same in both cases. Similar findings in lactating sheep were reported by Thomas and Ali (1983) who also provided, in hypoalbuminaemia, some evidence that effects on protein metabolism occur in the susceptible adult as well as in the young.

OTHER INTERNAL PARASITES

Parasitism of the respiratory tract by *Dictyocaulus* spp. and of the liver by *Fasciola* spp. reduces livestock performance. With both helminth species hyperglobulinaemia and hypoalbuminaemia feature and in the latter case anaemia (Dargie 1980, 1986). There is exaggerated production of cellular tissue and leakage of serum proteins into tissues, in common with infections in the gastrointestinal tract. There is no evidence for impaired digestive function or for significant change in efficiency of energy utilization in such infections (Sykes *et al.* 1980; Kroonen *et al.* 1986) although overall efficiency of feed utilization for weight gain in both states is reduced by reductions in feed intake. Rate of infection is undoubtedly important in influencing tissue damage and degree of anorexia.

DIRECT AND INDIRECT EFFECTS OF DIETARY NUTRIENT CONCENTRATIONS ON LARVAL ESTABLISHMENT AND ON THE PATHOGENESIS OF INFECTION

Nutrient intake or nutrient density may affect development of larvae directly within the tract, or indirectly through impact on the ability of the host to develop immunity. It may also affect the ability of the host to meet the additional nutrient demands of infection and therefore moderate impairment of metabolic effectiveness, i.e. the pathogenicity of infection. In this latter context the apparently greater resistance of the mature non-productive animal, compared to the growing or lactating animal may reflect, in part at least, a less precarious nutrient status and greater ability to use surplus dietary nutrients or body reserves to meet the nutritional penalty of parasitism.

Anecdotal and some experimental evidence suggests that $CuSO_4$ has anthelmintic properties in the abomasum where acid conditions will ensure dissociation to the toxic Cu ion. Copper sulphate did form the basis of some early anthelmintics.

Other data suggest that low dietary P concentration (1.5 g P /kg DM) may predispose sheep infected with *T.colubriformis* to greater worm burdens than one containing 2.5 g P /kg DM (Coop and Field 1983). Whether the effect occurred directly on the parasite in the tract or indirectly through the immune system of the host is difficult to judge.

Greater larval establishment has been attributed to low dietary protein concentration (Preston and Allonby 1978) but concurrent

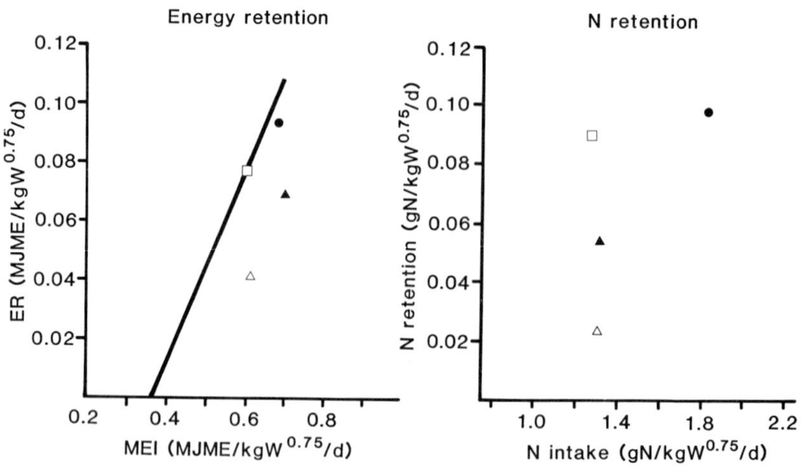

Figure 3. Effect of infection with *T.colubriformis* on energy retention and N retention of infected (△) sheep, and the effect of infusion of 50 g /d casein (●) or isocaloric amounts of glucose (▲); control (□).

changes in concentrations of other nutrients make these studies unreliable. The recent well-controlled studies of Abbott *et al.* (1986a), on the other hand, failed to demonstrate an effect of short term (4 week) or long term (8 month) feeding of a diet with low crude protein concentration (88 g /kg DM) to sheep on faecal egg count or worm establishment 12 weeks after single infection with *H.contortus*. Development of immunity in general is, however, affected by level of nutrition (Halliday 1980). Moreover, Wagland *et al.* (1984) showed a negative correlation between weight gain during a vaccination challenge against *T.colubriformis* and serum antibody titre at subsequent challenge and with evidence for improved resistance to parasite development.

More recently, an attempt has been made to establish the importance of protein *per se* for establishment of infection and the ability of the host to withstand infection (Bown 1986; Bown *et al.* 1986). Confounding effects of diet composition were avoided by infusion of casein or isocaloric quantities of glucose directly into the abomasum of sheep infected daily with 3000 larvae of *T.colubriformis*. Protein infusion rate was calculated to supply protein (50.5 g CP /d) to meet the likely additional protein synthesis in the alimentary tract in infected animals, calculated from previous experiments with similar rates of infection. Control animals were either not infected or infected but infused with saline. Gross efficiency of use of digested energy and rate of body protein deposition was fully restored to that in uninfected controls by the additional protein but not by energy (Fig. 3). The implication is that parasitism induces a protein, as opposed to energy deficiency, but that provided this can be overcome metabolic performance may be normal. A reservation must exist that the protein infusion moderated establishment and rejection of the parasite population, through enhancement of the host's immune system, rather than by supplying nutrients to limit pathogenicity of infection. However, no difference was observed between groups in the establishment of larvae after 6 weeks when mean worm burdens were about 7000 $\pm$ 2000. On the other hand, infusion into the abomasum of casein, but not glucose, led to significantly lower worm burdens after 12 weeks when numbers were 39,300, 17,800 and 45,700 $\pm$ 6,250 respectively in saline-, casein- and glucose-infused groups. These data, perhaps, extend the findings of Wagland *et al.* (1984) in indicating the importance of current protein status for the development of immunological competence during challenge. Whether this difference in response was associated with differences in pathological damage in the intestine and therefore nutrient demand is not clear, since similar rates of plasma protein loss into the tract were observed in all groups at all times.

It must now be apparent that there are extreme technical difficulties in answering the question as to whether nutrition affects the ability of the animal to withstand the effects of established infection precisely, because there is no method by which equivalence in pathogenicity of infection can be ensured to test effects of nutrition. Nutrition itself may change parasite establishment. However, the general consensus would be that diets with low protein concentration limit the ability of the host to withstand the metabolic demands of infection with helminths (Bawden 1969b; Dargie 1980, 1986; Abbott *et al.* 1986b).

HERBAGE AS A SOURCE OF INFECTIVE LARVAE

The epidemiology or ecology of endoparasite populations on pasture is extremely complex. The major sources of infective larvae are eggs passed in the faeces of a susceptible or naive animal, predominantly the young or lactating member of a species or those members subject to such nutritional or other stress that their immune responses have broken down. The subsequent appearance of larvae on pasture depends on temperature, moisture, light and the nature of the dung pat. The optimum temperature for larval development ranges from 13-25°C and varies with species, for example outbreaks of haemonchosis occur more frequently in periods when mean maximum temperature is above 18°C while those of trichostrongylosis are more likely when mean maximum temperature is 13-18°C. On the other hand *O.circumcincta*, for example, have better cold tolerance than do *H.contortus*. The infective 3rd stage larvae are particularly susceptible to dessication, but there is variation between species. For example, the larvae of *H.contortus* are

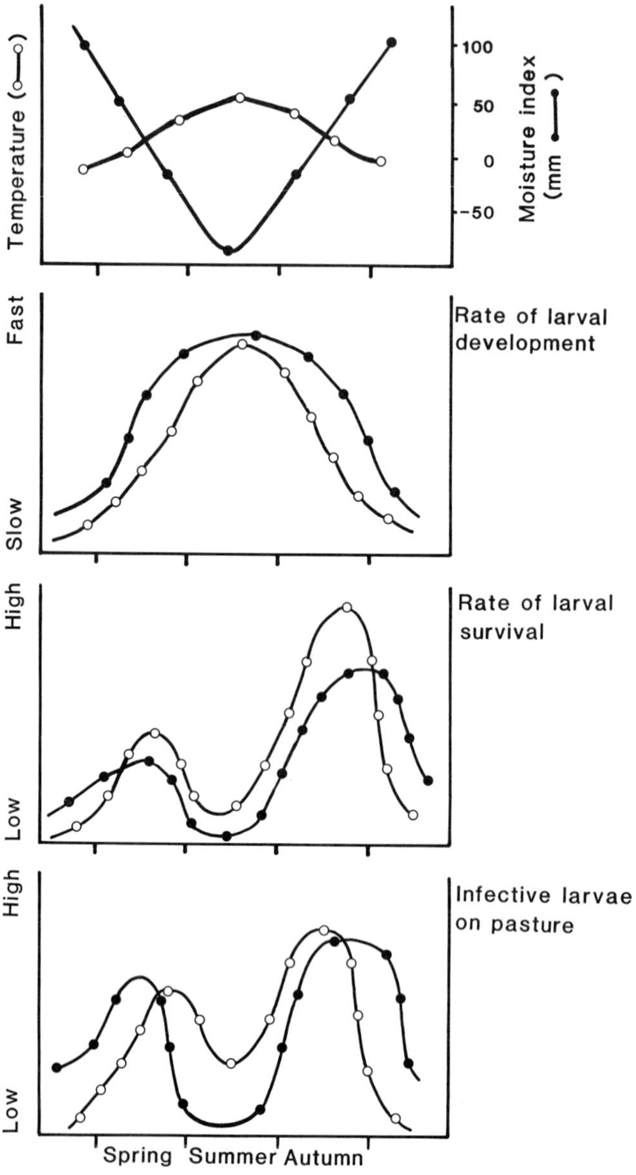

Figure 4. Effect of climate on rate of larval development and survival and the consequence for number of infective larvae on pasture (from Vlasoff 1982). ○ and ● represent species with relative heat and cold tolerance, respectively.

considered to survive repeated dessication more successfully than do *T.colubriformis* (Todd *et al*. 1970). Moisture is also required to allow migration of larvae from the dung pat onto a part of the herbage where their chances of prehension by the grazing animal are enhanced. Migration of larvae is probably largely a random process. They are known to migrate up to 100 cm from a dung pat (Skinner and Todd 1980) but 90% will be found within 10 cm. Larvae of *Ostertagia* spp. which infect cattle have been found at a soil depth of 7–15 cm (Al Saqur *et al*. 1982). It is not, at this stage, clear what controls their movement between soil and herbage (Rose and Small 1985) or the precise significance of the soil as a larval reservoir. Soil type influences migration and survival; while sandy soils promote greater migration than do clay soils (Furman 1944a) the latter have been associated with greater survival of larvae in the surface vegetation, presumably through maintenance of more favourable conditions of moisture (Furman 1944b). Larvae are vulnerable to ultraviolet light and greater survival may be expected in shaded areas of the sward.

Adequate aeration of the dung pat, essential for larval development, may be influenced by the nature of the feed and the structure of the undigested carbohydrate moeity in which eggs are contained.

In short, moist sheltered conditions with moderate temperatures are required for optimum development and survival of larvae on pasture, but the conditions vary for different helminth species. A theoretical pattern of distribution on pasture of two nematode species with differing cold and moisture tolerance is given in Fig. 4. The specific epidemiology of nematode parasites is very much influenced by local climate and understanding of risk to local livestock can only be developed with access to such data (Michel 1976; Vlassoff 1982; Young and Anderson 1981; Anderson 1983; Waller et al. 1981; Callinan 1978). The environmental demands of the larvae mean that the greatest accumulations of infective larvae occur in the lower parts of dense rapidly growing swards. Vlassoff's (1982) summary of this distribution in relation to site of deposition and sward height is given in Fig. 5.

There have been many excellent reviews of programmes for control of helminths in grazing systems (Nansen 1986). The essential principles involve control of egg deposition by naive animals by frequent anthelmintic treatment or grazing with resistant animals of the same or of another species prior to grazing by susceptible animals. In practice this is coupled with anthelmintic treatment to slow down the rate of accumulation of mature worms in the susceptible group of animals, accepting that they will inevitably be exposed to infective larvae. This practice invites and accepts any losses incurred by the animal resulting from the stimulation of the immune system and from early tissue damage by developing larvae. These latter losses could be considerable especially with those nematode species whose immature stages have a particularly close association with the tissues of the host.

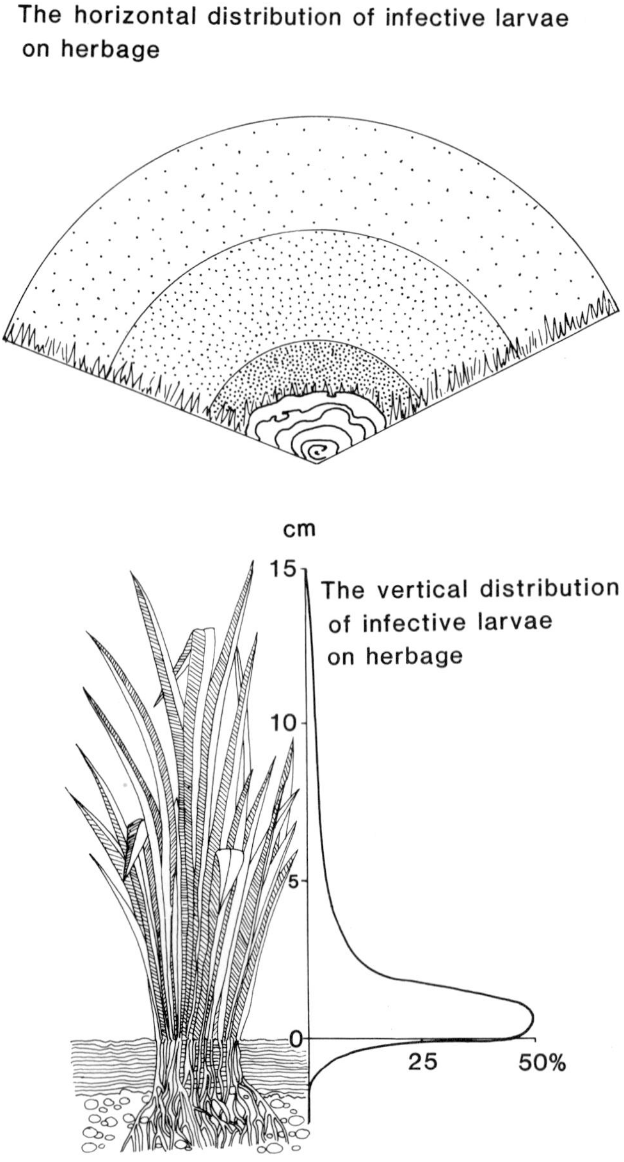

Figure 5. The vertical and horizontal distributions of larvae on pasture (from Vlassoff 1982).

For example, it was demonstrated by Coop *et al.* (1982) that under a situation of continuous infection with 5000 *O.circumcincta* per day, anthelmintic treatment at 21 day intervals, while suppressing development of mature worms and therefore eggs in faeces, restored only 20% of a 50% loss of growth rate, reflecting the damage caused by immature larval stages.

This emphasises the need to control parasitism by limiting larval availability on pasture. The impact of grazing strategies and the importance of herbage species in determining intakes of infective larvae have, however, received comparatively little attention. The distribution of larvae demonstrated in Fig. 4 has several implications, given knowledge of the grazing behaviour of the susceptible animal. Firstly, it may be possible to graze heavily contaminated pasture with naive animals if grazing pressure is not sufficient to promote grazing in the lower horizons of the sward. Secondly, swards in which much of the leaf material is carried on a high pseudostem exposed to ultraviolet light and to dessication may have a more uneven larval distribution, most remaining in the lower regions of the sward. For example, Furman (1944b) demonstrated very clearly that clover leaflets, which are readily accessible and consumed by grazing livestock as they graze down through the sward, provide a much better environment than do grass stems and leaves for larval development, migration and survival, as judged by their relative content of larvae. In contrast, larvae are found predominantly in the lower stem areas of lucerne, a part of the plant which is only selected by the animal when under extreme grazing pressure. Furman (1944b) also showed that the upper leaves of ryegrass had much lower larval counts than the lower leaves and stem. Irrigation and particularly rainfall and dew may induce short term changes in this pattern through facilitating larvae migration. It is well known that some animal species, for example, the equine, influence the lateral distribution of larval by their peculiar, but beneficial, habit of localizing defaecation to a small area of their range.

Of crucial importance is rate and seasonality of herbage production and this concept is demonstrated graphically in Fig. 6. Those pastures, typical of temperate regions, such as in New Zealand, which have persistent year-round herbage growth, and consistency of herbage production between seasons, attract high stocking rates and high rates of year-round utilization of pasture (Fig. 6a). Not only are the lower horizons of the swards intensively utilized, predisposing to high infection rates, but the high stocking rates ensure high rates of re-infection. The problem is further compounded by high rates of larval survival in such climatic conditions. Where pasture production is more seasonal in nature and less predictable, and stocking rates as a consequence more conservative (Fig. 6c), it is generally assumed that parasite contamination of herbage is much reduced. Moreover, utilization of pasture during the period when growth is rapid and conditions favour larval availability is likely to be low, because of the low stocking rate. This should lead to lower rates of larval ingestion.

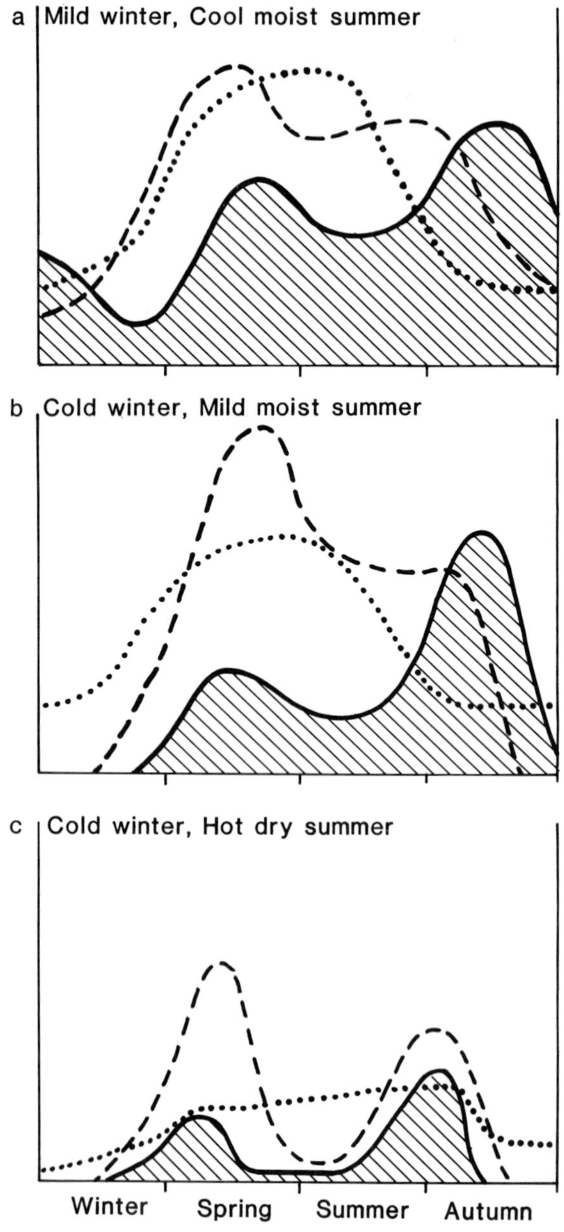

Figure 6. Effect of climate on relative seasonal pasture production (▬▬▬) feed demand of a sheep flock (••••) and availability of larvae (hatched area).

While this would be true in a general sense, it is rare for there not to be more favoured areas or strategically irrigated areas on which stock congregate in unfavourable periods and which can provide larval contamination rates closer to those described in Fig. 6 a or b.

Exploration of these aspects of the problem of endoparasite – nutrition interactions in the grazing herbivore are worthy of much more detailed study than heretofore, in view of the major nutritional penalties encountered when herbivores are forced to use their own adaptive mechanisms to survive exposure to endoparasites. Indeed the concern about the development of anthelmintic-resistant populations of parasites will help to ensure that this more ecologically sound approach to parasite control does attract research funding.

Abbott, E.M., Parkins, J.J., and Holmes, P.H. (1986a). *Vet. Parasitol.* **20**, 275-89.

Abbott, E.M., Parkins, J.J., and Holmes, P.H. (1986b). *Vet. Parasitol.* **20**, 291-306.

Al Saqur, I., Bairden, K., Armour, J., and Gettinby, G. (1982). *Res. Vet. Sci.* **32**, 332-7.

Altaif, K.I., and Dargie, J.D. (1978). *Parasitology.* **77**, 161-75.

Anderson, N. (1983). *Aust. J. Agric. Res.* **34**, 583-92.

Anderson, N., Hansky, J., and Titchen, D.A. (1981). *Parasitology.* **82**, 401-10.

Barger, I.A., and Southcott, W.H. (1975). *Aust. J. Exp. Agric. Anim. Husb.* **15**, 167-72.

Barker, I.K. (1973). *Parasitology.* **67**, 307-14.

Barker, I.K., and Titchen, D.A. (1982). *Int. J. Parasitology.* **12**, 345-56.

Bawden, R.J. (1969a). *Aust. J. Agric. Res.* **20**, 589-99.

Bawden, R.J. (1969b). *Aust. J. Agric. Res.* **20**, 601-6.

Bown, M.D. (1986). Ph.D. Thesis. University of Canterbury.

Bown, M.D., Poppi, D.P., and Sykes, A.R. (1984). *Can. J. Anim. Sci.* **64** (Suppl.) 187-98.

Bown, M.D., Poppi, D.P., and Sykes, A.R. (1986). *Proc. N.Z. Soc. Anim. Prod.* (in press).

Callinan, A.P.L. (1978). *Int. J. Parasit.* **8**, 233-7.

Care, A.D., Barlet, J.P., and Abdel-Hafeez, H.M. (1980). In "Digestive Physiology and Metabolism in Ruminants." (Eds Y. Ruckebush and P. Thivend.) pp. 429-46. (MTP: Lancaster, UK.)

Coop, R.L., and Field, A.C. (1983). *Res. Vet. Sci.* **35**, 175-81.

Coop, R.L., Angus, K.W., and Sykes, A.R. (1979). *Res. Vet. Sci.* **26**, 363-71.

Coop, R.L., Sykes, A.R., and Angus, K.W. (1982). *J. Agric. Sci.* **99**, 247-55.

Dargie, J.D. (1980). "Digestive Physiology and Metabolism in Ruminants". (Eds Y. Ruckebusch and P. Thivend.) pp. 349-71. (MTP: Lancaster UK.)

Dargie, J.D. (1982). In "Animal Models in Parasitology" (Ed. D.G. Owen.) pp. 17–51. (Macmillan: London.)
Dargie, J.D. (1986). *Proc. 6th Int. Congr. Parasitol.* (Ed. M.J. Howell.) pp. 453–63. (Aust. Acad. Sci.: Canberra.)
de Jong, A. (1986). In "Control of Digestion and Metabolism in Ruminants." (Eds L.P. Milligan, W.L. Grovum and A. Dobson.) pp. 459–78. (Prentice-Hall: New Jersey.)
Dineen, J.K., and Wagland, B.M. (1966). *Parasitology*. **56**, 639–50.
Dobson, C. (1967). *Parasitology*. **57**, 201–19.
Forbes, J.M. (1986). In "Control of Digestion and Metabolism in Ruminants." (Eds L.P. Milligan, W.L. Grovum, and A. Dobson.) pp. 420–35. (Prentice Hall: New Jersey.)
Furman, D.P. (1944a). *Amer. J. Vet. Res.* **5**, 79–86.
Furman, D.P. (1944b). *Amer. J. Vet. Res.* **5**, 147–53.
Gibson, T.E., and Parfitt, J.W. (1972). *Res. Vet. Sci.* **13**, 529–35.
Gregory, P.C. (1985). In "Veterinary Research Communication Vol. 1. The Ruminant Stomach." (Eds L.A.A. Oams, A.D. Degryse and R. Marsboom.) (Janesson Research Foundation.)
Halliday, R. (1980). In "Growth in Animals" (Ed. T.L.J. Lawrence.) pp. 65–82. (Butterworths: London.)
Holmes, P.H. (1986). Proc. 6th Inter. Cong. Parasitol. (Ed. M.J. Howell.) pp. 203–8. (Aust. Acad. Sci.: Canberra.)
Hucker, D.A., and Yong, W.K. (1986). *Vet. Parasitol.* **19**, 67–76.
Johnston, J.T., Familton, A.S., McAnulty, R.W., and Sykes, A.R. (1984). *N.Z. Vet. J.* **32**, 177–9.
Jones, D.G. (1982). *Res. Vet. Sci.* **32**, 316–23.
Jones, W.O., and Symons, L.E.A. (1982). *Int. J. Parasitol.* **12**, 295–301.
Kroonen, J.E.G.M., Verstegen, M.W.A., Boon, J.H., and van der Hel, W. (1986). *Br. J. Nutr.* **55**, 351–60.
Leyva, V., Henderson, A.E., and Sykes, A.R. (1982). *J. Agric. Sci.* **99**, 249–59.
MacRae, J.C., Smith, J.S., Sharman, G.A.M., and Corrigal, W. (1982). In "Energy Metabolism of Farm Animals." (Eds A. Ekern and F. Sundstl.) (E.A.A.P., Publ., 19: Agric. Univ., Norway.)
Michel, J.F. (1976). *Adv. Parasitol.* **14**, 355–97.
Miller, H.R.P. (1984). In "Advances in Veterinary Immunology, 1983." (Eds F.J. Bourne and N.T. Gorman.) (Elsevier: Amsterdam.)
Miller, T.A. (1979). *Adv. Parasitol.* **17**, 315–84.
Murray, M., Jennings, F.W. and Armour, J. (1970). *Res. Vet. Sci.* **11**, 417–27.
Nansen, P. (1986). Proc. 6th Int. Cong. of Parasitol. (Ed. M.J. Howell). pp. 425–33. (Aust. Acad. Sci.: Canberra.)
Poppi, D.P. (1986). Proc. Int. Symp. on use of nuclear techniques in studies of animal production and health in different environments. (FAO/IAEA: Vienna) (in press).
Poppi, D.P., MacRae, J.C., Brewer, A.C., Dewey, P.J.S., and Walker, A. (1985). *J. Comp. Path.* **95**, 453–64.

Poppi, D.P., Macrae, J.C., Corrigall, W., and Coop, R.L. (1981). *Proc. Nutr. Soc.* **40**, 116A.

Preston, J.M., and Allonby, E.W. (1978). *Vet. Rec.* **103**, 509–12.

Preston, J.M., and Allonby, E.W. (1979a). *Res. Vet. Sci.* **26**, 134–9.

Preston, J.M., and Allonby, E.W. (1979b). *Res. Vet. Sci.* **26**, 140–4.

Prichard, R.K., Hennessy, D.R., and Griffiths, D.A. (1974). *Res. Vet. Sci.* **17**, 182–7.

Richardson, R.I., and Jouan, A.R.P. (1986). *Br. J. Nutr.* **55**, 149–56.

Rose, J.H., and Small, A.J. (1985). *J. Helminthol.* **59**, 127–35.

Roseby, F.B. (1977). *Aust. J. Agric. Res.* **28**, 155–64.

Rowe, J.B., Abbott, E.M., Dargie, J.D., and Holmes, P.H. (1982). *Proc. Nutr. Soc.* **41**, 74A.

Skinner, W.D., and Todd, K.S. (1980). *Amer. J. Vet. Res.* **41**, 395–8.

Smith, W.D., Jackson, F., Jackson, E., and Williams, J. (1983). *J. Comp. Path.* **93**. 471–8.

Steel, J.W. (1978). In "Recent Advances in Animal Nutrition" (Ed. D.J. Farrell.) p. 98. (Univ. New England: Armidale, N.S.W.).

Steel, J.W., and Hogan, J.P. (1972). Ann. Rep., C.S.I.R.O., Division of Animal Health, Sydney. Project 6.4, p. 75.

Steel, J.W., Jones, W.O., and Symons, L.E.A. (1982). *Aust. J. Agric. Res.* **33**, 131–40.

Sykes, A.R. (1982). In "Biology and Control of Endoparasites." McMaster Laboratory 50th Anniversary Symposium. (Eds L.E.A. Symons, A.D. Donald and J.K. Dineen.) pp. 217–34. (Academic Press: Sydney.)

Sykes, A.R. (1983). In "Sheep Production." Nottingham 35th Easter School Proceedings. (Ed. W. Haresign.) pp. 317–34. (Butterworths: London.)

Sykes, A.R., and Coop, R.L. (1976). *J. Agric. Sci.* **86**, 507–15.

Sykes, A.R., and Coop, R.L. (1977). *J. Agric. Sci.* **88**, 671–7.

Sykes, A.R., Coop, R.L., and Angus, K.W. (1979). *Res. Vet. Sci.* **26**, 372–7.

Sykes, A.R., Coop, R.L., and Rushton, B. (1980). *Res. Vet. Sci.* **28**, 63–70.

Symons, L.E.A. (1965). *Gastroenterology.* **49**, 158–68.

Symons, L.E.A. (1978). *J. Parasitol.* **64**, 958–9.

Symons, L.E.A. (1985). *Adv. Parasitol.* **24**, 103–33.

Symons, L.E.A., and Jones, W.O. (1970). *Exp. Parasitol.* **27**, 496–506.

Symons, L.E.A., and Jones, W.O. (1983). *Int. J. Parasitol.* **13**, 309–12.

Thomas, R.J., and Ali, D.A. (1983). *Int. J. Parasitol.* **13**, 393–8.

Todd, K.S., Levine, N.D., and Whiteside, C.C. (1970). *J. Nematol.* **2**, 330–3.

Vlassoff, A. (1982). In "Internal Parasites of Sheep." Technical Handbook Animal Industries Workshop. (Ed. A.D. Ross.) (Lincoln College: Canterbury.)

Wagland, B.M., Steel, J.W., Windon, R.G., and Dineen, J.K. (1984). *Int. J. Parasitol.* **14**, 39–44.

Waller, P.J., Dobson, R.J., Donald, A.D., and Thomas, R.J. (1981). *Int. J. Parasitol.* **11**, 359–67.

Webster, A.J.F. (1983). In "Mammalian Thermogenesis." (Eds L. Girardier and M.J. Stock.) pp. 176–207. (Chapman and Hall: London.)

Wilson, W.D., and Field, A.C. (1983). *J. Comp. Path.* **93**, 61–71.

Young, R.R., and Anderson, N. (1981). *Aust. J. Agric. Res.* **32**, 371–88.

11

MICROBIAL DIGESTION OF FORAGES IN HERBIVORES

by R.I. Mackie

The microflora of the digestive tract can affect the nutritional status of the host animal in many and often complex ways. The provision of nutrients and regulation of the environment by the host seems to have obvious advantages for the intestinal flora, whereas the effects of microflora on the host animal are not well defined. Research on herbivore nutrition is essentially a study of the interactions between the host, microorganisms present, substrates available and end-products of digestion. Furthermore, the interaction of the normal intestinal flora with the host can be manipulated to improve the efficiency of nutrient utilization in animals. The present paper presents evidence in support of the theme that not only is it advantageous for animals, especially herbivores, to have a resident microbial population in the gut, but it is also essential from a nutritional standpoint. In short, the host and its associated microflora form a closely integrated ecological unit.

RELATIONSHIPS BETWEEN ORGANISMS

The ultimate aim of ecologists is to understand the relationships between organisms and their natural habitat or environment (Hungate 1969, 1975). Organisms can live together in associations which range from competitiveness to complete dependency. Interspecies interactions in a two-membered association can be either positive (beneficial), negative (harmful) or zero (neutral) and nine different states are possible (Table 1). This is a gross over-simplification, since it concerns only one type of interaction, under a single set of environmental conditions (nutrients, pH, temperature, etc.). If a third organism is added the total number of possibilities would be 81. Fortunately, as the number of interacting organisms increases, the chances of any organism drastically affecting another one decreases, because of the lower probability of frequent and consistent encounters between the two organisms. Thus it is possible for 10-12 interacting species to exist in ecological balance as occurs in most compartments of the gastro-intestinal tract.

TYPES OF ANIMAL-MICROBE RELATIONSHIP

The complexity of animal-microbe relationship varies tremendously, ranging from competition to cooperation. The animal alimentary tract

NUTRITION OF HERBIVORES
ISBN 0 12 750052 9

Table 1. Analysis of two species (A,B) population interactions (after Odum 1983).

Type of interaction		Species* A	B	Nature of interaction
Positive	Commensalism	+	0	Species A, the commmensal, benefits while B, the host, is not affected
	Mutualism	+	+	Interaction favourable to both but not obligatory
	Protocooperation	+	+	Interaction favourable to both but not obligatory
Negative	Competition: direct interference	–	–	Direct inhibition of each species by the other
	Competition: resource	–	–	Indirect inhibition when common resource is limited
	Amensalism	–	0	Species A inhibited, B not affected
Both	Parasitism	+	–	Species A, the parasite, generally smaller than B, the host
	Predation (incl. herbivory)	+	–	Species A, the commensal, benefits, while B, the host is not affected.
None	Neutralism	0	0	Neither species affects the other

* 0 indicates no significant effect; + indicates growth, survival or other characteristic enhanced; – indicates growth or other characteristic inhibited.

has evolved as an adaptation enabling the animal to secure food and limit consumption by other animals. This allows the retention and digestion of ingested food, followed by absorption and metabolism of digestion products, whilst feeding and other activities continue. Since microorganisms grow rapidly under favourable conditions in the gut they could be serious competitors for the animal's food. This microbial challenge has modified the course of evolution in animals, resulting in selection for different types of animal–microbe relationship (Hungate 1976, 1984).

Competition model
In the competition model, exemplified by carnivorous animals, host and microbe are competitors for the same food. Immunological and other adaptations delay consumption of microbes by animals and prevent invasion of animal tissues by microorganisms in the gastro-intestinal tract. Microbicidal concentrations of acid are secreted preventing attack on ingested feed, allowing the host's digestive enzymes to act followed by absorption of its enzymic digestion products. However, slower rate of passage, together with rapid growth, results in a large microbial population in the hindgut.

Cooperation model
The abundance of carbohydrate in plant cell walls (cellulose and hemicellulose) is the basis for the evolution of the cooperation model. The carbohydrate polymers of plant cell walls are indigestible by most animals (including all mammals), but can be hydrolysed and fermented by the microbial partner, with the resultant end-products of fermentation plus microbial cells being utilized by the host animal. The most widely known and economically important example of a cooperative animal-microbe relationship is found in ruminants in which the capacious, continuously fermenting rumen delays passage of digesta, allowing time for solubilization of fibre components by microorganisms. One problem is the breakdown or sacrifice of dietary protein by rumen microorganisms before digestion by the animal's own enzymes. However, poor quality protein is upgraded through conversion to microbial protein (Tamminga 1979).

Combination model
The combined competition-cooperation model for animal-microbe interaction avoids this difficulty since the animal's enzymes function and the breakdown products are absorbed before the microbial fermentation takes place. Thus, the host obtains not only the nutrients digested by its own enzymes but also fermentation products from materials its enzymes cannot digest. This type of interaction occurs in horses, elephants, hyraxes, rodents, and lagomorphs (hares and rabbits) but is probably best exemplified in the termites.

The difference between this and the competition model lies in the extent to which anatomical modifications of the host allow longer retention of the digesta in the hindgut with consequent increased solubilization and fermentation. In both models the microbes act after the host has absorbed the nutrients made available through its own enzymes and in both models there is marked microbial activity in the hindgut. A disadvantage of the combination model is that, although the host absorbs the fermentation end-products, the microbial cells themselves cannot be used. Some animals have overcome this deficiency by consuming the faeces containing the microbes, that is through coprophagy or caecotrophy.

ROLE OF PROTOZOA IN GASTRO-INTESTINAL ECOSYSTEMS

Some of the genera of ciliate protozoa are widely distributed in the gastro-intestinal tract of herbivorous mammals whereas others are limited to one or a few hosts (Giesecke 1970; Clarke 1977). However, the problem of determining host specificity at species level is complicated by problems in identification. Specificity may be determined by a number of factors including diet and its associated effects, physiological stress and microbial antagonisms (Clarke 1977).

Despite a long history of ecological investigation, factors that regulate the appearance and even the numbers and types of particular protozoan species are only superficially understood. *Buetschlia parva* was first reported in the rumen by Schuberg (1888) and by many others since. The monograph of Dogiel (1927) is probably still the most authoritative work on the classification of ophryoscolecid ciliates although the work of Kofoid and MacLennan (1930, 1932, 1933) and Lubinsky (1957a, b; 1958a, b) are useful in species identification. Hungate's (1966) book remains the most complete account of the rumen protozoa. However, protozoal taxonomy remains confused. This problem can be partially resolved in a number of ways.

The first approach would be the use of functional or trophic groups of protozoa based on their nutritional or substrate requirements. The advantages of using functional groups when compared to a detailed taxonomic study in a particular ecosystem are as follows:

1. reduced difficulty in identifying poorly described taxa,
2. simplification of information on ecosystem structure,
3. elimination of taxa with redundant functions,
4. less specific observer expertise required,
5. increased sampling precision due to decrease in analysis time per sample and increase in number of samples taken, and
6. identification of functional groups decreases numbers of species enumerated but increases precision by increasing enumeration of individuals or relative biomass in each category (Pratt and Cairns 1985).

Various factors would need to be taken into account when developing a functional group classification based on nutritional strategy or feeding type. There is likely to be some overlap between groups since protozoa are unlikely to restrict themselves to a particular feeding type and could also switch feed sources. Microspatial distribution patterns would also reduce the possibility of a particular species encountering a specific feed type. Under nutritional stress, some species may expand normal feeding tolerances although this normally involves switching to the next lower trophic level (Pimm 1980).

In the rumen, and probably most other gut compartments, it is possible to define six different functional groups of ciliate protozoa based on nutritional strategy (Table 2). An example of overlap is the bacteriovores and herbivores where no differentiation can be made between those forms which are selective bacterial feeders and those that simply ingest articles in appropriate size ranges. With existing knowledge, it should be possible to determine in a wide variety of ecosystems the species distribution in the different functional groups.

Table 2. Nutritional strategy of different functional groups of ciliate protozoa in the rumen.

Group	Apparent nutritional strategy
Predators	Feed primarily on other protozoa (e.g. *Polyplastron multivesiculatum*, *Eudiplodinium maggii*, *Enoploplastron confluens*, *Epidinium gigas*, *Entodinium bursa*).
Fungivores	Feed on zoospores of phycomycete fungi and flagellate protozoa of similar small size (e.g. *Enoploplastron confluens*, *Entodinium lobosospinosum*).
Bacteriovores	Feed on bacteria and small particles with attached bacteria. Possibly the largest group (e.g. *Entodinium simplex*, *E. caudatum*, *E. bursa*, *Polyplastron multivesiculatum*).
Herbivores	Feed specifically on plant particles such as chloroplasts, particulate protein and fibre particles which may have an adherent bacterial population (e.g. *Epidinium ecaudatum*, *Eudiplodinium neglectum*, *E. maggii*, *Diplodinium dentatum*, *Ostracodinium obtusum*, *Cycloposthium* spp.)
Saprotrophs	Feed on dissolved organic compounds (soluble sugars, peptides, amino acids, nucleic acids, acetate, etc.) in the surrounding milieu. These soluble compounds are released into the medium by other micro-organisms (bacteria and protozoa) and also by the host animal (e.g. *Isotricha prostoma*, *I. intestinalis*, *Dasytricha ruminatium*).
Non-selective omnivores	Ingest particulate matter both plant and bacteria in appropriate size range without apparent preference (e.g. *Polyplastron multivesiculatum*, *Entodinium caudatum*, *Ophryoscolex* spp.

This would provide information relating species numbers and diversity of resources. A wide diversity of resources would reduce the strong interactions or competition in the ecosystem allowing more species to occupy similar functional roles. It should also be possible to assess changes in functional groups after changing the diet, during initial acquisition of fauna or refaunation, and after any perturbation of the ecosystem. The grouping of species according to functional characteristics would be the first step towards a more complete understanding of the role of protozoa in the ruminal ecosystem.

The second approach aimed at clarifying the taxonomy of ciliate protozoa would be the use of modern nucleic acid methodology, serology or chemotaxonomy depending on which is most suitable for the arrangement or classification of organisms into taxonomic groups. Subsequently it would be necessary to identify new isolates or reclassify existing organisms. Classifications based on genetic relatedness tend to be more stable and reliable and not subject to frequent or radical changes. The simplest method is the determination of DNA base ratios (mol% G+C). However, this does not take into account the linear arrangement of the nucleotides in the DNA molecule. Therefore DNA homology experiments are used to detect similarities between closely related organisms and RNA homology is used to detect similarities between more distantly related organisms (Johnson 1984). The genetic approach using rRNA oligonucleotide sequences can be used for the elucidation of phylogenetic, evolutionary relationships. Serological and chemical techniques are based on the premise that chemical constituents either on the cell surface or within the cell, such as proteins or enzymes, are genetically determined and that relatedness can be measured by comparing "fingerprints" or profiles. This would be more precise and reliable than current morphological criteria such as size, spination, number and position of contractile vacuoles, shape and size of macronucleus, position of micronucleus and shape and number of skeletal plates, all of which are clearly influenced by nutrition, predation, intracellular parasites or commensals and other factors. Disadvantages of the genetic approach are the contribution of DNA from contaminating bacteria essential for the nutrition of the ciliate protozoa and the fact that genetic material is present in both the macronucleus and micronuclei. The macronucleus controls all "vegetative" functions, most notably the synthesis of RNA, whereas the micronuclei are largely inactive in RNA synthesis but are responsible for transmitting the chromosomes during conjugation (Gall 1984).

Role of the ciliates

Depending on diet the ciliate protozoa can account for up to 40% of total microbial nitrogen in the rumen and produce as much as 60% of the total microbial fermentation products (Clarke 1977; Coleman 1980; Williams 1986). The role played by the gut protozoa in ruminants in particular has long been debated and present feeling is that they are not essential to the survival of the host. Under many conditions it appears that defaunation has no adverse effect and ruminants can live indefinitely without ciliates. However, a recent report demonstrated that ciliate protozoa can alleviate chronic copper toxicity in sheep (Ivan *et al.* 1986).

Rumen VFA concentrations are often higher in faunated than

in defaunated animals although this is not always the case. VFA proportions also vary with faunation resulting in increased butyrate and often propionate. The effects depend on diet with those high in concentrates (either finely ground or pelleted) which lower ciliate populations tending to increase the amount of propionate produced. However, the situation with barley-based diets may be different. Animals fed restricted amounts of barley develop large populations of ciliate protozoa with a resulting high production of butyrate; this is also found in defaunated animals on *ad libitum* barley feeding (Eadie and Mann 1970). Clearly the relationships between the ciliate protozoa, the bacteria and VFA production are complex, being influenced by the nature, physical form and feeding level of the diet. There is no consensus of opinion that defaunation decreases weight gains by lowering VFA production (Veira 1986).

The instability of rumen microbial populations on high concentrate diets is more pronounced in ciliate-free animals (Eadie and Mann 1970). The ciliates control fermentation rate to a certain extent by ingesting starch grains (entodiniomorphs) and soluble sugars (holotrichs) thereby preventing rapid bacterial fermentation. Rapid fermentation leads to the accumulation of lactate and a detrimental reduction in ruminal pH (Mackie *et al.* 1978). Furthermore, ciliates rapidly assimilate soluble sugars forming storage polysaccharide (amylopectin) (Coleman 1979, 1980). The subsequent fermentation of ingested starch and storage polysaccharide over a prolonged period would prevent an "explosive" fermentation and create a levelling effect helping to maintain a balanced ruminal fermentation.

The contribution of protozoa to the protein economy of the host animal has been the subject of much research. In general, there is an inverse relationship between protozoal and bacterial pool sizes since the protozoa engulf large numbers of bacteria and also compete for available nutrients (Leng and Nolan 1984). Large populations of protozoa which are selectively retained in the rumen result in an increased recycling of nitrogen. Ruminal degradation of protein and amino acids of microbial origin reduces the efficiency of microbial growth and the net yield of microbial amino acids available to the host animal (Leng 1984). The ciliates in the hindgut of non-ruminant herbivores would not be able to supply protein except through coprophagy. It should be remembered that, although pathways for the breakdown and synthesis of protein in the rumen are well understood, estimates of N pool sizes and flux rates are often based on meagre evidence obtained in a single laboratory (Chalupa 1984).

It would be premature to suggest that large increases in productivity, particularly in grazing animals, could be obtained by defaunation unless further studies on a wide range of diets demonstrate marked improvement in animal productivity over prolonged periods. Furthermore, there are practical difficulties associated with establishing and maintaining defaunated herds. At present, the defaunating agents are unsatisfactory and may endanger animal health.

A sustained research effort is required to fully understand and appreciate the contribution of ciliate protozoa to animal productivity. This should be aimed at understanding the relationship or interactions between protozoa and anaerobic fungi. At present this is limited to visual observations and no quantitative estimates are available on the relative biomass of anaerobic fungi or their contribution to fibre or protein degradation. Pure culture studies indicate that they have high cellulase and protease activity (Orpin and Letcher 1979; Pearce and Bauchop 1985; Wallace and Joblin 1985; Wood *et al.* 1986). Care should be taken in interpreting experimental results after defaunation when rumen fungi "bloom" and expand to fill the competitive niche vacated by the protozoa or simply by removal of predation. It may be possible to design experiments with animals (i.e. free of all microbes or with organisms derived from pure cultures) to study these problems. More information on the growth physiology of protozoa, particularly growth yields, efficiency and maintenance energy requirements, is required.

Do protozoa prey differentially on bacteria? *Entodinia* spp. engulf cellulolytic species more rapidly than other ruminal or non-ruminal bacterial species (Coleman and Sandford 1979). Selective engulfment could have an important influence not only on microbial ecology in the rumen but could also lead to alteration of ruminal fermentation patterns. It is unlikely that protozoal feeding is sufficiently selective in most species to distinguish bacteria from other particulate material of similar size (dead bacteria, plant particles).

What role does bacterial motility play in controlling predation? It seems that bacteria can outcompete protozoa for soluble organic nutrients at dilute concentrations found in the rumen and other ecosystems. This may be explained by affinity or simply a surface area phenomenon. Thus protozoa have evolved as effective ingestors of bacteria on which they prey and particulate matter of similar size. Protozoa control bacterial numbers but do not reduce them below a certain level. The reduction in bacterial biomass would allow increased substrate levels and faster growth rates.

Symbiosis in protozoa

Many protozoa harbour symbionts of their own although the significance of such close relationships have been largely neglected. Corliss (1985) proposes the use of the word "xenosome" for all kinds of endosymbionts which are organisms or bodies within a eukaryotic cell that contains DNA and is bounded by at least one membrane. It is evident that an indispensable feature of endosymbiosis is the mutualistic transfer or movement of metabolites between host and xenosome (Corliss 1985). Only in rare instances does the endosymbiont become so tightly enmeshed in the internal economy of the cell that it becomes a genuine cell organelle (mitochondria, chloroplasts). This contrast demonstrates that protozoa are well adapted to serve as hosts for endosymbionts but that the process of conversion of symbionts into organelles requires complex genetic and molecular adaptations (Cavalier-Smith and Lee

1985). There is often some kind of specificity since a certain host is usually associated with a limited number of partners.

Bacterial symbionts often play an important role in protozoal nutrition or provide some essential functions for their host protozoa. Bacterial endosymbionts provide haem compounds for their host trypanosomatid protozoa, *Crithidia oncopelti* (Cheng and Dave 1980). Epibiotic spirochaetes are responsible for the movement of their flagellate hosts (Ball 1969). *Mixotricha paradoxa*, an intestinal flagellate found in the termite gut, is a striking example of this type of association with motility supplied by attached spirochaetes (Cleveland and Grimstone 1964).

Cellulose digestion by ruminal ciliates and termite flagellates

In the rumen, the ciliate protozoa are largely associated with plant material and may be responsible for some 30 – 40% of microbial fibre digestion (Demeyer 1981). Recent research confirms that the protozoal fraction contains appreciable cellulolytic activity (Williams and Strachan 1984). In a survey (Coleman 1985), highest activities were found in *Eudiplodinium maggii*, *Epidinium ecaudatum caudatum* and *Ostracodinium obtusum bilobum* with little or no activity in five *Entodinium* spp. on both carboxymethylcellulose and microcrystalline cellulose (Sigmacell). Further studies have shown the presence of the glucosidases involved in cellulose and hemicellulose breakdown in holotrich and entodiniomorphid ciliates isolated from the rumen or cultures *in vitro* (Williams *et al.* 1984). Preliminary subcellular fractionation confirmed that enzymes which degrade polysaccharides in the plant cell wall were sedimentable in fractions that contained membrane-bound subcellular organelles (Williams and Ellis 1985; Williams *et al.* 1986). In a well planned experiment in which *Polyplastron multivesiculatum* was maintained free of extra- and intra-cellular cellulolytic bacteria the presence of endo-1,4-β-glucanase (EC 3.2.1.4) and β-glucosidase (EC 3.2.1.2) was demonstrated (Bonhomme *et al.* 1986). However, this organism showed low activity on native cellulose (cotton powder).

According to the accepted model, a complete cellulase system comprises three main enzymatic components acting synergistically during hydrolysis of crystalline cellulose. These components are endo-1,4-β-glucanase, exo-1,4-β-glucanase (EC 3.2.1.91) and β-1,4-glucosidase. Many ruminal bacteria, including non-cellulolytic species, have been shown to possess endoglucanase and β-glucosidase activity (Williams and Withers 1981, 1982; Russell 1985; MacKenzie and Patel 1986). The presence of exoglucanase activity is critical for complete hydrolysis and utilization of native cellulose (Ljungdahl and Eriksson 1985). Further study of the origin and nature of the different components of the cellulase complex in ciliate protozoa is warranted before stating that ciliate protozoa do not rely on cellulases from endosymbiotic bacteria.

Cellulose digestion in termite flagellates may also rely on cellulases from endosymbiotic bacteria. However, recent work (Yamin 1978, 1981) with axenic cultures of major cellulolytic protozoa

Trichomitopsis termopsidis and *Trichonympha sphaerica* from the hindgut of the wood-eating "lower" termite (*Zootermopsis* sp.) demonstrated that hindgut protozoa were vital in cellulose degradation. For *T. termopsidis* final axenic cultures appeared to be free of live extracellular bacteria as well as endosymbiotic bacteria and were able to ferment cellulose to acetate, CO_2 and H_2 (Yamin 1980, 1981). Crude extracts of *T. termopsidis* possessed endo-glucanase and cellobiase activity (Yamin and Trager 1979; Odelson and Breznak 1985). Cell extracts also hydrolysed microcrystalline cellulose with glucose as the sole product liberated. Cellulolytic activity in *T. termopsidis* appeared to be entirely cell-associated and is probably localised in food vacuoles (Odelson and Breznak 1985). Secretion of cellulase from the salivary glands of some termites could also augment the action of the microbial enzymes (O'Brien and Slaytor 1982).

Symbiosis of protozoa with hydrogen-utilizing methanogens

Associations of epi- and endo-symbiotic methanogenic bacteria with host protozoa have recently been described. The methanogens are a unique group of bacteria and in the rumen produce CH_4 from CO_2 and H_2 (CO_2 reduction). Some of the cofactors (F_{342} and F_{420}) required in this series of biochemical reactions show fluorescence when irradiated with short wavelength blue light and thus methanogens can be easily recognized using epifluorescence microscopy (Doddema and Vogels 1978). Using this method numerous methanogenic bacteria (probably *Methanobrevibacter ruminantium*) were seen attached to entodiniomorphid ciliates. Between 8 and 55% of the cells of each ciliate species showed this episymbiotic association with methanogens (Vogels *et al.* 1980). The frequency of association varied not only between species but also within one species obtained from different rumens. When this was investigated further using fistulated sheep, it was found that after fasting for 14 hours the frequency of association was high but decreased rapidly after feed intake. When a fasted rumen was flushed with H_2 gas the frequency of association decreased. In contrast, flushing with N_2 gas 2.5 hours after feeding restored association frequency to the same level as that of fasting sheep (Stumm *et al.* 1982). A positive correlation between methanogenic activity and the fraction of strained rumen fluid containing the highest number of ciliates confirms the episymbiosis (Krumholz *et al.* 1983).

Freshwater and marine sediments (the sapropel) also form an important habitat for methanogenic bacteria. The numbers of bacteria and protozoa in the sapropel are much lower than those in the rumen. Fluorescent bacteria were detected inside cells of sapropelic protozoa, ciliates and the amoebae *Pelomyxa palustris* and *Mastigella* sp. (Van Bruggen *et al.* 1983, 1984, 1985). Electron microscopy confirmed the presence of endosymbiotic bacteria which were characterized as methanogens based on typical fluorescence; extraction, identification and quantitation of specific coenzymes; methane production by washed protozoa; isolation and identification of methanogens.

Although there is little solid experimental evidence it is reasonable to assume that the methanogenic bacteria and host protozoa form an integral metabolic consortium. Methanogenic bacteria utilize a small spectrum of substrates (acetate, H_2 plus CO_2, formate, methanol and methylamines) for growth. It therefore seems obvious that methanogenic symbionts obtain one or more of these substrates from the ciliates. In the case of protozoa known to harbour methanogens it is likely that H_2 originates from hydrogenosomes in the protozoa (Yarlett *et al.* 1984). Interspecies hydrogen transfer is well documented in anaerobic bacteria (Bryant 1979; Wolin 1979; McInerney and Bryant 1981) but not in protozoa. Until now, no endosymbiotic methanogen with the ability to utilize acetate has been isolated but this seems likely in the future with the correct nutritional and growth conditions. From the observations with rumen protozoa it can be concluded that the symbiosis is facultative in these organisms, especially if partial pressure of H_2 is high as occurs shortly after feeding. When partial pressure of H_2 is low the eukaryotic cell becomes the predominant hydrogen donor. The interaction between sapropelic protozoa and their endosymbionts is more obligatory since they always contain methanogens, substrate concentrations are low and the association guarantees the most advantage in terms of mass transfer and competition. The advantage to the ciliate protozoa is less obvious since they can survive without associated methanogens, not only in the rumen but also when cultivated *in vitro* (Coleman 1980). Furthermore, the holotrich species are never associated with methanogens in spite of H_2 production during metabolism of soluble carbohydrates (Williams 1986). It is unlikely that the methanogens excrete an essential metabolite for the eukaryotic cells, with the possible exception of certain growth factors. However, by maintaining thermodynamically favourable low H_2 partial pressures, the methanogens enable the ciliate to obtain a higher energy yield per mole of glucose fermented. Further research is required to clarify the exact function of the symbiotic association.

NUTRITIONAL SIGNIFICANCE OF BACTERIA IN THE GASTRO-INTESTINAL TRACT

The gastro-intestinal tract of herbivorous mammals and birds harbours a complex mixed culture of bacteria. This flora develops as a result of the interaction between intestinal physiology and the bacteria which colonize the gut. The initial inoculum is usually derived from the mother at birth and during suckling and grooming. Coprophagy may also be necessary for the establishment and maintenance of intestinal flora. The population changes as the host develops and the climax flora differ with animal species. Each structural and functional region of the gut will have a distinct ecology although diet has an important influence on the gastro-intestinal flora.

The intestinal flora may be regarded either as the most

intimate part of the environment or as an organ of the body. If the flora is considered as an organ it is potentially the most metabolically adaptable and rapidly renewable portion of the body. Although investigation of the gut flora and its metabolic activity provide important insights one must remember that the bacteria growing in the gastro-intestinal tract form an integrated ecological unit. The rumen microbial ecosystem has been the most thoroughly studied gut system, particularly the quantitative aspects. It is a dynamic system in a continuous state of flux with about 30–34% of bacteria free in rumen fluid while the majority (60–70%) are associated with, or firmly attached to, the feed particles (Forsberg and Lam 1977; Nuzback *et al.* 1983). A distinct bacterial population of facultative anaerobes adherent to the ruminal epithelium comprises about 1% of the total population (Cheng and Wallace 1979; Cheng and Costerton 1980). A further 1% is probably adherent on other organisms, both protozoa and bacteria.

The contribution of the rumen to the host's nutrition has been studied in considerable depth (Bauchop 1977; Prins 1977; Prins *et al.* 1984). The significance of hindgut fermentation in non-ruminant herbivores has not been studied in the same detail (McBee 1977; Stevens *et al.* 1980; Argenzio and Stevens 1984). The ruminant forestomach and the mammalian large intestine show some remarkable similarities with respect to microbial digestion, secretion and absorption (Stevens *et al.* 1980).

Energy metabolism

Rumen microorganisms ferment a wide variety of carbohydrates including cellulose, hemicellulose, pectins, starches and sugars; the principal end-products are three volatile fatty acids (VFA), acetate, propionate and butyrate plus CO_2 and CH_4. A small proportion of VFA may be produced from other sources such as amino acids and lipids.

Microbial digestion of carbohydrates in the hindgut produces VFA in proportions and concentrations similar to those found in the ruminant forestomach (Stevens *et al.* 1980; Argenzio and Stevens 1984). The nature of the hindgut fermentation is determined by the relative amounts of undigested plant structural polysaccharides, microbial protein and endogenous substrates such as host enzymes, sloughed epithelial cells and mucins. This means that, in general, the source and supply of substrate for hindgut fermentations is more constant, leading to a stable fermentation in this portion of the gastro-intestinal tract. Estimates of the efficiency of VFA absorption and utilization are higher than 90% in pigs and horses (Argenzio *et al.* 1974; Argenzio and Southworth 1975). The estimated daily absorption of VFA from the colon and rectum of the dog, goat and pig would provide approximately 7, 30 and 76% of their respective basal energy requirement (Stevens *et al.* 1980; Argenzio and Stevens 1984). In another study it was found that VFA could provide only 11.6 and 9.6% of the maintenance energy of pigs fed a low starch or high starch diet respectively and that a major proportion of the VFA absorbed from the large intestine was metabolized by its mucosas (Imoto and Namioka 1978). Recent work showed that 37.7% of the maintenance energy requirements of rabbits were supplied by VFA produced in the hindgut (Marty and Vernay 1984), considerably more than the 12% reported earlier (Hoover and Heitmann 1972). Microbial fermentation is important in herbivorous marsupials. Foregut fermenters such as the kangaroos and wallabies (Hume 1977, 1982) are similar to ruminants in which VFA provide 70–80% of the animals daily ME intake (Annison and Armstrong 1970). The hindgut fermenters include the koala (*Phascolarctos cinereus*) and other marsupial folivores (gliders, possums) where 9–10% of the energy is provided by VFA absorption from the gut (Hume 1984). Although many of the above estimates are based on *in vitro* fermentation procedures which clearly underestimate VFA production they provide valuable comparative information. It appears that VFA can provide a substantial portion of the metabolizable energy requirements of hindgut fermenters.

Fermentation in the gastro-intestinal tract of birds

Birds have not made as much use of fermentative fibre digestion as have mammals. Research with poultry suggested that, although birds can digest plant fibre this is of little nutritional value. Instead of utlizing fibre, many avian herbivores consume large quantities of plant material but extract only the readily digestible components. In such birds, the bulk of the plant cell wall constituents are rapidly expelled without significant microbial fermentation. Fibre digestion is significant in some birds, notably in the family Tetraonidae (grouse and ptarmigan) and these birds utilize enlarged caeca as fermentation chambers. Even in these birds the contribution of fibre digestion to the total energy expenditure appears to be low with less than 20% of basal energy metabolism being derived from this source. The weight of digesta in the fermentation compartments of the gut in the ptarmigan were lower than those found in similar sized mammalian herbivores. It has been suggested that this is related to the weight reduction necessary for flying.

Thus, it would be expected that the birds which should be able to make most use of fermentative digestion are the large flightless species such as the emus and ostriches. This is not only because the constraints associated with flying are absent but also because in large animals it is possible to have large fermentation chambers relative to metabolic rate. A recent investigation into the dietary energy requirements of the Australian emu (*Dromaius novaehollandiae*) showed that they digested 35–45% of the neutral detergent fibre (NDF) in their diets (NDF content 26–36%). Energy from the digestion of NDF contributed up to 63% of standard metabolism and 50% of maintenance requirements on the highest fibre diet. This was achieved despite the fact that the rate of passage of feed residues through the tract was rapid and the gastro-intestinal tract is simple in structure with a small caecum and short colon (Herd and Dawson 1984). In contrast the ostrich (*Struthio camelus*) has paired caeca 1.29 m long and a long colon (5.4–8.6 m). Rate of passage experiments using ^{51}Cr-mordanted samples of the diet as marker were carried out on ostriches of different liveweight. In the group with mean liveweight of 6.8 kg, mean retention time averaged 39 hours while in the group with mean liveweight 45.8 kg the value was 48 hours. This rate of passage is similar to values of 38 and 41 hours for the sheep and kangaroo respectively which are both foregut fermenters. The mean retention time in the emu was found to be very short (only 5.5 hours). The results obtained with the ostrich indicate significant retention of feed in the gut of this species. This is most likely to occur in the hindgut although the proventriculus may also be involved to a lesser extent.

The mean concentration of total volatile fatty acids was high in the gizzard and proventriculus of the ostrich. The concentration decreased in the intestine and then increased again in the hindgut (Swart and Mackie unpublished). The mean concentration in the caeca was 140.7 mM and increased further from the proximal colon to the distal colon (171 to 195 mM). This was closely related to the increase in DM (Fig. 1). Quantitatively, acetate was the most important of the individual VFA and in the proventriculus, gizzard and small intestine was the only VFA. In the hindgut, propionate and butyrate were also present together with trace amounts of iso-butyrate, iso-valerate and valerate. High molar proportions of acetate and low proportions of propionate are indicative of the fermentation of fibre or material high in plant cell walls.

Although lactic acid may be released from the gastric mucosa, it is also produced by microbial fermentation of carbohydrates and at lower pH bacteria producing lactic acid would be at an advantage over those producing VFA. The results (Fig. 1) show a low concentration in the proventriculus and gizzard. However, in the small intestine the concentration increased dramatically to 66 and 51 mM with 95–96% present as the L-isomer. In the caeca, the concentration dropped although the proportion of the D-isomer increased to *ca* 30%. In the colon the concentration of lactic acid ranged from 19–26 mM with 35–40% as the D-isomer. The presence of the L-isomer as the predominant form in the small intestine indicates that the gastric mucosa could be the source of some of the acid while the high proportion of the D-isomer in the caeca and colon indicates the contribution being mainly of microbial origin.

Production of acetate in the large intestine was relatively high, ranging from 26–32.5 mM /h per 100 g DM. The VFA production rates (65–81.3 mM /l per hour) were considerably higher than those reported for the ileum (14.0) and rectum (16.8 mM /l per hour) of the emu. In general, they are much higher than the values obtained from the hindgut of mammals including sheep and wallabies (Hume 1977). The levels are characteristic of those found in the forestomach of ruminants and some of the ruminant-like mammals with foregut fermentation.

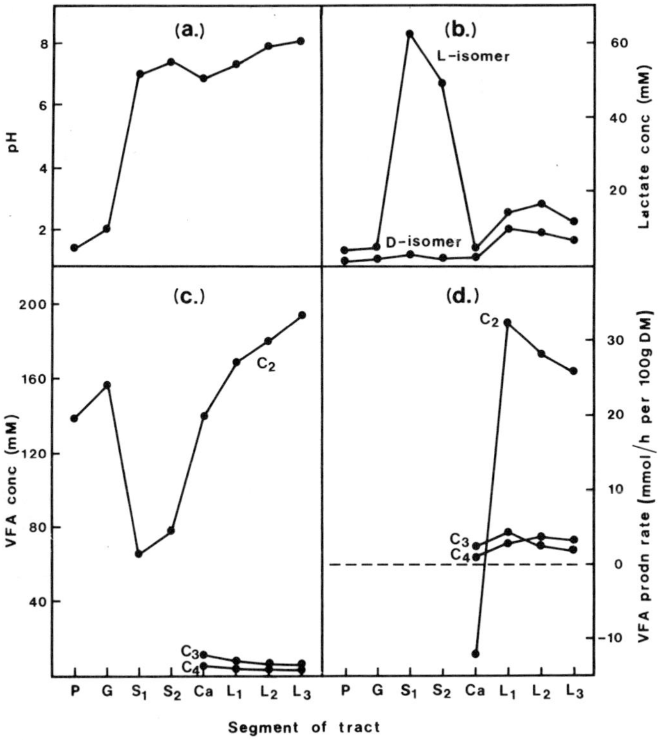

Figure 1. Measurement of (a) pH, (b) lactic acids, (c) VFA concentration, and (d) VFA production rate in different segments of the gastro-intestinal tract of the ostrich (n = 12) where P = proventriculus, G = gizzard, S_1 = proximal small intestine, S_2 = distal small intestine, Ca = caeca, L^1 = proximal colon, L_2 = mid-colon, and L_3 = distal colon and C_2 = acetate, C_3 = propionate and C_4 = butyrate.

Microbial colonization of the gastro-intestinal tract of the ostrich
Almost all research on the avian gut microflora has been done using the domestic fowl (*Gallus domesticus*) as an experimental model. In the domestic fowl the two main sites of bacterial activity are the crop and the caeca. Broadly speaking it can be said that the flora of the anterior gut from gizzard to terminal ileum is composed of facultative anaerobes whereas the caeca contain strict anaerobes as the predominant organisms (Fuller 1984). Thus the intestinal microflora is likely to be a complex mixture of many different types of bacteria which affect the nutrition and growth of birds.

Microbial ecology in the gut of ostriches was studied by relating the numbers and kinds of bacteria present to environmental conditions in each of the gut segments (Mackie and Swart, unpublished data). Ciliate protozoa were not present in any of the gut segments. Preliminary results from the caeca and proximal colon showed that viable (cultural) counts were similar to those obtained in the rumen of ruminants. The counts in the proximal colon were generally 15–40% higher than in the caeca with the exception of the lactate-utilizers and the hemicellulolytics (Table 3). The functional groups fermenting the different carbohydrate substrates contributed a large proportion of the total culturable counts. This is particularly interesting in the case of the lactate-utilizers which would be able to utilize as substrate the high endogenous production of lactate from the small intestine, resulting in a reduction of lactate concentration in the caeca. The presence of a diverse bacterial population in the hindgut was also observed using transmission electron microscopy (Fig. 2).

Table 3. Viable counts (no /g ingesta) of different functional groups of bacteria in the hindgut of the ostrich (Mackie and Swart, unpublished)

Functional group of bacteria	Segment	
	Caeca	Colon
Total culturable	12.9×10^8	14.6×10^8
Glucolytics	7.2×10^8	9.6×10^8
Amylolytics	9.2×10^8	11.5×10^8
Proteolytics	3.3×10^8	4.6×10^8
Lactilytics	6.3×10^8	6.0×10^8
Hemicellulolytics	7.5×10^8	$8.1 \times 10_8$
Cellulolytics	10^4	10^5

Protein synthesis and nitrogen recycling

Ammonia is an important source of nitrogen for the growth of rumen microorganisms. Depending on diet, 60–90% of the daily N intake by the ruminant is converted to ammonia and from 50–75% of bacterial-N is synthesized from ammonia (Mathison and Milligan 1971; Pilgrim *et al.* 1970; Nolan and Leng 1972). This ammonia is derived mainly from proteolysis of dietary and endogenous proteins and rapid deamination of the resulting amino acids. Non-protein nitrogen in the form of urea synthesized by the liver and secreted into the gut from the blood in addition to that recycled in saliva provides a further source of ammonia. This enables the ruminant to conserve nitrogen as well as the water no longer required for urea excretion. Most important, this valuable biosynthetic activity of rumen bacteria can be utilized to

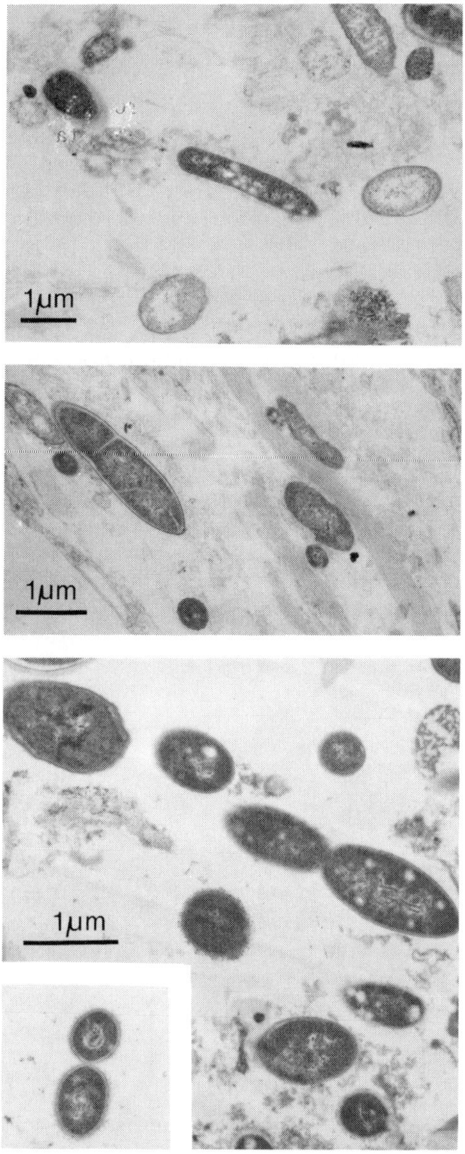

Figure 2. Transmission electron micrographs showing the diverse anaerobic bacterial flora present in the hindgut of the ostrich. Fermentative digestion results in the production of VFA, mainly acetate, which contribute to the metabolisable energy requirements of this unique avian herbivore.

provide a cheap source of fermentable N to ruminants in the form of urea supplements under conditions of N deficiency.

It is important to note that urea recycling is not confined to the rumen. In fact 50–80% of endogenous urea in sheep enters the gut post-ruminally including both small and large intestines (Huntington 1986). It has also been established that significant urea recycling occurs in the horse (Prior *et al.* 1974), rabbit (Regoeczi *et al.* 1965; Forsythe and Parker 1985a), rock hyrax (*Procavia abyssinica*; Hume *et al.* 1980), ringtail possum (*Pseudocheirus peregrinus*; Chilcott and Hume 1984), and tammar wallaby (*Macropus eugenii*; Mason 1984).

Nitrogen metabolism in the mammalian hindgut

This is highly developed in herbivores whose diets, post weaning, ensure a reasonably constant supply of fermentable materials to the hindgut. However, the extent and complexity varies considerably between species and is influenced by diet and physiological age of the animal. In most animals a substantial part (40–80%) of the ileal-N passing to the hindgut consists of proteinaceous materials of dietary, endogenous and bacterial origin with little urea-N (<2%). The wall of the large gut supplies N in the form of urea, mucins and sloughed epithelial cells. A considerable portion of the urea-N influx in the caecum and colon would be hydrolysed to ammonia by bacteria living in close association with the gut-wall although little quantitative information is available.

There is considerable protease, deaminase, decarboxylase and urease activity in the hindgut. Animal enzymes probably contribute little to these processes since there is no evidence for enzyme secretion by the mucosa in this region of the gut. Microbial involvement in these processes has been largely inferred from the large number of viable organisms in the hindgut. As in the rumen, NH_3 plays a central role in N metabolism with some 65% or more of ammonia being derived from degradation of bacterial, endogenous and dietary residues from the upper gut in sheep (Nolan *et al.* 1976). The hindgut flora synthesize their nitrogenous cell components and secretions from simpler molecules such as NH_3 and amino acids or peptides using carbohydrates as the energy source. The primary N source for bacterial protein synthesis in the hindgut is NH_3. The quantitative significance of these processes of catabolism and anabolism is indicated by the presence of 50–78% and 62–76% bacterial-N in the faecal-N of sheep (Mason 1969, 1979) and pig (Mason *et al.* 1976, 1982).

The important point relating to this activity in the hindgut is the nutritional significance to the host animal since this occurs caudal to the main site of protein digestion and amino acid absorption in the small intestine and many animals do not salvage these products by coprophagy. Although digestion of microbial protein has been demonstrated in the proximal colon of the rabbit (Bonnafous and Raynaud 1970; Wootton and Argenzio 1975; Forsythe and Parker 1985b), and the large intestine of both the horse (Slade *et al.* 1970, 1971) and pig (Niiyama *et al.* 1979) have been shown to absorb microbially synthesized amino acids, the amounts are small relative to NH_3 absorption. The possibility of amino acid or peptide absorption thus deserves further study (Webb 1986). The most important function is the uptake of ammonia which serves as N source for the synthesis of non-essential amino acids and urea in the gut-wall and liver. The urea is recycled to the rumeno-reticulum to assist in supporting microbial growth, especially under survival conditions, and to the hindgut to support a large population of viable bacteria which actively ferment plant cell wall material. The bacterial population converts osmotically active materials to absorbable molecules assisting in the absorption of water, sodium and other essential minerals while at the same time providing the host with an extra supply of energy in the form of VFA. Absorption of Na and VFA account for most of the H_20 removed during passage of digesta through the large intestine (Argenzio and Stevens 1984) and the presence of hindgut fermentation may be obligatory for survival under arid conditions. Certainly water conservation, body size and hindgut fermentation are intimately related and warrant further research.

Nitrogen metabolism in the caeca of birds

There are many similarities between nitrogen metabolism in the hindgut of mammals and birds. There are three different routes for the entry of nitrogenous compounds. A large part of caecal-N is derived from dietary protein which escapes digestion and absorption in the small intestine together with intestinal secretions. Urinary-N gains access to the caeca by a filling mechanism involving retrograde peristalsis from the cloaca (Fenna and Boag 1974). In the turkey up to 35% of the urine is transported to the caeca (Bjornhag and Sperber 1977). It is thought that during the squeezing of colonic contents, material of higher protein content than the coarse fibres remaining in the colon enters the caeca and that the carbohydrate is also more susceptible to microbial degradation. Urea can enter the caeca directly through the caecal wall (Mortensen 1984).

Evidence thus far indicated that all nitrogenous compounds entering the caeca are degraded to ammonia by the action of proteases (Salter and Coates 1971; Hanssen 1979; Bedbury and Duke 1983), uricase (Barnes and Impey 1974; Mortensen and Tindall 1981a) and urease (Okumura *et al.* 1976). Ammonia can be used as a nitrogen source for synthesis of non-essential amino acids by the bird (Salter 1973). This presupposes that the diet contains sufficient amounts of essential amino acids and a deficit of non-essential amino acids which is unlikely especially under normal dietary conditions. Within the caeca the complex microbial population assimilates ammonia mainly by the glutamate dehydrogenase pathway but when ammonia becomes limiting the high affinity glutamine synthetase-glutamate synthetase system is utilized (Mortensen and Tindall 1981b). The enzymes of ammonia assimilation and their regulation by ammonia level are similar to those operating in the rumen (Hespell 1984). The rates of fermentation and ammonia assimilation are positively correlated and the fact that activity of caecal bacteria in pigs can be increased by addition of protein (Just *et al.* 1981) suggests that the growth of caecal microflora is limited by N. The relationship between the caecal microflora and the host bird therefore represents a well developed and functional symbiosis. The caecal flora provides energy-rich fermentation end-products (VFA) but they require N to do this and this is supplied by the bird. This system is likely to have developed because of the mutual detoxification which occurs during this mutual exchange of metabolic end-products (Mortensen 1984).

Synergism between different species of proteolytic rumen bacteria
It is possible to consider bacterial interaction as the enhancement or amplification of activities when they are grown together as compared to the activity of individuals grown in pure culture. Recent work by Wallace (1985) has demonstrated considerable synergism in protein degradation by defined mixtures of ruminal bacteria. *Selenomonas ruminantium* and *Streptococcus bovis* grew slowly on casein in pure culture because of their low activity but when mixed together, growth was rapid (Fig. 3a).

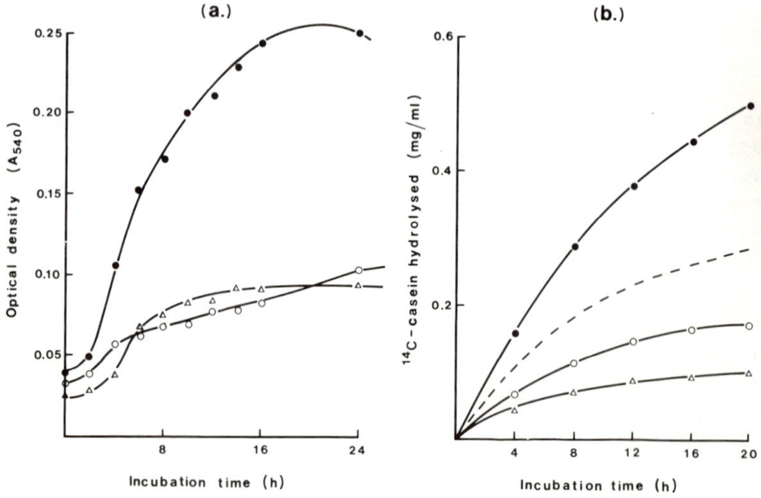

Figure 3. (a) Growth of proteolytic rumen bacteria in medium containing casein (1 mg /ml) as sole N source. (b) Hydrolysis of ^{14}C-labelled casein (2 mg /ml) by proteolytic rumen bacteria. The broken line represents the sum of individual activities of the two species. (O, *Selenomonas ruminantium*; Δ, *Streptococcus bovis*; ●, combined growth or activity of the two species in the mixture) (after Wallace 1985).

Presumably this was due to cooperativity during proteolysis (Fig. 3b) by complementary endopeptidase and exopeptidase activity of *S. ruminantium* and *S. bovis* respectively although this has not been proved as yet (Wallace 1986). Synergistic activity of proteolytic strains of *Butyrivibrio alactacidigens* and other proteolytic bacteria was not due to cooperativity but probably to nutritional interdependence (Wallace 1985). Cooperative proteolysis also occurred in mixtures of *Selenomonas ruminantium* and *Bacteroides ruminicola* which is capable of growth on casein in monoculture.

These results help to explain (i) why weakly active or seemingly insignificant bacteria may nevertheless play an important role in the overall proteolytic activity in the rumen, (ii) the poor digestion of protein in pure cultures of rumen bacteria (Laycock *et al.* 1985; R.I. Mackie unpublished), and (iii) why highest proteolytic activity is often obtained with cultures containing more than one organism (Blackburn and Hobson 1962). These findings clearly highlight and further document interspecies interactions with regard to proteolysis, cellulose digestion and fermentation (Wolin and Miller 1983) and the formation and utilization of H_2 (Wolin 1975, 1979).

Detoxification of phytotoxins and mycotoxins

Phytotoxins occur in many common feeds, including grains, protein supplements and forages (Kingsbury 1980). They range from tannins, alkaloids, goitrogens, gossypol, saponins, glucosinolates, mimosine and cyanogens to nitrate and oxalate. In many instances the ruminant forestomach provides a protective function and the ruminal flora effectively degrade a wide variety of toxic compounds. In some cases the opposite can occur with the production of toxic metabolites from innocuous compounds. In terms of ruminal metabolism, the formation of 3-methylindole from tryptophan and the induction of bovine pulmonary emphysema has been well documented (Carlson *et al.* 1972; Hammond *et al.* 1978, 1979).

Oxalate detoxification by bacteria

Many plants contain soluble oxalates but only a few have amounts sufficient (> 50g oxalic acid /kg DM) to be considered toxic (James *et al.* 1975). Toxicity of oxalic acid is due to the formation of calcium oxalate crystals in tissues such as the kidney, and in acute cases, tetany, where circulating, ionized calcium is complexed with oxalate. The main concern here is the interaction which occurs between gut microbial populations and dietary oxalate.

Ruminants fed gradually increasing amounts of oxalate adapt to and develop a tolerance to concentrations of oxalate that would be lethal to unadapted animals (James and Butcher 1972). This adaptation is mediated by the development of increased capacity for oxalate degradation by the ruminal microflora (Allison *et al.* 1977). The adaptation occurred over a 3 - 4 day period and was not accompanied by a change in the predominant ruminal bacteria. Furthermore, none of the cultures obtained from an extensive collection of predominant ruminal anaerobes, nor from the rumen of a sheep adapted to a diet containing the high-oxalate plant halogeton (*Halogeton glomeratus*), were found to degrade oxalate. When an *in vitro* continuous culture system inoculated with rumen contents was fed gradually increasing amounts of oxalate, the rates of oxalate degradation increased 100-fold (Allison *et al.* 1981). Gradual adaptation of the fermenter could be obviated by inoculation with rumen contents of an adapted sheep. This adaptation was the result of selection for increased numbers of bacteria which specifically degraded oxalate rather than a metabolic adaptation of the existing predominant flora.

An enrichment culture containing 45 mM oxalate as the principal energy source was established from an adapted continuous culture (Dawson *et al.* 1980a). When the methanogenic enrichment containing a simplified mixed population was established in continuous culture the first isolation of an obligately anaerobic, oxalate-degrading bacterium was obtained (Dawson *et al.* 1980b). This Gram-negative, obligate anaerobe was unable to use any of a wide variety of substrates except oxalate. Growth rates as high as 0.3 h^{-1} and cell yields of 1.1 g /mol of oxalate were observed. Selection for increased numbers of oxalate degraders would be closely regulated by oxalate availability in view of the limited substrate range of these organisms. The relationship between CH_4 production and oxalate degradation in mixed culture was partially explained by the formation of 1 mol formate per mol oxalate degraded. These oxalate degraders do not appear to fit into any existing taxonomic group and have therefore been assigned to a new genus and species, *Oxalobacter formigenes* (Allison *et al.* 1985). Measurements with one strain indicated that 1 mol of protons (H^+) was consumed per mol of oxalate degraded to produce approximately 1 mol of CO_2 and 0.9 mol of formate.

Further studies indicated that oxalate degraders could be detected in anaerobic roll tubes on the basis of clearings around colonies when the culture medium contained calcium oxalate. Numbers of oxalate degraders in adapted sheep were 1 - 5 x 10^5 /g rumen contents with an oxalate degradation rate of 0.8 - 3.3 μmol /g per h (Allison and Reddy 1984). The roll tube technique has facilitated the isolation of anaerobic oxalate-degrading bacteria from caecal contents of pigs and from human faeces (Allison *et al.* 1985, 1986). Oxalate degradation also occurs in the hindgut of horses, guinea-pigs and rabbits (Allison and Cook 1981), desert rodents (Shirley and Schmidt-Nielsen 1967) and in anaerobic aquatic sediments (Smith *et al.* 1985).

Mycotoxin degradation by rumen bacteria

Mycotoxins are secondary metabolites produced by fungi and normally have potent biological effects. Mycotoxins cause a broad spectrum of effects, both acute and chronic, and these are usually exerted on specific organs and tissues at the subcellular level resulting in a variety of biochemical changes which may or may not be accompanied by pathological changes (Shull and Cheeke 1983). It is important to note that many mycotoxins are produced by ubiquitous fungi which contaminate or invade common feeds such as maize, peanuts, rice and others which are utilized directly by livestock and humans. The importance of these mycotoxins cannot be overemphasized, especially in animal feedstuffs where a large proportion of an animal's diet in many parts of the world is comprised of products that are naturally infected with Fusaria and related fungal species.

Ruminants are generally less susceptible to mycotoxin poisoning than are monogastrics (Kurmanov 1977). Preparations of whole ruminal ingesta have been shown to degrade T-2 toxin (Kiessling *et al.* 1984), deoxynivalenol (King *et al.* 1984), zearalenone (Kiessling *et al.* 1984) and ochratoxin (Hult *et al.* 1976; Kiessling *et al.* 1984). Thus the rumen is probably responsible for conferring a degree of mycotoxin resistance in ruminants.

The trichothecenes are a chemically related group of biologically active metabolites of which the most important is T-2 toxin. The reactivity of trichothecenes is due to the 12,13-epoxide and 9,10-double bond in the trichothecane ring system and relative potency among the more than 40 trichothecenes is determined by the R-group substitution on the ring skeleton (Ueno 1977). A series of experiments was conducted to study the degradation of trichothecenes in the rumen. Preliminary experiments showed T-2 toxin, HT-2 toxin, deoxynivalenol and diacetoxyscirpenol were all degraded to varying extents by whole ovine ruminal contents. The rate of T-2 toxin degradation was 2 - 2.2 mg /l per h. From these results it appeared that T-2 toxin was first deacetylated to HT-2 toxin which was then further deacetylated to T-2 triol (R. Westlake, R.I. Mackie and M.F. Dutton, unpublished data). Toxin degrading activity was slightly higher in the protozoal than the bacterial fraction and absent from cell-free preparations. However, T-2 toxin caused an 83% decrease in protozoal protein over a 6 h incubation period whereas bacterial protein increased by 14% over the same period. This effect on eukaryotic cells is due to binding to ribosomes and inhibition of protein synthesis (Wei *et al.* 1974) whereas trichothecenes gave little effect on bacteria.

Since the detoxification of T-2 toxin involved deacetylation, further experiments were carried out with pure cultures of rumen bacteria known to have esterase activity (Westlake *et al.* 1987a). T-2 toxin (10 μg /ml) had no inhibitory effect on lag time or growth rate of pure cultures of *Butyrivibrio fibrisolvens*, *Selenomonas ruminantium* and *Anaerovibrio lipolytica* (Table 4). Even at a concentration of 1 mg /ml there was no effect on growth or lag time of *B. fibrisolvens*. This organism was shown to hydrolyse T-2 toxin to HT-2 toxin, T-2 triol and neosolaniol whereas the other two organisms only produced HT-2 toxin and T-2 triol. This suggested enzymatic hydrolysis at C-3 and C-4 to yield HT-2 toxin and T-2 triol, while a single hydrolytic cleavage at C-8 would lead to production of neosolaniol (Westlake *et al.*

Table 4. Effect of T–2 toxin on specific growth rate (μ), lag time (t), degradation products and esterase activity of pure cultures of predominant ruminal bacteria (after Westlake, Mackie and Dutton 1987).

Bacterial species and strain	Control[a] μ(/h)	Control[a] t(h)	T–2 toxin[a] μ(/h)	T–2 toxin[a] t(h)	% degraded T–2 in HT–2	% degraded T–2 in T–2 triol	% degraded T–2 in Neosolaniol	Esterase[b] activity
Butyrivibrio fibrisolvens								
CE 46	0.69	5.5	0.90	6.3	52	10	29	nd
CE 51	0.64	4.4	0.76	6.2	51	7	23	474
CE 52	0.60	0.6	0.61	1.6	63	23	0	nd
CE 56	0.54	0.3	0.51	0.8	53	13	19	nd
Selenomonas ruminantium	0.80	1.4	0.90	1.3	47	26	0	240
Anaerovibrio lipolytica	0.52	3.3	0.52	4.2	54	17	0	136

a Glucose (1% w/v) as growth substrate. T–2 toxin added to final concentration of 10 μg/ml.
b μmoles p–nitrophenol formed/min/mg protein using p–nitrophenylacetate as substrate.

1987a). A membrane preparation isolated from *B. fibrisolvens* was able to degrade T-2 toxin to HT-2 toxin and T-2 triol. This protein fraction had an approximate molecular weight of 65,000 and showed esterase activity (396 µmoles p-nitrophenol formed /mg protein per minute with p-nitrophenylacetate as substrate). A certain degree of enzyme-substrate specificity is evident since this esterase preparation did not produce neosolaniol and thus two different pathways are probably involved.

Further experiments with *B. fibrisolvens* showed that, although aflatoxin B_1 was not inhibitory, there was no degradation. However, there was degradation to a greater or lesser extent of zearalenone, verrucarin A, diacetoxyscirpenol, deoxynivalenol, and acetyl T-2 toxin (Westlake *et al.* 1987b). Thus this predominant species of rumen bacterium plays an important role in increasing the resistance of ruminants to trichothecene toxicity. Detoxification of deleterious compounds would be an important mechanism for assisting the survival of the host animal.

Vitamin synthesis and absorption

In the past there has been much experimental work on this subject and it has been found that rats can rely completely on their intestinal microflora for their requirements of vitamin K, folic acid, the vitamin B complex, biotin and vitamin E. Avitaminosis can be induced by oral administration of antibiotics. The adult ruminant obtains most of its vitamin requirements and essential growth factors as a result of microbial activity in the rumen. In fact, the biosynthetic ability of the rumen bacteria makes the adult ruminant completely independent of a dietary source of the B-vitamins and vitamin K. Although experimental evidence is lacking it appears that most, if not all, foregut fermenters share this ability. Hindgut fermenters rely on microbial synthesis in the small intestine which is also the major site of absorption. There is some evidence of B-vitamin absorption in the hindgut (Sorrel *et al.* 1971). However, some of the B-vitamins (riboflavin, nicotinic acid, folic acid and vitamin B_{12}) are largely intracellular (Phillipson 1977) and if the microbial cells are not digested in the hindgut these vitamins will be largely excreted. In small hindgut fermenters faecal loss is countered by coprophagy or caecotrophy (McBee 1977; Hornicke and Bjornhag 1980). The degree to which synthesized vitamins are available to the host in the case of monogastric animals or non-ruminant herbivores which do not practise coprophagy is worthy of further consideration.

Manipulation of rumen fermentation using recombinant DNA techniques

No paper would be complete without some mention of genetic engineering of ruminal bacteria. Most interest has been focused on biochemical and genetic aspects of gene manipulation (Smith and Hespell 1983; Teather 1985; Hespell 1985; Forsberg *et al.* 1986). Little consideration seems to have been given to the ecological limitations of engineered organisms once released back into an open ecosystem or indeed in the research required before deliberately releasing a genetically engineered organism into the environment. However, it may be less of a problem assessing the potential impact on the environment than obtaining a suitably engineered organism.

With new recombinant DNA technology it is now possible to confer new properties on a competitive wild type strain to increase its genetic potential without reducing its biological efficiency. Thus it is now possible in principle to introduce new genetic information into a natural ecosystem and reasonably expect it to be maintained and expressed. The level of expression cannot be high since this would adversely affect the competitive fitness of the organism carrying the information but many possibilities can be envisaged in which a low level of expression of new genetic information could have a major effect on the system as a whole (Teather 1985). This is largely speculative and obviously only subtle changes to existing ruminal bacteria will have any chance of success considering the intense selection pressure for organisms best fitted for survival in their particular ruminal niche. This does not imply that it cannot be done or that we should not try, rather that we should be more realistic in our assessment of the risks and realities of this exciting new technology.

Hungate (1984) pointed out that two basic axioms of bacterial ecology – (a) one cannot change the bacterial numbers or composition of an open natural ecosystem by inoculating with exotic species, and (b) the numbers and composition can be changed by altering the environment, – are only approximations. An example of this is the recent finding that detoxification of mimosine in *Leucaena leucocephala* can be transferred from Hawaiian cattle to Australian ruminants by rumen inoculation (Jones 1981). Furthermore the numbers and composition of the gut microflora can change with time without any discernible change in environmental conditions although these changes seem to be fluctuations rather than definite trends. Mutants constantly appear, some finding a niche and displacing other strains and later, in turn, being displaced themselves. These changes may be mediated by viral transfer of genetic material or through random transfer of genetic information within the population. In other words transfer of genetic material occurs naturally in the population and research with recombinant DNA technology seeks to direct this transfer towards some goal identified by the researcher.

Before deliberately releasing an engineered organism one would have to determine if the acquisition of new properties gives the engineered organism an undesirable selective advantage over unaltered members of the species. Apart from being able specifically to identify and detect the engineered organism in the open ecosystem or environment one would like to know if the engineered genes or even other genes in the engineered organism can be transferred in nature to other organisms and if genes from other organisms can be transferred to genetically engineered organisms more readily than to unaltered ones.

The preceding comments are intended to serve as a note of caution - with an indication of some of the ecological problems which will have to be overcome before practical application of the new technology becomes a reality - rather than an exhaustive discussion on the topic.

CONCLUSION

The gastro-intestinal flora can be considered as the most metabolically adaptable and rapidly renewable organ of the body which plays a vital role in the normal nutritional and physiological functions of the host animal. The types of animal-microbe relationships vary tremendously. However, the fermentative processes which occur in the rumen and mammalian large intestine are similar in many respects, in particular the production of VFA, synthesis of microbial protein and vitamins and the degradation or modification of toxins. The extent and complexity of this type of digestive strategy and the nutritional significance of these processes require considerable research activity in order to increase our understanding so that the microflora can be manipulated to the ultimate benefit of the host, whether it be one of the domestic animals, protozoa, or man himself. These problems will be most productively resolved by multidisciplinary groups of scientists answering basic questions relating to mechanisms and their regulation and control.

The author wishes to thank Dr A. Kistner and Dr F.M.C. Gilchrist for helpful discussion and constructive criticism during the preparation of the manuscript.

Allison, M.J., and Cook, H.M. (1981). *Science* **212**, 675-6.

Allison, M.J., and Reddy, C.A. (1984). In "Current Perspectives in Microbial Ecology." (Eds M.J. Klug and C.A. Reddy.) pp. 248-56. (Am. Soc. Microbiol.: Washington DC.)

Allison, M.J., Cook, H.M., and Dawson, K.A. (1981). *J. Anim. Sci.* **53**, 810-6.

Allison, M.J., Cook, H.M., Milne, D.B., Gallagher, S., and Clayman, R.V. (1986). *J. Nutr.* **116**, 455-60.

Allison, M.J., Dawson, K.A., Mayberry, W.R., and Foss, J.G. (1985). *Arch. Microbiol.* **141**, 1-7.

Allison, M.J., Littledike, E.T., and James, L.F. (1977). *J. Anim. Sci.* **45**, 1173-9.

Annison, E.F., and Armstrong, D.G. (1970). In "Physiology of Digestion and Metabolism in the Ruminant." (Ed. A.T. Phillipson.) pp. 423-37. (Oriel: Newcastle-upon-Tyne, UK.)

Argenzio, R.A., and Southworth, M. (1975). *Am. J. Physiol.* **228**, 454-60.

Argenzio, R.A., and Stevens, C.E. (1984). *Proc. Nutr. Soc.* **43**, 13-23.

Argenzio, R.A., Southworth, M., and Stevens, C.E. (1974). *Am. J. Physiol.* **226**, 1043-50.

Ball, G.H. (1969). *Res. Protozool.* **3**, 565–718.
Barnes, E.M., and Impey, C.S. (1974). *J. Appl. Bact.* **37**, 393–409.
Bauchop, T. (1977). In "Microbial Ecology of the Gut." (Eds R.T.J. Clarke and T. Bauchop.) pp. 223–50. (Academic: London.)
Bedbury, H.P., and Duke, G.E. (1983). *Poult. Sci.* **62**, 675–82.
Bjornhag, G., and Sperber, I. (1977). *Swed. J. Agric. Res.* **7**, 57–66.
Blackburn, T.H., and Hobson, P.N. (1962). *J. Gen. Microbiol.* **29**, 69–81.
Bonhomme, A., Fonty, G., Foglietti, M.J., Robic, D., and Weber, M. (1986). *Can. J. Microbiol.* **32**, 219–25.
Bonnafous, R., and Raynaud, P. (1970). *Experientia* **26**, 52–3.
Bryant, M.P. (1979). *J. Anim. Sci.* **48**, 193–201.
Carlson, J.R., Yokoyama, M.T., and Dickinson, E.O. (1972). *Science* **176**, 298–9.
Cavalier-Smith, T., and Lee, J.J. (1985). *J. Protozool.* **32**, 376–9.
Chalupa, W. (1984). *J. Dairy Sci.* **67**, 1134–46.
Chang, K.-P., and Dave, C. (1980). In "Endocytobiology, Endosymbiosis and Cell Biology, a Synthesis of Recent Research." (Eds W. Schwemmler and H.E.A. Schenk.) pp. 349–59. (Walter de Gruyter:Berlin.)
Cheng, K.-J., and Costerton, J.W. (1980). In "Digestive Physiology and Metabolism in Ruminants." (Eds Y. Ruckebusch and P. Thivend.) pp. 227–50. (MTP: Lancaster, UK.)
Cheng, K.-J., and Wallace, R.J. (1979). *Br. J. Nutr.* **42**, 553–7.
Chilcott, M.J., and Hume, I.D. (1984). *Aust. J. Zool.* **32**, 615–22.
Clarke, R.T.J. (1977). In "Microbial Ecology of the Gut." (Eds R.T.J. Clarke and T. Bauchop.) pp. 251–75. (Academic: London.)
Cleveland, L.R., and Grimstone, A.V. (1964). *Proc. R. Soc. Lond. B.* **159**, 668–86.
Coleman, G.S. (1979). In "Biochemistry and Physiology of Protozoa." 2nd Ed. Vol. 2. (Eds M. Levandowsky and S.H. Hutner.) pp.381–408. (Academic: New York.)
Coleman, G.S. (1980). *Adv. Parasitol.* **18**, 121–73.
Coleman, G.S. (1985). *J. Agric. Sci.* **104**, 349–60.
Coleman, G.S., and Sandford, D.C. (1979). *J. Agric. Sci.* **92**, 729–42.
Corliss, J.O. (1985). *J. Protozool.* **32**, 373–6.
Dawson, K.A., Allison, M.J., and Hartman, P.A. (1980a). *Appl. Environ.* Microbiol. **40**, 833–9.
Dawson, K.A., Allison, M.J., and Hartman, P.A. (1980b). *Appl. Environ. Microbiol.* **40**, 840–6.
Demeyer, D.I. (1981). *Agric. Environ.* **6**, 295–337.
Doddema, H.J., and Vogels, G.D. (1978). *Appl. Environ. Microbiol.* **36**, 752–4.
Dogiel, V.A. (1927). *Arch. Protistenk.* **59**, 1–228.
Eadie, J.M., and Mann, S.O. (1970). In "Physiology of Digestion and Metabolism in the Ruminant." (Ed. A.T. Phillipson.) pp. 335–47. (Oriel: Newcastle-upon-Tyne, UK.)
Fenna, L., and Boag, D.A. (1974). *Can. J. Zool.* **52**, 537–40.

Forsberg, C.W., and Lam, K. (1977). *Appl. Environ. Microbiol.* **33**, 528–37.
Forsberg, C.W., Crosby, B., and Thomas, D.Y. (1986). *J. Anim. Sci.* **63**, 310–25.
Forsythe, S.J., and Parker, D.S. (1985a). *Br. J. Nutr.* **53**, 183–90.
Forsythe, S.J., and Parker, D.S. (1985b). *J. Appl. Bact.* **58**, 363–9.
Fuller, R. (1984). *Proc. Nutr. Soc.* **43**, 55–61.
Gall, J.G. (1984). *Nature* (Lond.) **310**, 453–4.
Giesecke, D. (1970). In "Physiology of Digestion and Metabolism in the Ruminant." (Ed. A.T. Phillipson.) pp. 306–18. (Oriel: Newcastle-upon-Tyne, UK.)
Hammond, A.C., Bradley, B.J., Yokoyama, M.T., Carlson, J.R., and Dickinson, E.O. (1979). *Am. J. Vet. Res.* **40**, 1398–401.
Hammond, A.C., Carlson, J.R., and Breeze, R.G. (1978). *Science* **201**, 153–5.
Hanssen, I. (1979). *Acta Vet. Scand.* **20**, 365–71.
Herd, R.M., and Dawson, T.J. (1984). *Physiol. Zool.* **57**, 70–84.
Hespell, R.B. (1984). In "Herbivore Nutrition in the Subtropics and Tropics." (Eds F.M.C. Gilchrist and R.I. Mackie.) pp. 346–58. (Science Press: Craighall, South Africa.)
Hespell, R.B. (1985). *Rev. Rural Sci.* **6**, 95–100.
Hoover, W.H., and Heitmann, R.N. (1972). *J. Nutr.* **102**, 375–9.
Hornicke, H., and Bjornhag, G. (1980). In "Digestive Physiology and Metabolism in Ruminants." (Eds Y. Ruckebusch and P. Thivend.) pp. 707–30. (MTP: Lancaster, UK.)
Hult, K., Teiling, A., and Gatenbeck, S. (1976). *Appl. Environ. Microbiol.* **32**, 443–4.
Hume, I.D. (1977). *Comp. Biochem. Physiol. A* **56**, 299–304.
Hume, I.D. (1982). "Digestive Physiology and Nutrition of Marsupials." (Cambridge Univ.: Cambridge.)
Hume, I.D. (1984). *Bioscience* **34**, 435–40.
Hume, I.D., Rubsamen, K., and von Engelhardt, W. (1980). *J. Comp. Physiol. B* **138**, 307–14.
Hungate, R.E. (1960). *Bact. Rev.* **24**, 353–64.
Hungate, R.E. (1966). "The Rumen and its Microbes." (Academic: New York.)
Hungate, R.E. (1975). *Ann. Rev. Ecol. Syst.* **6**, 39–66.
Hungate, R.E. (1976). In "Fiber in Human Nutrition." (Eds G.A. Spiller and R.J. Amen.) pp. 131–49. (Plenum: New York.)
Hungate, R.E. (1984). *Proc. Nutr. Soc.* **43**, 1–11.
Huntington, G.B. (1986). *Fed. Proc.* **45**, 2272–6.
Imoto, S., and Namioka, S. (1978). *J. Anim. Sci.* **47**, 467–78.
Ivan, M., Veira, D.M., and Kelleher, C.A. (1986). *Br. J. Nutr.* **55**, 361–7.
James, L.F., and Butcher, J.E. (1972). *J. Anim. Sci.* **35**, 1233–8.

James, L.F., Allison, M.J., and Littledike, E.T. (1975). In "Digestion and Metabolism in the Ruminant." (Eds I.W. McDonald and A.C.I. Warner.) pp. 576–90. (Univ. of New England: Armidale, NSW.)

Johnson, J.L. (1984). In "Bergey's Manual of Systematic Bacteriology." (Eds N.R. Krieg and J.G. Holt.) pp. 8–11. (Williams and Wilkins: Baltimore, USA.)

Jones, R.J. (1981). *Aust. Vet. J.* **57**, 55.

Just, A., Jorgensen, H., and Fernandez, J.A. (1981). *Br. J. Nutr.* **46**, 209–19.

Kennedy, P.M., and Hume, I.D. (1978). *Comp. Biochem. Physiol. A* **61**, 117–21.

Kiessling, K.H., Pettersson, H., Sandholm, K., and Olsen, M. (1984). *Appl. Environ. Microbiol.* **47**, 1070–3.

King, R.R., McQueen, R.E., Levesque, D., and Greenhalgh, R. (1984). *J. Agric. Food Chem.* **32**, 1181–3.

Kingsbury, J.M. (1980). In "Toxicology: The Basic Science of Poisons." (Eds J. Doull, C.D. Klaassen and M.O. Amdur.) pp. 578–90. (MacMillan: New York.)

Kofoid, C.A., and MacLennan, R.F. (1930). *Univ. Calif. Publ. Zool.* **33**, 471–544.

Kofoid, C.A., and MacLennan, R.F. (1932). *Univ. Calif. Publ. Zool.* **37**, 53–152.

Kofoid, C.A., and MacLennan, R.F. (1933). *Univ. Calif. Publ. Zool.* **39**, 1–34.

Krumholz, L.R., Forsberg, C.W., and Veira, D.M. (1983). *Can. J. Microbiol.* **29**, 676–80.

Kurmanov, I.A. (1977). In "Mycotoxic Fungi, Mycotoxins and Mycotoxicoses." (Eds T.D. Wyllie and L.G. Morehouse.) pp. 85–110. (Marcel Dekker: New York.)

Laycock, K.A., Hazlewood, G.P., and Miller, E.L. (1985). *Proc. Nutr. Soc.* **44**, 54A.

Leng, R.A. (1984). In "Ruminant Physiology – Concepts and Consequences." (Ed. S.K. Baker.) pp. 161–73. (Univ. of W. Australia: Nedlands.)

Leng, R.A., and Nolan, J.V. (1984). *J. Dairy Sci.* **67**, 1072–89.

Ljungdahl, L.G., and Eriksson, K.E. (1985). *Adv. Microb. Ecol.* **8**, 237–99.

Lubinsky, G. (1957a). *Can. J. Zool.* **35**, 135–40.

Lubinsky, G. (1957b). *Can. J. Zool.* **35**, 141–59.

Lubinsky, G. (1958a). *Can. J. Zool.* **36**, 819–35.

Lubinsky, G. (1958b). *Can. J. Zool.* **36**, 937–59.

MacKenzie, C.R., and Patel, G.B. (1986). *Arch. Microbiol.* **145**, 91–6.

Mackie, R.I., Gilchrist, F.M.C., Robberts, A.M., Hannah, P.E., and Schwartz, H.M. (1978). *J. Agric. Sci.* **90**, 241–54.

Marty, J., and Vernay, M. (1984). *Br. J. Nutr.* **51**, 265–77.

Mason, V.C. (1969). *J. Agric. Sci.* **73**, 99–111.

Mason, V.C. (1979). *Z. Tierphysiol. Tierernahr. Futtermittelk.* **41**,

140-9.
Mason, V.C. (1984). *Proc. Nutr. Soc.* **43**, 45-53.
Mason, V.C., Just, A., and Bech-Andersen, S. (1976). *Z. Tierphysiol. Tierernahr. Futtermittelk.* **36**, 310-24.
Mason, V.C., Kragelund, Z., and Eggum, B.O. (1982). *Z. Tierphysiol. Tierernahr. Futtermittelk.* **48**, 241-52.
Mathison, G.W., and Milligan, L.P. (1971). *Br. J. Nutr.* **25**, 351-66.
McBee, R.H. (1977). In "Microbial Ecology of the Gut." (Eds R.T.J. Clarke and T. Bauchop.) pp. 185-222. (Academic: London.)
McInerney, M.J., and Bryant, M.P. (1981). In "Biomass Conversion Processes for Energy and Fuels." (Eds S. Sofer and O.R. Zaborsky.) pp. 277-96. (Plenum: New York.)
Mortensen, A. (1984). In "Current Perspectives in Microbial Ecology." (Eds M.J. Klug and C.A. Reddy.) pp. 273-8. (Am. Soc. Microbiol: Washington DC.)
Mortensen, A., and Tindall, A. (1981a). *Acta Physiol. Scand.* **III**, 129-33.
Mortensen, A., and Tindall, A. (1981b). *Acta Physiol. Scand.* **113**, 465-9.
Niiyama, M., Deguchi, E., Kagota, K., and Namioka, S. (1979). *Am. J. Vet. Res.* **40**, 716-8.
Nolan, J.V., and Leng, R.A. (1972). *Br. J. Nutr.* **27**, 177-94.
Nolan, J.V., Norton, B.W., and Leng, R.A. (1976). *Br. J. Nutr.* **35**, 127-47.
Nuzback, D.E., Bartley, E.E., Dennis, S.M., Nagaraja, T.G. Galitzer, S.J., and Dayton, A.D. (1983). *Appl. Environ. Microbiol.* 46, 533-8.
O'Brien, R.W., and Slaytor, M. (1982). *Aust. J. Biol. Sci.* **35**, 239-62.
Odelson, D.A., and Breznak, J.A. (1985). *Appl. Environ. Microbiol.* **49**, 622-6.
Odum, E.P. (1983). "Basic Ecology." (Holt-Saunders: Tokyo.)
Okumura, J., Hewitt, D., Salter, D.N., and Coates, M.E. (1976). *Br. J. Nutr.* **36**, 265-72.
Orpin, C.G., and Letcher, A.J. (1979). *Curr. Microbiol.* **3**, 121-4.
Pearce, P.D., and Bauchop, T. (1985). *Appl. Environ. Microbiol.* **49**, 1265-9.
Phillipson, A.Tb. (1977). In "Dukes' Physiology of Domestic Animals." 9th Ed. (Ed. M.J. Swenson.) pp. 250-86. (Comstock: Ithaca, NY.)
Pilgrim, A.F., Gray, F.V., Weller, R.A., and Belling, C.B. (1970). *Br. J. Nutr.* **24**, 589-98.
Pimm, S.L. (1980). *Ecology* **61**, 219-25.
Pratt, J.R., and Cairns, J. (1985). *J. Protozool.* **32**, 415-23.
Prins, R.A. (1977). In "Microbial Ecology of the Gut." (Eds R.T.J. Clarke and T. Bauchop.) pp. 73-183. (Academic: London.)

Prins, R.A., Lankhorst, A., and van Hoven, W. (1984). In "Herbivore Nutrition in the Subtropics and Tropics." (Eds F.M.C. Gilchrist and R.I. Mackie.) pp. 408–34. (Science Press: Craighall, South Africa.)

Prior, R.L., Hintz, H.F., Lowe, J.E., and Visek, W.J. (1974). *J. Anim. Sci.* **38**, 565–71.

Regoeczi, E., Irons, L., Koj, A., and McFarlane, A.S. (1965). *Biochem. J.* **95**, 521–32.

Russell, J.B. (1985). *Appl. Environ. Microbiol.* **49**, 572–6.

Salter, D.N. (1973). *Proc. Nutr. Soc.* **32**, 65–71.

Salter, D.N., and Coates, M.E. (1971). *Br. J. Nutr.* **26**, 55–69.

Schuberg, A. (1888). *Zool. Jarhrb. Jena, Abt. Syst.* **3**, 365–418.

Shirley, E.K., and Schmidt-Nielsen, K. (1967). *J. Nutr.* **91**, 496–502.

Shull, L.R., and Cheeke, P.R. (1983). *J. Anim. Sci.* **57** (Suppl. 2), 330–54.

Slade, L.M., Bishop, R., Morris, J.G., and Robinson, D.W. (1971). *Br. Vet. J.* **127**, xi–xiii.

Slade, L.M. Robinson, D.W., and Casey, K.E. (1970). *J. Anim. Sci.* **30**, 753–60.

Smith, C.J., and Hespell, R.B. (1983). *J. Dairy Sci.* **66**, 1536–46.

Smith, R.L., Strohmaier, F.E., and Oremland, R.S. (1985). *Arch. Microbiol.* **141**, 8–13.

Sorrell, M.F., Frank, O., Thomson, A.D., Aquino, H., and Baker, H. (1971). *Nutr. Rep. Int.* **3**, 143–8.

Stevens, C.E., Argenzio, R.A., and Clemens, E.T. (1980). In "Digestive Physiology and Metabolism in Ruminants." (Eds Y. Ruckebusch and P. Thivend.) pp. 685–706. (MTP: Lancaster, UK.)

Stewart, C.S., Paniagua, C., Dinsdale, D., Cheng, K.-J., and Garrow, S.H. (1981). *Appl. Environ. Microbiol.* **41**, 504–10.

Stumm, C.K., Gijzen, H.J., and Vogels, G.D. (1982). *Br. J. Nutr.* **47**, 95–9.

Tamminga, S. (1979). *J. Anim. Sci.* **49**, 1615–30.

Teather, R.M. (1985). *Can. J. Anim. Sci.* 65, 563–74.

Ueno, Y. (1977). In "Mycotoxins in Human and Animal Health." (Ed. J.V. Rodricks.) p. 189. (Pathotox: Illinois.)

Van Bruggen, J.J.A., Stumm, C.K., and Vogels, G.D. (1983). *Arch. Microbiol.* 136, 89–95.

Van Bruggen, J.J.A., Stumm, C.K., Zwart, K.B., and Vogels, G.D. (1985). *FEMS Microbiol. Lett.* **31**, 187–92.

Van Bruggen, J.J.A., Zwart, K.B., van Assema, R.M., Stumm, C.K., and Vogels, G.D. (1984). *Arch. Microbiol.* **139**, 1–7.

Vogels, G.D., Hoppe, W.F., and Stumm, C.K. (1980). *Appl. Environ. Microbiol.* **40**, 608–12.

Wallace, R.J. (1985). *Curr. Microbiol.* **12**, 59–64.

Wallace, R.J. (1986). In "Comparative Aspects of the Physiology of Digestion in Ruminants." (Ed. A. Dobson.) (Cornell Univ.: Ithaca, NY.)

Wallace, R.J., and Joblin, K.N. (1985). *FEMS Microbiol. Lett.* **29**, 19–25.
Webb, K.E. (1986). *Fed. Proc.* **45**, 2268–71.
Wei, C.-M., Campbell, I.M., Laughlin, C.S., and Vaughan, M.H. (1974). *Mol. Cell Biol.* **3**, 215–9.
Westlake, K., Mackie, R.I., and Dutton, M.F. (1987a). *Appl. Environ. Microbiol.* **53**, in press.
Westlake, K., Mackie, R.I., and Dutton, M.F. (1987b). *Appl. Environ. Microbiol.* **53**, in press.
Williams, A.G. (1986). *Microbiol. Rev.* **50**, 25–49.
Williams, A.G., and Ellis, A.B. (1985). *Curr. Microbiol.* **12**, 175–82.
Williams, A.G., and Strachan, N.H. (1984). *Curr. Microbiol.* **10**, 215–20.
Williams, A.G., and Withers, S.E. (1981). *J. Appl. Bact.* **51**, 375–85.
Williams, A.G., and Withers, S.E. (1982). *J. Appl. Bact.* **52**, 389–401.
Williams, A.G., Ellis, A.B., and Coleman, G.S. (1986). *Curr. Microbiol.* **13**, 139–47.
Williams, A.G., Withers, S.E., and Coleman, G.S. (1984). *Curr. Microbiol.* **10**, 287–94.
Wolin, M.J. (1975). In "Digestion and Metabolism in the Ruminant." (Eds I.W. McDonald and A.C.I. Warner.) pp. 134–48. (Univ. of New England: Armidale, NSW.)
Wolin, M.J. (1979). *Adv. Microb. Ecol.* **3**, 49–77.
Wolin, M.J., and Miller, T.L. (1983). *Fed. Proc.* **42**, 109–13.
Wood, T.M., Wilson, C.A., McCrae, S.I., and Joblin, K.N. (1986). *FEMS Microbiol. Lett.* **34**, 37–40.
Wootton, J.F., and Argenzio, R.A. (1975). *Am. J. Physiol.* 229, 1062–7.
Yamin, M.A. (1978). *J. Protozool.* **25**, 535–8.
Yamin, M.A. (1980). *Appl. Environ. Microbiol.* **39**, 859–63.
Yamin, M.A. (1981). *Science* **211**, 58–9.
Yamin, M.A., and Trager, W. (1979). *J. Gen. Microbiol.* **113**, 417–20.
Yarlett, N., Coleman, G.S., Williams, A.G., and Lloyd, D. (1984). *FEMS Microbiol. Lett.* **21**, 15–9.

12

NON-MICROBIAL DIGESTION OF FORAGES BY HERBIVORES

by E.R. Ørskov and R.N.B. Kay

Herbivores digest their food in part by their own digestive secretions and in part by playing host to microbes which can ferment fibre as well as the more tractable nutrients the host animal could otherwise degrade unaided. It has proved difficult to establish the nutrient requirements of the herbivore independently of those of its microbes, especially in species that ferment their food before it reaches the absorptive small intestine. These requirements have now been studied in ruminants in which microbial fermentation has been reduced to negligible proportions and maintained by infusion of liquid diets into the rumen and abomasum.

FORESTOMACH *VS* HINDGUT FERMENTATION

Selective pressure will cause the host animal to adapt its metabolism to the supply of nutrients absorbed from its digestive tract. One may therefore anticipate substantial differences in the requirement and utilization of nutrients between forestomach fermenters and hindgut fermenters. The former obtain most of their energy from volatile fatty acids (VFA) and most of their amino acids from microbial protein; their microbes also supply B vitamins and may degrade some dietary toxins. The hindgut fermenters can make little or no use of microbial protein and vitamins, unless they practice coprophagy, but such starch, sugar and protein as their diet may contain is available for digestion in their small intestine and little but the fibrous residue and resistant starch granules are fermented to VFA in the large intestine (Stevens *et al.* 1980).

Although perhaps a rather naive approach, one may consider what is the most appropriate construction of the alimentary tract to achieve greatest productivity from a given type of herbage. Ruminant animals selectively retain fibrous matter in the rumen for prolonged fermentation and show quite a wide range of morpho-physiological adaptations of the digestive tract which suit them to diets ranging from succulent leaves and fruits to coarse fibrous grasses (Hofmann 1973). Yet those selecting the most digestible diets tend to reduce rumen fermentation by shortening retention time with a corresponding increase in the importance of the large intestine (Hofmann 1983) and those relying on the least digestible diets have to enlarge their rumen capacity greatly to permit sufficient intake despite prolonged retention in the rumen (Mould *et al.* 1982). Perhaps it is the herbivores practising hindgut fermentation that are best adapted to both extremes

NUTRITION OF HERBIVORES
ISBN 0 12 750052 9

of the quality range of herbage. At one end of the spectrum are pigs which in Europe and North America receive diets based largely on grains and in China and South East Asia are fed mainly on good quality herbage and kitchen offal. At the other end of the spectrum are horses which in some regions of France will gain in weight when consuming poor quality roughage on which cattle lose weight. When forage quality becomes very poor on the rangelands of Africa the zebra (*Equus* sp.) often remains in good condition while ruminants become severely malnourished, presumably because, with no rumen to impede digesta flow, it can increase intake as forage quality falls, so maintaining an adequate rate of nutrient absorption even though rapid passage may depress efficiency of fibre digestion. The giant panda (*Ailuropoda melanoleuca*), another hindgut fermenter, adopts the same strategy, the very rapid passage of digesta allowing it to consume huge amounts of its poor quality bamboo diet.

SECRETION OF NITROGENOUS SUBSTANCES INTO THE GUT

Some nutrients of value to the gut microbes, especially urea and sulphur, pass from plasma to gut contents through the gut wall and in secretions. Forestomach fermenters have an advantage here in having such nutrients provided in the saliva. In the case of urea, however, plasma concentration tends to be highest when ammonia absorption from the rumen is greatest, so that urea recycling is most pronounced when nitrogen supply to the rumen microbes already exceeds their requirement.

A more helpful control over urea flow to the rumen was postulated by Wallace *et al.* (1979). They showed that the urease activity of the bacteria adhering to the rumen wall increased as the rumen ammonia concentration decreased and suggested that the creation of high-ammonia low-urea concentrations in the rumen wall would promote the movement of urea-N to the rumen in situations when rumen N requirements were most urgent. The regulation of urea movement into the rumen is not yet adequately understood, however. F.G. Whitelaw (unpub. observ.) has recently shown that the movement of urea into the rumen of sheep nourished only by intraruminal infusion of VFA was little different from that in normally-fed sheep.

Plasma urea also passed readily into the intestinal contents and can provide the N required by the microbes in the large intestine. Although there is a net absorption of N between terminal ileum and rectum of sheep under normal conditions (Clarke *et al.* 1966), when fermentation in the hindgut was enhanced by infusion of starch into the caecum, microbial activity and faecal N excretion were greatly increased with corresponding reduction in urinary urea excretion (Ørskov *et al.* 1970). However, while the N secreted into the rumen may be incorporated into microbial protein and thus become available as amino N to the host animal, it seems unlikely that significant amounts of microbial protein synthesised in the hind gut can be utilized except by coprophagy, although further research on this subject is required (Hornicke and Bjornhag 1980).

Table 1. The flow of non-ammonia N through different segments of gut in cows, steers and lambs nourished by N-free nutrients given by intragastric infusion (Ørskov and McLeod 1982; Storm *et al.* 1983; Ørskov *et al.* 1986).

	Dairy cows	Steers	Lambs
Rumen outflow			
mg /kg$^{0.75}$LW	51	90	79
g /d	8.3	6.0	1.3
Abomasal outflow			
mg /kg$^{0.75}$LW		196	194
g /d		13.4	2.9
Ileal outflow			
mg /kg$^{0.75}$LW			42
g /d			0.54
Faecal output			
mg /kg$^{0.75}$LW	34	29	36
g /d	4.1	2.2	0.44

The excellent account of N recycling in ruminants recently compiled by Egan *et al.* (1986) makes it clear that urea is not the only important vehicle carrying endogenous N to the rumen. Shed epithelial cells and mucus from the mouth, oesophagus, rumen and respiratory tract may also be appreciable. Ørskov *et al.* (1986) measured the quantity of this contribution in lambs, steers and dairy cows maintained by infusion of liquid diets containing no nitrogen. It can be seen from Table 1 that the non-ammonia N leaving the rumen, which would not include urea N as this is completely hydrolysed to ammonia, accounts for about a half of the non-ammonia N leaving the abomasum. The increment supplied by the abomasum is probably largely derived from abomasal enzymes together with further epithelial debris and mucus. The amino acid composition of both non-ammonia N fractions was similar to that of animal protein (except for a high cystine content that presumably arises from the keratin component) and was indicative of a high biological value. Animals sustained by intragastric nutrition lack the physical and chemical stimuli to digestive secretions and epithelial replacement provided by ingestion and fermentation of normal food and so probably show a subdued level of endogenous N production. Here one may note, however, that the mitotic index of rumen epithelial tissue is little different in liquid-infused and in normally-fed animals.

The point remains that even in liquid-infused animals given a N-free diet a substantial flow of protein is maintained from the stomach to the intestines.

ABSORPTION OF VOLATILE FATTY ACIDS

Although the VFA produced by fermentation of food in the rumen are very largely in their neutralized, ionic form at rumen pH values of 6 or more, the permeability of the rumen wall is such that they are largely absorbed in their acid, un-ionised form. Therefore only a fraction, around one third, of the VFA produced needs to be neutralized by salivary bicarbonate and phosphate to maintain rumen pH within its normal range (Kay 1966). Equally only a part of the VFA given into the rumen of animals sustained by infusion of liquid diets needs to be neutralized, thus reducing the external load of sodium and alkali that such animals have to excrete in their urine. However, infusion of inadequately neutralized VFA can readily produce acidosis in such animals, although preliminary trials in which salivary secretion was augmented by oral infusion of VFA indicate a way in which alkali infusion could be further reduced (E. Storm unpublished).

Under normal feeding conditions it has been found that the molar proportions of VFA in the rumen reflect fairly closely the rates of production (Bergman *et al.* 1965; Whitelaw *et al.* 1970). It has long been known that, under acute experimental conditions, butyric and propionic acids are absorbed more readily than acetic acid at low rumen pH, and at very high rumen pH the converse relationship exists (Danielli *et al.* 1946). Results indicating selective VFA absorption have also been obtained by Thorlacius and Lodge (1973) using rumen pouches and by Sutton and Morant (1978) using isotopically-labelled VFA.

The intragastric infusion technique offers a new approach to this question as "production rate" (i.e. infusion rate) is precisely known and controlled. A comparison of VFA proportions in infusate and rumen content at rumen pH values ranging from 5.5 to 6.8 was made by MacLeod *et al.* (1984). The molar proportions in the rumen clearly differed from the proportions in the infusate. At pH 5.6 the molar proportion of acetate in the rumen was about 10% higher than in the infusate, indicating preferential absorption of butyrate and propionate under acid conditions, whereas at pH 6.8 little difference between rumen content and infusate was seen.

UTILIZATION OF VOLATILE FATTY ACIDS

Much controversy still surrounds this subject. Are individual VFA utilized with equal efficiency or not? Part of the uncertainty stems from disagreements about how much glucose or glucose precursor is needed for efficient oxidation of acetate through the citric acid cycle and also to provide for other processes, such as oxidative metabolism in the brain and in foetal tissues and through the pentose phosphate pathway for the formation of NADPH used in fatty acid synthesis. This is coupled with the clear knowledge that the digestible energy from the poor-quality fibrous roughage diets, that give rise to a high proportion of acetate in rumen VFA, is almost always utilized with less efficiency than that from good-quality roughages and concentrates. Thus it is tempting to suggest a cause and effect relationship. However, except in earlier experiments (Armstrong and Blaxter 1957), when pure VFA or their salts were given, no significant differences in net efficiency of utilization of various VFA, alone or in combination, have been demonstrated (Ørskov and Allen, 1966; Hovell *et al.* 1976).

Again, maintenance of ruminants by infusion of liquid diets allows a critical assessment of the question. Table 2, from Ørskov *et al.* (1979), shows the efficiency with which various mixtures of fatty acids were used for energy retention when infused intraruminally into sheep maintained by intragastric infusion. It can be seen that, if one disregards the infusion containing the lowest proportion of acetate (which would never be found in normal rumen contents), there was no consistent variation of k_f with VFA proportions (SE of mean, 0.04 to 0.11). MacRae and Lobley (1982) suggested that a plentiful supply of dietary protein may suppress an effect of VFA proportion on efficiency of energy utilization by providing abundant glucogenic amino acids. Yet in the experiment illustrated in Table 2 the protein:energy ratio resembled that expected when a carbohydrate source is fermented in the rumen to microbial protein and VFA. It seems therefore that differences in utlization of metabolisable energy between diets are not to be explained by differences in VFA proportions but by other factors such as activity associated with eating, work of digestion, etc. (Graham 1986).

Table 2. The effect of different proportions of volatile fatty acid (VFA) on efficiency (k_f) for energy retention in sheep using a common estimate of energy for maintenance (0.45 kJ /kg$^{0.75}$LW) (Ørskov *et al.* 1979).

VFA proportions (mmol/mol)			VFA proportions (kJ/MJ)			k_f
Acetic	Propionic	Butyric	Acetic	Propionic	Butyric	
350	550	100	200	620	160	0.78
450	450	100	300	530	170	0.64
550	350	100	390	430	180	0.57
650	250	100	480	330	190	0.61
750	150	100	590	210	200	0.61
850	50	100	720	70	210	0.59

Dairy cows require glucose not only for the metabolic processes mentioned above but also for synthesis of lactose and other requirements of milk secretion. A study was made by Ørskov *et al.* (1977) on Friesian cows to see how far the supply of glucose or glucose precursors places constraints on milk production. The cows were fed in early lactation to yield only 10 kg of milk daily and mixtures of glucose and casein were infused into the abomasum. The results are shown in Table 3. It can be seen that replacement of glucose by casein progressively stimulated milk yield and increased the protein and fat content of the milk. This increased production occurred despite the reduced supply of total glucose plus glucogenic amino acids in the infusate, the increasing energy deficit presumably being met by mobilisation of fat reserves. This effect has recently been substantiated in experiments reported by Whitelaw *et al.* (1986) in which heat production was also determined.

Table 3. Effect of casein and glucose infusion into the abomasum on milk yield and milk composition in cows in early lactation and negative energy balance (Ørskov *et al.* 1977).

Casein (g /d)	Glucose (g /d)	Milk yield (kg /d)	Milk fat (g /kg)	FCM* (kg /d)	Milk Protein (g /kg)	Net energy deficit (MJ /d)
0	750	16.8	48.2	18.9	25.2	20.5
250	500	19.8	49.8	22.7	28.4	30.5
500	250	21.6	51.0	25.2	29.6	38.6
750	0	21.4	54.0	26.1	31.5	41.0
S.E. of means		0.8	1.2	-	0.6	-

*Fat corrected milk yield

In order to see if it would be possible to drive lactating animals into a primary glucose deficiency and so cause ketosis, dairy cows were given a high silage diet mixed with various amounts of fishmeal which largely escaped from the rumen in an undegraded state (Ørskov *et al.* 1982). The cows were calculated to be consuming sufficient metabolisable energy to yield from 12 to 15 kg of milk. At the high level of fishmeal inclusion (12% of dietary dry matter) the cows yielded 35 kg milk daily, about 20 kg of which was produced at the expense of body tissue, and they developed primary ketosis. At the lower level of fishmeal (8%) the cows yielded rather less milk, 28kg, but still a half of it depended on fat mobilisation and the cows now showed no sign of ketosis. In this experiment virtually no glucose would be absorbed from the gut and the only exogenous sources of glucogenic material would be the dietary fishmeal and ruminal microbial protein, yet this was sufficient to support lactation where a half of the yield was fuelled from body fat.

Ruminants that are subjected to fasting or low level nutrition have to mobilise body fat at a time when little or no nutrients are being absorbed from the gut. This is therefore another condition in which glucose precursors may be limiting metabolic processes. During fasting there is a sharp increase in N excretion due to oxidation of body protein which may yield glucogenic metabolites as well as energy. Studies on ruminants maintained by intragastric infusions have shown that urinary excretion of N is always greater during fasting than when N-free nutrients are infused at an estimated maintenance energy level. Unpublished data from a recent experiment by J.C. Ku Vera, E.R. Ørskov and F.D. Hovell are summarised in Table 4. Mixtures of VFA containing high levels of either acetate or propionate were infused into the rumens of sheep to provide about 20% or 40% of the energy estimated to be needed for energy balance. It can be seen there was almost no difference between the two levels of intake nor between the high acetate and high propionate mixture in their ability to reduce urinary N excretion to well below that recorded in the fasting state.

Table 4. The effect on urinary N excretion by sheep of infusing volatile fatty acids containing a high or low proportion of acetic acid at 2 levels of infusion, 75 or 150 kJ/kg$^{0.75}$LW, compared to fasting animals.

Molar proportion of infused acids (Acetic:propionic:butyric)	Level of infusate (kJ)	Urinary excretion (mg N /kg$^{0.75}$LW)	
15 : 3 : 2	75	250	
15 : 3 : 2	150		224
9 : 9 : 2	75	234	
9 : 9 : 2	150		215
Fasting		325	
Fasting			372
SE		20	14

In both these examples of severely submaintenance nutrition, one in the lactating female the other in the non-productive adult, depot fat is being mobilised and there is also, in the former case, a massive synthesis of milk constituents. Yet the need for glucose precursors to support these metabolic processes appears to be low and may only become critical in lactating or pregnant animals that become inappetant for reasons of acidosis or other stress conditions. The need for glucose and glucose precursors seems to be met adequately by the glucogenic products of rumen fermentation, propionate and microbial matter (nucleic acids and carbohydrate as well as non-essential amino acids), which are not used directly for protein synthesis.

ABSORPTION OF ENERGY-YIELDING NUTRIENTS FROM THE SMALL INTESTINE

The most important energy-yielding substrates presented to the small intestine of ruminants are dietary lipids (which are generally associated with particulate digesta), small amounts of soluble sugars that evade rumen fermentation, starch under some circumstances such as with feeding of maize or sorghum grain, and microbial matter. Potentially digestible and indigestible fibre also enters the small intestine and a little may be digested by microbial fermentation in the terminal ileum. The capacity for digestion and absorption of sugar, starch and protein is seldom exceeded when roughages are eaten, but can easily be overwhelmed when large amounts of grain or molasses are consumed. The capacity for digestion of sucrose is very limited; some forms of starch (such as the flinty starch of maize grain and raw potato starch) are not readily attacked by amylase and the capacity for glucose absorption is not great (Ørskov 1980). Such undigested carbohydrate then provides a substrate for increased fermentation in the large intestine.

The capacity for digestion of protein is seldom exceeded in adult ruminants when ruminal microorganisms provide the main source of protein (Clarke *et al.* 1966). Digestion by the powerful proteolytic secretions of the abomasum, pancreas and intestinal mucosa is aided by the fact that there is a fairly constant flow of protein from the rumen. The capacity can be exceeded when non-clotting proteins are used in milk replacers (Soliman *et al.* 1982).

DIGESTION IN THE LARGE INTESTINE

Digestion in the large intestine is of minor importance in ruminants consuming coarse roughages but may be somewhat greater in species that are adapted to selection of high quality diets (Hofmann 1983) and propel food rapidly through their gut. The main functions of the large intestine are the absorption of water and salts and the storage and elimination of food residues (Bruce *et al.* 1966). However, fermentation in the large intestine is enhanced when the capacity for digestion and absorption of carbohydrate in the small intestine is exceeded, as mentioned above. The faeces then become moist and the apparent digestibility of N is reduced because increased quantities of microbial protein are formed in the large intestine which are not subsequently digested (Ørskov *et al.* 1970).

DIGESTION OF DIETARY PROTEIN

The amount of dietary protein that evades fermentation in the rumen and so enters the small intestine depends both on its rate of degradation in the rumen and on how long it is retained in the rumen. Since the dietary proteins that are most digestible in the small intestine may tend to have the most rapid rate of degradation in the rumen, and in any case the truly indigestible N in the diet will still be present in the smaller amount of undegraded dietary protein that reaches the intestine, one might expect the digestibility of dietary protein entering the small intestine to be rather low. Hvelplund (1983) assessed the digestibility of undegraded protein in the small intestine in relation to the degradability in the rumen. His results, summarized in Table 5, confirm the expected tendency for intestinal digestibility of undegraded protein to fall as rumen degradability increases.

Table 5. Rumen degradability and the digestibility in the small intestine of proteins which escape rumen degradation (Hvelplund 1983).

Protein source	Degradability (g/g)	Digestibility in small intestine of undegraded protein (g/g)
Coconut cakes	0.33	0.85
Fish meal	0.44	0.86
Cottonseed cake	0.59	0.70
Soyabean meal	0.75	0.81
Rapeseed meal	0.77	0.63
Sunflower cakes	0.87	0.67

ABSORPTION AND UTILIZATION OF MICROBIAL PROTEIN

In ruminants, by far the most important source of protein is normally the microbial protein produced as a result of fermentation of carbohydrate in the rumen. When this was recognized many workers isolated small amounts of rumen microbial protein and determined its digestibility and biological value in rats. These results and others are summarized in Table 6 (Storm *et al.* 1983a, b).

It can be seen that the values reported by different workers vary considerably. Due to variation in the quantity of microbial protein produced it is very difficult to estimate directly the value of microbial protein to intact ruminants and it is not easy to obtain quantities of separated rumen microbial matter to allow feeding experiments with large animals. Two developments that allowed more controlled studies of the digestion and utilization of microbial protein by ruminants have been a technique for isolating large amounts of microorganisms and nutrition by intragastric infusion. Storm and Ørskov (1983) isolated 75kg of freeze-dried microbial matter from slaughter house rumen contents. This, being their only scource of protein, was then infused into lambs maintained by intragastric infusion protein. Collection of digesta at the terminal ileum allowed the digestibility of each amino acid in the small intestine and the biological value of microbial amino acids to be determined.

Table 6. Published values of true nitrogen digestibility, biological value and net N utilization of mixed rumen microorganisms (RMO).

Method used	True N digest-ibility	Biol. value	Net N utili-zation	Source
Sheep given purified diets with urea as the main N source	0.89	0.62	0.55	Harris & Mitchell (1941)
	0.55	0.66	0.36	Johnson *et al.* (1944)
Rats given semi-purified diets with isolated RMO as main source of protein	0.64	0.79	0.51	Reed *et al.* (1949)
	0.73	0.88	0.64	McNaught *et al.* (1950)
	0.74	0.81	0.60	McNaught *et al.* (1954)
	0.75	0.85	0.64	Bergen *et al.* (1968)
	0.79	0.72	0.57	Mason & Palmer (1971)
Intragastric infusion of RMO	0.81	0.67	0.54	Storm *et al.* (1983b)

Some of the more important nutritive characteristics of this microbial matter are given in Table 7. Only about 80% of the microbial N was amino acid N but the biological value of this fraction was high. The amount of microbial N required for the lambs to achieve maintenance of N balance was 843 mg N /kg$^{0.75}$LW. If the requirement of metabolisable energy (ME) for maintenance of energy balance in sheep is 0.4 MJ /kg$^{0.75}$LW and the yield of rumen microbial N is 1.37 g N /MJ ME (1985) then at maintenance 548 mg microbial N /kg$^{0.75}$LW will be produced. The amount of microbial protein produced by fermentation of a diet that allows maintenance of energy balance in a mature sheep thus appears insufficient to allow maintenance of N balance unless supplemented (as it normally would be) by undegraded dietary protein. When fed below maintenance energy intake the production of microbial protein will fall still further behind requirement, though when energy intake and protein of dietary origin are restored the animals will show the compensatory growth that is observed in the many areas of the world which have a seasonal fluctuation in the availability of good forage.

Table 7. Nutritive values of microbial matter in ruminants when infused into the abomasum of lambs sustained by intragastric infusion (Storm *et al.* 1983 a,b).

Composition of microbial dry matter (g /kg DM)	
Lipids	92
Carbohydrate	93
Crude protein	640
Composition of microbial N (g /kg N)	
RNA-N	114
DNA-N	43
Amino acid N	806
Digestibility of microbial fractions	
Dry matter	0.711
Apparent N	0.774
True N	0.813
Digestibility in small intestine	
Dry matter	0.742
Total N	0.815
Amino acid N	0.847
RNA-N	0.871
DNA-N	0.812
Utilisation of microbial fractions	
Total N	0.543
Truly digested N	0.663
Absorbed amino acid N	0.801
N equilibrium requirement	
Microbial N (mg /kg$^{0.75}$LW	843

The requirement of the animal for amino acids from its diet (that is those amino acids limiting in microbial protein), will be influenced by the amino acid requirement of the animal product concerned. Thus the optimum composition of absorbed amino acids will depend on whether the main product is wool, milk or lean tissue. Although the various microorganisms differ somewhat in amino acid composition there is no evidence that total microbial composition varies to any great extent.

Storm and Ørskov (1984) devised a new method for estimating the order and extent of limitation of the amino acids of rumen microbial protein in growing lambs. They assumed that if amino acids were present in a dietary protein in optimal proportions, utilisation of amino N would be 1.00. The lower value of about 0.80 (Table 7), obtained for microbial protein, was due to a deficiency of one or more amino acids. If 25% more microbial protein were infused into the sheep, the deficiency would be repaired and the N balance would represent that achieved if the lower intake of protein had contained an optimal balance of amino acids. Instead of adding 25% more microbial protein Storm and Ørskov (1984) added 25% more of a mixture of essential amino acids. After ascertaining that the expected N balance was achieved, each amino acid was removed from the mixture in turn. Methionine removal returned the N balance to that found with unsupplemented microbial protein, showing that methionine was the first limiting amino acid. Removal of other amino acids in turn gave a N balance indicative of the point at which each one became limiting. For the growing lambs concerned it was found that the limiting amino acids in microbial protein, in order of importance, were methionine, lysine, histidine and arginine. There is clearly an opportunity for many further studies in this area of investigation.

Agricultural Research Council (1985). "The Nutrient Requirements of Ruminant Livestock." Supplement No. 1. (Commonwealth Agricultural Bureaux: Farnham Royal, UK.)

Armstrong, D.G., and Blaxter, K.L. (1957). *Br. J. Nutr.* **11**, 413–25.

Bergen, W.G., Purser, D.B., and Cline, J.H. (1968). *J. Anim. Sci.* **27**, 1497–501.

Bergman, E.N., Reid, R.S., Murray, M.G. Brockway, J.M., and Whitelaw, F.G. (1965). *Biochem. J.* **97**, 53–8.

Bruce, J., Goodall, E.D., Kay, R.N.B., Phillipson, A.T., and Vowles, L.E. (1966). *Proc. Roy. Soc. Lond. B* **166**, 46–62.

Clarke, E.M.W., Ellinger, G.M., and Phillipson, A.T. (1966). *Proc. Roy. Soc. Lond. B* **166**, 63–79.

Danielli, J.F., Hitchcock, M.W.S., Marshall, R.A., and Phillipson, A.T. (1946). *J. Expt. Biol.* **22**, 75–84.

Egan, A.R., Boda, K., and Varady, J. (1986). "In Control of Digestion and Metabolism in Ruminants." (Eds L.P. Milligan, W.L. Grovum, and A. Dobson.) pp. 386–402. (Prentice Hall: New Jersey.)

Graham, N.McC. (1986). *Br. J. Nutr.* **56**, 515.

Harris, L.E., and Mitchell, H.H. (1941). *J. Nutr.* **22**, 167–82.

Hofmann, R.R. (1973. "The Ruminant Stomach." East African Monographs in Biology No. 2. (East African Literature Bureau: Nairobi.)

Hofmann, R.R. (1983). *Roy. Soc. N.Z. Bull.* **20**, 51–8.

Hornicke, H., and Bjornhag, G. (1980). In "Digestive Physiology and Metabolism in Ruminants." (Eds Y. Ruckebusch and P. Thivend.) pp. 707–30. (MTP: Lancaster, UK.)

Hovell, F.D., Greenhalgh, J.F.D., and Wainman, F.W. (1976). *Br. J. Nutr.* **35**, 343–63.

Hvelplund, T. (1983). IV Int. Symposium, Protein Metabolism and Nutrition. pp. 283–6 (INRA Publications: France.)
Johnson, B.C., Hamilton, T.S., Robinson, W.B., and Garey, J.C. (1944). *J. Anim. Sci.* **3**, 287–8.
Kay, R.N.B. (1966). *Wld Rev. Nutr. Diet.* **6**, 292–325.
MacLeod, N.A., Ørskov, E.R., and Atkinson, T. (1984). *J. Agric. Sci.* **103**, 459–62.
MacRae, J.C., and Lobley, G.E. (1982). *Livestock Prod. Sci.* **9**, 447–56.
Mason, V.C., and Palmer, R. (1971). *J. Agric. Sci.* **76** 567–72.
McNaught, M.L., Owen, E.C., Henry, K.M., and Kon, S.K. (1954). *Biochem. J.* **56**, 151–6.
McNaught, M.L., Smith, J.A.B., Henry, K.M., and Kon, S.K. (1950). *Biochem. J.* **46**, 32–6.
Mould, F.L., Saadullah, M., Haque, M., Davis, D., Dolberg, F., and Ørskov, E.R. (1982). *Trop. Anim. Prod.* **7**, 174–81.
Ørskov, E.R., and Allen, D.M. (1966). *Br. J. Nutr.* **20**, 519–32.
Ørskov, E.R., Fraser, C., Mason, V.C., and Mann, S.O. (1970). *Br. J. Nutr.* **24**, 671–82.
Ørskov, E.R., Grubb, D.A., and Kay, R.N.B. (1977). *Br. J. Nutr.* **38**, 397–405.
Ørskov, E.R., Grubb, D.A., Smith, J.S., Webster, A.J.F., and Corrigall, W. (1979). *Br. J. Nutr.* **41**, 541–51.
Ørskov, E.R. (1980). In "Digestive Physiology and Metabolism in Ruminants." (Eds Y. Ruckebusch, and P. Thivend.) pp. 309–24. (MTP: Lancaster, UK.)
Ørskov, E.R., and MacLeod, N.A (1982). *Br. J. Nutr.* **47**, 625–36.
Ørskov, E.R., Reid, G.W., and Tait, C.A.G. (1982). *Anim. Prod.* **34**, 361.
Ørskov, E.R., MacLeod, N.A., and Kyle, D.J. (1986). *Br. J. Nutr.* **56**, 241–8.
Reed, F.M., Moir, R.J., and Underwood, E.J. (1949). *Aust. J. Sci. Res. B.* **2**, 304–17.
Soliman, H.S., Ørskov, E.R., Davies, N.T., and McDonald, I. (1982). *J. Agric. Sci.* **98**, 377–89.
Stevens, C.E., Argenzio, R.A., and Clemens, E.T. (1980). In "Digestive Physiology and Metabolism in Ruminants." (Eds Y. Ruckebusch and P. Thivend.) pp. 685–706. (MTP: Lancaster, UK.)
Storm, E., and Ørskov, E.R. (1983). *Br. J. Nutr.* **50**, 463–70.
Storm, E., and Ørskov, E.R. (1984). *Br. J. Nutr.* **52**, 613–20.
Storm, E., Brown, D.S., and Ørskov, E.R. (1983b). *Br. J. Nutr.* **50**, 479–85.
Storm, E., Ørskov, E.R., and Smart, R. (1983a). *Br. J. Nutr.* 471–78.
Sutton, J.D., and Morant, S.V. (1978). In "Ruminant Digestion and Feed Evaluation." (Eds D.F. Osborn, E.E. Beever, and D.J. Thomson.) (Agric. Res. Counc.: London.)
Thorlacius, S.O., and Lodge, G.A. (1973). *Can J. Anim. Sci.* **53**, 279–88.
Wallace, R.J., Cheng, K.J., Dinsdale, D., and Ørskov, E.R. (1979). *Nature*, Lond. **279**, 424–6.

Whitelaw, F.G., Milne, J.S., Ørskov, E.R., and Smith, J.S. (1986). *Br. J. Nutr.* **55** 537–56.

Whitelaw, F.G., Hyldgaard-Jensen, J., Reid, R.S., and Kay, M.G. (1970). *Br. J. Nutr.* **24**, 179–95.

13

THE TRANSITION FROM MILK TO FORAGE DIETS

by P.A. Janssens and J.H. Ternouth

Mammals first appeared in the fossil record about 200 million years ago and marsupials and eutherians became differentiated during the Cretaceous Period. Initially, the mammals remained a small and insignificant group and only began to assume their present dominant position about 65 million years ago when the dinosaurs finally became extinct. During the early part of the Tertiary Period, mammals showed extensive adaptive radiation, particularly into zones which had previously been occupied by the dinosaurs (Keast 1972).

Mammals are thought to have entered Australia by a southern route about 60 - 70 million years ago, when the continent was still part of Gondwanaland (Keast 1972) or to have originated in Australia at an even earlier age (Kirsch 1984). However, the fossil record is poor and the oldest known Australian fossil mammals date from about 23 million years ago (Archer 1984). The subsequent geographical isolation of Australia has lead Rich (1982) to say, "During the Cainozoic Period at least, Australia's terrestrial mammalian fauna evolved in virtually complete isolation from their nearest relatives on other continents". Thus, it is possible that all Australian terrestrial mammals descended from a single, or at the most a very few, species which entered the continent in the late Mesozoic or early Cainozoic Periods (Rich 1982). The superfamily Macropodoidea, which includes the macropodids (kangaroos, wallabies, wallaroos and euros) and the potoroids (rat-kangaroos), first appeared in the fossil record some 15 million years ago (Flannery 1984). Clearly then, the eutherian and marsupial herbivores have had distinct lineages for many millions of years.

Eutherian herbivores, especially domestic ruminants, have been extensively studied for many years but it is only recently that interest in the marsupials has burgeoned and a body of information on the physiology of their digestion has been gathered together. When this information is examined, there are striking similarities between the digestive physiology of the eutherian and marsupial herbivores. It is our aim to review the information currently available on the transition from a milk to a herbage diet in marsupial animals and to compare and contrast the picture in marsupials with that in eutherian animals. We are hampered in this aim by the patchiness of the studies on marsupials, and, for that matter, on the transition period in eutherian animals. In the case of the marsupials, by far the greatest amount of information is available on the tammar wallaby, *Macropus eugenii*. Some scattered data are available for a variety of other macropodids, the common brushtail possum, *Trichosurus vulpecula*, and the carnivorous eastern quoll, *Dasyurus viverrinus*. The tammar is

NUTRITION OF HERBIVORES
ISBN 0 12 750052 9

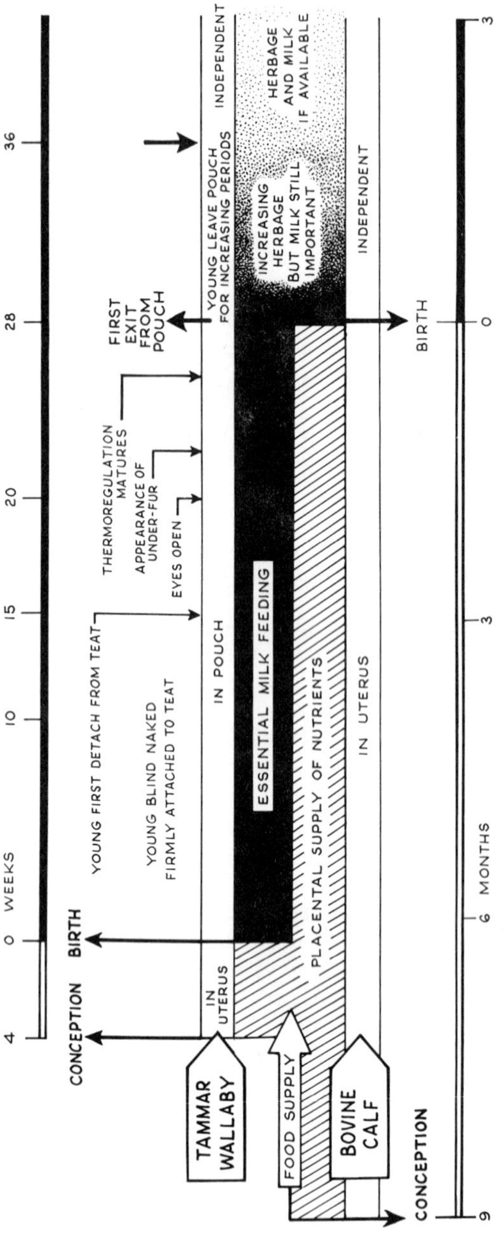

Figure 1. The patterns of pre- and post-natal development of the tammar wallaby and the bovine calf.

pre-eminent in studies of marsupial physiology; a large colony of them was established by CSIRO in Canberra 15 years ago. Many of the studies discussed in this paper have used animals originating from this colony. It is unlikely that studies on the physiology of the Australian marsupial fauna will be significantly broadened until self-sustaining colonies of other species have been established, a time-consuming and expensive undertaking.

The tammar wallaby is a small macropodid with adult females weighing 4 - 6 kg and males 5 - 8 kg. The young weighs about 400 mg when born, crawls to the pouch in the first few minutes after birth, and begins to suck from one of the four teats to which it remains attached for about 15 weeks. During the period from 15 to 28 weeks it remains in the pouch but periodically detaches from the teat. The eyes open at about 20 weeks, the animal becomes homeothermic at about 26 weeks and the first exit from the pouch is made at about 28 weeks. The young then spends increasing periods out of the pouch until permanent pouch exit, at about 36 weeks. For several further weeks, the young returns to the mother at intervals to suck, and then finally becomes independent at 45 - 50 weeks of age. In this chapter we define weaning as a process which lasts from the time the young first ingests significant amounts of food other than milk, until the time when milk is no longer essential for the survival of the young. Using this definition, the process of weaning in the tammar wallaby lasts from about week 28 of lactation until about week 36 (Fig. 1). Although the young put their heads out of the pouch and nibble grass from about week 25, examination of their stomach contents indicates that herbage is a negligible component of their diet at this time.

The eutherian herbivores that have been most thoroughly studied are the calf and the lamb which have intra-uterine lives of 280 and 150 days respectively. The pre- and post-natal life of the calf has been compared with that of the tammar in Fig 1. In both lambs and calves, small quantities of dry food are consumed from about two weeks after birth, and from this time the animals are considered to have entered a period of ingestive, digestive and metabolic transition (Wardrop and Coombe 1960; Santhirasegaram and Watts 1970). Herbage forms a significant part of the diet from about 4 weeks in lambs (Wardrop and Coombe 1960; Dove and Freer 1979) and 4-6 weeks in calves (Santhirasegaram and Watts 1970) and milk ceases to be essential between 6 and 9 weeks. In this paper, comparison has been made using the calf as a model eutherian animal. However, other eutherian animals do not have the same independent mobility immediately after birth; for instance, many polytocous mammals including the rabbit are born at a much more immature stage. The ready availability of bovine milk as a basic ingredient for diets has resulted in large numbers of young ruminants and other young mammals, particularly dairy cattle calves, being hand-fed. This has resulted in man regulating the times at which young ingest their first dry food and cease to drink milk; in other words, man regulates the age at which weaning occurs (Fig. 1).

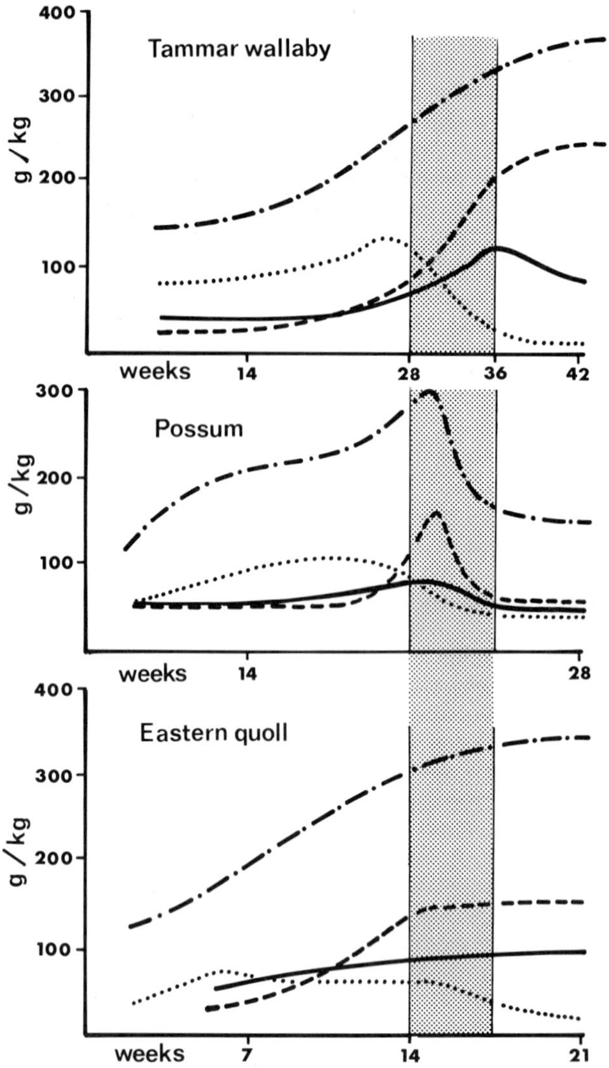

Figure 2. Variation in the milk composition of the tammar wallaby, the common brushtail possum and the eastern quoll; ▬ ▪ ▬, total solids; • • • • •, carbohydrates; ▬▬▬, lipid; ———, protein concentration. The horizontal time axis is adjusted so that the weaning periods (shaded areas) are synchronized.

MILK COMPOSITION

The composition of the milk of marsupials so far studied changes dramatically during the period of lactation. Green *et al.* (1980) have found that the tammar wallaby, which is a fore-gut fermenter, has milk solid levels which increase from about 120 g /kg at the start of lactation to approach 400 g /kg at the end (Fig 2a). A similar change in milk solids occurs in the eastern quoll and in the common brushtail possum; milk solids increase from 100 g /kg to about 300 g /kg at the start of weaning after which they fall to about 160 g /kg (Fig. 2).

The milk of the cow contains about 130 g DM /kg throughout lactation, composed of 30 – 55 g fat, 31 – 39 g protein and 47 – 51 g carbohydrate (Schmidt 1971). Only relatively small changes in the concentrations of these constituents occur during a lactation except for the high levels of non-casein proteins secreted in the first few days of the lactation. Other eutherian herbivores including the sheep (Cook *et al.* 1970; Peart *et al.* 1972, 1975), red deer (*Cervus elephas*; Arman *et al.* 1974), goat (Parkash and Jenness 1968), water buffalo (*Bubalus bubalis*; Laxminarayana and Dastur 1968) and guinea-pig (*Cavia porcellus*; Mepham and Beck 1973) have milks with relatively constant composition throughout lactation.

In the first week of lactation, the carbohydrate content of the milk of the tammar is *ca* 40 g /kg (Messer *et al.* 1984) after which there is a gradual increase to a maximum of about 130 g /kg between 26 and 28 weeks of lactation (Fig 2a). During the next 4 to 6 weeks, during the weaning period, the milk carbohydrate content falls to less than 20 g /kg and remains at this level until the end of the lactation (Messer and Green 1979). At the start of lactation, lactose is the only carbohydrate present in the milk but higher oligosaccharides begin to appear in the first week and thereafter their concentration gradually increases whereas that of lactose falls (Messer *et al.* 1984). Between the fifth and 28th weeks of lactation, the milk carbohydrate contains relatively little lactose and is primarily composed of oligosaccharides containing up to eight hexose moieties together with a large oligosaccharide fraction containing sialic acid (Messer and Green 1979). These oligosaccharides are the trisaccharide, 3'-galactosyl lactose (Messer *et al.* 1980) and the series of homologues produced by the addition of (1->3)-linked β-D-galactosyl residues to the non-reducing end of the molecule (Collins *et al.* 1981). After 28 weeks of lactation, the oligosaccharides are gradually lost from the milk so that from 36 weeks until the end of lactation, milk carbohydrate is almost entirely in the form of free monosaccharides.

In most eutherians, the carbohydrate is almost entirely in the form of lactose and usually has a concentration less than 60 g /kg. However, in the milk of the domestic rabbit (Cowie 1969) there is a 90% decrease in the concentration of lactose and a 100% increase in the concentration of fat and protein in mid-lactation. The concentration of lactose actually falls from 20 to 2 g /kg. The findings of Cowie (1969) may be peculiar to domestic rabbits which have very long lactations as the carbohydrate content of wild rabbit milk does not change markedly (B.Green, K.Merchant and J.Newgrain, unpublished).

Early in lactation, the lipid content of tammar milk is low, remaining below 50 g /kg until at least 26 weeks (Fig 2a). Thereafter, there is a rapid increase in concentration reaching a maximum of about 250 g/kg at 38 weeks, at which level it remains for the remainder of the lactation (Green *et al.* 1983). This increase in lipid concentration coincides with the period of weaning, the decrease in carbohydrate concentration, and a change in the composition of milk lipids. At 26 weeks of age, oleic, stearic and palmitic acids are the predominant fatty acids and form 400, 250 and 200 g/kg total fatty acids. By 38 weeks of lactation, these concentrations have changed to 500, 200 and 60 g/kg respectively. The proportion of saturated acids thus falls from 50 to 30% during the period of weaning, with a reciprocal increase in the proportion of unsaturated fatty acids (Green *et al.* 1983).

The protein concentration of tammar milk is about 40 g /kg during the first 20 weeks of lactation, rises to about 120 g /kg at 36 weeks and then falls to 80 g /kg at 45 weeks (Fig 2a; Green *et al.* 1980). The proteins in tammar milk are predominantly in the whey fraction and both the number and amounts of individual proteins increase during the lactation. However, there is no specific change associated with weaning (Green and Renfree 1982).

In raw bovine milk, the concentration of whey proteins and immunoglobulins are 10.3 and 0.9 g /kg respectively (Roy 1980). However, the concentrations in the first post-partum milk (colostrum) may exceed 95.0 and 68.0 g/kg respectively. This results from the high concentration of the immunoglobulins IgG_1, IgM and IgA and albumin, lactalbumin and lactoglobulin in the colostrum. Over the period of the first four 12-hourly milkings after calving, the concentration of total proteins in the colostral milk falls from 190 to 45 g /kg (Parrish *et al.* 1948).

As well as in the tammar, the pattern of low lipid/high carbohydrate before weaning, switching to low carbohydrate/high fat after weaning has also been found in four other groups of marsupials. These include four macropodids, the red, eastern grey and western grey kangaroos and the red necked wallaby (*Macropus rufus*, *M. giganteus*, *M. fuliginosus* and *M. rufogriseus;* Lemon and Barker 1967; Poole *et al.* 1982; Green 1984) and the potoroo (*Potorous tridactylus*; H.M.Crowley, D.R.Woodward and R.M.Rose, unpublished), all of which are fore-gut fermenters. The other groups of marsupials with this pattern include the common brushtail possum (Gross and Bolliger 1959), which is a hind-gut fermenter, the American opossum (*Didelphis virginiana* – Bergman and Housley 1968), which is omnivorous and the Eastern quoll (Green 1984), which is a carnivore. The absolute amounts of carbohydrate and fat do, however, differ between species (Fig. 2). As this switch is found in such a wide range of species, it is highly unlikely that it has any direct association with the diet of the adult animal. Rather it appears to be a characteristic of the marsupial animal, as also is the high concentration of carbohydrate in the milk. It is perhaps worth restating here that, although some early papers reported low levels of carbohydrate in marsupial milk (Gross and Bolliger 1959; Lemon and Barker 1967), these carbohydrates were not subjected to hydrolysis before analysis, and, because the oligosaccharides had a lower reducing power than monosaccharides the concentration of carbohydrate was under-estimated (Messer and Green 1979). Recent determinations of the carbohydrate content of the milk of the common brushtail possum give values more than double those previously reported (P. Cowan, unpublished; Fig. 2).

In discussing the unusual form of carbohydrate found in marsupial milk, Messer and Green (1979) pointed out that a solution containing 110 g lactose /kg is iso-osmotic with plasma. Thus, the lactose in bovine milk represents almost half the osmotically active moieties in milk. It follows that to concentrate the carbohydrates in milk, without making the milk hyperosmolar, the hexoses must be polymerised. It is significant in this regard that at the end of lactation, when the concentration of carbohydrate is low, the sugars are in the form of free hexoses (Messer and Green 1979). The question of why the marsupial, particularly the herbivorous marsupial, has milk which has such high levels of carbohydrate is of considerable interest. It may be related to the ability of the pouch marsupial to digest milk lipids. It is of note that bovine milk contains substantial quantities of butyric acid esterified at the sn-3 position on the glycerol (Pitas *et al.* 1967) whereas macropodid milk contains no short-chain fatty acids. Edwards-Webb and Thompson (1977) and Ternouth *et al.* (1980) have found that short-chain acids are much more easily hydrolysed from bovine milk triglyceride than the longer chain acids by both salivary and pancreatic lipases (EC 3.1.1.3) so such triglycerides are easily digested.

The elevated lipid content of tammar milk in the later stages of lactation results in a large increase in the energy level of the milk and coincides with the emergence of the young from the pouch (Green *et al.* 1980). Those authors have suggested that this increase in the energy content of the milk is adaptive as the energy requirements of the young increase after emergence from the pouch, because of the energetic demands of maintaining its own body temperature and independent mobility. The coincident fall in milk carbohydrate content may well be associated with the appearance of fore-gut microflora. In eutherian herbivores it is well known that high levels of soluble carbohydrate in the rumen promote the growth of *Streptococcus* spp. which increase lactic acid production and cause ruminal acidosis (Schwartz and Gilchrist 1975). Although the milk of eutherian animals contains substantial quantities of lactose, ruminal acidosis is prevented by the ingested milk bypassing the rumen and entering the abomasum. The change in the concentration of particular fatty acids in the milk may be the result of this adaptive mechanism and indicate that the mothers are in negative energy balance. Stobbs and Brett (1974) have found similar increases in oleic and stearic acids and decreases in palmitic acid concentrations in bovine milk when the cows were underfed due to the greater proportion of milk fatty acids being derived from body reserves.

There is a large increase in the concentration of Fe in the milk of the tammar at weaning, but the concentrations of Ca, P, Mg, Cu, and Zn do not change (Green 1984). Similar changes in Fe concentration have been reported in the milks of the quokka (*Setonix brachyurus*), the red-necked wallaby and the eastern quoll (Green 1984). The concentrations of both Fe and Cu are higher in marsupial than in eutherian milks (Green 1984). It is of note that if calves are fed on milk diets alone for extended periods, it is recommended that supplementary Fe and Cu are added to the milk (Roy 1980).

MILK INTAKE

The rate of growth of marsupial pouch young is far less than that of comparable eutherians (Russell 1982) and, although comparisons are difficult because of the paucity of the data, milk intake is also low (Green 1984; Cork and Dove 1986). The peak milk intake of the tammar wallaby occurs when the pouch young are starting to consume dry food and they weigh *ca* 1 kg. At this time the young consumes 56 ml /d (504 kJ /d), which is equivalent to a daily energy intake of about 162 kJ /kg maternal liveweight ($kg^{0.75}$). These estimates are low when compared with those from most eutherian mammals, especially eutherian herbivores (Oftedal 1984). As the rate of weight increase relative to milk consumption is similar in the tammar and eutherian species (Green 1984), it appears that the growth of tammar pouch young is limited by the supply of milk from the mother rather than by the intake of milk by the young. Thus, marsupial young will grow faster when transferred to pouches in which they receive an increased milk supply (Green 1984).

This limitation of growth by the milk supply is not seen in eutherians where the supply of milk by the mother increases in response to demand by the young such that lactating eutherian mothers produce milk in excess of that required by a single young. This difference in regulation of the size of the nutrient intake probably accounts for the slower growth rates of marsupial pouch young compared with their eutherian counterparts.

DIGESTION

There are major structural and functional changes in the gastro-intestinal tract of both the macropodid marsupials and the eutherian mammals during the transition from milk to herbage as the major constituent of their diets. The stomach of the adult red kangaroo is divided into two major regions, an anterior cardiac region representing 80% of the total stomach and a posterior gastric pouch (Griffiths and Barton 1966). The mucosa of the anterior cardiac region secretes no enzymes and the pH of this region is close to neutrality. In all macropodids that have been studied, this region of the stomach acts as a fermentative chamber and therefore is analogous in size and function to the rumeno-reticulum of the major domestic ruminants (Moir *et al.* 1956; Hume 1982). The gastric pouch is analogous to the abomasum of ruminants; the contents of both are acid and the mucosas contain both chief and parietal cells which secrete pepsinogen (EC 3.4.23.1) and hydrochloric acid (Griffiths and Barton 1966).

The stomach of the red kangaroo pouch young is very different. The whole stomach contents have an acid pH and proteolytic activity is high in both the anterior and posterior regions of the stomach. As in some other species, the gastric gland cells of the whole stomach wall are able to secrete both acid and enzyme (Griffiths and Barton 1966). After about 200 days of pouch life, mucous-secreting cardiac glands start to appear in the anterior cardiac stomach and the beginnings of gastric glands with both chief and parietal cells are seen in the gastric pouch. By 236 days, which is the time when the young red kangaroo is weaned and vacates the pouch, the gastric pouch contains gastric glands with both types of secretory cells. Both regions have a pH approaching adult values by 250 days of age (Griffiths and Barton 1966).

Comparable changes in stomach structure and physiology have been reported in the tammar wallaby (Janssens 1984), two pademelons (*Thylogale stigmata and T. thetis*; Langer 1979), and the longnosed potoroo (P. Frappell and R. Rose, unpublished), although the details vary between species. It is likely that a similar pattern is present in both the macropodids, which are true fore-stomach fermenters and in the potoroids (Hume 1982). Coincident with the increase in pH towards neutrality in the cardiac chamber of the tammar stomach (Table 1), herbage and micro-organisms appear in the contents (R.G. Warner - unpublished). In the potoroo, micro-organisms and fungal hyphae appear simultaneously (P. Frappell and R. Rose, unpublished). Thus, in all these marsupials, the anterior portion of the stomach changes its function from digestion using endogenous enzymes to microbial fermentation at the time when dry food becomes a significant part of the diet.

Post-natal changes in the morphology and size of the gastro-intestinal tract of eutherian mammals have been detailed for both the calf and the lamb (Wardrop and Coombe 1960; Stobo *et al.* 1966). The weight of the rumeno-reticular tissue in the newborn is 20 - 30% of total stomach weight, increasing to over 60% in the weaned calf or lamb. The increase is associated with a great increase in the volume of the rumeno-reticulum, and is the result of increases in the weight of the musculature together with the mucosa and submucosa of the mucosal papillae. Whilst there are these large changes in the size and weight of the rumeno-reticulum, the weight of the empty abomasum of the lamb is constant at 0.5% of liveweight. If the calves or lambs are maintained on a milk diet then less development of the rumeno-reticulum occurs, especially the papillae, so that the weight of the rumeno-reticulum and the abomasum are similar (Wardrop 1960). The feeding of high-concentrate, as distinct from high-roughage diets, is associated with more rumeno-reticular development but the volume of contents is smaller than for the roughage-fed calves (Stobo *et al.* 1966).

Table 1. The milk hexose concentration and the pH and contents of the fore-stomach of tammar wallabies at various ages.

Pouch age (Weeks)	Milk hexose (g/kg)	pH#	Forestomach contents#
18		2.5, 3.5	Milk only
24		2.0, 3.5	Milk only
28	120	3.5, 3.5	Milk + fragments of herbage
30	90	4.5, 6.0	Milk + some herbage
32	60	6.5, 7.0	Herbage + some milk
34	30	6.5, 7.0	Herbage + little milk

\# Based on 2 pouch-young at each age.

In ruminant animals the reflex closure of the reticular groove and the discrete nature of the forestomachs allows ingested milk to by-pass the rumeno-reticulum. The milk enters the abomasum, where coagulation occurs, prior to efficient digestion in the abomasum and small intestine by endogenous secretions (Black 1971). The reflex closure of the reticular groove has been described by Comline and Titchen (1951) and occurs whenever milk is drunk (Ørskov *et al.* 1970; Abe *et al.* 1979).

In the abomasum of the new-born calf and lamb, the pH of the contents is close to neutrality. In lambs, Hill (1956) has found that there is a 10-fold increase in the number of parietal cells in the first 72 hours of post-natal life. There is a further increase with age in the secretion of acid and a decrease in the pH of the digesta entering the duodenum (Ternouth *et al.* 1976). The limited acid secretion of the newborn ruminant does not adversely affect the abomasal proteolytic activity as chymosin (EC 3.4.23.4) rather than pepsin is the predominant abomasal enzyme. Chymosin has optimal proteolytic activity at a higher pH than pepsin. A similar proteolytic enzyme is found in the secretion of the stomach of the rabbit (Henschel 1973). With increasing age and with weaning there is an increase in pepsin and a decrease in chymosin in the abomasal mucosa (Guilloteau *et al.* 1984, 1985). Similar changes have been recorded in the mucosal wall of the lamb (Guilloteau *et al.* 1983).

The role of coagulation in digestion is to extend the time ingested protein and lipid remain in the abomasum so that abomasal digestion is enhanced and there is a slow constant outflow of milk protein and lipid. Within 6 hours of drinking 3–4 kg of milk, 80–90% of the whey fluids have passed out of the abomasum into the small intestine of the calf but only 25–30% of the proteins and, by calculation, less than 10% of the casein (Ternouth *et al.* 1975). Commencing 2–4 hours after feeding, partially hydrolysed casein and lipids are sloughed from the surface of the coagulum and pass into the small intestine. This process regulates abomasal outflow so that the outflow is constant for many hours (Ternouth *et al.* 1974, 1975). Irregular passage of digesta into the duodenum has been found to reduce the overall digestion of the diet (Guilloteau *et al.* 1981).

The enzymes present in the small intestine of the tammar pouch young alter dramatically at the time of the change from a milk to a herbage diet (Fig. 3). The activities of lactase (EC 3.2.1.23), neuraminadase (EC 3.2.1.18), β-D-acetylglucosaminidase (EC 3.2.1.30) and α-L-fucosidase (EC 3.2.1.51) are high during the first 34 weeks of pouch life but subsequently decline to negligible levels (Walcott and Messer 1980). In contrast, the activities of the disaccharidases maltase (EC 3.2.1.20), sucrase (EC 3.2.1.48), isomaltase (EC 3.2.1.10) and trehalase (EC 3.2.1.28) are all very low before 34 weeks of pouch life (Fig 3d) but then increase rapidly to adult levels (Walcott and Messer 1980). Lactase activity is higher in the distal portion of the intestine and its pH optimum is 4.6, which is lower than the intestinal lactase of most eutherians. Furthermore, the neutral lactase of eutherians is located in the brush border of the mucosa whereas both histochemical and biochemical studies show that acid lactase of the tammar is intra-cellular (Messer and Czolij 1984; Crisp and Messer 1986). Thus, in the tammar the lactose enters the mucosal cells by a pinocytotic process prior to intra-cellular digestion. These changes in enzyme activity are clearly associated with the change in diet. Although glucose is not present in the stomach of the adult tammar, it can be detected in the small intestine at concentrations which increase distally (Janssens 1984). To our knowledge, there are no comparable data for other non-macropodid marsupial species.

In the young milk-fed calf the pancreas secretes increasing quantities of trypsin (EC 3.4.21.4), chymotrypsin (EC 3.4.21.1), amylase (EC 3.2.1.1) and lipase with age (Ternouth *et al.* 1976; Guilloteau *et al.* 1984). The magnitude of the change in enzyme secretion is different for various enzymes, being greatest for amylase and lipase and only small for trypsin. Weaning further increases the concentration of all the enzymes except lipase (Guilloteau *et al.* 1985). In the milk-fed calf, the secretion of pancreatic ribonuclease (EC 3.1.27.5) also increases with age (Ternouth *et al.* 1975) and the levels are high in the adult ruminant (Barnard 1969). Generally similar changes have been observed

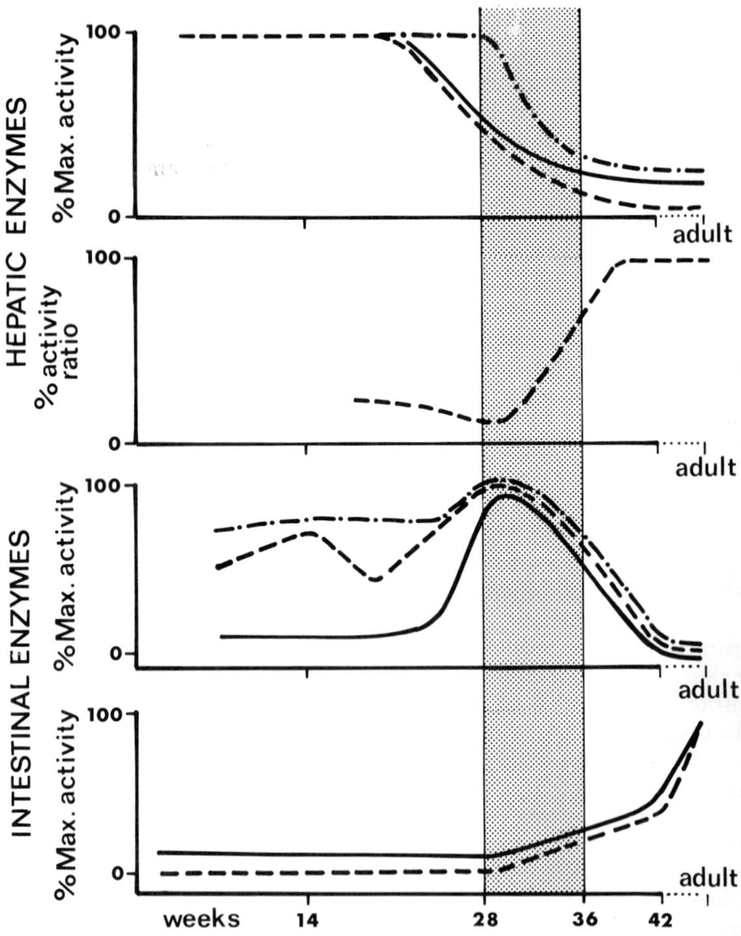

Figure 3. Changes in the activities of enzymes involved in carbohydrate metabolism in the intestine and liver of the young tammar wallaby. The weaning area is shaded.

(a) Hepatic enzymes ——, galactokinase; — · — ·, hexose-1-P uridyl transferase; — — —UDP-glucose-4-epimerase.

(b) Hepatic PEPCK/pyruvate kinase ratio.

(c) Intestinal enzymes — — — , lactase; — · — · — · , acetyl-glucose - aminidase; ———, neuramididase.

(d) Intestinal enzymes ——, maltase; — —, sucrase, isomaltase and trehalase.

in the pancreatic tissue of the lamb (Guilloteau *et al.* 1983). In the mucosal wall of the young calf, there are significant levels of lactase, glucosidase, trehalase and maltase but the concentrations of these enzymes decrease rapidly after weaning, except for maltase which increases further (Toofanian *et al.* 1974).

As in the marsupials, it is easy to see the physiological significance of these changes in enzyme activity in the eutherian animals. During early life when the milk contains lactose and fats, enzymes which can digest these substances are present. After weaning, the diet contains no lactose and only very small amounts of lipid but the digesta passing to the small intestine are likely to contain a variety of microbial products including proteins, nucleoproteins and a range of carbohydrates. During weaning, the concentrations of the spectrum of abomasal and intestinal enzymes alter with age, or are induced to alter with dietary change to digest the available feed particles. In particular, nucleases may be important in the conservation of nitrogenous bases and phosphate (Barnard 1969), both of which are likely to be in short supply in the diet.

A question of some importance is the origin of the gut micro-organisms and the preparation of the young for a herbage diet. Adult folivores have to tolerate and detoxify the high concentrations of phenolic compounds in the eucalyptus leaf diet. At the start of weaning, the common brushtail possum increases the concentration of these compounds in the milk from low levels to those found in the plasma. This is believed to condition the pouch young to these phenolic compounds (MacLennan *et al.* 1983). In the koala (*Phascolarctos cinereus*), the young have been observed to ingest maternal faeces during a period of a few days before they begin to eat eucalyptus leaves (Fleay 1937) and this may serve to expose the young to compounds not present in the milk but encountered when leaves enter the diet.

Such eating of faeces or regurgitates from the mother may also serve to ensure complete inoculation of the young's gut with micro-organisms which are essential for the digestion of herbage. In herbivorous marsupials, colonisation of the gut with microrganisms does not seem to occur until close to the time of weaning. In eutherians, colonisation of the gastro-intestinal tract of non-ruminant animals is complete within 48 hours of birth, both in terms of ecological balance and numbers (Ducluzeau 1983). Large populations of bacteria are present in the rumeno-reticulum of calves by ten days of age (Smith 1965). These bacteria may enter the rumeno-reticulum from the abomasum with the backflow of milk or with saliva following oral contact between calves, between calves and their dams or contaminated objects in their habitat. The results of Winter (1985) suggest that the rumeno-reticular microbial population is able to digest diets effectively by the time calves are four weeks of age.

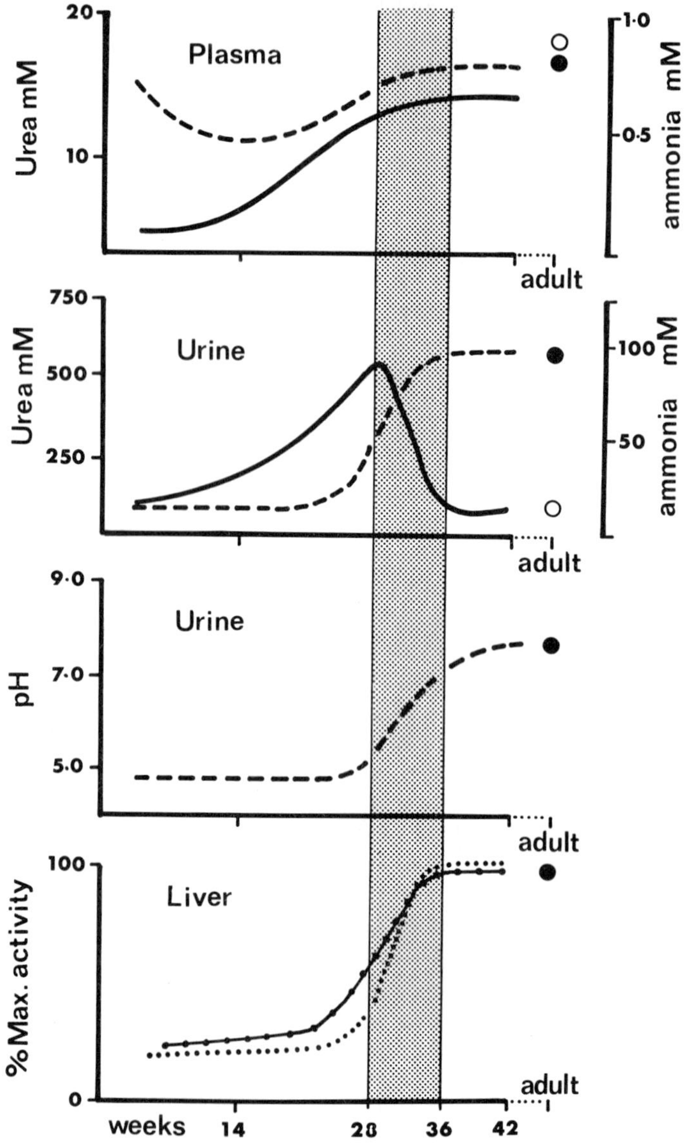

Figure 4. Changes in plasma and urine ammonia (——— ○) and urea (— — — ●) concentrations, urine pH and hepatic enzymes, carbamoyl phosphate synthetase (EC 6.3.4.16) (•—•—•—•) and argininosuccinate synthetase (EC 6.3.4.5.) (••••••••), in the young tammar wallaby. The weaning area is shaded.

THE SUPPLY OF GLUCOSE

The alterations in structure and function of the gastro-intestinal tract of herbivorous mammals during weaning result in a substantial change in the types and quantities of organic moieties absorbed by the animal. In the tammar, during the first 28 weeks of pouch life, carbohydrate is the principal source of energy, as the amount of lipid is low. After weaning has begun in the tammar, energy is supplied mainly in the form of lipids, either from the fermentative hydrolysis of carbohydrate to form volatile fatty acids (VFA) or from the milk lipids. As weaning is completed, the energy supply to the animal switches more strongly to VFA and glucose is essentially absent from the anterior stomach of the tammar (Janssens 1984). Ballard (1965) has found that hepatic glucokinase (EC 2.7.1.2) has a very low level of activity in the liver of the tammar. Fore-gut fermenters do not require enzymes for the rapid uptake of glucose from the portal circulation but the enzyme is required by hind-gut fermenters such as the rabbit and the common brushtail possum. This change in the form of the energy supply has profound consequences for the young. Although blood glucose levels are lower in adult tammars than in the pre-weaning young (Janssens 1984), glucose is still necessary for the maintenance of normal brain function. In addition, glucose is probably essential for other tissues such as the renal medulla, which relies largely upon anaerobic metabolism because of the low blood flow necessary to maintain the medullary concentration gradient. It has been found that the concentration of urea in the urine of the tammar increases at the time the young exit from the pouch (Fig. 4; Wilkes and Janssens 1986b). This change coincides with the reduction in water and carbohydrate intake in the form of milk and suggests an increased requirement for glucose precisely at the time when the supply of glucose from the diet is decreasing. Therefore, glucose must be supplied by endogenous synthesis from non-carbohydrate sources. The most likely substrates for gluconeogenesis are the VFA, particularly propionate, and amino acids. Several authors, including Moir *et al.* (1956), Ballard *et al.* (1969) and Hume (1982), have remarked upon the similarity between the metabolism of VFA in eutherian animals and macropodids and Barker (1961) has demonstrated the capacity of the adult quokka to synthesise glucose from propionate *in vivo*. However, except in the tammar wallaby, there is little information on the development of systems associated with the switch from carbohydrate to VFA as the principal energy source.

In isolated hepatocytes prepared from livers from fed pre-weaning tammar pouch young, the rate of gluconeogenesis from substrates such as lactate, propionate or alanine is relatively low and is significantly increased after 24 hours fasting (Janssens *et al.* 1977; Paton and Janssens 1981). After fasting, the concentration of malate in the liver decreases by over 60% whereas that of phosphoenolpyruvate (PEP) increases about 3-fold (Janssens *et al.* 1977). Therefore there is a cross-over (Rolleston 1972) between malate (=oxaloacetate) and PEP which suggests that increased activity of phosphoenolpyruvate carboxy-kinase (PEPCK) (EC 4.1.1.38) may be responsible for the increased rate of gluconeogenesis. When the activities of the specific gluconeogenic enzymes pyruvate carboxylase (EC 6.4.1.1.), PEPCK, fructose-bis-phosphatase (EC 4.1.1.38) and glucose-6-phosphatase (EC 3.1.3.9.) were determined, cytosolic PEPCK had increased about 10-fold in activity while no other enzyme had changed significantly (Janssens *et al.* 1977; Paton and Janssens 1981). In pre-weaning pouch young, therefore, gluconeogenesis is inducible by fasting and the rate-limiting enzyme is PEPCK.

In adult tammars, the rate of gluconeogenesis by isolated hepatocytes is not increased by fasting and the concentration of malate in the liver does not fall in fasted adults. As these findings would suggest, cytosolic PEPCK activity in both fed and fasted adults is similar to that in fasted pouch young (Paton and Janssens 1981). In adult tammars, therefore, gluconeogenesis is a constitutive pathway and cytosolic PEPCK is not inducible by fasting. Recently, we determined the relative activities of cytosolic PEPCK and pyruvate kinase in the liver of the tammar pouch young at intervals through the process of weaning. Between 28 and 36 weeks of pouch life, the ratio of cytosolic PEPCK activity to pyruvate kinase activity increased some 10-fold (Fig. 3b; Wilkes and Janssens 1986b). This corresponds very precisely with the period of weaning and it is concluded that the change in gluconeogenic regulation from inducibility to constitutiveness is part of the weaning process.

The regulation of glucose supply in adult tammars is very similar to that observed in adult eutherian ruminants. Adult ruminants characteristically have low blood glucose levels and glucokinase is absent from the liver (Ballard 1965; White and Leng 1980; Lindsay and Oddy 1986), indicating that the liver is not adapted for uptake of glucose from the hepatic portal vein. Lindsay and Oddy (1986) have concluded that twice as much carbon dioxide was derived from the oxidation of glucose in milk-fed than grass-fed sheep. Nevertheless, ruminants must continually synthesise glucose (Leng 1970) and, in contrast with non-ruminant herbivores, the activity of PEPCK in livers of fed ruminants is relatively high and is not increased with fasting (Ballard *et al.* 1969; Filsell *et al.* 1969). As the liver of ruminants receives large quantities of VFA, the coenzyme A derivatives of these fatty acids can act as activators of pyruvate carboxylase. Thus, in contrast with other species, bovine pyruvate carboxylase is activated by butyryl-CoA in addition to acetyl-CoA and propionyl-CoA (Ballard *et al.* 1969; Clark *et al.* 1976).

The ontogeny of gluconeogenesis and its enzymes differs between ruminant and non-ruminant species, the enzymes generally appearing earlier in the ruminants (Ballard *et al.* 1969; Ballard 1971; Arinze 1975) and in no case is the pattern observed in the tammar duplicated in eutherians. In foetal sheep liver gluconeogenic enzymes are present, but gluconeogenesis from lactate is not activated until after birth (Warnes *et al.* 1977a, b). Furthermore, Edwards *et al.* (1975) have reported an increase in the ratio of PEPCK to pyruvate kinase in the last few days of gestation and after birth. However, Prior and Christenson (1977) have reported that the rate of gluconeogenesis from alanine in foetal sheep is about double that in suckling lambs. Clark *et al.* (1976) and Lindsay and Oddy (1986) report that gluconeogenesis is active in both pre-ruminant and ruminant lambs although alanine and serine are better substrates before rumination develops whereas propionate is a better substrate after rumination has begun. The lambs studied by Clark *et al.* (1976) were kept without food for 24 hours before hepatocyte preparation so no comment can be made on the effect of fasting and, while glucagon stimulated gluconeogenesis in pre-ruminant lambs, ruminant lambs were not tested.

In the guinea-pig, the gluconeogenic pathway is active before birth, increases in activity during the first few days after birth and then declines to adult levels (Jones and Ashton 1976; Raghunathan and Arinze 1977). The gluconeogenic enzymes pyruvate carboxylase, soluble and particulate PEPCK, fructose-bis-phosphatase and glucose-6-phosphatase, which are low in activity early in gestation, increase in activity by 40 or 50 days to a plateau which is maintained until birth. Immediately after birth all these enzymes, but especially soluble PEPCK and glucose-6-phosphatase, increase rapidly in activity for a few days after which they decline to the adult level (Jones and Ashton 1976; Raghunathan and Arinze 1977).

In all cases, these developmental patterns differ in timing from that seen in the tammar but a common thread can be discerned. Gluconeogenesis becomes operative at the stage of development when the young obtains an insufficient supply of glucose from the mother. In some eutherian foetuses, ample glucose is supplied by the mother throughout gestation and gluconeogenesis is turned on at birth, for example, in the rat. In other species, such as the guinea-pig, the glucose supply from the mother may become inadequate during the later stages of gestation when the young is relatively large. Many eutherian species produce milk which has a relatively low carbohydrate content and in these species, which include the calf, the sheep and the guinea-pig, gluconeogenesis is relatively high in activity during lactation. The tammar, in contrast, gives birth to a very small young and glucose is unlikely to be limiting during gestation (Renfree 1973). The milk is high in carbohydrate and, although the pouch young is capable of gluconeogenesis when it is only ten days old, the activity of the pathway in the fed young is low. It is only when milk is replaced by herbage and the population of micro-organisms in the fore-stomach becomes mature that the young requires a high rate of endogenous glucose synthesis and gluconeogenesis becomes constitutive. The common thread then, in the emergence of gluconeogenesis as an important pathway, is the requirement for glucose by the young. In the tammar this occurs at weaning. In most eutherians it happens earlier when the glucose supply from the mother is either inadequate or irregular. In the first case, gluconeogenesis may be constitutive, in the second, inducible.

A second function of the liver is the metabolism of nutrients derived from the diet. Vernon *et al.* (1981) have shown that three enzymes of galactose metabolism, galactokinase (EC 2.7.1.6), hexose-1-phosphate uridyl transferase (EC 2.7.7.12) and UDP glucose 4-epimerase (EC 5.1.3.2) are very active in the livers of tammar pouch young until 24–30 weeks of lactation. Thereafter, these three enzymes decrease in activity, reaching the low levels characteristic of the adult by about 36 weeks (Fig. 3). The decrease in activity in these three enzymes parallels the decrease in galactose content of the milk. In pre-weaning pouch young, the activities of galactokinase and uridyl transferase are similar to those in rat sucklings. UDP glucose 4-epimerase, on the other hand, has more than double the activity in rat liver, which is consistent with this enzyme being rate-limiting in galactose metabolism (Segal 1978) since the galactose content of pre-weaning tammar milk is so much higher than that of rat milk. Vernon *et al.* (1981) also determined the activity of these three enzymes of galactose metabolism in the livers of early pouch young of four other macropodids and in the Eastern quoll. In most cases, the enzymes were lower in activity than in similarly aged or staged tammar pouch young, although they were invariably higher than in adult tammars. These findings indicate that marsupial pouch young have a metabolic capacity appropriate to the composition of their diet.

Although the milk of the tammar contains protein, the growth rate of the pouch young appears to be limited by the rate of production of milk by the mother (Merchant and Sharman 1966; Green 1984). This suggests that most milk protein is used in growth and not as a gluconeogenic substrate so that urea production would be low. This conclusion is supported by the finding that the plasma urea concentration of the pre-weaning tammar young is only half that found in the adult (Fig. 4). Indeed, Wilkes and Janssens (1986a) have argued that the inability of the tammar pouch young to produce a concentrated urine is, at least in part, due to the lack of available urea. This inability of the young tammar to produce sufficient urea to maintain a concentration gradient in the renal medulla attests to the low rate of deamination of proteins during this stage of development.

HOMEOSTASIS AND THE DISPOSAL OF WASTE

When the diet of the young tammar is composed exclusively of milk, the metabolism is acidotic. The pH of the urine is between 4.5 and 5.5 (Fig. 4) and the ammonia content of the urine progressively increases from less than 20 mM at 6 weeks to a maximum of about 100 mM by 26 weeks of age. During the same time period, the urinary urea content is consistently below 200 mM. During the few weeks after herbage is first detectable in the stomach and the process of weaning begins, there are dramatic changes in all these parameters. The pH of the urine begins to rise and approaches adult levels of around pH 8 by week 36 (Fig. 4); the ammonia content of the urine falls rapidly to the adult level of about 18 mM by week 34, despite the rise in plasma ammonia concentration which occurs at this time, and there is a reciprocal increase in the concentration of urea in the urine which approaches the adult level of 550 mM by 34 weeks. The activity of the five enzymes of the ornithine-urea cycle each increase steeply in activity to the adult level in parallel with the increase in urinary urea content (Fig. 4). Clearly these events are inter-related and are associated with the process of weaning which is occurring at the same time.

It is well-documented that adult fore-gut fermenters tend to be alkalotic, that is they do not produce large amounts of metabolic acid. It is notable that such species have low concentrations of ammonia in the urine and the urinary pH is usually rather high. In contrast, animals which have a high intake of milk are likely to need to remove strong acids such as sulphate and phosphate, which are produced in protein catabolism, from the tissues. The changes which occur in the urine of the tammar pouch young at the time of commencement of weaning exemplify these differences rather well.

Until 28 weeks, the tammar pouch young drink increasing amounts of milk which supports an increasing rate of growth (Green 1984). The urine pH is low and the urine also contains increasing amounts of ammonia which together suggest that the animal is acidotic (Pitts 1973). Atkinson and Camien (1982) have argued that urinary ammonia is not an indication of proton excretion but their arguments have been largely refuted by Halperin and Jungas (1983) and, for the purposes of this argument, we shall accept the current dogma that urinary ammonia is used for buffering excess protons. On this assumption, the pre-weaning tammar must excrete large amounts of protons but the origin of these excess protons is not, at present, clear.

Protons can be produced during protein catabolism when phosphate and sulphate ions are released as strong acids. However, it is considered that very little protein is being catabolised. The rate of urine formation in tammar pouch young is not known and it is not possible to calculate the rate of protein catabolism from the rate of nitrogen excretion at this time. However, tammars gain about 0.35 g in liveweight for every 1 ml of milk ingested (Green 1984), and, if we assume that 15% of the weight gain is protein, then about 5% of the weight of milk consumed must be incorporated into tissue protein. At 26 weeks when urine ammonia is at its highest, tammar milk contains about 60 g protein /kg and so most of the protein in the milk contributes to tissue protein. The remainder is probably deaminated and gives rise to the ammonia and urea in the urine. However, this remainder is small and although it is possible that the young is acidotic because of production of strong acids from protein metabolism, this does not seem especially likely. An alternative possibility is that the tissues become acidotic as a result of the breakdown of the carbohydrate component of the milk. As Messer and Green (1979) have shown, pre-weaning tammar milk contains relatively large amounts of sialic acid which accounts for between 4 and 5.6% of the carbohydrate. As the carbohydrate fraction of the milk increases up to 28 weeks and as the intestine of the pouch young has been shown to contain enzymes capable of breaking down the sialic acid-containing carbohydrates (Walcott and Messer 1980), free sialic acid may well be released in increasing quantities as lactation proceeds. As the sialic acids are relatively strong with pKa's of 2.60 to 2.75 (Svennerholm 1963), they may be the cause of the acidosis seen in pre-weaning pouch young. Certainly all suggestion of acidosis has disappeared by 36 weeks when the carbohydrate, and the sialic acid, content of the milk has been considerably reduced. (A study of the sialic acid content of the urine during the process of weaning might prove worthwhile.)

The development of the urea cycle in eutherians varies considerably from species to species but has not been associated with the development of herbivory. Many eutherian sucklings differ from marsupials in that they produce a concentrated urine well before the start of weaning (Dalton 1968 b, c) but this may be diluted after drinking milk (Dalton 1968a). The five urea-cycle enzymes are present in foetuses of the human (Raiha and Suihkonen 1968a), pig (Kennan and Cohen 1959) and sheep (Rattenbury *et al.* 1980). The foetal sheep can synthesize urea at 50 days of gestation at rates comparable with the adult (Rattenbury *et al.* 1980). In the rat, the urea cycle is inactive in the foetus but appears at birth and soon reaches maximal activity (Kretchmer *et al.* 1966; Raiha and Suihkonen 1968b). There is thus no association between weaning and urea-cycle activity in any eutherian which has so far been studied. Marsupials such as the tammar produce a dilute urine until the start of the weaning period (Wilkes and Janssens 1986a) and this urine is consumed by the mother. Hence, the mother recovers all water lost in the urine of the young and concentrates it in her own kidneys and so there is little pressure on the tammar pouch young to produce a concentrated urine until the time it begins to vacate the pouch. This is the time when urea-cycle activity increases markedly (Fig. 4).

CONCLUSIONS

At weaning, the herbivorous mammal undergoes a digestive metamorphosis which, although not as obvious to the casual observer, is as significant as the transformation of a tadpole to a frog. The gastro-intestinal tract changes in both structure and enzyme complement to accommodate the considerable change in the composition, ease of digestion and quality of the diet. At the same time, enormous microbial populations become established in either the fore- or hind-gut which have a vital symbiotic relationship with the animal. There are equally large changes in the metabolic processes in the liver to accommodate the change in the supply of nutrients from the gastro-intestinal tract. In addition, in many species there are large changes in the function of the kidney.

Both marsupial and eutherian herbivores pass through this metamorphosis. There are considerable differences in the timing of these changes, which relate largely to the very short gestation period in the marsupial mammals. Nevertheless, there are great similarities in the changes which occur in these two infra-classes of herbivorous mammals.

We thank Jim Merchant, Division of Wildlife and Rangelands Research, CSIRO, Phil Cowan, Ecology Division, DSIR New Zealand, and Helen Crowley, Peter Frappell and Randy Rose, Department of Zoology, University of Tasmania, for allowing us to quote their unpublished findings; and Frank Carrick, Hugh Dove, Hugh Tyndale-Biscoe and Peter Woodall for reading and commenting on the manuscript. We thank Hugh Tyndale-Biscoe, Lyn Hinds, Brian Green and Terry Fletcher of the Division of Wildlife and Rangelands Research, CSIRO, for much valuable support with and discussion of research on the tammar wallabies.

Abe, M., Iriki, T., Kondoh, K., and Shibui, H. (1979). *Br. J. Nutr.* **41**, 175.

Archer, M. (1984). In "Vertebrate Zoogeography and Evolution in Australasia." (Eds M. Archer and G. Clayton.) pp. 633-806. (Hesperian: Carlisle, WA.)

Arinze, I.J. (1975). *Biochem. Biophys. Res. Comm.* **65**, 184-9.

Arman, P., Kay, R.N.B., Goodall, E.D., and Sharman, G.A.M. (1974). *J. Reprod. Fert.* **37**, 67-84.

Atkinson, D.E., and Camien, M.N. (1982). *Top. Cell. Reg.* **21**, 261-302.

Ballard, F.J. (1965). *Comp. Biochem. Physiol.* **14**, 437-43.

Ballard, F.J. (1971). *Biochem. J.* **124**, 265-74.

Ballard, F.J., Hanson, R.W., and Kronfeld, D.S. (1969). *Fed. Proc.* **28**, 218-31.

Barker, J.M. (1961). *Quart. J. Exp. Physiol.* **46**, 54-68.

Barnard, E.A. (1979). *Nature*, Lond. **221**, 340-4.

Bergman, H.C., and Housley, C. (1968). *Comp. Biochem. Physiol.* **25**, 213-8.

Black, J.L. (1971). *Br. J. Nutr.* **25**, 31-55.

Clark, M.G., Filsell, O.H., and Jarrett, I.G. (1976). *Biochem J.* **156**, 671-80.

Collins, J.G., Bradbury, J.H., Trifonoff, E., and Messer, M. (1981). *Carbohydr. Res.* **92**, 136-40.

Comline, R.S., and Titchen, D.A. (1951). *J. Physiol.*, Lond. **115**, 210-26.

Cook, H.W., Pearson, A.M., Simmons, N.M., and Baker, B.E. (1970). *Can. J. Zool.* **48**, 629-33.

Cork, S.J., and Dove, H. (1986). *Proc. Nutr. Soc. Aust.* **10**, (in press.)

Cowie, A.T. (1969). *J. Endocr.* **44**, 437–50.

Crisp, E.A., and Messer, M. (1986). *Proc. Aust. Biochem. Soc.* **18**, 98.

Dalton, R.G. (1968a). *Br. Vet. J.* **124**, 371–81.

Dalton, R.G. (1968b). *Br. Vet. J.* **124**, 451–9.

Dalton, R.G. (1968c). *Br. Vet. J.* **124**, 498–502.

Dove, H., and Freer, M. (1979). *Aust. J. Agric. Res.* **30**, 725–39.

Ducluzeau, R. (1983). *Ann. Rech. Vet.* **14**, 354–9.

Edwards, E.M., Dhand, U.K., Jeacock, M.K., and Shepherd, D.A.L. (1975). *Biochim. Biophys. Acta* **399**, 217–27.

Edwards-Webb, J.D., and Thompson, S.Y. (1977). *Br. J. Nutr.* **37**, 431–40.

Filsell, O.H., Jarrett, I.G., Taylor, P.H., and Keech, D.B. (1969). *Biochim. Biophys. Acta* **184**, 54–63.

Flannery, T. (1984). In "Vertebrate Zoogeography and Evolution in Australasia." (Eds M. Archer and G. Clayton.) pp. 817–35. (Hesperian: Carlisle, W.A.)

Fleay, D. (1937). *Aust. Zoologist* **9**, 68–80.

Green, B. (1984). *Symp. Zool. Soc. Lond.* **51**, 369–87.

Green, B., Griffiths, M., and Leckie, R.M.C. (1983). *Aust. J. Biol. Sci.* **36**, 455–61.

Green, B., Newgrain, K., and Merchant, J. (1980). *Aust. J. Biol. Sci.* **33**, 35–42.

Green, S.W., and Renfree, M.B. (1982). *Aust. J. Biol. Sci.* **35**, 145–52.

Griffiths, M., and Barton, A.A. (1966). *CSIRO Wildl. Res.* **11**, 169–85.

Gross, R., and Bolliger, A. (1959). *Am. J. Dis. Child.* **98**, 768–75.

Guilloteau, P., Corring, T., Garnot, P., Martin, P., Toullec, R., and Durand, G. (1983). *J. Dairy Sci.* **66**, 2373–85.

Guilloteau, P., Corring, T., Toullec, R., and Guilhermet, R. (1985). *Reprod. Nutr. Devel.* **25**, 481–93.

Guilloteau, P., Corring, T., Toullec, R., and Robelin, J. (1984). *Reprod. Nutr. Devel.* **24**, 315.

Guilloteau, P., Toullec, R., Patureau-Mirand, P., and Prugnaud, J. (1981). *Reprod. Nutr. Devel.* **21**, 885–99.

Halperin, M.L., and Jungas, R.L. (1983). *Kidney Int.* **24**, 709–13.

Henschel, M.J. (1973). *Br. J. Nutr.* **30**, 285–96.

Hill, K.J. (1956). *Q. J. Exp. Physiol.* **41**, 421–3.

Hume, I.D. (1982). "Digestive Physiology and Nutrition of Marsupials." (Cambridge Univ.: Cambridge.)

Janssens, P.A. (1984). *Proc. Nutr. Soc. Aust.* **9**, 68–75.

Janssens, P.A., Jenkinson, L.A., Paton, B.C., and Whitelaw, E. (1977). *Aust. J. Biol. Sci.* **30**, 183–95.

Jones, C.T., and Ashton, I.K. (1976). *Arch. Biochem. Biophys.* **174**, 506–22.

Keast, A. (1972). In "Evolution, Mammals and Southern Continents." (Eds A. Keast, F.C. Erk, and B. Glass.) pp. 23–87. (Suny: Albany, NY.)

Kennan, A.L., and Cohen, P.P. (1959). *Dev. Biol.* **1**, 511–25.

Kirsch, J. (1984). In "Vertebrate Zoogeography and Evolution in Australasia." (Eds M. Archer and G. Clayton.) (Hesperian: Carlisle, WA.)

Kretchmer, N., Hurwitz, R., and Raiha, N. (1966). *Biol. Neonat.* **9**, 187–96.

Langer, P. (1979). *Zoomorphologie* **93**, 137–51.

Laxminarayana, H., and Dastur, N.N. (1968). *Dairy Sci. Abstr.* **30**, 177–86.

Lemon, M., and Barker, S. (1967). *Aust. J. Exp. Biol. Med. Sci.* **45**, 213–9.

Leng, R.A. (1970). *Adv. Vet. Sci. Comp. Med.* **14**, 209–60.

Lindsay, D.B., and Oddy, V.H. (1986). *Proc. Nutr. Soc.* **45**, 33–8.

MacLennan, D.G., Blume, L.M., and Johnson, D.E. (1983). *Toxicon Supp. 3*, pp. 261–64.

Mepham, T.B., and Beck, N.F.G. (1973). *Comp. Biochem. Physiol.* **45A**, 273–81.

Merchant, J.C., and Sharman, G.B. (1966). *Aust. J. Zool.* **14**, 593–609.

Messer, M., and Czolij, R. (1984). *Proc. Nutr. Soc. Aust.* **9**, 114.

Messer, M., and Green, B. (1979). *Aust. J. Biol. Sci.* **32**, 519–31.

Messer, M., Griffiths, M., and Green, B. (1984). *Aust. J. Biol. Sci.* **37**, 1–6.

Messer, M., Trifonoff, E., Stern, W., Collins, J.G., and Bradbury, J.H. (1980). *Carbohydr. Res.* **83**, 327–34.

Moir, R.J., Somers, M., and Waring, H. (1956). *Aust. J. Biol. Sci.* **9**, 293–304.

Oftedal, O.T. (1984). *Symp. Zool. Soc. Lond.* **51**, 33–85.

Ørskov, E.R., Benzie, D., and Kay, R.N.B. (1970). *Br. J. Nutr.* **24**, 785–95.

Parkash, S., and Jenness, R. (1968). *Dairy Sci. Abstr.* **30**, 67–87.

Parrish, D.B., Wise, G.H., Hughes, J.S., and Atkeson, F.W. (1948). *J. Dairy Sci.* **31**, 889.

Paton, B.C., and Janssens, P.A. (1981). *Comp. Biochem. Physiol.* **70B**, 105–13.

Peart, J.N., Edwards, R.A., and Donaldson, E. (1972). *J. Agr. Sci.* **79**, 303–13.

Peart, J.N., Edwards, R.A., and Donaldson, E. (1975). *J. Agr. Sci* **85**, 315–24.

Pitas, R.E., Sampugna, J., and Jensen, R.G. (1967). *J. Dairy Sci.* **50**, 1332–6.

Pitts, R.F. (1973). In "Handbook of Physiology." Section 8. Renal Physiology. (Eds J. Orloff and R.W. Berliner.) pp. 455–96. (American Physiology Society: Washington DC.)

Poole, W.E., Sharman, G.B., Scott, K.J., and Thompson, S.Y. (1982). *Aust. J. Biol. Sci.* **35**, 607–15.

Prior, R.L., and Christenson, R.K. (1977). *Amer. J. Physiol.* **233**, E462–8.

Raghunathan, R., and Arinze, I.J. (1977). *Int. J. Biochem.* **8**, 737–43.

Raiha, N.C.R., and Suihkonen, J. (1968a). *Acta Pediat. Scand.* **57**, 121-4.

Raiha, N.C.R., and Suihkonen, J. (1968b). *Biochem. J.* **107**, 793-7.

Rattenbury, J.M., Jeacock, M.K., and Shepherd, D.A.L. (1980). *Biochim. Biophys. Acta* **630**, 210-9.

Renfree, M.B. (1973). *Dev. Biol.* **33**, 62-79.

Rich, T.H. (1982). In "The Fossil Vertebrate Record of Australasia." (Eds P.V. Rich and E.M. Thompson.) pp. 385-478. (Monash Univ.: Melbourne.)

Rolleston, F.S. (1972). *Curr. Top. Cell. Reg.* **5**, 47-75.

Roy, J.H.B. (1980). "The Calf." Edition 4. (Butterworths: London.)

Russell, E.M. (1982). *Biol. Rev.* **57**, 423-86.

Santhirasegaram, K., and Watts, J.I. (1970). *Proc. Aust. Soc. Anim. Prod.* **8**, 258-62.

Schmidt, G.H. (1971). "The Biology of Lactation." (Freeman: San Francisco.)

Schwartz, H.M., and Gilchrist, F.M.C. (1975). In "Digestion and Metabolism in the Ruminant." (Eds I.W. McDonald and A.C.I. Warner.) pp. 165-79. (University of New England: Armidale, NSW.)

Segal, S. (1978). In "The Metabolic Basis of Inherited Disease." (Eds J.B. Stanbury, J.B. Wyngaarden and D.S. Frederickson.) 4th Edn. (McGraw-Hill: New York.)

Smith, H.W. (1965). *J. Path. Bact.* **90**, 495-513.

Stobbs, T.H., and Brett, D.J. (1974). *Aust. J. Agric. Res.* **25**, 657-66.

Stobo, I.J.F., Roy, J.H.B., and Gaston, H.J. (1966). *Br. J. Nutr.* **20**, 171-88.

Svennerholm, L. (1963). In "Methods in Enzymology." (Eds S.P. Colowick and N.O. Kaplan.) Vol. 6, pp. 453-9. (Academic: New York.)

Ternouth, J.H., Roy, J.H.B., and Shotton, S.M. (1976). *Br. J. Nutr.* **36**, 523-35.

Ternouth, J.H., Roy, J.H.B., and Siddons, R.C. (1974). *Br. J. Nutr.* **32**, 37-45.

Ternouth, J.H., Roy, J.H.B., Thompson, S.Y., Toothill, J., Gillies, C.M., and Edwards-Webb, J.D. (1975). *Br. J. Nutr.* **33**, 181-96.

Ternouth, J.H., Thompson, S.Y., and Edwards-Webb, J.D. (1980). *Br. J. Nutr.* **44**, 141-50.

Toofanian, F., Kidder, D.E., and Hill, F.W.G. (1974). *Res. Vet. Sci.* **16**, 382-92.

Vernon, J.K., Messer, M., and Green, B. (1981). *Aust. J. Biol. Sci.* **34**, 401-9.

Walcott, P.J., and Messer, M. (1980). *Aust. J. Biol. Sci.* **33**, 521-30.

Wardrop, I.D., and Coombe, J.B. (1960). *J. Agric. Sci.* **54**, 140-13.

Wardrop, I.D. (1960). *J. Agric. Sci.* **55**, 127-31.

Warnes, D.M., Seamark, R.F., and Ballard, F.J. (1977a). *Biochem. J.* **162**, 617-26.

Warnes, D.M., Seamark, R.F., and Ballard, F.J. (1977b). *Biochem. J.* **162**, 627–34.

White, R.G., and Leng, R.A. (1980). *Comp. Biochem. Physiol.* **67A**, 223–9.

Wilkes, G.E., and Janssens, P.A. (1986a). *J. Comp. Physiol. B.* **156**, 573–82.

Wilkes, G.E., and Janssens, P.A. (1986b). *J. Comp. Physiol. B.* **156**, 829–37.

Winter, K.A. (1985). *Can. J. Anim. Sci.* **65**, 445–50.

14

THE PARTITIONING OF NUTRIENTS BY HERBIVORES

by R.F. Thornton

Food for the mammalian herbivore is the supply of energy and nutrients which together with oxygen and water support the chemistry of life. The energy of food is expressed in terms of joules but such an expression does not specify the actual substrates that yield ATP, in other words glucose, amino acids and fatty acids. Similarly the term "nutrients" refers to specific substrates – amino acids, fatty acids, glucose, minerals and electrolytes, and vitamins – which become the structural and functional constituents of body tissues. To separate food into these broad compartments, i.e. energy and nutrients, is purely an arbitrary definition that does not account for the fact that some of the same substrates are utilised for both fuel and tissue constituents. For example, among the amino acids those with branched-chains are oxidized while others are glycogenic, yet these same amino acids are found in the enzymatic, structural and contractile proteins of tissues. Similarly, the long-chain fatty acids are readily oxidised, yet form prostaglandins. Carbohydrates are not commonly associated with structural and functional lipids, for example cellular and plasma membranes and mammalian cell structures, but the glycoproteins contain carbohydrate moieties as do some of the lipoproteins, while glycogen is found in liver, muscle and adipose tissues.

The electrolytes Na and K are usually thought of in terms of osmotic balance but they are also an integral part of the plasma membrane Na^+K^+ ATPase enzyme system, the activity of which is a major component of the energy expenditure of animals. Similarly Ca is vital to bone, horn and antler growth, but is equally imperative for milk protein secretion, the regulation of muscle contraction through the Ca^{++} activated ATPase system, and other enzymes such as adenyl cyclase, which catalyses the synthesis of cyclic AMP from ATP. Thus even in the most static time frame imaginable for cellular metabolism, there is considerable interchange or partitioning of substrates between the fuels which drive cellular metabolic processes and the anabolism and catabolism of cellular constituents.

The complexity of metabolic regulation is increased with the realization that the molecular components of every cell are turning over at a variety of rates. The turnover rates for proteins have half lives from minutes to months, and the same protein may have a variety of turnover rates depending upon its location. Thus the substrates for metabolism are exogenous (from the diet) or endogenous (from turnover of tissue components). Although the substrates are undifferentiated in a metabolic sense, the utilization of endogenous substrates has energic

NUTRITION OF HERBIVORES
ISBN 0 12 750052 9

advantages. Furthermore, the products of digestion, in the lumen of the gut, can be modified and/or utilised by the gut during absorption into the portal blood or subsequent metabolism in the liver.

The complexity of nutrient partitioning is further enhanced by the physiological state of the herbivore, for example, daily grazing, drinking, resting and rumination patterns, seasonal migration and mating schedules, seasonal wool or hair growth, as well as pregnancy, lactation, growth and fattening, and disease. Realization of the complexity and dynamic nature of metabolism led Bauman to coin the term "homeorhesis" to describe "the orchestrated changes for the priorities of a physiological state" (Bauman and Currie 1980; Bauman 1984). Bauman (1984) states "Homeorhetic control involves the coordination of metabolism, resulting in the direct partitioning of nutrients to support requirements specific for each physiological state. Thus homeorhetic mechanisms provide chronic regulation, while homeostatic controls prevail in the short term and, in life-threatening situations, may even override the long-term regulation to preserve vital functions".

Simplified schematic models of homeostatic (Model a) and homeorhetic (Model b) regulation were presented by Bauman *et al.* (1982) and Bauman (1984). However, from the above general discussion of the coordinated dynamic exchange of substrates and the turnover of cellular components, it is evident that there is no stasis in metabolism. Bauman's separation of terms and models (homeorhetic *vs* homeostatic) is in essence based on time (chronic *vs* acute) rather than on the mechanisms of metabolic regulation. Constancy is not a feature of metabolic processes, be they biochemical or physiological, and time frames are arbitrary. Thus the concept of homeorhetic regulation is more apt to describe the control of metabolism, and the term homeostasis, in a regulatory sense, could well be abandoned. Homeostasis can be viewed as the result of homeorhetic regulation.

Since the mid-1970s there has been a growing interest in the metabolism of ruminant tissues as is indicated by the fact that the phrase "protein metabolism" no longer refers only to "protein digestion in the rumen". This new interest has spawned a variety of review articles and whole symposia have addressed the subject of partitioning of nutrients. Some relevant reviews are those of Baldwin *et al.* (1984), Bauman (1984), Robelin and Geay (1984), Annison (1984), Van der Walt (1984), Buttery (1984), Bergman and Pell (1984), Armstrong and Blaxter (1984) and Black (1984) from the first Herbivore Nutrition Symposium; Bauman, *et al.* (1982), Prior and Smith (1982), Lindsay (1982), Hood (1982) and Byers (1982) from a symposium on "Partitioning of Nutrients between Muscle and Adipose Tissue Growth in the Ruminant", Weekes (1983), Buttery (1983), McDowell (1983), Robinson *et al.* (1983), Hart (1983) from a symposium on "Nutrient-Hormone Interactions in Animal Production"; Milligan and Summers (1986), Weekes (1986), Lobley (1986), Moe and Tyrrell (1986) and Gill (1986) from the symposium on "Assessing Responses to Nutrients by Ruminants, together with Bauman and Currie (1980), McCutcheon and Bauman (1983), Hart and Johnsson (1985), Bauman and McCutcheon (1986), MacRae and Lobley (1986) and Johnsson and Hart (1986).

Rather than attempt to be encyclopaedic and cover this entire subject, the intention of this article is to highlight some concepts for the regulatory control of nutrient partitioning for wool growth, lactation, growth and fattening, together with the role and mode of action of repartitioning agents.

WOOL GROWTH

It has long been recognised that under field conditions, physiological state contributes to significant variations in wool growth and this subject has been comprehensively reviewed by Williams (1978), Black and Reis (1979), Corbett (1979) and Oddy and Annison (1979). Of particular interest are the depressions in wool growth associated with pregnancy and lactation. Corbett (1979) summarised these effects thus: "Wool growth is reduced by about 30% during the last two months of pregnancy or about 3-10% in terms of annual fleece production. Lactation generally reduces the annual fleece by 5-8%. The full cycle of reproduction reduces the annual fleece production by 10-14% and the efficiency of its production by about 60% of that of non-breeding ewes". Behind these generalizations is a large literature of variable but reasonably consistent results from various breeds of sheep and pasture conditions. It does appear that Merino ewes, which produce the greatest amounts of wool when not pregnant and not lactating, show the greatest depressions in wool growth when they are pregnant and lactating (Oddy and Annison 1979; Williams 1978).

More recently Oddy (1985) attempted to quantify the effects of pregnancy and lactation on wool growth of Merino ewes in a carefully designed experiment. The results of this experiment indicate that for each litre of milk production, clean wool growth was depressed by 12 g (Figure 1). This relationship clearly indicates that there has been a "repartitioning" of available nutrients from wool growth to milk production. Oddy (1985) calculated that the amount of sulphur lost in milk closely approximated that which did not go into wool production. Similar calculations for uterine and foetal sulphur deposition *versus* wool sulphur were not so close but these calculations depended upon literature assumptions rather than actual estimates. These shifts in sulphur deposition together with the observed changes in plasma cysteine concentration suggested that the supply of sulphur- containing amino acids was limiting wool growth during lactation. As supplementation with sulphur amino acids has been shown to stimulate wool growth (Reis 1979) such an interpretation would seem reasonable.

However, Oddy (1985) discounts such an interpretation on the basis of the findings of Williams *et al.* (1978) who reported that supplementation with sulphur amino acids did not lead to increased wool growth during lactation and concluded "that diminution of supply of substrate to the wool follicle is not the prime cause of reduced wool growth in pregnancy and lactation." The findings of Williams *et al.* (1978) warrant further investigation as casein supplementation during lactation increased wool production by 33% but sulphur amino acid supplementation actually depressed wool production by 8% relative to controls. Furthermore, wool growth of the lactating sheep was 41% greater than when they were dry. It is unlikely that a protein/energy imbalance occurred as the protein intake during lactation can be calculated as ranging from 15 g /MJ ME in the unsupplemented ewes to 18 g /MJ ME in the supplemented ewes. Kempton (1979) calculated that some digested protein per (12 g /MJ ME) is required to support maximum wool growth and an additional 2-3 g /MJ ME is required for lactation. Thus Oddy (1985) concluded very generally "that inhibition of wool growth in pregnancy and lactation is the result of interactions between many factors, and hence not clearly correlated with the concentration of any specific hormone or with the availability of any one specific nutrient".

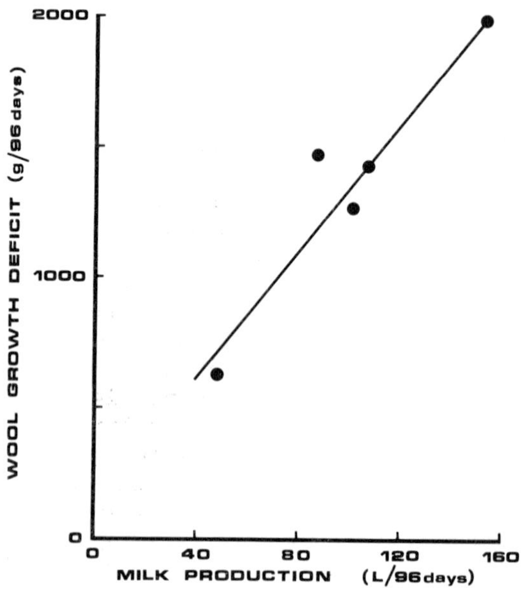

Figure 1. The relationship between wool growth deficit and milk production of lactating Merino ewes. The equation describing this relationship is $y = 12.2x + 136.5$ ($r = 0.95$, $p < 0.01$; Oddy *et al.* 1985).

In this context the role of the growth hormone should not be overlooked. It is generally considered that exogenous ovine growth hormone depresses wool growth during the period of its application (Ferguson *et al.* 1965; Wheatley *et al.* 1966; Wallace 1979). Exceptions to this generalization are the recent studies of Johnsson *et al.* (1985, 1986) in which the application of 0.1 mg of bovine pituitary growth hormone and 0.1 mg of recombinant bovine somatotrophin /kg LW, respectively, increased greasy fleece weight of young lambs fed a concentrate diet. Notwithstanding, during lactation in sheep, growth hormone levels have been shown to increase by 4-5 times, i.e. to levels of 17 ng /ml (Vernon *et al.* 1981). It can be hypothesised that growth hormone levels during lactation stimulate milk production, thus diverting nutrients to milk synthesis and away from wool growth, with consequent reduction in wool growth. Recently it has been reported that exogenous growth hormone increased circulating levels of insulin (Johnsson *et al.* 1985; P.C. Wynn, A.L.C. Wallace and E.F. Annison unpublished) prolactin (Johnsson *et al.* 1985) and somatomedins (Wynn *et al. loc. cit.*) and depressed methionine levels in plasma by 20% (Wynn *et al. loc. cit.*. These responses suggest a "preferential uptake of amino acids for protein synthesis in muscle tissue may account for the growth hormone induced depression in wool growth" (Wynn *et al. loc. cit.*).

LACTATION

Milk production is one of the most metabolically demanding activities of female herbivores. The magnitude of this demand is dependent upon the number of offspring; for example in sheep or cattle, twins result in close to double the milk production when compared to levels produced to support a single offspring. However, even the milk production demands of a single offspring can cause significant reductions in fat reserves and loss of bodyweight of lactating females, particularly if herbage quality or quantity is low. The ability of some female herbivores to lactate has been exploited by man who has selected sheep, goats and cows for milk production. The high-producing dairy cow is a metabolic marvel, capable of producing as much as 14 tonnes of milk in a *ca* 300 day lactation, which at 3.8% fat, 3.5% protein and 4.8% lactose is a daily loss of 1.77, 1.63 and 2.24 kg of fat, protein and lactose respectively. The demands of early lactation are so great in high-producing cows that they are unable to ingest sufficient energy and so mobilise body fat extensively, and to a lesser extent protein, to meet this demand. Such changes are quantitatively the greatest example of nutrient conversion and partitioning in the herbivores.

The regulation of the metabolic processes involved in lactation has been extensively studied in the 1980s with the consequent demonstration of the major role of growth hormone in orchestrating the orderly repartitioning of exogenous and endogenous energy and nutrient supplies (Bauman 1984; Bauman and McCutcheon 1986; Johnsson and Hart 1986). The key to this discovery has been the advent of recombinant DNA technology which has enabled the production of sufficient purified bovine growth hormone for complete lactation studies.

The results of this research are best illustrated by the long term study of Bauman *et al.* (1985) in which dairy cows were treated with bovine somatotrophin (Figure 2). Several aspects of these results deserve comment.

1. The composition of the milk was not affected by treatment; fat was 36 g /kg, protein 34 g /kg and lactose was 48 g /kg and the concentrations of Ca and P were also not affected (Eppard *et al.* 1985). Short term experiments have resulted in similar findings when cows were in positive energy and nitrogen balance.
2. Although there were changes from high to medium to low energy diets in the course of the experiment, the net energy intake of the treated groups was higher than that of controls, but the controls were always in positive energy balance.
3. Gross lactational efficiency (kg FCM /Mcal NE intake) was dose-dependent and ranged from 0.83 in controls to 1.03 in the 40.5 mg /day group, and when corrected for bodyweight, efficiency ranged from 0.94 to 1.12.
4. The responses in milk yield were dose-dependent and were 23 to 41% above the production of the control animals of (27.9 kg /day), that is the responses were from, and above, those of high-producing cows.

As growth hormone production is positively correlated with milk yield, both within a lactation of the same cow and across cows of varying yields (Hart *et al.* 1980), it would appear that these massive responses are an extension of the natural regulatory mechanism(s). What is truly remarkable is that the application of growth hormone alone could induce these responses. This indicates the key regulatory role of growth hormone in lactation and nutrient partitioning of cows.

There is no evidence that growth hormone has a direct effect on food intake, rather food intake increases as a result of the metabolic responses - increased growth or increased milk yield - to growth hormone application. Similarly, there is no evidence that growth hormone affects the apparent digestibility of energy or nitrogen, or maintenance requirements (Bauman *et al.* 1985).

Increased mammary blood flow is associated with increase in milk yield during growth hormone treatment (Johnsson and Hart 1986). There can be little doubt that increased cardiac output and blood flow to the mammary gland are mechanistic in the greater milk production achieved with exogenous growth hormone application. In the Bauman *et al.* (1985) study, steady-state concentrations of blood metabolites were maintained which suggests that either a greater blood flow and/or greater extraction rates of substrates from the blood must have occurred. Other studies on nutrient partitioning in lactation (McDowell *et al.* 1984; Oddy *et al.* 1984; Leenanuruksa and McDowell 1985) indicate that blood flow rate across the mammary gland, rather than nutrient transport rates across cell membranes, is regulatory for milk production. Johnsson and Hart (1986) cite evidence that "an increase in cardiac output and peripheral blood flow now appears to be a more generalised action of exogenous growth hormone in both lactating and non-lactating ruminants", and point out that "the role of growth hormone on local vascular regulation and membrane transport in the mammary gland is unknown."

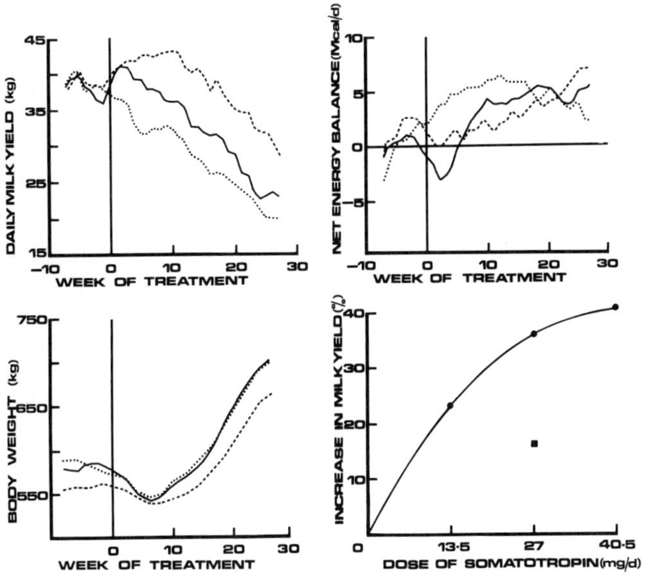

Figure 2. The effects of exogenous bovine somatotrophin on daily milk yield (kg /day), net energy balance (cal /day) and bodyweight (kg) throughout the experimental period.
▬ ▬ ▬ ▬ 27.0 mg of methionyl bovine somatotrophin (MBS) per day;
———— 27.0 mg of pituitary bovine somatotrophin (PBS) per day;
•••••••• controls (average production 27.9 kg /day). Also shown is the relationship between increase in milk yield (%) and dose of somatotrophin (mg/day) for MBS (●) and PBS (■) (Bauman *et al.* 1985).

It is evident that exogenous growth hormone must:

1. alter the metabolism of body tissues so that more nutrients, be they exogenous or endogenous, are partitioned to the mammary glands, and
2. enhance the ability of the mammary glands to secrete milk.

The major recent reviews of this subject, Bauman and McCutcheon (1986) and Johnsson and Hart (1986), are at variance in their interpretation of the responses to the exogenous growth hormone. This is largely due to the interpretation of short term *vs* long term experiments in which growth hormone has resulted in increases in milk yield. Thus Johnsson and Hart (1986) conclude "that growth hormone does exert lipolytic and diabetogenic activities in lactating ruminants", but Bauman and McCutcheon (1986) deliberately avoid such terms. It is thus important to clarify the specific effects of growth hormone as distinct from the stimulus of lactation *per se*, which in turn is often confounded with energy balance. The stimulus of lactation may involve other hormones, for example, prolactin (Bauman and McCutcheon 1986).

The well documented effects of lactation on adipose tissue lipid metabolism (for example, increased lipolysis and therefore mobilization of long-chain fatty acids and glycerol), decreased adipocyte volume, decreased lipogenesis from acetate, decreased lipoprotein lipase activity (and thus a reduced uptake of fatty acids) and the increased lipolytic response of adipose tissue to adrenalin), are all consistent with negative energy balance (Smith and Walsh 1984). The claim of Vernon (1980) that special mechanisms operate during lactation (as opposed to negative energy balance and weight loss) to regulate the partitioning of nutrients is undoubtedly true; all that is needed is the evidence. The data of Bauman *et al.* (1985) illustrate the complexity of metabolic regulation during lactation. In this experiment (Figure 2) the effect of growth hormone was to change the form of the lactation curve. Whereas milk production of the control animals declined almost from the start of the experiment, that of the treated cows rose for some 10 weeks after treatment began, and then declined at about the same rate as that of controls. Concomitant changes in energy intake and energy balance reflected milk production and the period of energy deficit for the treated cows was not long. However, all cows regained similar and substantial amounts of bodyweight (100 - 150kg). It is likely that muscle protein and reserves of adipose tissue lipid were being replenished during this period of weight gain in all groups even though exogenous growth hormone maintained milk production at some 10 kg /day over that of controls. In other words growth hormone was promoting increased milk yield when adipose tissue was lipolytic in early lactation and lipogenic in late lactation. Thus the partitioning of nutrients related to lipid metabolism must be markedly different during the early and later stages of lactation.

Similar circumstances and considerations apply to the regulation of protein metabolism during lactation. However, the role of growth hormone is likely to be even more complex in protein metabolism than it is in lipid metabolism as it could be expected to enhance synthesis rather than catabolism of muscle protein (Johnsson and Hart 1986). Treatment of cows with growth hormone when they are in negative energy and nitrogen balance causes the concentration of milk fat to increase while that of protein decreases. Whether this "additional" milk fat is derived from adipose tissue, and/or if this reduced protein secretion is a "protein sparing" response to negative balances, has not been ascertained.

Glucose, apart from being an energy source, is the essential precursor for lactose synthesis. Thus the effects of growth hormone on glucose metabolism and partitioning are critical in understanding the increased milk yield of cows treated with growth hormone. In the long term study of Bauman *et al.* (1985) the cows were fed a corn meal based diet so there was some likelihood of glucose being derived directly from the diet. However, in another long-term study, with twin cows on high quality pasture, daily treatment with pituitary growth hormone (50 mg /day) resulted in a 17.7% increase in milk yield (Peel *et al.* 1985). In this experiment gluconeogenesis would have provided virtually all of the glucose requirements of the lactating cows. This glucose supply from gluconeogenesis is maintained through both positive and negative energy balance states. It is not known whether growth hormone stimulates gluconeogenesis from amino acids, glycerol, propionate, or lactate, and/or repartitions glucose from other tissues to the mammary gland. Recent short-term studies on cows by McDowell *et al.* (1984) have shown that the uptake of glucose by the hind limb muscle was significantly decreased and the mammary gland increased its uptake of free fatty acids, glucose and acetate. Only further demonstrative experiments like that of McDowell *et al.* (1984) will indicate the direct and indirect effects of growth hormone which result in increased milk production. The most likely indirect effects are through the somatomedins, which are known to increase, in both activity and circulating levels, in response to growth hormone (Johnsson and Hart 1986).

The complexity of these metabolic processes in lactation is magnified out of all proportion by the realization that the dairy cow begins to ovulate some 4-6 weeks after calving and lactation initiation. At this time she is usually inseminated, naturally or artificially, and thus a new pregnancy begins. Although foetal, uterine and placental growth are not immediately significant in early lactation, they certainly become significant towards the end of lactation when the cow is replenishing her lipid and protein reserves. Thus, at least in the high-producing dairy cow, lactation and pregnancy are concurrent for most of their terms. However, particularly in circumstances of poor nutrition, the demands of lactation can be such that ovulation is suppressed and pregnancy does not ensue until some reserves of lipid and protein are regained. Under these conditions lactation and pregnancy, together with the utilization and replenishment of tissue reserves, become cyclic rather than concurrent (Vernon and Flint 1984).

GROWTH AND FATTENING

Fattening is the colloquial term for weight gain or improvement in "condition" and is applied to all herbivores. An often propounded thesis is that "domestication" of wild herbivores has resulted in selection for excess fat deposition. This is certainly true of some British breeds of both cattle and sheep, but is not true of all domestic cattle and sheep. It is even less correct when applied across the range of herbivores that have been domesticated for man's use, for example, water buffaloes, camels and horses, all of which are noted for their leanness. Where the selection of animals has been primarily for meat production, there seems to be an associated "overfatness", as in some British breeds of cattle and sheep. If, however, selection has been for draught or meat production, then the resultant breeds, or types of animals, are not so fat, as occurs in some European and Zebu cattle, water buffaloes, camels and horses. Across the spectrum of herbivores is a range of animals from extremely lean species such as the marsupials, to extremely fat. For instance, Tribe and Peel (1963) reported that kangaroos contained only 1% fat (most being in the tail pad) and we have had apparently normal, healthy sheep that were 70% fat, on an empty bodyweight basis (Thornton *et al.* 1983).

To generalise about the relative leanness or fatness of genera and species, and breeds within species, across the entire spectrum of mammalian herbivores is fraught with exceptions, some of which are well documented. Dairy-type cattle are leaner but have more internal fat (omental, perirenal and pelvic) and less subcutaneous fat than beef cattle (Williams 1978; Truscott *et al.* 1983). Similarly, some breeds of sheep which are noted for both fecundity and milking ability, have heavier internal fat deposits than those breeds which are noted for their meat characteristics (Wood *et al.* 1980).

Within a breed, sex is an obvious determinant of fatness and in most herbivores, females are fatter than castrates which are fatter than entire males. As these sex effects can be demonstrated when animals are eating the same diet, it is clear that the sex hormones influence the partitioning of dietary nutrients between muscle and fat. Indeed this is the scientific basis of some of the steroid-based growth promotants as males are also larger than females. However, no such specific reason(s) are known or can be applied to the obvious repartitioning of dietary nutrients between muscle and fat in across-species comparisons, for example, the relative fatness of horses as compared with sheep grazing the same swards, or eating the same diet. Growth promotants generally result in carcasses which are heavier and leaner (Buttery 1985) and thus must effect nutrient partitioning. Recently Hunter and Vercoe (1986) have shown that growth promotants reduced the metabolic rate of cattle by 10%. Similarly exogenous growth hormone has been shown to promote growth as well as increase muscle and reduce fat deposition in both sheep and cattle (Hart and Johnsson 1985).

The composition of adipose tissue is extremely variable and is largely dependent upon plane of nutrition. For instance, in a study by Vickery (1977) the fat content of the perirenal adipose tissue of "emaciated to fat" cows ranged from 2.7% to 80.4% and that of the subcutaneous adipose tissue from 2.7% to 69.5%, respectively. On the other hand, the composition of muscle is relatively constant and particularly constant if the intramuscular fat content is disregarded. In Vickery's (1977) study, intramuscular fat levels ranged from 0.9% in the brisket muscle of emaciated cows to 4.8% in fat cows. The intramuscular fat content of the *longissimus dorsi* muscle from 450 kg Santa Gertrudis steers ranged from 1% to 12% (Thornton *et al.* 1974) but values as high as 20% are not uncommon in grain-fed cattle. It is important to note that the intramuscular fat content of lean from cattle is as low as that from so called "game species", e.g. 1.0% to 1.4% for kangaroos and wallabies (Tribe and Peel 1963; Ford and Fogarty 1982), or 1.8% to 3.9% for deer (Tan and Fennessy 1981).

The composition of mammalian muscle is remarkably constant at about 75% water and 21% protein, regardless of species, sex or plane of nutrition (Callow 1947; Lawrie 1974; Thornton *et al.* 1981). However, the neck muscles of the rutting male deer are an exception to this generalization. These muscles become enlarged by a factor of about two during the rut and the protein content of these muscles (*rhomboideus, splenius* and *longissimus capitus*) is reduced to about 19% and the water content increased to about 78%. Associated with this enlargement of the neck muscles is a 1000-fold increase in plasma testosterone levels and a doubling in area of Type I muscle fibres and an increase in type IIA fibre area by a factor of 2.6 (Field and Young 1985). Despite this exception, which only represents about 1% of the total muscle of the deer, there seem little ground(s) for suggesting that nutrient partitioning has major effects on protein accumulation in muscle.

It is quite remarkable that both domesticated cattle and sheep show similar basic patterns of protein and fat deposition (Haecker 1920; Moulton *et al.* 1922; Hedrick 1968; Searle *et al.* 1972; Searle and Graham 1975). Fig. 3 shows that protein accumulation is virtually linear but fat deposition increases dramatically when empty bodyweights of approximately 300 kg for cattle and 30 kg for sheep are reached. There are several variations of this basic pattern. For instance, young suckling or milk-fed animals are fatter than Fig. 3 depicts (Searle and Griffiths (1976). Breed effects are reflected by differing slopes of the fat accumulation relationships and/or the bodyweight at which rapid fat accumulation begins, in other words larger late-maturing animals begin to fatten at heavier bodyweight than smaller early-maturing animals (Searle *et al.* 1982). Finally, Fig. 3 is based on data from castrates and does not represent the accumulation patterns of males or females, which are generally considered to be leaner and fatter, respectively, than castrates. It should also be noted that the bodyweight at which rapid fat accumulation begins (termed physiological maturity or the break point) is not associated with weaning or any other dietary transition, nor with puberty. Clearly this increased fat accumulation represents a marked repartitioning of dietary nutrients.

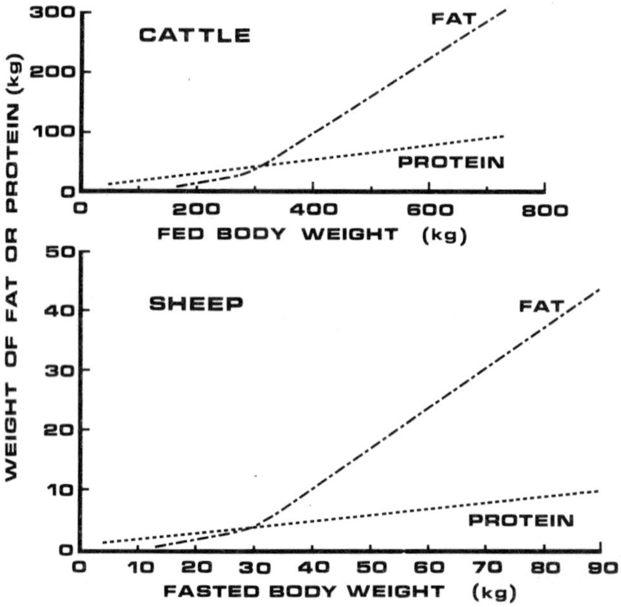

Figure 3. The relationship between the weight of fat and protein and body weight (kg) of cattle and sheep (Searle *et al.* 1972).

The physiological and biochemical changes which accompany this transition to rapid fat deposition have not been identified. In fact, likely explanations have not been confirmed by experiment. For instance, there is no clear evidence that this increase in adipose tissue mass is directly associated with an increase in adipocyte number and/or differentiation of pre-adipocytes into adipocytes. In a study on growing sheep, adipocyte number increased until the animals were about 11 months of age and some 44 kg in liveweight (Hood and Thornton 1979). Similarly, cattle attained a maximum adipocyte number at 14 months of age and 470 kg liveweight (Hood and Allen 1973). In both these studies, adipocytes were fixed with osmium tetroxide and counted in a coulter counter. This method does not adequately measure the smaller cells (< 20 µ; Hood 1982) but there is no indication in either of these studies, on sheep or cattle, that there was a significant hyperplasia and/or differentiation of adipocytes associated with the extent of fattening. Some workers have reported bimodal distribution of adipocyte size in adipose tissues of fat cattle (Allen 1976; Cianzio *et al.* 1985) which implies that there has been a stimulus to hyperplasia during excess fat deposition. Both these studies used the osmium fixation/coulter counter technique with statistical methods of removing "electrical noise". In contrast, using direct observation, we found mostly normal statistical distributions of adipocyte size in the perirenal, omental and subcutaneous adipose tissues, of both thin and fat adult sheep (Thornton *et al.* 1984b).

With such a dramatic change in rate of fat deposition at physiological maturity, it might be hypothesised that changes in the rates of lipogenesis, long-chain fatty acid incorporation and lipolysis would be occurring. However, direct evidence to support this hypothesis has not been forthcoming. For instance, we could not detect changes in the rate of lipogenesis (Hood and Thornton 1980), nor in the activity of lipoprotein lipase, nor in lipolysis rates (R.F. Thornton and R.K. Tume unpublished) in the subcutaneous adipose tissues of sheep fed the same diet and growing through this transition period of fat deposition. Other workers have associated changes in acetate utilization for fatty acid synthesis and the activities of enzymes involved in NADPH generation (Ingle *et al.* 1972a, b), or lipoprotein lipase activity (Haugebak *et al.* 1974) in the adipose tissues of growing lambs and mature sheep, with differential rates of fat deposition. However, these studies have not been systematic investigations of growth and have used different sheep and variable diets to illustrate shifts in adipose tissue metabolism. A further problem is the adequate expression of biochemical activity when adipocyte size is enlarging. Lipogenesis is highly related to cell size, although cell size is not the only determinant of lipogenic activity (Hood 1982).

The regulatory role(s) of insulin in carbohydrate and lipid metabolism of non-ruminants are well documented (Etherton 1982). In some mammals excess fat deposition is associated with enhanced insulin production but whether greater insulin secretion causes excess fat deposition, or is a result of increased insulin resistance, has not been resolved. Furthermore, the role of insulin in ruminant adipose tissue metabolism remains equivocal.

The effect of insulin on the metabolism of rat adipocytes was clearly demonstrated in a study by Green and Newsholme (1979). Increasing the insulin concentration in the media from 1 to 100 μU /ml increased glucose uptake almost fourfold and decreased glycerol release sixfold. However, when we repeated this experiment with ovine adipocytes, lipogenesis and lipolysis rates were unaffected by insulin concentration as high as 1000 μU /ml. (Table 1; Thornton *et al.* 1982). These same preparations of ovine adipocytes responded to adrenalin as lipolysis rates were markedly increased. Rat adipocytes, isolated and incubated under identical conditions to the ovine adipocytes, responded in terms of glucose incorporation and glycerol release in a similar manner to those studied by Green and Newsholme (1979). These findings on ovine adipocytes, and similar observations on isolated bovine adipocytes (Prior and Smith 1982), led to the conclusion that insulin has a negligible role in the regulation of ruminant adipose tissue metabolism. Recently Etherton and Evock (1986) have reported that insulin markedly stimulates lipogenesis of isolated bovine adipocytes and that this response is dependent upon the source of bovine serum albumin used in the medium.

Table 1. Effect of adrenalin and insulin on lipogenesis in isolated ovine subcutaneous, perirenal and omental adipocytes. The tissues (n = 8) were from sheep gaining liveweight at 100g /day; subcutaneous incorporation of acetate was significantly less ($P < 0.001$) than perirenal and omental; adrenalin significantly ($P < 0.001$) reduced lipogenesis but insulin was without effect (Thornton *et al.* 1982).

		Acetate incorporation (mol /g lipid /h)			
	Adrenalin (μM)	Insulin (μU/ml)			
		0	10	100	1000
Subcutaneous	0	0.70	0.72	0.75	0.79
	1	0.42	0.47	0.54	0.58
Perirenal and omental	0	1.97	1.88	2.00	2.05
	1	0.66	0.81	1.10	1.18

In vivo experiments on cattle have also indicated that insulin does not have a major regulatory role in fat deposition. The tolbutamide model was used in experiments on Hereford and Friesian steers by Gregory *et al.* (1982). Although serum insulin concentration rose in response to tolbutamide treatment, the insulin response was not correlated with percentage body fat within or across breeds. Smith *et al.* (1983) used alloxan-diabetic cattle, insulin injections and glucose infusions, to study the interrelationships between insulin and lipid metabolism. They concluded: "It appears that insulin may have only a limited effect on the processes (lipogenesis and lipolysis) in the bovine". In these same experiments alloxan caused significant increases, and insulin caused significant decreases, in the plasma concentration of branched-chain amino acids, indicating an important role for insulin in bovine amino acid metabolism (Prior and Smith 1982).

Recent reports from New Zealand have indicated that infusion of insulin into early-weaned lambs causes a further reduction in the percentage fat in the empty body (Munro *et al.* 1985) and that increasing the circulating levels of insulin is associated with an increase in the rate of backfat synthesis in lambs (Davey *et al.* 1985). In this latter study, use of the euglycaemic insulin clamp technique increased the plasma insulin concentration from 13.5 to 160 μU /ml while glucose turnover and glucose converted to glyceride glycerol were both doubled.

The production of large quantities of purified insulin(s) by recombinant DNA technology should permit long term *in vivo* studies to resolve the role of insulin, and insulin-like growth factors, in lipid metabolism of ruminants.

REPARTITIONING AGENTS

The consumer demand for leaner meat has provided a stimulus for research aimed at reducing the fat content of meat by altering the regulatory control systems which determine the accumulation of fat in adipose tissue. A significant outcome of this research has been the finding that the β_2-adrenergic agonists, clenbuterol and its analogue cimaterol, dramatically reduce the accumulation of fat in the adipose tissues with a concomitant increase in the muscle mass of meat-producing animals (Ricks *et al.* 1984a). These findings have been confirmed and extended in sheep (Baker *et al.* 1984; Thornton *et al.* 1985a; Beermann *et al.* 1986), cattle (Ricks *et al.* 1984b; Williams *et al.* 1986), poultry (Dalrymple *et al.* 1984) and pigs (Moser *et al.* 1986).

These reciprocal responses of adipose tissue and muscle to β_2-adrenergic agonists invoked the term "repartitioning agents" to broadly

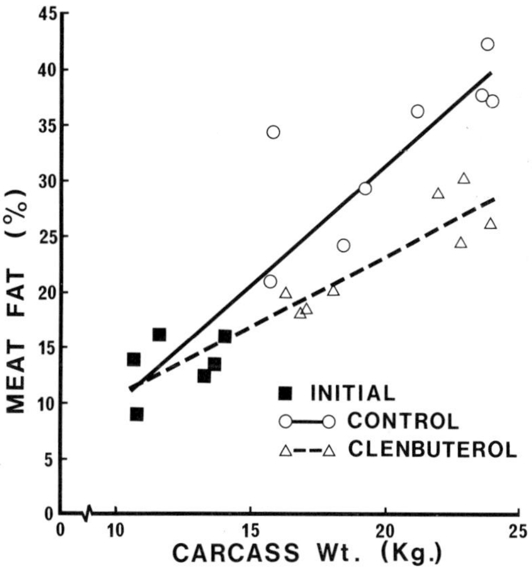

Figure 4. The relationship between the percentage of fat in meat and carcass weight (kg) of control and clenbuterol-treated (100µg /day) sheep (Thornton *et al.* 1985).

describe their effects (Ricks *et al.* 1984a). The extent to which these repartitioning agents affect carcass meat composition is well illustrated in the experiments on lambs described by Thornton *et al.* (1985a). In a study on pen-fed lambs both control and clenbuterol-treated animals grew at approximately 140 g /day. Treatment resulted in significantly less fat in the meat and consequently more lean (Fig. 4). For a 12 kg gain in carcass weight some 11.3 kg of meat was gained of which 42% was lean and 58% was fat in control animals *vs* 62% lean and 38% fat in the treated animals. The results from the final slaughter groups in this experiment are summarised in Table 2 which in quantitative terms shows a repartitioning of some 2 kg of fat for 2 kg of lean in 19 kg of meat.

Table 2. The effects of clenbuterol on the carcass meat of pen-fed lambs.

Variable	Clenbuterol (μg/head/day)		Change
	0	100	%
Carcass wt (kg)	23.2	23.0	
Meat wt (kg)	19.5	19.3	
Meat fat (%)	38.3	27.3	-28.7
Meat fat wt (kg)	7.5	5.2	-30.6
Lean meat wt (kg)	12.0	14.1	+17.5
Meat protein (%)	12.8	14.8	+15.6
Meat protein wt (kg)	2.5	2.9	+16.0

In another experiment, suckling-grazing lambs at foot with their mothers gained at approximately 200 g /day irrespective of treatment (control *vs* clenbuterol at 2.5 μg /kg liveweight/day) or sex (ewes, wethers or rams). The results of this experiment are summarised in Table 3 which shows that treatment reduced the weight of fat by 1.3, 1.1 and 0.7 kg or 33, 33 and 29% in the meat of ewes, wethers and rams, respectively.

It is clear from these two experiments that small quantities (50-100 μg/day) of clenbuterol can reduce the fat content of lamb carcass meat by some 30%. Furthermore, there was a consistent increase of 3 to 4 units in dressing percentage, from the lambs studied by Baker *et al.* (1984) and Beermann *et al.* (1985, 1986). Other fat deposits, for instance perirenal and omental fat (R.F. Thornton *et al.* unpublished), or kidney and pelvic fat (Beermann *et al.* 1986), are reduced by similar proportions to the reductions in meat fat. The gain in lean is not a gain in water only, as both protein and water are gained. The composition of the meat in terms of fat, water and protein from clenbuterol-treated animals (Thornton *et al.* 1985a, 1987) was not different from that of studies on boneless sheep meat (Thornton *et al.* 1981). Detailed studies indicate that the feeding of cimaterol to lambs has no effect on concentrations of protein, DNA, and RNA of the *semitendinosus* muscle (Beermann *et al.* 1985). The question remains as to how these β_2-adrenergic agonists exert their repartitioning effects.

Table 3. The effects of clenbuterol on the carcass meat of suckling-grazing lambs.

Variable	Ewes		Wethers		Rams	
	C	T	C	T	C	T
Carcass wt (kg)	16.2	15.8	16.9	15.8	15.7	15.6
Meat wt (kg)	12.9	12.4	13.4	12.4	12.2	12.0
Meat fat (%)	30.1	20.2	24.3	17.4	19.5	14.0
Meat fat (kg)	3.9	2.6	3.3	2.2	2.4	1.7
Lean meat wt (kg)	9.0	9.8	10.1	10.2	9.8	10.3

C = Control T = Treatment (2.5μg clenbuterol/kg liveweight/day)

As previously mentioned the growth rates of control and clenbuterol-treated lambs in the Australian studies (Thornton *et al.* 1985a) were not different, but some American studies have found greater growth rates and improved efficiency of feed conversion (Baker *et al.* 1984; Beermann *et al.* 1986). Improved feed conversion in clenbuterol-treated animals seems unlikely when the findings of short term experiments are considered. Abomasal infusion of clenbuterol (1.5 mg /day) caused an initial abrupt increase in heart rate and rectal temperature of a 35 kg lamb (Herbert *et al.* 1985) and 1.5 mg of clenbuterol in the diet of sheep fed dried grass at twice maintenance levels increased heat production by 0.3 to 0.8 MJ /day (MacRae *et al.* 1986). Further studies with larger numbers of animals in closely controlled experiments are required to resolve these questions.

Ricks *et al.* (1984a) proposed a scheme in which β_2-adrenergic agonists increased lipolysis rates in adipose tissue while decreasing protein catabolism and/or increasing protein synthesis in muscle, and suggested that these effects may be mediated or amplified by increased serum growth hormone levels (Ricks *et al.* 1984b).

Agonists can be described as compounds which promote function through their interaction with receptors on cell membranes. The β_2-adrenergic agonists promote a response similar to that of adrenalin (catacholamine) in the tissues and organs of the animals. The classification of the catacholamine receptors (α_1 β_1 β_2) of adipose tissue in different species has been reviewed by Mersmann (1979).

Clenbuterol has mainly β_2-receptor activity (Ricks *et al.* 1984b). That clenbuterol stimulates lipolysis has been demonstrated *in vitro* by Thornton *et al.* (1984a, 1985a) who showed that clenbuterol increased lipolysis rates of isolated ovine subcutaneous adipocytes (Fig. 5), and *in vivo* by Beermann *et al.* (1985) and Thornton *et al.* (1987) who showed that circulating levels of free fatty acids were increased in cimaterol-treated and clenbuterol-treated lambs respectively. The mechanism involved in the stimulation of lipolysis is probably through enhanced adenyl cyclase activity with consequent increased cyclic AMP production, resulting in activation of hormone-sensitive lipase by phosphorylation, that is the typical adrenalin response, but adenyl cyclase responses to β_2-adrenergic agonists have not been reported.

Although increased lipolysis could have been predicted as a response to clenbuterol, the finding that incorporation of acetate into lipid (lipogenesis) is markedly decreased in isolated ovine adipocytes by clenbuterol (Thornton *et al.* 1984a; see Fig. 5) was unexpected. A mechanism for this response is unknown and such a response has not been demonstrated *in vivo*. However, adrenalin reduced lipogenesis in isolated ovine adipocytes (Thornton *et al.* 1982). Furthermore, clenbuterol decreased the incorporation of a long-chain fatty acid (^{14}C stearate) into lipid of isolated ovine adipocytes (Thornton *et al.* 1985b). Again the mechanism of this response is unknown but it has been shown that clenbuterol at concentrations of up to 10 μg /ml does not influence the activity of lipoprotein lipase, the key enzyme in the

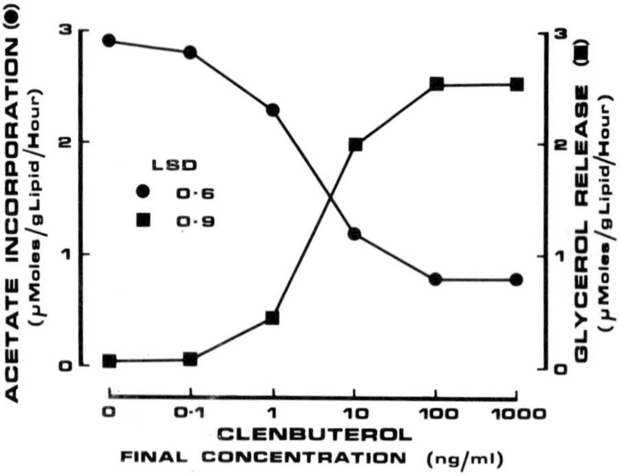

Figure 5. The relationships between lipogenesis (acetate incorporation (μmol /g lipid/h) and lipolysis (glycerol release; μmol /g lipid/h) *versus* the final concentration of clenbuterol in the incubation medium (0–1000 ng /ml) of isolated ovine subcutaneous adipocytes. Each point is the mean of 6 observations).

control of fatty acid uptake by tissues, in either isolated ovine adipocytes or in purified form (R.K. Tume and R.F. Thornton unpublished).

It is clear that β_2-adrenergic agonists exert direct effects on the adipocyte and both lipid deposition (lipogenesis and long-chain fatty acid incorporation) and mobilization (lipolysis) are affected. These responses suggest mechanisms for decreased fat deposition in adipose tissue which would enhance the availability of both acetate and long-chain fatty acids for metabolism in other tissues.

The role of growth hormone in the repartitioning effects of the β_2-adrenergic agonists is yet to be delineated. Treatment of sheep with clenbuterol increased serum growth hormone levels from 4.0 in controls to 5.1 ng /ml (Thornton *et al.* 1987; see also Ricks *et al.* 1984 and Beermann *et al.* 1985). However, only unnaturally high concentrations of ovine growth hormone (>10 ng /ml) depressed lipogenesis and were without significant effect on lipolysis rates of isolated ovine adipocytes (Duquette *et al.* 1984; Thornton *et al.* 1986). Thus the *in vivo* responses of ovine adipose tissue to chronic β_2-adrenergic agonist treatment appear to be a direct consequence of treatment rather than mediated through growth hormone levels.

Circulating levels of some hormones and associated metabolites in the blood of lambs fed 10 ppm cimaterol were reported by Beermann *et al.* (1985). In that study insulin levels were more than halved by cimaterol (3.7 to 1.6 ng /ml) but glucose levels (79 *vs* 78 mg /100ml) were not affected; thyroxine levels were increased (6.3 *vs* 8.1 µg /100ml) as were free fatty acids (287 *vs* 468 µeq /l) but cortisol levels (14.1 *vs* 15.5 ng /ml), prolactin levels (225 *vs* 195 ng /ml) and triiodothyronine levels (1.20 *vs* 1.34 ng /ml) were unaffected. These observations were interpreted as indicating "a maintenance of glucose homeostasis and mobilization of lipid stores which would be expected to be associated with the greater use of free fatty acids as an energy source". Recently, we have found that clenbuterol significantly reduced ($P < 0.001$) circulating insulin levels in sheep (from 2.0 to 1.4 ng /ml, controls *versus* treated animals; Thornton *et al.* 1987). The critical question raised by this data, and that of Beermann *et al.* (1985), is whether or not such a reduction in insulin levels would inhibit lipogenesis and stimulate lipolysis rates in adipose tissue.

The muscular hypertrophy of sheep receiving β_2-adrenergic agonists has been associated with increased size of muscle fibre (in terms of both cross sectional area and longitudinal growth) and a marked reduction in the proportion of Type 1 fibres in the *semitendinosus, semimembranosus* and *longissimus dorsi* muscles (Beermann *et al.* 1985). Recently, N.J. Rothwell, M.J. Stock and D.K. Sudera (personal communication) have shown that chronic clenbuterol treatment of rats increased blood flow to adipose tissues but reduced blood flow and β_2-adrenoreceptor density in skeletal muscle. In contrast to findings on rats (Emery *et al.* 1984) it has now been shown that protein catabolism of sheep is reduced and amino acid oxidation is increased, while whole body protein synthesis is unaffected, by clenbuterol treatment (MacRae *et al.* 1986). Whether this reduced muscle protein catabolism is a direct effect of the β_2-adrenergic agonist and/or is induced by some associated hormonal response, such as by increased growth hormone level, or metabolic intermediate, for example 2-ketoisocaproate concentration (Flakoll and Nissen 1986), remains to be clarified.

CONCLUSIONS

In this essay no real attempt has been made to integrate what is obviously a most integrated subject. Headings have been used on an arbitrary basis and although they correspond with physiological states, they should not be thought of as metabolically discrete. To embrace Bauman's concept of homeorhesis requires acknowledgement of the fact that few, if any, metabolic reactions occur or are regulated independently. However, to elucidate the roles of specific regulatory factors by experimentation, adequate control of all other variables is essential. It is equally important to integrate the findings of carefully controlled experiments and this can only be effectively achieved by computer modelling.

It is most unlikely that a single regulatory agent, or system, could be dominant in the cellular metabolism of herbivores. The emphasis on growth hormone in this essay is not meant to convey such an impression. On the contrary, the multifaceted roles of growth hormone in several physiological states well portrays the integrated nature of the regulation of cellular metabolism. However, it is also evident that the manipulation of growth hormone levels, by exogenous application or endogenous manipulation, eg. immunization against somatostatin, has considerable potential for the manipulation of physiological functions of domestic animals for the further benefit of mankind.

I am grateful to my colleague Ron Tume for helpful discussions and to Vicki Ford for typing the first draft.

Allen, C.E. (1976). *Fed. Proc.* **35**, 2302-7.

Annison, E.F. (1984). In "Herbivore Nutrition in the Subtropics and Tropics." (Eds F.M.C. Gilchrist and R.I. Mackie.) pp. 549-70. (Science Press: Craighall, South Africa.)

Armstrong D.G., and Blaxter, K.L. (1984). In "Herbivore Nutrition in the Subtropics and Tropics." (Eds F.M.C. Gilchrist and R.I. Mackie.) pp. 631-47. (Science Press: Craighall, South Africa.)

Baker, P.K., Dalrymple, R.H., Ingle, D.L., and Ricks, C.A. (1984). *J. Anim. Sci.* **59**, 1256-61.

Baldwin, R.L., Bywater, A.C., Taylor, J., and Murphy, M. (1984). In "Herbivore Nutrition in the Subtropics and Tropics." (Eds F.M.C. Gilchrist and R.I. Mackie.) pp. 478–502. (Science Press: Craighall, South Africa.)

Bauman, D.E. (1984). In "Herbivore Nutrition in the Subtropics and Tropics." (Eds F.M.C. Gilchrist and R.I. Mackie.) pp. 505–24. (Science Press: Craighall, South Africa.)

Bauman, D.E., and Currie, W.B. (1980). *J. Dairy Sci.* **63**, 1514–29.

Bauman, D.E., and McCutcheon, S.N. (1986). In "Control of Digestion and Metabolism in Ruminants." (Eds L.P. Milligan, W.L. Grovum and A. Dobson.) pp. 436–55. (Prentice-Hall: Englewood Cliffs, NJ.)

Bauman, D.E., Eisemann, J.H., and Currie, W.B. (1982). *Fed. Proc.* **41**, 2538–44.

Bauman, D.E., Eppard, P.J., De Getter, M.J., and Lanza, G.M. (1985). *J. Dairy Sci.* **68**, 1352–62.

Beermann, D.H., Campoin, D.R., and Dalrymple, R.H. (1985). *Proc. Recip. Meat Conf.* **38**, 105–17.

Beermann, D.H., Hogue, D.E., Fishell, V.K., Dalrymple, R.H., and Ricks, D.A. (1986). *J. Anim. Sci.* **62**, 370–80.

Bergman, E.N., and Pell, J.M. (1984). In "Herbivore Nutrition in the Subtropics and Tropics." (Eds F.M.C. Gilchrist and R.I. Mackie.) pp. 613–28. (Science Press: Craighall, South Africa.)

Black, J.L. (1984). In "Herbivore Nutrition in the Subtropics and Tropics." (Eds F.M.C. Gilchrist and R.I. Mackie.) pp. 648–71. (Science Press: Craighall, South Africa.)

Black, J.L., and Reis, P.J. (1979). In "Physiological and Environmental Limitations to Wool Growth." pp. 269–94. (Eds J.L. Black and P.J. Reis.) (Univ. New England: Armidale, NSW.)

Buttery, P.J. (1983). *Proc. Nutr. Soc.* **42**, 137–48.

Buttery, P.J. (1984). In "Herbivore Nutrition in the Subtropics and Tropics." (Eds F.M.C. Gilchrist and R.I. Mackie.) pp. 597–612. (Science Press: Craighall, South Africa.)

Buttery, P.J. (1985). *Reviews in Rural Science* **6**, 141–9.

Byers, F.M. (1982). *Fed. Proc.* **41**, 2562–66.

Callow, E.H. (1947). *J. Agric. Sci.* **37**, 113–29.

Cianzio, D.S., Topel, D.G., Whitehurst, G.B., Beitz, C.D., and Selt, H.E. (1985). *J. Anim. Sci.* **60**, 970–6.

Corbett, J.L. (1979). In "Physiological and Environmental Limitations to Wool Growth." (Eds J.L. Black and P.J. Reis.) pp. 79–98. (Univ. New England: Armidale, NSW.)

Dalrymple, R.H., Baker, P.K., Gingher, P.E., Ingle, D.L., Pensack, J.M., and Ricks, C.A. (1984). *Poultry Sci.* **63**, 2376.

Davey, H.W., Bickerstaffe, R., Munro, J.M., and O'Connell, D. (1985). *Proc. Nutr. Soc. NZ.* **10**, 79.

Duquette, P.F., Scones, C.G., and Muir, L.A. (1984). *J. Anim. Sci.* **58**, 1191–7.

Emery, P.W., Rothwell, D.J., Stock, M.J., and Winter, P.D. (1984). *Biosci. Rep.* **4**, 83–91.

Eppard, P.J., Bauman, D.E., Bitman, J., Wood, D.L., Akers, R.M., and House, W.A. (1985). *J. Dairy Sci.* **68**, 1047–54.

Etherton, T.D. (1982). *J. Anim. Sci.* **54**, 58–6[illegible]

Etherton, T.D., and Evock, C.M. (1986). *J. Anim. Sci.* **62**, 357–63.

Ferguson, K.A., Wallace, A.L.C., and Lindner, H.R. (1965). In "Biology of the Skin and Hair Growth." (Eds A.G. Lyne and B.F. Short.) pp. 655–77. (Angus and Robertson: Sydney.)

Field, R.A., and Young, O.A. (1985). *Proc. Recip. Meat Conf.* **38**, 141–5.

Flakoll, P., and Nissen, S. (1986). *Fed. Proc.* **45**, 232.

Ford, G.L., and Fogerty, A.C. (1982). *CSIRO Food Res. Quart.* **42**, 57–60.

Gill, M. (1986). *Proc. Nutr. Soc.* **45**, 221–9.

Green, A., and Newsholme, E.A. (1979). *Biochem. J.* **180**, 365–70.

Gregory, N.G., Truscott, T.G., and Wood, J.D. (1982). *J. Sci. Food Agric.* **33**, 276–82.

Haecker, T.L. (1920). Univ. Minnesota Agric. Expt. Stat. Res. Bulletin No. **193**, 1–99.

Hart, I.C. (1983). *Proc. Nutr. Soc.* **42**, 181–94.

Hart, I.C., Bines, J.A., and Morant, S.V. (1980). *Life Sci.* **27**, 1839–47.

Hart, I.C., and Johnsson, I.D. (1985). In "Control and Manipulation of Animal Growth." Proc. 43rd Easter School in Agric. Sci., Univ. Nottingham. (Eds P.J. Buttery, D.B. Lindsay and N.B. Haynes.) pp. 135–59 (Butterworths: London.)

Haugebak, C.D., Hedrick, H.B., and Asplund, J.M. (1974). *J. Anim. Sci.* **39**, 1026–31.

Hedrick, H.B. (1968). Univ. Missouri Agric. Expt. Stat. Res. Bulletin No. **928**, 1–96.

Herbert, F., Hovell, F.D. De B., and Reeds, P.J. (1985). *Proc. Nutr. Soc.* **44**, 105A.

Hood, R.L. (1982). *Fed. Proc.* **41**, 2555–61.

Hood, R.L., and Allen, C.E. (1973). *J. Lipid Res.* **14**, 605–10.

Hood, R.L., and Thornton, R.F. (1979). *Aust. J. Agric. Res.* **30**, 153–61.

Hood, R.L., and Thornton, R.F. (1980). *Aust. J. Agric. Res.* **31**, 155–61.

Hunter, R.H., and Vercoe, J.E. (1986). *Proc. Nutr. Soc. Aust.* **11**, 151.

Ingle, D.L., Bauman, D.E., and Garrigus, U.S. (1972a). *J. Nutr.* **102**, 609–16.

Ingle, D.L., Bauman, D.E., and Garrigus, U.S. (1972b). *J. Nutr.* **102**, 617–24.

Johnsson, I.D., and Hart, I.C. (1986). In "Manipulation of Milk Yield with Growth Hormone." 20th Nutrition Conf. Feed Manufacturers, Univ. Nottingham.

Johnsson, I.D., Butler-Hogg, B.W., Hathorn, D.J., and Wilde, R. (1986). Anim. Prod. **42**, 433.

Johnsson, I.D., Hart, I.C., and Butler-Hogg, B.W. (1985). *Anim. Prod.* **41**, 207–17.

Kempton, T.J. (1979). In "Physiological and Environmental Limitations to Wool Growth." (Eds J.L. Black and P.J. Reis.) pp. 209–22. (Univ. New England: Armidale, NSW.)

Lawrie, R.A. (1974). In "Meat." Proc. 21st Easter School in Agric. Sci., Univ. Nottingham. (Eds D.J.A. Cole and R.A. Lawrie.) pp. 249–68. (Butterworths: London.)

Leenanuruksa, D., and McDowell, G.H. (1985). *Aust. J. Biol. Sci.* **38**, 197–208.

Lindsay, D.B. (1982). *Fed. Proc.* **41**, 2550–4.

Lobley, G.E. (1986). *Proc. Nutr. Soc.* **45**, 203–14.

MacRae, J.C., and Lobley, G.E. (1986). In "Control of Digestion of Metabolism in Ruminants." (Eds L.P. Milligan, W.L. Grovum and A. Dobson.) pp. 367–85. (Prentice-Hall: Englewood Cliffs, NJ.)

MacRae, J.C., Lobley, G.E., Skene, P.A., and James, S. (1986). *J. Anim. Sci.* (Suppl.). In press.

McCutcheon, S.N., and Bauman, D.E. (1983). *Reviews in Rural Science* **6**, 131–40.

McDowell, G.H. (1983). *Proc. Nutr. Soc.* **42**, 149–67.

McDowell, G.H., Leenanuruksa, D., Gooden, J.M., Jois, M., and Annison, E.F. (1984). *Proc. Nutr. Soc. Aust.* **9**, 164.

Mersmann, H.J. (1979). *Proc. Recip. Meat Conf.* **32**, 93–8.

Milligan, L.P., and Summers, M. (1986). *Proc. Nutr. Soc.* **45**, 185–93.

Moe, P.W., and Tyrrell, H.F. (1986). *Proc. Nutr. Soc.* **45**, 215–20.

Moser, R., Dalrymple, R.H., Cornelius, S.G., Pettigrew, J.E., and Allen, C.E. (1986). *J. Anim. Sci.* **62**, 21–6.

Moulton, C.R., Trowbridge, P.F., and Haigh, L.D. (1922). Uni. of Missouri Agric. Expt. Stat. Res. Bulletin No. **55**, 1–89.

Munro, J.M., Bickerstaffe, R., Geenty, K.G., and Willis, J.A. (1985). *Proc. Nutr. Soc. NZ* **10**, 78.

Oddy, V.H. (1985). *J. Agric Sci.* **105**, 613–22.

Oddy, V.H., and Annison, E.F. (1979). In "Physiological and Environmental Limitations to Wool Growth." (Eds J.L. Black and P.J. Reis.) pp. 295–309. (Univ. of New England: Armidale, NSW.)

Oddy, V.H., Gooden, J.M., and Annison, E.F. (1984). *Aust. J. Biol. Sci.* **37**, 375–88.

Oddy, V.H., Gooden, J.M., Hough, G.M., Teleni, E., and Annison, E.F. (1985). *Aust. J. Biol. Sci.* **38**, 95–108.

Peel, C.J., Sandles, L.D., Quelch, K.J., and Herrington, A.C. (1985). *Anim. Prod.* **41**, 135–42.

Prior, R.L., and Smith, S.B. (1982). *Fed. Proc.* **41**, 2545–49.

Reis, P.J. (1979). In "Physiological and Environmental Limitations to Wool Growth." (Eds J.L. Black and P.J. Reis.) pp. 223–42. (Univ. New England: Armidale, NSW.)

Ricks, C.A., Baker, P.K., and Dalrymple, R.H. (1984). *Proc. Recip. Meat Conf.* **37**, 5–11.

Ricks, C.A., Dalrymple, R.H., Baker, P.K., and Ingle, D.L. (1984). *J. Anim. Sci.* **59**, 1247–55.

Robelin, J., and Geay, Y. (1984). In "Herbivore Nutrition in the Subtropics and Tropics." (Eds F.M.C. Gilchrist and R.I. Mackie.) pp. 525–48. (Science Press: Craighall, South Africa.)
Robinson, C.J., Hall, J., and Beshir, S.O. (1983). *Proc. Nutr. Soc.* **42**, 169–80.
Searle, T.W., and Graham, N. McC. (1975). *Aust. J. Agric. Res.* **26**, 355–61.
Searle, T.W., and Griffiths, D.A. (1976). *J. Agric. Sci.* **86**, 483–93.
Searle, T.W., Graham, N. McC., and Donnelly, J.B. (1982). *J. Agric. Sci.* **98**, 241–5.
Searle, T.W., Graham, N. McC., and O'Callaghan, M. (1972). *J. Agric. Sci.* **79**, 371–82.
Smith, R.W., and Walsh, A. (1984). *Res. Vet. Sci.* **37**, 320–3.
Smith, S.B., Prior, R.L., and Mersmann, H.J. (1983). *J. Nutr.* **113**, 1002–15.
Tan, G.Y., and Fennessy, P.F. (1981). *NZ. J. Agric. Res.* **24**, 1–3.
Thornton, R.F., Hood, R.L., Rowe, R.W.D., and Jones, P.N. (1983). *Aust. J. Agric. Res.* **34**, 447–52.
Thornton, R.F., Husband, P.M., and Larsen, T.W. (1981). *Food Tech. Aust.* **33**, 468–73.
Thornton, R.F., Shaw, F.D., and Hood, R.L. (1974). *Aust. J. Exp. Agric. Anim. Husb.* **14**, 281–5.
Thornton, R.F., Tume, R.K., and Larsen, T.W. (1982). *Proc. Nutr. Soc. Aust.* **7**, 115.
Thornton, R.F., Tume, R.K., Larsen, T.W., and Johnson, G.W. (1984a). *Proc. Nutr. Soc. Aust.* **9**, 185.
Thornton, R.F., Tume, R.K., Larsen, T.W., and Johnson, G.W. (1984b). *Proc. Aust. Soc. Anim. Prod.* **15**, 758.
Thornton, R.F., Tume, R.K., Larsen, T.W., and Johnson, G.W. (1985b). *Proc. Nutr. Soc. Aust.* **10**, 205.
Thornton, R.F., Tume, R.K., Larsen, T.W., Johnson, G.W., and Wynn, P.C. (1986). *Proc. Nutr. Soc. Aust.* **11**, 152.
Thornton, R.F., Tume, R.K., Payne, G., Larsen, T.W., Johnson, G.W., and Hohenhaus, M.A. (1985a). *Proc. NZ. Soc. Anim. Prod.* **45**, 97–101.
Thornton, R.F., Tume, R.K., Wynn, P.C., Larsen, T.W., and Johnson, G.W. (1987). *Proc. Aust. Asian Anim. Prod. Conf. Con.* **4**, 486.
Tribe, D.E., and Peel, L. (1963). *Aust. J. Zool.* **11**, 273–89.
Truscott, T.G., Wood, J.D., and MacFie, H.J.H. (1983). *J. Agric. Sci.* **100**, 257–70.
Van der Walt, J.G. (1984). In "Herbivore Nutrition in the Subtropics and Tropics." (Eds F.M.C. Gilchrist and R.I. Mackie.) pp. 571–93. (Science Press: Craighall, South Africa.)
Vernon, R.G. (1980). *Prog. Lipid Res.* **19**, 23–106.
Vernon, R.G., and Flint, D.J. (1984). *Symp. Zool. Soc. Lond.* **51**, 119–45.
Vernon, R.G., Clegg, R.A., and Flint, D.J. (1981). *Biochem. J.* **200**, 307–14.

Vickery, J.R. (1977). CSIRO Div. Food Res. Tech. Paper. No. 42, pp. 1–17.

Wallace, A.L.C. (1979). In "Physiological and Environmental Limitations to Wool Growth." (Eds J.L. Black and P.J. Reis.) pp. 257–68. (Univ. New England: Armidale, NSW.)

Weekes, T.E.C. (1983). *Proc. Nutr. Soc.* **42**, 129–36.

Weekes, T.E.C. (1986). *Proc. Nutr. Soc.* **45**, 195–202.

Wheatley, I.S., Wallace, A.L.C., and Bassett, J.M. (1966). *J. Endocrinol.* **35**, 341–53.

Williams, A.J. (1979). In "Physiological and Environmental Limitations to Wool Growth." (Eds J.L. Black and P.J. Reis.) pp. 337–54. Univ. New England: Armidale, NSW.)

Williams, A.J., Tyrrell, R.N., and Gilmour, A.R. (1978). *Aust. J. Expt. Agric. Anim. Husb.* **18**, 52–7.

Williams, D.R. (1978). In "Patterns of Growth and Development in Cattle." (Eds H. DeBoer and J. Martin.) pp. 219–29. (Martinus Niijhogg: The Hague.)

Williams, P.E.V., Paglioni, L., Innes, G.M., and Pennie, K. (1986). *Anim. Prod.* **42**, 434.

Wood, J.D., MacFie, H.J.H., Pomeroy, R.W., and Twinn, D.J. (1980). *Anim. Prod.* **30**, 135–52.

15

THE ABSORPTION, RETENTION AND FUNCTION OF MINOR NUTRIENTS

by N.F. Suttle

Knowledge of the needs of herbivores for nutrients other than those which supply energy and protein (i.e. the minor nutrients), is astonishingly incomplete. The principal reasons for this have been the pre-occupation of research workers with the mechanisms which animals possess for controlling minor nutrient metabolism and the use of experimental models which bear little resemblance to the herbivore's natural environment. The emphases of this paper will therefore be on mass rather than mechanism and on grazing rather than housed animals. It will try to establish some principles and models whereby minor nutrient deficiencies can be anticipated and prevented. Quantitative information concerning absorption, retention and function will be of primary concern.

The plant and animal factors which impinge upon absorption, retention and function are illustrated in Fig 1. The importance of particular factors and interactions varies from nutrient to nutrient, from feed to feed and from one herbivore species to another. However, it will generally be impossible to predict the functional outcome of a given minor nutrient supply from herbage using single measures of status in either the source or the recipient. Examples relating to the macro-minerals Ca, P and Mg and to the micro nutrients Cu, Co and Se will principally be used to illustrate the primary areas of ignorance concerning the herbivore's requirements for minor nutrients. We know that they are needed and for the most part we know why they are needed: we rarely know how much is needed and cannot therefore explain why deficiencies occur in particular animals on particular pastures.

MICROBIAL SYMBIOTES

The dependence of the herbivore on populations of fermentative microorganisms in the fore- and hind-gut brings a unique dimension to their minor-nutrient nutrition.

Microbial needs

The microbes themselves have nutrient requirements which must be satisfied from the host's diet and digestive secretions. Requirements for P are low and are normally met by the abundant flow of P recycled via the saliva. Although the microbial requirement for P increases with

NUTRITION OF HERBIVORES
ISBN 0 12 750052 9

the amount of organic matter they digest (Durand *et al.* 1983), it takes extreme depletion of the host to reduce salivary P flow into the rumen to a sufficient extent to interfere with microbial function (Breves *et al.* 1985). By diverting salivary flow, Milton and Ternouth (1985) depressed the digestibility of neutral detergent fibre (Fig. 1.1) without depleting the sheep of P and without affecting appetite.

There may, however, be circumstances in which appetite is reduced as a consequence of microbial P depletion (Fig. 1.3). In their study of Ca and P deficiencies in growing lambs, Field *et al.* (1975) found an immediate loss of appetite in the lamb given a diet low in P, long before the fall in plasma P and in contrast to all other studies. Important features of their study were the use of a pelleted, complete diet which would presumably have been a poor stimulator of salivary flow and the fact that removal of a $CaCO_3$ supplement largely prevented the loss of appetite. It is possible that the $CaCO_3$ caused insoluble P salts to form, thus exacerbating the poor supply of P to the microbes (Fig. 1.4) from both the diet and saliva.

The importance of salivary P flow raises an interesting question. Could water deprivation also accelerate the P-depletion of

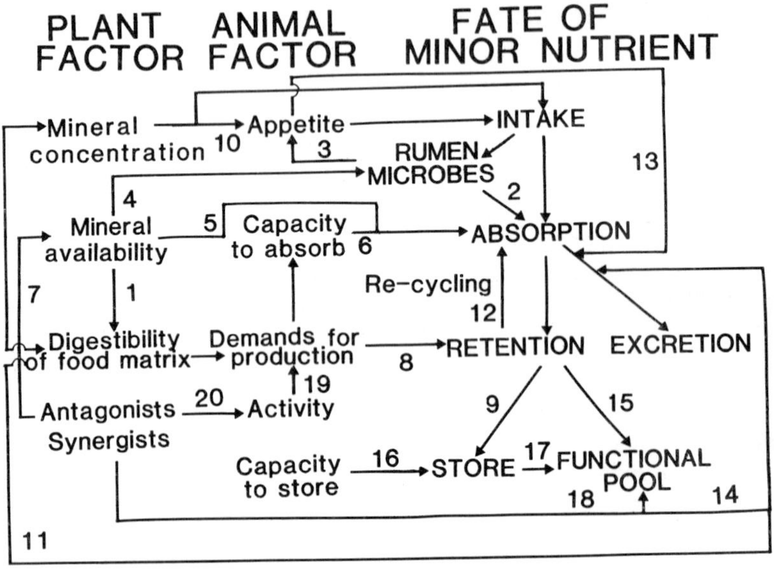

Figure 1. An illustration of the complex actions and interactions of plant and animal factors in determining the absorption, retention and function of minor nutrients in herbivores.

rumen microbes by restricting the flow of salivary P? It is common for forage P concentrations to be low in regions with arid climates to go together and this coincidence may depress feed digestibility and possibly intake.

The addition of fat to the ruminant's diet often depresses fibre digestibility but the effect can be prevented by concomitantly adding Ca and Mg or adding the fat as a Ca soap (Jenkins and Palmquist 1984). The inference is that either the Ca and Mg requirements of the rumen microorganisms are not met when the cations are removed from the liquid phase of the rumen by reacting with fat to form soaps, or the added soaps provide fat in a form which does not physically obstruct digestion by the microbes.

Microbial products

Microbial populations transform the food substrates and some of the minor nutrients ingested by the host. Cobalt, for example, is incorporated into a coenzyme (cobalamin) and analogue forms of vitamin B_{12}. The latter become increasingly predominant in the rumen as Co intakes increase (Gawthorne 1970) but the cobalamins are selectively absorbed through their affinity for intrinsic factor. There have been claims that the substitution of concentrates for roughage in the diet increases the proportion of analogues synthesised by the rumen microbes (ARC 1980) but it is harder to deplete the plasma cobalamin pool with a Co-deficient barley than a Co-deficient hay (Fig. 2). Nevertheless, for cobalt and other minor nutrients, the end products of their metabolism by microbes will have implications for herbivore nutrition.

Microbial by-products

By-products of rumen metabolism can also influence the minor-nutrient status of the host. Ivan *et al.* (1986) found that the elimination of the protozoal population from the sheep's rumen increased the hepatic accumulation of Cu sufficiently to cause death from Cu poisoning: it was suggested that the protozoa generated high ruminal sulphide concentrations through the rapid breakdown of soluble proteins and thus lowered Cu absorption (Fig. 1.2).

Experimental models

Experimental models for assessing the minor nutrient requirements of grazing herbivores should attempt to provide a rumen milieu and microbial population which match those of the grazing species. The use of one species of herbivore as a model for another, even within the fore-gut fermenter group, is likely to create anomalies in view of the 4 to 5-fold differences between species in the 'retention times' of both the liquid and particulate compartments of the rumen (Engelhardt *et al.* 1985). An example of species differences in rumen metabolism with effect on minor nutrient metabolism is provided by the relatively poor synthesis of thiomolybdates in the rumen of the deer compared with sheep (J. Mason pers. comm.).

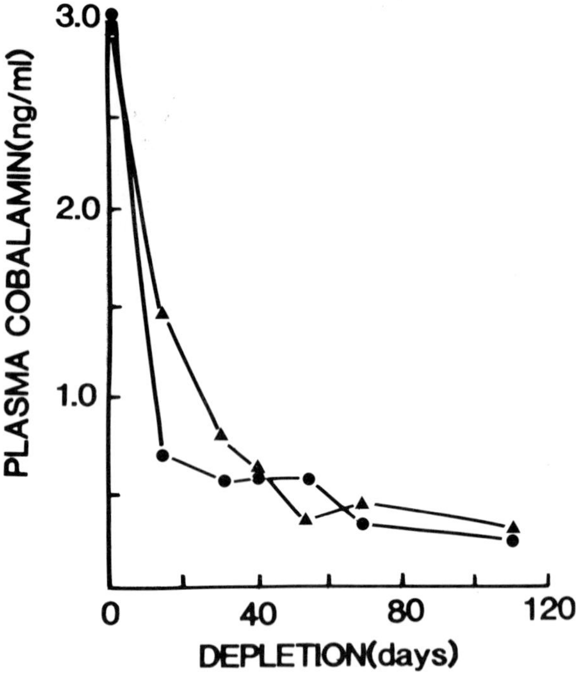

Figure 2. Depletion of plasma vitamin B_{12} in ewes on a cobalt-deficient barley diet (●; 0.03 mg Co /kg DM) is similar to that on a hay of higher cobalt content (▲; 0.05 mg Co /kg DM) (N.F. Suttle unpublished)

ABSORPTION

Absorption is influenced by both dietary and animal factors i.e. nutrient intake, the proportion which is available for absorption from the diet ('availability'; Suttle 1986; Fig. 1.5) and how much the animal can absorb or needs to absorb (Fig. 1.6). If availability is uniform then nutrient concentration and some idea of food intake may be all that is required to predict absorption but if availability is variable, then nutrient concentrations may be downright misleading. If the animal influences absorption, the validity of estimates of 'availability' as a dietary attribute is questionable. Until the range of availability is known, one cannot describe the minor nutrient value of a particular feed source.

Phosphorus availability

The data on the availability of P is summarised in Table 1 and shows that we know next to nothing about fresh pasture and only a little about conserved pasture. More is known about mixed diets and the constituents of compounded feeds when fed to sheep: with cattle, only data for mixed diets have been reported (Playne 1976). The range for P availability in grass or legume products is wide (0.54–0.95) but it is impossible to separate experimental bias from true differences in availability. The fact that a simple inorganic P source such as $CaHPO_4$ has been credited with availbility values ranging from 0.49 (Bruggemann *et al.* 1959) to 0.87 (Dingwall *et al.* 1987) suggests that experimental bias is important.

Table 1. Availability or absorption coefficients for P in roughages, concentrates and mixed diets fed to sheep.

	Roughages		Concentrates		Mixed diets
Fresh	Ruanui ryegrass	0.62(1)**			
	Maize silage	0.71(2)			
Dried	Grass hay	0.64(3)			
		0.64(4)	Wheat	0.78(4)	
		0.75(4)	Barley	0.78(4)	
		0.74(5)	Rapeseed		0.63 ± 0.06(10)
			meal	0.84(4)	0.66 ± 0.04(11)
					0.67 ± 0.12(12)
	Legume hay	0.72(4)	Soya bean		
			meal	0.81(4)	
		0.91(6)	Maize gluten		
			meal	0.71(4)	
		0.95(7)	Fish meal	0.80(4)	
		0.54(1)			
		0.55(8)			
	Grass pellets	0.75(4)			
	Lucerne pellets	0.74(6)			
	Bran	0.25(9)			
		0.63(4)			

** References 1. Grace (1981); 2. Gueguen and Durand (1976); 3. Bruggemann *et al.* (1959); 4. Field *et al.* (1984); 5. Field *et al.* (1982); 6. Lofgreen and Kleiber (1975); 7. Lofgreen and Kleiber (1954); 8. Gueguen (1962); 9. Ellis and Tillman (1961); 10. Braithwaite (1985); 11. Braithwaite (1983b); 12. Playne (1976).

The availability of P may reflect the heritable capacity of the selected animals to absorb P, the P intake chosen for the experiment or interactions with other minerals. Field *et al.* (1984) noted marked differences in P availability between, but not within, chimaera-derived sheep: such differences may be responsible for breed differences in urinary P excretion (Field *et al.* 1986). Braithwaite (1984) has reported that over a 10-fold increase in P intake the efficiency of P absorption decreases from 0.8 to 0.4. Field *et al.* (1983) found that an increase in dietary Ca from 3.4 to 5.4 g /kg DM reduced the availability of P by 0.18 $\pm$ 0.04. The addition of K to the diet increases urinary P excretion in sheep (Suttle and Field 1967) and cattle (Field and Suttle 1979), possibly by increasing P absorption (Greene *et al.* 1983).

The data of Field *et al.* (1984) is particularly useful insofar as several different feed sources were compared within the same experiment at similar moderate P intakes and with the influence of genotype minimised by the skilful use of chimaera. However, they used only one sample of each feed type and it is not known how representative each one was. There is clearly a need for a systematic study of the availability of P from fresh and conserved herbage, using a common protocol. Once systematic variation has been shown, then the reasons for that variation and *in vitro* methods for its prediction should be pursued.

There are probably differences in the availability of P in feeds fed to sheep and cattle. In the data cited by Playne (1976) for mixed diets, the mean ($\pm$ sd) values for sheep and cattle were 0.67 $\pm$ 0.12 (n=7) and 0.59 $\pm$ 0.14 (n=15), respectively, and ARC (1980) used absorption coefficients of those orders in calculating the respective requirements for dietary P. The apparent species difference must be confirmed by direct comparisons within experiments in adults fed on forage diets, since it may help to explain why aphosphorosis is a problem in cattle but not in sheep (Anon 1976).

Calcium availability

The Technical Committee on Responses to Nutrients Working Party on Ca and P requirements of ruminants (TCORN 1987) has concluded that little is known of the capacity of most feeds and supplements to provide Ca because the experimental conditions for measuring Ca availability have usually been inappropriate. Most species absorb Ca according to needs (ARC 1980; Fig. 1.6), the modulation of Ca absorption being a primary homeostatic mechanism alongside the control of a net Ca accretion in bone. Unless Ca availability is measured with Ca intakes at or below requirement, the potential of the feed to provide Ca is underestimated. In trying to allow for this phenomenon, ARC (1980) scaled the published estimates of Ca availability in balance trials with lactating dairy cows according to the relationship between intake (I) and requirement (R) i.e. according to I/R. They concluded that the availability of Ca pleateaued at a maximum of 0.68. The diets involved would have been mostly hay/concentrate mixtures with inorganic Ca supplements. In examining the recent, though still sparse, literature and extending the ARC (1980) approach to lambs, TCORN (1987) found no convincing evidence that Ca availability cannot reach 0.68 for any feed source including forages.

Theoretically, the availability of Ca in fresh or conserved herbage could be limited by incomplete release from the feed or interactions with other dietary components which render Ca insoluble. Ca is extensively released from forages in the rumen, although more rapidly from silage than from tropical hays (Rooke *et al.* 1983; Playne *et al.* 1978). Ward *et al.* (1975) claimed that the high oxalate content of lucerne hay was responsible for its low Ca 'availability' but it is more likely to reflect a high Ca content. Blaney *et al.* (1982) report good Ca absorption by cattle fed tropical grasses high in oxalate. Herbage lipids are unlikely to suppress Ca availability through soap formation in the rumen because the soaps are subsequently digested (Jenkins and Palmquist 1984).

The extent to which vitamin D enhances Ca absorption in herbivores is unknown but the lack of insolation at extreme latitudes in winter results in exceedingly low plasma vitamin D concentrations in outdoor ewes (Smith and Wright 1984) and may reduce Ca absorption.

Table 2. Mean (±s.d.) availability of magnesium in forage-based diets for sheep

Diet	Estimation technique: Measured with ^{28}Mg	ARC (1980) derivation
Fresh grass	0.25 ± 0.09 (n=5)	0.23 ± 0.023 (n=3)
Hay	0.25 ± 0.066 (n=7)	-
Hay + concentrates	-	0.43 ± 0.032 (n=5)
Silage	-	0.43 ± 0.032 (n=4)

From ARC (1980); Powley *et al.* (1977); Suttle and Field (1966, 1967); Ivan *et al.* (1983).

Magnesium availability

Of the macro-minerals, Mg shows the widest variation in availability between and within feed sources (Table 2). The availability values for hay and grass, measured with ^{28}Mg, indicate no difference between them but a three-fold variation within each material. The values for hay and concentrate mixtures derived from stable balance data by the ARC (1980) method, are nearly twice as high as those for grass, indicating a much higher Mg availability in concentrates than fresh grass or hay. Various explanations for these differences have been put forward based on differences in the chemical composition of the feeds.

Potassium - magnesium interaction. The spring pastures on which hypomagnesaemic tetany often occurs are rich in K and since the addition of K to cereal diets lowers Mg availability (Fig. 1.7), K has long been viewed as a factor contributing to the disorder (Suttle and Field 1967; Greene *et al.* 1983). McLean *et al.* (1985) recently found no effect of K on the availability of Mg when it was added to a pelleted grass diet and suggested that the earlier effects may have been a feature of the concentrate diets used. However, their unsupplemented diet was already high in K (20 g /kg DM). Magnesium availability was found to be greater in a grass high in Na and low in K than in one with the opposite attributes (Powley *et al.* 1977), supporting the concept that K interferes with Na-dependent absorptive mechanisms in the epithelium of the rumen, the primary site of absorption (Field 1983).

Formation of soaps. The high lipid content of spring pasture and the reduction in Mg availability following the addition of animal fat to a hay + concentrate diet led to the suggestion that formation of unavailable Mg soaps (Fig. 1.7) in the gastro-intestinal tract contributed to the incidence of hypomagnesaemia (Kemp *et al.* 1966). Studies with vegetable oils, which simulate the predominantly unsaturated herbage lipids more closely, have given conflicting results. Wilson *et al.* (1969) found that daily bolus doses of peanut oil in grazing cows lowered their plasma Mg concentrations. However, Grace and Body (1979), found that a continuous infusion of corn oil had no effect on the apparent availability of Mg in penned sheep fed twice daily on freshly cut white clover, and that soaps formed in the large intestine rather than at the primary absorptive site, the rumen. Whether the form of lipid (micelle size) and contrasting entry rates of oil and Mg were conducive to soap formation in the rumen in their study is a matter for speculation.

Rumen pH and the precipitation of salts. The early hypothesis that high ammonia concentrations in spring grass reduced Mg availability through the formation of insoluble $MgNH_4PO_4$ was dismissed (Field 1983) after the mineral was found to be a good source of Mg and P for lambs (Hemingway and Brown 1967). However, the lamb experiments were carried out with cereal diets and therefore at a low rumen pH and it has been suggested that at pH's of the higher order prevailing on grass diets, insoluble $MgNH_4PO_4$ may form and remain insoluble (Axeford *et al.* 1982); such a reaction (Fig. 1.7) would also reduce P availability and a reduction in P retention in sheep given a large supplement of magnesium oxide has been reported (Hjerpe 1968).

Hypotheses concerning the aetiology of hypomagnesaemia in herbivores have been postulated and criticised, largely on the basis of inaccurate simulations of the diet and rumen environment of grazing herbivores. The fact remains that Mg availability is inconsistently low in diets based on dried or fresh herbage and the reasons remain largely unknown. Since the difference in Mg availability between cereals and hay is conferred upon an added inorganic Mg salt (Dingwall and Field; cited by Suttle 1983a), this and other differences (for example the high availability of Mg in silage; Table 2) may relate to the chemical conditions established in the rumen rather than the form in which Mg is present in the food, with pH an important determinant (Fig. 1.2).

Copper availability

Variations in the availability of Cu have received more attention and are probably more nutritionally significant to herbivore nutrition than are those affecting any other micro-nutrient. The narrow base from which most of the data has been generated (Suttle 1974) must however be a cause of concern in view of the influence of experimental technique. Suttle and Peter (1985) found that when giving $CuSO_4$ in a semi-purified diet, rich in S, to sheep, the availability of the Cu was reduced by one third by simply changing from a single feed to continuous feeding. Their suggestion that changes in rumen sulphide 'profiles' after feeding were an important determinant of Cu availability were supported by within-treatment relationships between pre-prandial rumen sulphide concentrations and Cu availability in individual ewes. They were also supported by earlier reports of a negative curvilinear relationship between sulphide and soluble Cu outflows from the rumen (Bird 1970). The inference is that CuS is precipitated in the rumen and remains unavailable and that dietary S will be an important determinant of Cu availability.

The effect of sulphur. Natural forages show large differences in the inhibitory effects of S on Cu availability (Fig. 3). The effect of S

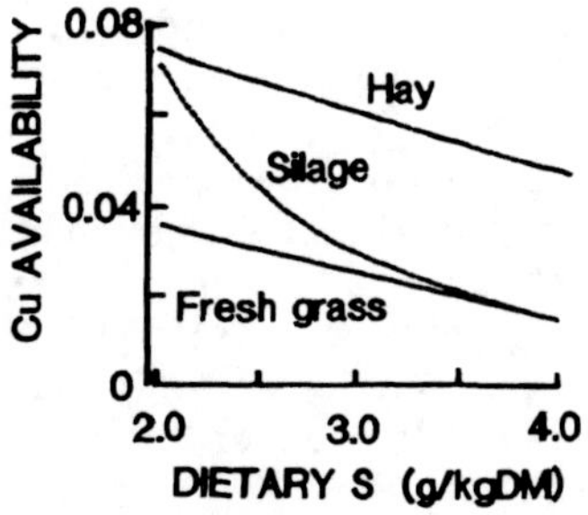

Figure 3. Sulphur concentration in forage is an important determinant of copper availability but the effects differ quantitatively in grass, hay and silage.

seems most pronounced in silages but the effect in fresh herbage may be masked because of the low availability even at the lowest S concentrations; the difference may involve differences in rumen sulphide metabolism arising from the continuous grazing of grass as opposed to the discontinuous intake of silage in the experiments.

Table 3. A comparison of the effects of molybdenum and sulphur concentrations on the availability coefficient of dietary copper to sheep as predicted from equations derived from two contrasting experimental sources (Suttle and McLauchlan 1976; Suttle 1983b).

Herbage Mo (mg /kg DM)	Herbage S (g /kg DM)	Prediction of Cu availability: Semi-purified diets	Prediction of Cu availability: Grazed herbages
1	1	0.062	0.045
	2	0.051	0.036
	4	0.035	0.015
3	1	0.057	0.020
	2	0.045	0.020
	4	0.028	0.003

The effect of molybdenum. The quantitative assessment of the inhibitory effects of Mo on the absorption of Cu (Fig. 1.7) was based initially on studies involving a semi-purified diet, a breed of sheep known to be particularly susceptible to Cu deficiency (Scottish Blackface; Woolliams *et al.* 1986a) and initially hypocupraemic animals (Suttle 1974; Suttle and McLauchlan 1976). However, the prediction equations inferred such major changes in Cu absorption for small changes in dietary Mo and S concentrations that a vast extrapolation was necessary to estimate Cu requirements for practical circumstances (ARC 1980).

It was not long before the predictions were tested under pastoral conditions in New South Wales and found to overestimate the effects of Mo and S on changes in liver Cu concentrations (Fig. 1.9) in the grazing Merino (Langlands *et al.* 1981). He suggested that the experimental conditions of Suttle and McLauchlan (1976) may be inappropriate. However application of their technique to a grazing situation showed that qualitatively at least it was relevant (Suttle 1983a). When sufficient experiments had been conducted to permit prediction equations for pasture to be generated (Suttle 1983a), there were clearly quantitative differences in the responsiveness of Cu absorption to changes in herbage Mo from the situation obtaining with semi-purified diets. However, the tendency was for the initial approach to have *underestimated* the inhibitory effects of Mo (Table 3). Furthermore on temperate, upland pastures, management practices which led to small increases in herbage Mo also induced hypocuprosis in sheep and cattle (Whitelaw *et al.* 1979; Woolliams *et al.* 1986a, b).

Effect of soil ingestion. Controversy surrounds the inhibitory effect of ingested soil on Cu absorption which was thought to involve a Fe x Cu antagonism (Suttle *et al.* 1984). Suttle *et al.* (1975) showed that the inclusion of soil in a semi-purified diet (at 10% DM) reduced the absorption of Cu by one third but Langlands *et al.* (1982) found that thrice weekly dosing with soil, *via* a rumen fistula, had no effect on liver Cu in the grazing Merino in New South Wales. However, in autumn in Scotland there is an exponential increase in herbage and faecal Fe with time due to soil contamination of ingested pasture and the best fit equation for the prediction of Cu absorption from autumn pasture contains a log Fe (mg /kg DM) component (N.F. Suttle unpublished).

The potential importance of soil ingestion on Cu absorption under grazing conditions is emphasised by the low levels of inclusion (1.75-3% DM) which have recently been found to inhibit Cu absorption in semi-purified diets (Brebner 1987); such levels of soil ingestion are common under grazing conditions. Others have shown that small supplements of soluble Fe salts enhance the depletion of Cu reserves in cattle (Brebner *et al.* 1983) but this may not simulate the effects of insoluble soil Fe which may adsorb Cu from the digesta (Brebner1987).

Sulphur, molybdenum and soil interactions. The separation of S from Mo and soil effects for the purposes of presentation cannot obscure the intimate relationships which determine how much absorbable Cu a forage can provide. The dependence of the Cu x Mo anatagonism on the presence of adequate dietary S is well known but the partial dependence of the soil x Cu antagonism on sulphur is a recent finding (Suttle *et al.* 1984). It is tempting to conclude that the contrasting responses of sheep in New South Wales and Scotland to both herbage Mo and ingested soil reflect the varying capacity for S to potentiate each interaction or the interactive nature of Mo and soil effects.

There are clearly limits to the accuracy with which one can predict the availability of the most-studied elements, P and Cu, in forages from current knowledge. It is impossible to contemplate such predictions for other minerals because availability has not been measured in herbivores even under convenient but inappropriate experimental conditions! The general lack of information on the absorption of minerals from forage is a major obstacle to the understanding and control of the related disorders of metabolism.

RETENTION

Once absorbed, the minor nutrients must be retained if the demands for production are to be met. Unavoidable losses *via* secretions entering faeces (Fig. 1.11) represent a drain on resources and an important source of variation in nutritional status through interactions between the animal and its food (Fig. 1).

Faecal endogenous loss of phosphorus

Phosphorus is actively secreted into saliva and secretion increases as salivary flow increases, although to a lesser extent, affording a considerable opportunity for recycling (Fig. 1.12) and economy, since P is subjected to a highly efficient (0.80; Field 1981) reabsorptive process in the small intestine. The unabsorbed (or endogenously lost) P in faeces constitutes the 'maintenance' need and this has conveniently been regarded as a constant per unit bodyweight (ARC 1980). Recent studies have, however, indicated that endogenous losses are influenced by the amount of food eaten and nature of the diet, in other words by factors which influence the flow of P in saliva.

Effect of food intake. In their efforts to redefine the maintenance requirements of ruminants for P, TCORN (1987) have used unpublished data of Field and Dingwall for sheep given pelleted, complete diets deficient in P and containing fine roughage (oat hulls, ground or fine), in which endogenous P (EP, g /d) loss was directly proportional to dry matter intake (DMI; kg /d; Fig. 1.13):

$$EP = 0.659\ DMI - 0.02$$

The negative balance of P increased as dry matter intake (and P intake) increased, confirming the increased demand for P. The maintenance needs arising from salivary secretion should therefore increase dramatically with the onset of lactation as DMI is increased to meet production needs (Fig. 1.8).

Effect of roughage. Since salivary flow is stimulated by the ingestion of coarse roughage, TCORN (1987) have suggested that an additional 'roughage factor' might be needed to predict endogenous P on P diets containing more than 20% roughage (Fig. 1.11). This recommendation stemmed from a study in which the inclusion of 20% chopped hay in a barley-soya bean meal diet caused a proportionate increase in endogenous P with no change in P intake (Field, Woolliams and Dingwall unpublished).

The precise values of endogenous P in animals on natural diets are, however, impossible to predict at present because the needs of the herbivore given forages have been almost entirely neglected in the studies of salivary P secretion and endogenous loss. The semi-purified diet has prevailed, often pelleted and sometimes lacking a roughage component, to the extent that only two measurements of endogenous P loss have been made in animals consuming fresh forage (Gueguen and

Durand 1976; Grace 1981). While there is information on herbage consumed as hay, the material was invariably chopped (often to unspecified proportions) for experimental convenience. The processing of roughage by milling decreases endogenous P loss (Tomas *et al.* 1967) and an effect of chopping is possible.

Effects of animal genotype. The maintenance needs of sheep and cattle for P are probably subject to heritable variation within and between species in P absorption (Fig. 1.6) because this determines the re-utilisation of secreted salivary P (Field 1981). Sheep with low genetically-determined efficiencies of P absorption have high endogenous P losses and hence high maintenance needs for P (Field *et al.* 1984). The lower efficiency of P absorption in cattle than in sheep could lead to higher endogenous losses and a widening difference between their maintenance needs when both species are on high intakes of long roughage. Extrapolation from sheep to cattle may not be valid and must be justified experimentally.

Effect of P intake. The relationship of endogenous P with P intake must be borne in mind when assessing data because high values may simply reflect the fact that an excessive supply was being excreted via saliva. Braithwaite (1984) has argued that since endogenous P increases as the intake and absorption of P increases, maintenance needs will increase with the production need, because higher production requires more P to be absorbed (Fig. 1.8). There remains the difficulty of distinguishing the obligatory component of endogenous P in such relationships and this can only be done by appealing to some 'external standard' of adequacy (e.g. the plasma P concentration or degree of bone mineralisation) associated with the different levels of production.

The daily endogenous P losses reported by Grace (1981) for sheep given fresh herbage (39 ± 5.5 (sd) mg /kg LW) were 2-3 times those predicted by the TCORN (1987) equations and were associated with plasma P concentrations of 1.87 ± 0.43 (sd) mmol /l, in other words, not particularly high and generally below the renal threshold. There is thus an urgent need for, first, agreement on the conditions under which the maintenance needs of herbivores for P should be measured and secondly, assessments of the effect of forage type and intake on endogenous P loss in order to predict the herbivore's needs for P and the adequacy of native pastures.

Faecal endogenous loss of calcium.

The maintenance requirement for Ca, like that for P, is predominantly to replace faecal endogenous loss (ECa) and regarded until recently as constant when scaled according to liveweight (ARC 1980). Braithwaite (1982) has challenged that assumption by showing that endogenous Ca was proportional to DMI /kg LW for sheep on a variety of dry diets (Fig. 1.3). TCORN (1987) have simplified that relationship to one involving unscaled DMI:

$$ECa = 0.623\ DMI + 0.228$$

The impact on the estimates of maintenance needs for lactating ewes (50 kg liveweight and 3 kg DMI) and rapidly growing lambs (30 kg liveweight and 1.8 kg DMI) is considerable, increasing ECa by 2–3 times the value given by the old constant (12 mg Ca /kg LW; ARC 1980). There is, however, an important question to be resolved for fresh forages; will endogenous Ca be determined by DMI or fresh food intake? If fresh food intake (i.e. FI) is the prime determinant, then maintenance needs for the grazing herbivore would have to be increased still further.

Faecal endogenous loss of other micronutrients

Effect of food intake. The fact that the first two nutrients to be examined, one (P) secreted predominantly in saliva and the other (Ca) post-ruminally, have shown a relationship between faecal endogenous loss and DMI suggests that this may be the rule rather than the exception as far as mineral requirements are concerned. The herbivore's maintenance requirements for many minor nutrients from forage diets may therefore be substantially underestimated by extrapolating from experiments with animals on low intakes of highly digestible diets. The relationship has important nutritional implications because a deficiency is likely to be exacerbated by high intakes of the deficient feed and reduction in appetite may be an important mechanism for limiting its impact. Milton and Ternouth (1985) have shown that the common, delayed form of anorexia in P deficiency is a humoral response and not a consequence of disturbed microbial function (Fig. 1.10).

Impaired recycling. Elements which are extensively recycled through the gastro-intestinal tract may be subject to impairment of reabsorption by herbage constituents which inhibit absorption. In cattle, biliary Cu secretion represents a major potential loss of Cu from the liver and hepatic loss is proportional to the size of the hepatic store (Simpson *et al.* 1981; Phillipo and Graca 1983). The negative slope of the relationship in sheep is increased by adding Mo to the diet as molybdate (Woolliams *et al.* 1983) or thiomolybdate (Suttle and Field 1983), suggesting impairment of Cu reabsorption. Since proportionately more Cu is secreted by cattle than by sheep with the same liver Cu concentration, exposure to diets rich in Mo may deplete cattle at a faster rate than sheep. Marked effects of Mo on hepatic Cu depletion in cattle were reported by Simpson *et al.* (1981). The lack of effect of the other dietary anatagonist, S, in that study could be explained in terms of a rumen based antagonism from sulphide which had no effect beyond that organ and thus could not impair the recycling of biliary Cu.

The impact of a dietary antagonist on the endogenous loss of a nutrient will therefore depend on its site of action relative to the site of reabsorption. The endogenous loss of Mg is unlikely to be increased on K-rich pasture (Powley *et al.* 1977) because K inhibits Mg absorption from the rumen i.e. prior to the site of most Mg secretion. The long term effect on retention of an antagonist which does reduce reabsorption will be less than that indicated by short-term studies because the amounts of the nutrient recycled will diminish with time. Thus the likelihood of Mo inducing hypocuprosis (i.e. clinical disease) in ruminants should be more accurately anticipated from studies in animals with low (Suttle and McLauchlan 1976) rather than high initial liver stores (Simpson *et al.* 1981; Langlands *et al.* 1981).

Increased recycling. Humoral influences on the efficiency of absorption, such as a high heritable capacity to absorb or an increased absorptive efficiency when nutrient intakes are low, should increase the reabsorption of the secreted nutrient and decrease its endogenous loss. The low efficiency of Zn recycling in sheep reported by Ivan *et al.* (1987) may thus reflect a generous dietary supply of Zn, and efficiency may increase when low Zn intakes induce a high absorptive efficiency (Suttle *et al.* 1982).

Urinary loss

The minor nutrients in urine usually represent excretion of an excess supply: they are an affordable loss and the obligatory component is normally small. There may, however, be situations in which retention is decreased and requirements are increased by antagonists which increase urinary losses (Fig. 1.14, 1.10). Harris *et al* (1987) have recently reported improvements in the fertility of grazing dairy cows following the use of Na supplements althougth the pasture contained more Na than was required according to ARC (1980; 1.5-2.5 *vs* 0.8-1.2 g /kg DM). In attempting to simulate the composition of spring pasture in studies of Mg metabolism in sheep, Suttle and Field (1966, 1967) found that the infusion of KCl and large volumes of water into the rumen of sheep on dry diets reduced their Na retention, largely through natriuresis. Thus, the ability of the grazing herbivore to retain Na may have been overestimated by inappropriate experimental models employing 'dry' diets low in K.

Distribution in the body

It is not sufficient for a nutrient to be retained, it must be present at functional sites in appropriate forms and amounts or concentrations (Fig. 1.15). Three examples are given to illustrate the importance of minor nutrient partition after absorption.

There have been suggestions that analogues of vitamin B_{12} are absorbed and retained by herbivores, especially cattle, so that concentrations of total B_{12} in plasma no longer reflect functional status with regard to cobalamin supply (Halpin *et al.* 1984). Furthermore the analogues may impair the action of cobalamins in the liver (Rickard and Elliot 1982). Technical problems may have led to overestimates of the extent of analogue absorption (Rothenberg and Marcoullis 1983) and further work is needed to establish the physiological importance of B_{12} analogue retention in herbivores.

It has been suggested that a shift in Mg from the extracellular to the intracellular pool occurs in animals on high K intakes (Larvor 1976); this could create a risk of hypomagnesaemic tetany because Mg^{2+} concentrations at the neuromuscular junction fall below the level necessary for normal transmission of nerve impulses. The recent study of McLean *et al.* (1985) supports this possibility; they reported a drop in plasma Mg concentration without a change in Mg balance or apparent Mg absorption in sheep given a dietary supplement of K.

An antagonist of mineral metabolism may occasionally act by altering the form in which the mineral is present in the tissues. Dietary Mo, with S, can alter the binding of Cu and Mo to plasma proteins through the systemic effects of thiomolybdates formed in the rumen (Suttle 1980). The trichloracetic acid (TCA) insoluble Cu:Mo complexes formed accumulate in the tissues and trithiomolybdate may be instrumental to their synthesis (Mason *et al.* 1982, Mason 1986). The importance of these systemic effects in the grazing herbivore has not been established, however. Thiomolybdates are absorbed onto fibre (Bray *et al.* 1982; Gawthorne *et al.* 1985) and their uptake is reduced by dietary Cu (Suttle and Field 1984): both reactions are likely to reduce the systemic effects of thiomolybdates. In sheep consuming Mo-rich hay or grass, TCA insoluble Cu in plasma increased slowly to a maximum of only 0.2 mg /l on forage containing 12 mg Mo /kg DM (Suttle 1983b).

In each of the above examples, total nutrient content of the body or even selected tissues may give misleading estimates of the functional status of the herbivore and current analytical and diagnostic approaches to the assessment of vitamin B_{12} and Cu deficiency in particular must be reconsidered.

Body stores

The capacity to accumulate nutrients (Fig. 1.16) in excess of the requirement for growth during periods of generous dietary supply, and mobilise them during periods of shortage (Fig. 1.17) is of particular importance to the grazing herbivore, which must cope with considerable fluctuations in the growth and composition of herbage. Initial body reserves can also influence the outcome of experiments and assessments of the adequacy of diets.

Skeletal reserves for growth. The most striking example of the impact of body stores relates to the reserve of Ca, P and Mg in well-mineralized bone. The ash concentration of bone increases substantially in lambs on upland pastures between birth and 15 months of age (Spence *et al.* 1985). It follows that the ability to tolerate a deficiency will increase with age and in the adult hill ewe, 25% of the complete reserve of Ca, P and Mg in bone may be removed during a period of winter deficit (Field *et al.* 1968). Studies with pregnant sheep (Little *et al.* 1978) and cattle (Call *et al.* 1978, 1986) have revealed a tolerance of diets which are low in P by all but the lowest standard (ARC 1980). This standard has been strongly criticised for its parsimony (Todd 1983). This tolerance of 'inadequate' diets almost certainly reflects the sample size of the skeletal reserve in relation to demands for maintenance and foetal growth (Table 4).

Table 4. Phosphorus requirements of sheep and cattle for daily maintenance, daily milk production and the whole of pregnancy as fractions of the skeletal reserve of P (ARC 1980).

	Ewe	Cow
Liveweight (kg)	50	500
Total P in body (g)	290	3500
Maintenance *	0.0034	0.0029
Pregnancy†	0.10	0.10
Lactation#	0.007	0.005

* At 20 mg P /kg LW.
† Gravid uterus of ewe and cow contains 29g and 342 g P, respectively.
\# For ewes 2 kg milk /d at 1.2 g P /kg: for cows 20 kg milk /d at 0.9 g P /kg.

The situation with young, rapidly growing animals is very different. Hypophosphataemia is readily induced by feeding diets deficient in P, and inorganic plasma P can readily reflect the dietary supply of P (Fig. 4). The use of plasma P as an index of nutritional status is widely criticised, but its value may be enhanced by sampling the younger members of the population at risk.

Skeletal reserves for lactation. With the onset of lactation reserves of Ca and P are likely to be mobilized at a greater rate (Table 4). Little *et al.* (1978) found that hypophosphataemia was induced by their low P diet once lactation commenced. However, Ca and P are deposited in bone in the fixed ratio necessary for mineralization (2.2:1) rather than for milk production where the required ratio in cattle and sheep is 1.3 to 1. Bone resorption during early lactation will therefore oversupply Ca. The difficulties of balancing supply and demand for Ca in the early lactation of sheep were indicated by Braithwaite (1983a) and by the persistence of milk-fever as a problem in the dairy cow. Mechanistic studies of parathyroid hormone (PTH), thyrocalcitonin and vitamin D metabolite concentrations in affected and unaffected animals

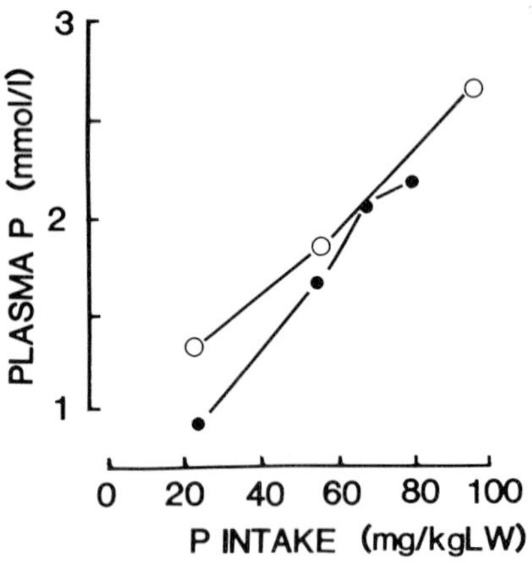

Figure 4. Plasma phosphorus concentrations in young lambs quickly reflect phosphorus intakes for a particular diet. Data for lambs after 2 (●, Fishwick 1978) or 4 weeks feeding (○, Braithwaite 1985).

have so far failed to identify a weak link in homeostatic control of Ca in milk fever. Could the problem be related to the concurrent hyphophosphataemia and an excessive supply of Ca from bone? There are reports of P-deficient animals failing to attain the high efficiencies of Ca absorption that should have attended the feeding of diets marginal in Ca, possibly because bone resorption was briefly compensating for the poor supply of Ca from the diet (Boxebeld *et al.* 1983; Braithwaite 1983a; 1986). B.W.S. Smith and H. Wright (pers. comm.) have noted that the feeding of a low P diet to lambs can decrease the plasma PTH concentration, whereas low Ca diets produce an increase.

A low Ca intake pre-partum, although hard to attain on natural diets, is an effective experimental technique for controlling milk fever (Goings *et al.* 1974) and it is thought to work by priming the homeostatic mechanism(s) for Ca. However, it may simply work by creating a balanced supply of Ca and P when coupled with the supply from bone. An equivalent effect has been obtained by raising P intake (Stott 1965) and it may have been achieved by allowing maximal Ca absorption to be attained.

Mobilisation of mineral reserves should not be relied upon in formulating Ca and P requirements and diets, but it should be considered in assessing the economics of supplementing the pastures grazed by herbivores. It is also important to ensure that macro-nutrient deficiencies do not limit the retention of minor nutrients. Sykes and Geenty (1986) have suggested that negative balances of Ca and P during lactation may be the result of a lack of protein for bone matrix formation.

FUNCTION

An awareness of the first limiting function of the minor nutrient in terms of animal health and the nutrient inputs needed to avoid malfunction is of crucial importance.

Vitamin E, selenium and their antagonists

Nutritional myopathy provides an excellent model to illustrate the interplay of synergists and antagonists in determining nutrient requirements. The limiting function is muscle membrane integrity in the face of peroxidative challenge. Since both vitamin E and the selenium-containing enzyme, glutathion peroxidase (GSHPx; EC 1.11.1.9), have antioxidant properties, they can act synergistically in preventing myopathy (Fig. 1.18). A low status of both elements may not, however, cause disease until a high intake of polyunsaturated fatty acids (PUFA) increases the endogenous concentration of toxic hydroperoxides (Fig. 1.18) and challenges the capacity to prevent membrane damage (McMurray and McEldowney 1977). Thus, myopathy develops acutely when cattle low in vitamin E and Se are turned out to graze pasture rich in the unsaturated fatty acid, linoleic acid, and is reproduced experimentally by adding protected linseed oil to a diet low in both micro-nutrients (McMurray *et al.* 1983).

The implications of such interactions in terms of herbivore nutrition have not been appreciated. Both the need for each nutrient and the biochemical criteria of adequacy (for example plasma tocopherol and blood GSHPx) will depend on the dietary concentration of the other and of their mutual antagonist, PUFA. Furthermore, since the antagonist must pass through the rumen without being hydrogenated and must be absorbed before it can cause damage to the muscle membrane, factors such as rate of passage through the rumen may determine the severity of challenge.

In a collaborative study with McMurray and his colleagues, we investigated some of these interrelationships by studying the effects of vitamin E and Se injection on the incidence of myopathy in lambs from ewes severely depleted of vitamin E during pregnancy and kept on a diet of marginal Se content (Table 5). Giving vitamin E to the lamb delayed the sharp decrease in plasma tocopherol but was insufficient to reduce myopathy, as indicated by increased plasma creatine kinase (CK; EC 2.7.3.3) activity. Se treatment was protective at 35 days even though GSHPx concentrations in untreated lambs were adequate by conventional standards (Anderson *et al.* 1979). With high GSHPx activity, the plasma tocopherol concentration could transiently reach undetectable levels without causing muscle damage. By 63 days, however, 60% of the Se-treated lambs had CK concentrations > 1000 U

/l despite a three-fold increase in GSHPx concentrations, suggesting that Se was only partially able to substitute for the lack of vitamin E. Together the supplements afforded complete protection.

Table 5. The effects of selenium and vitamin E injections on plasma creatine kinase (CK) in 5 week-old lambs from ewes given a diet low in vitamin E and marginal in Se during late pregnancy.

	Log plasma CK (IU /ml)		Blood GSHPx (U /g Hb)	Plasma tocopherol (μm /l)	
Dietary treatment	No. > 3	Mean (±se)	Mean (±se)	No. < 0.2	Mean of detectable values (±se)
0	4/7†	3.10 (0.28)	106 (65)	4/7	1.8
Se*	0/5	2.55 (0.24)	322 (200)	3/5	2.0
Vit E#	2/5	2.75 (0.32)	115 (77)	1/5	1.85 (0.18)
Vit E+Se	0/4	2.64 (0.24)	428 (79)	1/4	1.47 (0.18)

* Se - single injection of 5 mg Se as barium selenate at 7 d of age
\# Vit E - weekly injection of 100 mg tocopherol in vegetable oil given subcutaneously
† Proportion of animals with more than or less than stated value.

The outcome of such interactions in the face of a PUFA challenge has yet to be studied but, in the meantime, the temptation to increase the naturally low concentrations of essential fatty acids in the plasma of newborn herbivore (Soares 1986) by feeding PUFA during pregnancy should be resisted since growth may be retarded (Palmquist *et al.* 1977).

Activity. Challenges to nutrient function which influence nutritional and biochemical criteria of adequacy may come from sources other than the diet. In his attempts to induce myopathy in calves, Arthur (1981) found that the ingestion of PUFA-rich grass by calves deficient in Se- and vitamin E did not cause muscle damage when the grass was cut and fed to them indoors. However, when similar calves were turned out to graze the grass, CK concentrations increased to 500-5000 U /l. Myopathy in deer of low Se and vitamin E status is often precipitated by chasing and handling, giving rise to the term, 'capture myopathy' (Basson and Hofmeyr 1973; Ullrey *et al.* 1985). Thus exercise may compound the oxidative stress on the muscle membrane and increase requirements for Se and vitamin E (Fig. 1.19). Biochemical criteria of vitamin E and Se status which appear adequate in experiments with housed animals may be found wanting in the grazing situation.

Further examples of the possible effects of exercise are, first, the sudden onset of post-parturient haemoglobinuria which occurs when marginally Cu- and Se-deficient diary cows are turned out to graze spring pasture (Black 1981); secondly, the haemolytic, oxidative (or Heinz body) anaemia found in a large flock of Cu- or Se-deficient lambs, disturbed and harassed for 24 hours prior to blood sampling (Suttle *et al.* 1987a). Neither disorder has been reported in extensive studies of Cu and Se deficiency under the rarefied conditions of indoor experiments and they indicate that the erythrocyte membrane as well as the muscle membrane is vulnerable to challenge when activity increases.

Infection increases requirements for energy and protein and almost certainly does the same for the minor nutrients as their turnover and distribution is altered by lymphokines secreted during the immune response (Fig. 1.20). Most nutrient deficiencies will impair the multiplication of lymphocytes challenged *in vitro* with mitogens (for example, Se; Turner *et al.* 1985). However, the nutritional effects are often dependent on the dose of mitogen used and the source of the lymphocytes cultures (blood, spleen or thymus; Suttle and Jones 1986). It is therefore dangerous to assume that the ability of an animal to withstand an infection has been compromised by such artificial stimulation which may exaggerate the incidence of malfunction *in vivo* by inducing hyperactivity in the lymphocytes culture.

STRATEGIES FOR CONTROLLING MINOR NUTRIENT DEFICIENCIES

From what has been said so far about the number of factors which can influence the capacity of nutrients in herbage to satisfy the functional requirements of animals, strategies for identifying and preventing deficiencies will be beset with difficulties. The plant and animal variables summarised in Fig. 1 can influence minor nutrient requirements by at least an order of magnitude.

'Critical' concentrations

It will rarely be possible to approach a suspected deficiency situation with a table of minor nutrient concentrations in herbage in the one hand, and tables of animal requirements or biochemical criteria in the other, and define the scale of the animal health problem. Leaving aside the inadequacies of the description of the herbage supply, the definitions of animal requirements and critical tissue concentrations vary so widely between sources and experiments that the conclusion reached will depend on whose data were chosen: none is reliable for predicting field problems.

The danger of relying on unsubstantiated standards is illustrated by the attempts to define the cause of a wasting disease of Colombian cattle called 'Secadera'. Secadera has been described as 'primarily an induced thiamine deficiency, complicated in some geographic areas by deficiency or imbalance of Zn, Cu and less frequently Co and Se' (McDowell *et al.* 1985). Further studies implied that herbage S might be limiting and that bone indices of Ca and P status were generally low (McDowell *et al.* 1985). The disease was controlled by feeding a 'complete mineral mixture' with additional amounts of Zn, S, Cu and Se (Miles and McDowell 1983). 'Secadera' is probably a general term for ill-thrift which can arise from various minor-nutrient deficiencies, often singly, occasionally in combination but rarely in the abundance indicated by comparison with tables of 'critical' concentrations of the nutrients in herbage (McDowell *et al.* 1983). Impaired rumen microbial activity may be of primary importance and other changes may be secondary in 'Secadera'.

McDowell *et al.* (1983) cite 15 reports from eight developing countries in which mineral supplements improved calving percentages but in no case was a specific deficiency proven, although P deficiency seems likely in most instances and Na deficiency was ruled out by the use of NaCl-treated controls.

Table 6. Mean (± sd) trace element concentrations in Scottish pastures in relation to ARC (1980) estimates of requirements.

	Area	n	Grass (mg /kg DM)	Ovine requirement (mg /kg DM)
	East	99	5.9 (1.82)	Not less than 4.5
Cu	North	67	5.8 (2.04)	rising to 13.5 with Mo
	West	312	7.8 (2.35)	at 4 mg /kg DM
	East	20	0.095 (0.035)	
Co	North	46	0.072 (0.064)	0.08
	West	312	0.100 (0.100)	
	East	-	-	
Se	North	30	0.021 (0.010)	0.03 - 0.05
	West	82	0.060 (0.053)	

Indices of function

The proper assessment of minor-nutrient deficiencies is a challenge for developed as well as developing countries. What is the solution? Since there are few reliable predictors of functional deficiency in herbivores, they must be sought by careful observation of the responses to specific supplements in circumstances in which other elements are not limiting. Correlations between biochemical indices and improvements in performance will eventually establish whether critical threshold values

Table 7. An assessment of the determinants of Cu, Co and Se status in sheep on Scottish hill pastures and the relationship between status and growth limitation (Suttle *et al.* 1986).

Soil measurements	Herbage measurements	Animal (lamb) measurements	
		Biochemical Index	Functional Index
	Determinants of copper		
EDTA extractable Cu Ammonium acetate extractable Mo	Cu, Mo, S, Fe	Plasma Cu Blood SOD*	Growth response to CuO needles
	Determinants of cobalt		
Total and acetate extractable Co	Co, Ti**	Plasma B_{12}	Growth response to B_{12} injection Plasma MMA
	Determinants of selenium		
Se	Se	Blood GSHPx	Growth response to Se injection

* SOD Superoxide dismutase
** to indicate soil contamination.

exist, whether critical ranges are more appropriate, or whether response to treatment is the only reliable criterion.

An example of this 'dose : response' approach is proceeding in Scotland. Herbage analyses indicate that Cu, Co and Se concentrations were often well below the current recommendations in the UK (Table 6), but field problems were not widely recognised. In each of three years, the ability of one element to improve lamb growth when the others are not lacking is being examined (Suttle *et al.* 1986). Specific supplements are given in forms believed to be most certain of giving a response and new criteria of trace element status in the soil and animal are being assessed alongside the conventional parameters (Table 7). Measurements of the amount of herbage available are also being made. The pattern is not one which could or should be duplicated in developing countries: many of the analyses require sophisticated and expensive equipment and appropriate expertise in their use and maintenance: most of the analyses and even the supplements may turn out to be superfluous. However, the best service that developed nations can give the developing nations in defining minor nutrient problems in herbivore nutrition is to identify the key variables under conditions conducive to experimentation.

Already, the Scottish experience has led to the questioning of some important assumptions. In the first year when responses to copper were examined, growth responses were rare (1 in 20 sites) and hypocupraemia transient, if present at all. However, lamb growth rates varied widely between farms and were correlated with another criterion of Cu status, superoxide dismutase (SOD; EC 1.15.1.1) in blood. Although the Cu treatment improved both plasma Cu and SOD on all farms, its effect on the latter was small relative to between-farm variation. Was there an advantage in having a higher Cu status than the treatment could confer and above the accepted norm?

Sites had been chosen for likelihood of higher than normal herbage Mo concentrations, and the mean Mo value was 3.9 ± 2.3 mg /kg DM. Recent work at the Rowett Research Institute has suggested that Mo may have adverse effects on growth and fertility in cattle which are independent of total Cu concentration in plasma and liver (Humphries *et al.* 1983; Phillippo *et al.* 1985). There have been reports in the field of cattle in molybdeniferous areas responding to Cu even though blood Cu concentrations were scarcely below normal (Thornton *et al.* 1972). This phenomenon may also affect sheep: a recent study has shown that lamb carcase weight was decreased by adding 5 mg Mo /kg DM to the diet although plasma and liver Cu concentrations were 'normal' (Table 8), and similar relationships have been obtained in lambs grazing Mo-rich pastures (Hogan *et al.* 1971). It is possible that different criteria of Cu adequacy are needed when Mo intake is high.

Table 8. Effect of dietary molybdenum supplement on dressed carcase weight of lambs after 7 weeks on diet of marginal copper concentration (4 mg Cu /kg DM) and before plasma and liver Cu concentrations had reached 'critical levels' (0.6 mg /l and 20 mg /kg DM, respectively).

Mo (mg /kg DM)	Plasma Cu (mg /1)	Liver Cu (mg /kg DM)	Dressed carcase weight (kg)
0	0.6 (0.2)	142 (18.5)	15.9 (0.42)
5	0.7 (0.2)	35 (6.8)	14.7 (0.38)

N. F. Suttle (unpublished).

Indices of requirement

The accumulation of data on the minor nutrient composition of herbage in association with information on the most sensitive index of body function should generate values for the dietary requirement for that function under grazing conditions. This 'black box' approach ignores most of the interfaces and interactions in Fig. 1 but is likely to yield accurate values for requirement quicker than a compartmental or factorial approach: it will probably yield anomalies which only careful study of absorption, retention and function will resolve. The task is then to study each compartment under conditions more relevant to the grazing or forage-fed herbivore than those commonly used in the past.

ARC (1980). In "The Nutrient Requirements of Ruminant Livestock." (CAB: Farnham Royal, UK.)

Anderson, P.H., Berrett, S., and Patterson, D.S.P. (1979). *Vet. Rec.* **104**, 235-8.

Anon (1976). In "Reviews in Rural Science III." (Ed. G.D. Blair.) p.180 (Univ. New England: Armidale, NSW.)

Arthur, J.R. (1981). In "Proc. 4th Int. Symp. Trace Element Metabolism in Man and Animals." (Eds J. McC. Howell, J.M. Gawthorne and C.L. White.) pp. 356-9. (Aust. Acad. Sc.: Canberra.)

Axford, R.F.E., Hughes, A., and Evans, R.A. (1982). *Proc. Nutr. Soc.* **41**, 85A.

Basson, P.A., and Hofmeyr, J.M. (1973). In "The Capture and Care of Wild Animals." (Ed. E. Young.) pp. 151-60. (Human and Rousseau: Cape Town, S. Africa.)

Bird, P.R. (1970). *Proc. Aust. Soc. Anim. Prod.* **8**, 212-8.

Black, H. (1981). Proc. Sheep and Beef Soc. N.Z. Vet. Assoc. 11th Seminar. Palmerston North. pp. 11-4.

Blaney, B.J., Gartner, R.J.W., and Head, T.A. (1982) *J. Agric. Sci.* **99**, 533-9.

Boxebeld, A., Gueguen, L., Hannequart, G., and Durand, M. (1983). *Reprod. Nutr. Develop.* **23**, 1043-53.

Braithwaite, G.D. (1982). *J. Agric. Sci.* **99**, 355-8.

Braithwaite, G.D. (1983a). *Br. J. Nutr.* **50**, 711-22.

Braithwaite, G.D. (1983b). *Br. J. Nutr.* **50**, 723-36.

Braithwaite, G.D. (1984). *J. Agric. Sci.* **102**, 295-306.

Braithwaite, G.D. (1985). *J. Agric. Sci.* **105**, 67-72.

Braithwaite, G.D. (1986). *J. Agric. Sci.* **106**, 271-8.

Bray, A.A., Suttle, N.F., and Field, A.C. (1982). *Proc. Nutr. Soc.* **41**, 67A.

Brebner, J. (1987). Ph. D. Thesis. Imperial College, University of London.

Bremner, I., Phillippo, M., Humphries, W.R., Young, B.W., and Mills, C.F. (1983). In "Trace Elements in Animal Production and Veterinary Practice." Br. Soc. Anim. Prod. Occ. Publ. No. 7, 136-7.

Breves, G., Hoeller, H., and Lessmann, H.W. (1985). *Proc. Nutr. Soc.* **44**, 145A.

Bruggemann, J., Bronsch, K., Lorcher, K., and Seuss, H. (1959). *Zeitschr. Tierphysiol. Tierernahr. Futtermittelk* **14**, 224-41.

Call, J.W., Butcher, J.E., Blake, J.T., Smart, R.A., and Shupe, J.L. (1978). *J. Anim. Sci.* **47**, 216-25.

Call, J.W., Butcher, J.E., Shupe, J.L., Blake, J.T., and Olson, A.E. (1986). *Am. J. Vet. Res.* **47**, 475-81.

Dingwall, A.R., Munro, C.A., and Suttle, N.F. (1987). *J. Agric. Sci.*, (in press).

Durand, M., Beaumatin, P., and Dumay, C. (1983). *Reprod. Nutr. Develop.* **23**, 727-39

Ellis, L.C., and Tillman, A.D. (1961). *J. Anim. Sci.* **20**, 606-7.

Engelhardt, W.V., Dellow, D.W., and Hoeller, H. (1985). *Proc. Nutr. Soc.* **44**, 37-43.

Field, A.C. (1981). *Proc. Nutr. Soc.* **40**, 267-72.

Field, A.C. (1983). In "Proc. John Lee Pratt Int. Symp. on the Role of Magnesium in Animal Nutrition." Blacksburg, Virginia, 1980.

Field, A.C., and Suttle, N.F. (1979). *J. Comp. Path.* **89**, 431-9.

Field, A.C., Coop, R.L., Dingwall, R.A., and Munro, C.S. (1982). *J. Agric. Sci.* **99**, 311-7.

Field, A.C., Kamphues, J., and Woolliams, J.A. (1983). *J. Agric. Sci.* **101**, 597-602.

Field, A.C., Suttle, N.F., and Gunn, R.G. (1968). *J. Agric. Sci.* **71**, 303-10.

Field, A.C., Suttle, N.F., and Nisbet, D.I. (1975). *J. Agric. Sci.* **85**, 435-42.

Field, A.C., Woolliams, J.A., Dingwall, R.A., and Munro, C.S. (1984). *J. Agric. Sci.* **103**, 283-91.

Field, A.C., Woolliams, J.A., and Woolliams, C. (1986). *Anim. Prod.* **42**, 349-54.

Fishwick, G. (1978). *N.Z. J. Agric. Res.* **21**, 571-5.

Gawthorne, J.M. (1970). *Aust. J. Exp. Biol. Med. Sci.* **48**, 285-92.

Gawthorne, J.M., Allen, J.D., and Nader, C.J. (1985). In "Proc. 5th Int. Symp. Trace Element Metabolism in Man and Animals." (Eds C.F. Mills, I. Bremner and J.K. Chesters.) pp. 346-51. (CAB: Farnham Royal, UK.)

Goings, R.L., Jacobson, N.L., Beitz, D.C., Littledike, E.T., and Wiggers, K.D. (1974). *J. Dairy Sci.* **57**, 1184-8.

Grace, N.D. (1981). *Br. J. Nutr.* **45**, 367-74.

Grace, N.D., and Body, D.R. (1979). *N.Z. J. Agric. Res.* **22**, 405-10.

Greene, L.W., Webb, K.E., and Fontenot, J.P. (1983). *J. Anim. Sci.* **56**, 1214-21.

Gueguen, L. (1962). *Ann. Biol. Anim., Bioch. Biophys.* **2**, 143-9.

Gueguen, L., and Durand, M. (1976). *Ann. Zootech.* **25**, 543-9.

Halpin, C.G., Harris, D.J., Caple, I.W., and Petterson, D.S. (1984). *Res. Vet. Sci.* **37**, 249-51.

Harris, D., Caple, I., and Moate, P. (1987). Proc. 6th Int. Conf. Prod. Dis. Farm Animals (in press).

Hemingway, R.G., and Brown, N.A. (1967). *Proc. Nutr. Soc.* **26**, xxiv.

Hjerpe, C.A. (1968). *Am. J. Vet. Res.* **29**, 143-51.

Hogan, K.G., Money, D.F.L., White, D.A., and Walker, D.A. (1971). *N.Z. J. Agric. Res.* **14**, 687–701.

Humphries, W.R., Phillippo, M., Young, B.W., and Bremner, I. (1983). *Br. J. Nutr.* **49**, 77–86.

Ivan, M. (1987). *Br. J. Nutr.* **57** (in press).

Ivan, M., Ihnat, M., and Veira, D.M. (1983). *Anim. Feed Sci. Tech.* **9**, 131–42.

Ivan, M., Veira, D.M., and Kelleher, C.A. (1986) *Br. J. Nutr.* **55**, 361–7.

Jenkins, T.C., and Palmquist, D.L. (1984). *J. Dairy Sci.* **67**, 978–86.

Kemp, A., Deijs, W.B., and Kluvers, E. (1966). *Neth. J. Agric. Sci.* **14**, 290–5.

Langlands, J.P., Bowles, J.E., Donald, G.E., and Smith, A.J. (1982). *Aust. J. Agric. Res.* **33**, 313–20.

Langlands, J.P., Bowles, J.E., Donald, G.E., Smith, A.J., and Paull, D.R. (1981). *Aust. J. Agric. Res.* **32**, 479–82.

Larvor, P. (1976). *Cornell Vet.* **66**, 413–29.

Little, D.A., Siemon, N.J., and Moodie, E.W. (1978). *Aust. J. Exp. Agric. Anim. Husb.* **18**, 514–9.

Lofgreen, G.P., and Kleiber, M. (1953). *J. Anim. Sci.* **12**, 366–71.

Lofgreen, G.P., and Kleiber, M. (1954). *J. Anim. Sci*, **13**, 258–64.

Mason, J., Kelleher, C.A., and Letters, J. (1982). *Br. J. Nutr.* **48**, 391–7.

Mason, J. (1986). *Toxicol.* **42**, 99–102.

McDowell, L.R., Conrad, J.H., Ellis, G.L., and Loosli, J.K. (1983). Bulletin: Minerals for grazing ruminants in tropical regions. Co-operative Extension Service Univ. Florida, Inst. Food Agric. Sci. Florida.

McDowell, L.R., Vargas, R., Conrad, J.H., Martin, F.G., and Ellis, G.L. (1985). In "Proc. 5th Int. Symp. Trace Element Metabolism in Man and Animals." (Eds C.F. Mills, I. Bremner and J.K. Chesters.) pp. 862–4. (CAB: Farnham Royal, UK.)

McLean, A.F., Buchan, W., and Scott, D. (1985). *Br. J. Nutr.* **54**, 713–8.

McMurray, C.H., and McEldowney, P.K. (1977). *Br. Vet. J.* **133**, 535–42.

McMurray, C.H., Rice, D.A., and Kennedy, S. (1983). In "Trace Elements in Animal Production and Veterinary Practice." Br. Soc. Anim. Prod. Occ. Publ. No. 7, pp. 61–73.

Miles, W.H., and McDowell, L.R. (1983). *World Anim. Rev.* **46**, 2–10.

Milton, J.T.B., and Ternouth, J.H. (1985). *Aust. J. Agric. Res.* **36**, 647–54.

Palmquist, D.L., McClure, K.E., and Parker, C.F. (1977). *J. Anim. Sci.* **45**, 1152–9.

Phillippo. M., and Graca, D.S. (1983). *Proc. Nutr. Soc.* **42**, 46A.

Phillippo, M., Humphries, W.R., Bremner, I., Atkinson, T., and Henderson, G. (1985). In "Proc. 5th Int. Symp. Trace Element Metabolism in Man and Animals." (Eds C.F. Mills, I. Bremner and J.K. Chesters.) pp. 176–80. (CAB: Farnham Royal, UK.)

Playne, M.J. (1976). In "Reviews in Rural Science, 3. The Efficiency of Phosphorus Utilization." (Ed. G.D.J. Blair.) pp. 155–64. (Univ. New England: Armidale, NSW.)

Playne, M.J., Echevarria, M.G., and Megarrity, R.G. (1978). *J. Sci. Fd Agric.* **29**, 520–6.

Powley, G., Care, A.D., and Johnson, C.L. (1977). *Res. Vet. Sci.* **23**, 43–6.

Rickard, T.R., and Elliot, J.M. (1982). *J. Anim. Sci.* **55**, 168–73.

Rooke, J.A., Akinsoyinu, A.O., and Armstrong, D.G. (1983). *Grass and Forage Sci.* **38**, 311–6.

Rothenberg, S.P., and Marcoullis, G. (1983). *Am. Chem.* **29**, 1557–60.

SAC and SARI (1982). "Trace Element Deficiency in Ruminants." Report of a Scottish Agricultural Colleges (SAC) and Research Institutes (SARI) Working Party. (SAC: Edinburgh.)

Simpson, A.M., Mills, C.F., and McDonald, I. (1981) In "Proc. 4th Int. Symp. Trace Element Metabolism in Man and Animals." (Eds J. McC. Howell, J.M. Gawthorne and C.L. White.) pp. 133–6. (Australian Academy of Science: Canberra.)

Smith, B.W.S., and Wright, H. (1984). *Vet. Rec.* **115**, 537–8.

Soares, M.C. (1986). *J. Nutr.* **116**, 1473–9.

Spence, J.A., Sykes, A.R., Atkinson, P.J., and Aitchison, G.U. (1985). *J. Comp. Path.* **95**, 505–24.

Stott, G.H. (1965). *J. Dairy Sci.* **48**, 1485–9.

Suttle, N.F. (1974). *Br. J. Nutr.* **32**, 395–405.

Suttle, N.F. (1980). *Ann. N.Y. Acad. Sci.* **355**, 195–207.

Suttle, N.F. (1983a). In "Feed Information and Animal Production." Proc. 2nd Symp. Int. Network of Feed Information Centres. (Eds G.E. Robards, and R.G. Packham.) pp. 211–37. (CAB: Farnham Royal, UK.)

Suttle, N.F. (1983b). *J. Agric. Sci.* **100**, 651–6.

Suttle, N.F. (1986). "Proc. 13th Int. Cong. Nutr., Brighton." (Eds T.G. Taylor and N.K. Jenkins.) pp. 232–7. (John Libbey: London.)

Suttle, N.F., Abrahams, P., and Thornton, I. (1984). *J. Agric. Sci.* **103**, 81–6.

Suttle, N.F., and Field, A.C. (1966). *Br. J. Nutr.* **20**, 609–19.

Suttle, N.F., and Field, A.C. (1967). *Br. J. Nutr.* **21**, 819–31.

Suttle, N.F., and Field, A.C. (1983). *J. Comp. Path.* **93**, 379–89.

Suttle, N.F., and Jones, D.G. (1986). *Proc. Nutr. Soc.* **45**, 317–25.

Suttle, N.F., Jones, D.G., Woolliams, C., and Woolliams, J.A. (1987). *Br. J. Nutr.* **57**. (in press).

Suttle, N.F., Lloyd Davies, H., and Field, A.C. (1982) *Br. J. Nutr.* **47**, 105–12.

Suttle, N.F., and Peter, D.W. (1985) In "Proc. 5th Int. Symp. Trace Element Metabolism in Man and Animals." (Eds C.F. Mills, I. Bremner and J.K. Chesters.) pp. 367–70. (CAB: Farnham Royal, UK.)

Suttle, N.F., Alloway, B.J., and Thornton, I. (1975). *J. Agric. Sci.* **84**, 249–54.

Suttle, N.F., Wright, C., MacPherson, A., Harkess, R., Halliday, G., Miller, K., Phillips, P., and Evans, C. (1986). In "Proc. 6th

Int. Conf. Prod. Dis. in Farm Animals." pp. 100–3. (Veterinary Research Laboratory: Stormont, NI.)

Sutton, A.L., and Elliot, J.M. (1972). *J. Nutr.* **102**, 1341–6.

Sykes, A.R., and Geenty, K.G. (1986). *J. Agric. Sci.* **106**, 369–75.

TCORN (1987). A working party report. Nutr. Abstr. Rev. (In press).

Thornton, I., Kershaw, G.F., and Davies, M.K. (1972) *J. Agric. Sci.* **78**, 165–71.

Todd, J.R. (1983). Interdepartmental Working Party Report. Ministry of Agriculture, Fisheries and Food. London. p. 8.

Tomas, F.M., Moir, R.J., and Somers, M. (1967). *Aust. J. Agric. Res.* **18**, 635–45.

Turner, R.J., Wheatley, L.E., and Beck, N.G.F. (1985). *Vet. Immuno. Immunopath.* **8**, 119–24.

Ullrey, D.E., Schmitt, S.M., Cooley, T.M., Kic, P.K., and Whetter, P.A. (1985) In "Proc. 5th Int. Symp. Trace Element Metabolism in Man and Animals." (Eds C.F. Millar, I. Bremner and J.K. Chesters.) pp. 113–5. (CAB: Farnham Royal, UK.)

Ward, G., Harbers, L.H., and Blaha, J.J. (1979). *J. Dairy Sci.*, **62**, 715–22.

Whitelaw, A., Armstrong, R.H., Evans, C.C., and Fawcett, A.R. (1979). *Vet. Rec.* **104**, 455–60.

Wilson, G.F., Reid, C.S.W., Molloy L.F., Metson, A.J., and Butler, G.W. (1969). *N.Z. J. Agric. Res.* **12**, 4676–88.

Woolliams, J.A., Suttle, N.F., Wiener, G., Field, A.C., and Woolliams, C. (1982). *Anim. Prod.* **35**, 299–307.

Woolliams, J.A., Suttle, N.F., Wiener, G., Field, A.C., and Woolliams C. (1983). *J. Agric. Sci.* **100**, 441–9.

Woolliams, C., Suttle, N.F., Woolliams J.A., Jones, D.G., and Wiener, G. (1986a). *Anim. Prod.* **43**, 293–301.

Woolliams, J.A., Woolliams, C., Suttle, N.F., Jones, D.G., and Wiener, G. (1986b). *Anim. Prod.* **43**, 303–17.

16

ASSESSING THE NUTRIENT STATUS OF HERBIVORES

by J.P. Langlands

The grazing animal meets its needs for energy and other nutrients from the forage available and, when this is inadequate in quality or quantity or both, from body reserves laid down in times of plenty. Although higher animals have a requirement for energy, protein, at least 15 minerals, essential fatty acids, and vitamins, the range of nutrients which have been shown unequivocally to increase the productivity of grazing ruminants is relatively small (McDonald 1968) because of the ruminant mode of digestion, the low requirement or large reserves of some nutrients, or their abundance in the environment. Some responses are attributable to a reduced availability of a nutrient because of other dietary constituents (for instance response to Cu supplementation as a result of an elevated Mo intake), and some responses are restricted to a particular species or class of animals. Species differ in their requirements and in their sensitivity to deficiency. For example grazing (but not penned) sheep are less sensitive to P deficiency than cattle (McDonald 1968) which may in part reflect differences in grazing behaviour; grazing sheep frequently select a diet containing more P than cattle. In contrast, sheep appear to have a higher requirement for S perhaps as a result of the quantities deposited in the fleece, while cattle are more sensitive than sheep to Cu deficiency. An appreciation of the physiology and behaviour of the species under study is likely to be of value in determining its sensitivity to particular nutrient deficiencies.

Techniques for estimating nutrient intake, predicting nutrient reserves and identifying the more common nutrient deficiencies and excesses are presented. These techniques should ideally be suitable for research and diagnostic purposes with domesticated and wild herbivores under conditions varying from intensively managed sown pastures to rangeland. In practice most techniques are not universally applicable, and many have never been evaluated with wild herbivores. They are generally sensitive to biases which arise from errors of techniques, from physiological changes associated with mustering and sampling, and from interference with normal grazing behaviour. They range in complexity from visual assessment to assays of diet, digesta, body tissues or fluids. Techniques must provide meaningful answers to precisely specified questions. A systematic initial assessment of environmental and other processes which determine nutrient status is likely to be worthwhile in formulating these questions and in selecting appropriate techniques.

NUTRITION OF HERBIVORES
ISBN 0 12 750052 9

ENVIRONMENTAL ASSESSMENT

Climate and particularly rainfall can modify nutrient status through their effects on pedogenesis and soil drainage (Reid and Horvath 1980), on plant growth and senescence, on botanical and chemical composition, and on animal requirements and parasitism. For example pasture growth is sensitive to many factors including climate, and rapidly growing immature pastures are frequently of high digestibility and N content, but unless the rate of mineral uptake is correspondingly increased, mineral concentrations may be reduced by dilution. Rain frequently reduces the nutritive value of senesced pasture although the effect may be masked by new growth; it also leaches mobile ions from soil. High soil moisture concentrations and reduced aeration can impede root development and hence the uptake of some nutrients while increasing the availability and uptake of elements such as Co and Mo.

Grazing animals constrained within a given area must satisfy their nutrient requirements including their mineral requirements from that area, and are therefore sensitive to the characteristics of the soils and the underlying parent materials. Knowledge of any deficiencies or excesses which have been identified on similar soils in the region can be of assistance in the initial assessment of nutrient status. Stream sediment analysis, a geochemical technique developed for mineral exploration, has been used to survey large areas, and has been particularly useful in identifying areas of high Mo status and associated hypocupraemia in the United Kingdom (Webb *et al.* 1968). History of fertilizer usage may also be important. For example the application of lime may increase soil pH and the availability of Mo while depressing Co uptake. Changes in botanical composition may also occur and there is considerable variability within and between plant species in their nutritive value, and in their ability to extract and accumulate minerals from soil, and to translocate these nutrients into the aerial portion of plants where they are available to the grazing animal (Underwood 1981). For example clovers generally contain more Mg, Cu, Zn, Co, P and Ca than accompanying grasses, and are less sensitive to dilution effects as growth proceeds. Animals also ingest nutrients in drinking water and in soil. Water can be a significant source of F, Na, Ca, Mg and S, and soil contamination can provide a number of nutrients including Na and the essential trace elements I and Co; soil consumption can also reduce the availability of Cu (Suttle *et al.* 1975).

Information on the quality and availability of forage is required to distinguish situations in which intake is reduced because of insufficient pasture, from those where dietary concentrations of N, P, S or some vitamins or trace minerals are so low that the animal reduces its intake. Many techniques for estimating biomass have been published (Mannetje 1978) and recent emphasis has been on procedures for specific situations. For example, Vickery and Nicol (1982) have described further modifications to the electronic capacitance meter which provides a rapid non-destructive estimate of pasture yield with a low labour requirement. Another development has been the use of remote sensing from aeroplane or satellite which is particularly suitable for extensive and remote areas. Solar radiation reflected from vegetation is sensed at specific wavelengths, which is maximal in the infra-red region (750 - 1000 nm). Absorption is maximum for chlorophyll in the red (630 - 700 nm) and both bands can be used for estimating the quantity of green forage available. Correlation coefficients of *ca.* 0.9 have been reported (Myers 1983), and the technique appears to be most reliable when the vegetation is predominantly green and of low availability. Reflectance measurements are sensitive to cloud and to the angle of the sun particularly when vegetation is sparse and shading may occur; soil characteristics may also introduce signal noise and ratios of reflectance measurements at different wavelengths are sometimes calculated to minimise such effects (Hacker 1980). Many studies have been based on the LANDSAT satellite but new systems with improved spectral and spatial resolution have recently become available, and further progress is likely in both sensing and computer-interpretation of the signals received.

Liveweight and condition scores (Jefferies 1961; Russel *et al.* 1969) are cost-effective means of assessing nutrient status particularly when target values are available for comparable animals in a similar environment; for example, target values for New Zealand conditions have been suggested by Towers and Clark (1983). If liveweight or condition scores appear low or variability in liveweight is inexplicably high, more complex techniques may be required to identify the cause of the problem. The conclusions can frequently be confirmed by supplementation, but care is necessary. Animals of low nutrient status do not necessarily respond to supplementatiion. Multiple deficiencies occur frequently and productivity will only respond to supplementation to the extent that all deficiencies are corrected; a deficiency of a particular nutrient may be masked by other deficiencies. For example, Gartner *et al.* (1980) and others have observed that responses to P supplementation are most often observed in Northern Australia during the wet season when dietary P concentrations are higher than in the dry. The diet contains insufficient energy, N, P, S and other nutrients during the dry season, and P may no longer be the first limiting nutrient even though dietary P concentrations are at their seasonal minimum. Individual dietary components should not therefore be considered in isolation.

ESTIMATION OF NUTRIENT INTAKE AND DIETARY COMPOSITION

Many techniques for estimating the feed intake of grazing ruminants have been suggested during the last 40 years (Langlands 1975; Corbett 1978, 1981a), but the most successful have been based on estimating:

(a) the faecal output (F), and

(b) the digestibility coefficient of the diet (D), and thence deriving intake (I) and digestible nutrient intake (M):

$$I = (F/1 - D) \qquad \text{equation (1)}$$

$$M = (F * D/1 - D) \qquad \text{equation (2)}$$

If F and D are expressed on an organic matter basis, then I is the organic matter intake and M is the intake of digestible organic matter; one kg of digestible organic matter consumed as pasture provides approximately 15 to 16 MJ of metabolizable energy for ruminants.

Faecal output can be estimated by total collection using harnessed animals, or by regularly giving a known quantity of an inert marker such as chromium sesquioxide (Cr_2O_3), and determining its dilution in a sample of faeces. Typically the marker is given by mouth in a gelatine capsule twice daily for a ten day preliminary period to ensure that daily marker excretion is in equilibrium with marker intake, and thereafter faecal samples are collected either in bags secured to the animal by harnesses, or from the rectum or after excretion onto the pasture (Kotb and Luckey 1972). Two sources of error can be identified; the long term error arises from the failure of determined marker intake to equal marker output, and the short term error from differences in marker concentration between the sample collected and a representative sample of the faeces excreted. Both sources of error can result in biased and imprecise estimates of faeces output and hence intake (Langlands *et al.* 1963); long term errors can result from loss of marker at dosing, an inadequate preliminary period or incomplete recovery of marker during analysis. Short term errors typically arise because of diurnal variation in marker excretion arising from inadequate mixing of feed and marker in the rumeno-reticulum; the short (but not long) term errors can be reduced if an appropriate sampling regime is used or if the marker is given in a form in which it is only slowly released in the rumen (Cr_2O_3-paper). A recent development has been the incorporation of Cr_2O_3 into controlled release intra-ruminal devices (Laby *et al.* 1984). These have the potential to eliminate the need to repeatedly dose animals during either the preliminary or experimental period, with consequent saving in labour and reduction in the interference to grazing behaviour. Their use could also reduce the short term error because marker is continually released in the rumen, and should therefore mix more uniformly with ingested feed. Bird *et al.* (1984) used these devices with grazing cattle; they reported variability in device life ranging from 14 to 21 days, and discarded about 15% of the data because of irregularity in marker release rate. There was also evidence of bias when faecal samples were taken from the rectum because of diurnal variation in faecal Cr_2O_3 concentration which could preclude the use of rectal (grab) sampling strategies.

The digestibility of the diet consumed is usually estimated by either the faecal index technique or by *in vitro* digestibility determinations on extrusa collected from oesophageally fistulated animals. The major defect of the faecal index method is that a single relationship between digestibility and the indicator substance does not exist; it can vary with species of forage and species of animal, and with season of the year, and a relationship derived with forage cut from a single growth of pasture may not be appropriate for animals grazing that pasture. These differences result in biased estimates of intake of varying magnitude which is unfortunate because the technique has advantages in comparison with the oesophageal fistula procedure. For example estimates of digestibility are obtained for each animal and are not inferred from samples selected by surgically modified animals grazing with the experimental flock.

Procedures for establishing, maintaining and sampling from oesophageal fistulae were described by Langlands (1975). The sample is exposed to leaching and salivary contamination, and if dried at high temperatures may undergo non-enzymatic browning, which depresses digestibility; rapid freezing of the extrusa followed by freeze drying is to be preferred when facilities are available. Saul *et al.* (1986) reported a linear relationship with a residual standard deviation of $\pm$ 3.3 units between the predicted digestibilities of forages before and after mastication by oesophageally fistulated sheep. Digestibility of extrusa is usually estimated *in vitro* (Tilley and Terry 1963), and each batch should be calibrated against forage samples of known *in vivo* digestibility. The original method required rumen inoculum and therefore fistulated animals but cellulases are now available commercially (Jones and Hayward 1973); other approaches to predicting the digestibility of extrusa are ruminal incubation in nylon bags (Playne *et al.* 1978) and infrared reflectance spectroscopy. The precision of prediction is critical because intake is estimated from the complement of digestibility (equations 1 and 2), and an error of 0.01 in estimating digestibility is an error of 5, 3.3 and 2.5% in estimating intake when digestibility is 0.80, 0.70 and 0.60 respectively.

Digestibility can also be estimated from the concentrations of an indigestible reference substance occurring naturally in the diet and in the resulting faeces, but most markers examined (lignin, chromogens) have been found to be unsuitable because they were either digestible or difficult to determine chemically. Mayes and Lamb (1983) proposed the use of n-alkanes with odd numbered carbon atoms which are found in cuticular wax. Mayes *et al.* (1986) reported that the indigestibility of n-alkanes increased with chain length and for C_{33}-alkane was 0.89. They compensated for the bias in estimated feed digestibility, and hence intake, by regularly administering a known dose of C_{33}-alkane which does not occur in the diet but is similarly indigestible, and used this material to estimate faeces output. The biases in estimated digestibility and faecel output are then mutually compensating, and Mayes *et al.* (1986) estimated intakes of twelve penned sheep without bias and with a precision of *ca.* $\pm$ 3%. The estimation of C_{33}-alkane in the diet of grazing animals necessitates the use of oesophageally fistulated animals and is likely to inflate this error. Unbiased estimates of faecal output and hence digestibility can be derived if faecal output is simultaneously estimated using a truly indigestible marker such as Cr_2O_3.

An advantage of any oesophageal fistula procedure is that extrusa can be subjected to botanical (Hamilton and Hall 1975) and chemical analysis. Salivary contamination precludes direct chemical analysis for some nutrients (Na, P), and leaching can reduce the concentration of others (K). The losses and gains appear to balance for N, and neither appear to be a problem with Ca. Little *et al.* (1977) combined the oesophageal fistula technique with a radionucleide procedure to correct for salivary P contamination. Fistulated animals are injected with ^{32}P on the day preceding sampling, saliva is collected immediately prior to sampling and specific radioactivity of P is determined (B). The animals are then allowed to graze, and specific radioactivity of P in the extrusa (A) is determined. Little *et al.* (1977) estimated the P concentration of the diet (PD, g P/kgDM) from the concentration in the extrusa (PE, g P/kg DM) thus:

$$PD = (1 - A/B) * PE \qquad \text{equation (3)}$$

The accuracy of the procedure was examined by Little *et al.* (1977) using cattle and by the author (unpublished) using sheep. The results are presented in Figure 1. Shortcomings of the technique include legislative and safety requirements inevitable when any radionucleide is used, and the short half life of the isotope (14.3 days).

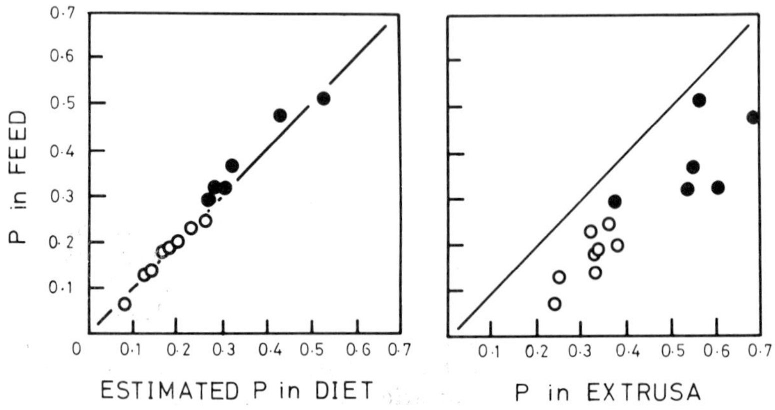

Figure 1. Use of oesophageally fistulated animals to predict dietary phosphorus content. Relationship between dietary P content and P estimated from extrusa composition using a radionucleide procedure and dietary P content and extrusa P content. O cattle (Little *et al.*, 1977) ● sheep (Langlands unpublished).

The oesophageal fistula technique can also be adapted to estimate the intake of sucking animals (Langlands 1972). Milk production of the dam must be estimated independently, perhaps by machine milking following oxytocin injection (Corbett 1968). The faecal output of the sucking animal is determined using one of the methods described earlier, and the estimate is reduced by the product of the organic matter consumed as milk and the indigestibility of the milk organic matter which is typically *ca.* 25 g /kg OM. Forage intake is estimated from equations (1) and (2) using adjusted faecal output, and the digestibility of the diet from extrusa collected by oesophageally fistulated animals. It is assumed that sucking animals select forage similar to that selected by the fistulated animals and that its digestibility is not affected by the quantity of milk consumed nor by the animal's age.

Fistulated animals are not always available, and plucking of pastures to simulate selective grazing is sometimes adopted as a means of estimating dietary composition or nutritive value. The reliability of the procedure is likely to vary with the experience of the operators, and the heterogeneity of the pasture. Three operators plucked samples from eight pastures in one study (Langlands 1974) in which the mean difference in estimated digestibility coefficient between samples ingested by fistulated sheep and by plucking ranged from −0.053 to +0.090, and in N content from −8.4 to +7.7 g N /kg OM on the different pastures. Plucking overestimated nutritive value on high quality pastures and underestimated nutritive value on poorer pastures. The pastures were relatively uniform and it is likely that errors would be even greater under range conditions. Faecal analysis has also been used to estimate dietary N and P concentrations (Moir 1960a, b), but the relationships between dietary N and faecal N concentrations, and between dietary P and faecal P concentrations vary with different diets and classes of animals (Holechek *et al.* 1985). Another problem is that P deficiency depresses intake and hence faecal output, which tends to increase faecal P concentration in severely deficient animals (Read *et al.* 1986).

Urine samples are rarely used to monitor the nutrient intake of grazing animals although urinary excretion of Na (Kemp 1964) and Mg (Field 1964) are closely correlated with intake, and Ca and P nutrition of horses have been assessed from urine samples (Caple *et al.* 1982). Sample concentrations vary widely with the pattern of water consumption, and can be equated to daily excretion by reference to osmolality or creatinine concentration. Creatinine occurs naturally as a waste product of muscle metabolism and is excreted in amounts which vary approximately with liveweight. The errors in estimating urinary excretion can be partitioned into long and short term errors in a manner analogous to that described earlier for estimating faecal output using Cr_2O_3 (Langlands 1966). The errors are large and diurnal variation occurs indicating a need for multiple sampling. Alternatively the animal can be fitted with equipment for the direct determination of urinary output. Total urine output may be transferred to a reservoir

(Beeston and Hogan 1960) or volume may be recorded and an aliquot retained (Betteridge and Andrewes 1986); some procedures may also involve surgical manipulation of the urinary tract but care is necessary to avoid kidney malfunction. All procedures generally require some training of the animal, may interfere with normal grazing behaviour and can be labour demanding.

Much of the progress in ruminant nutrition in the last two decades has been attributable to the development of techniques for studying energy and protein digestion in the gastro-intestinal tract (Faichney 1975), and these techniques have been adapted for use with the grazing ruminant (Corbett 1981b; Pickering *et al.* 1982). The techniques involve the infusion of radionucleides such as ^{103}Ru-phenanthroline chloride, ^{51}Cr-EDTA and ^{35}S into animals with multiple cannulae in the gastro-intestinal tract using portable infusion pumps. Samples are collected manually; diurnal variation in sample composition occurs and adjustments are applied using a Fourier relationship which implies that the diurnal pattern is repeatable between days. The appropriateness of this relationship should be confirmed if applied in situations other than those for which it was developed. There are also risks that grazing behaviour may be disturbed by the need to wear harnesses and carry apparatus such as infusion pumps, and by regular mustering for sample collection.

ESTIMATION OF BODY RESERVES AND THE OCCURRENCE AND DIAGNOSIS OF DEFICIENCIES IN ANIMALS

Analysis of the forage on offer can provide insights into the likely nutrient status of the animals, but soil contamination, variability in the availability of ingested nutrients and selective grazing, limit its usefulness. The analysis of animal tissue is generally more reliable. Samples may be collected from the main site of nutrient storage, from blood or other tissues or fluids (for example saliva and milk), and analysed for the nutrient (Se, Cu), a metabolite (serum urea), or biologically active forms such as enzymes (blood glutathione peroxidase for Se; EC 1.11.1.9), hormones (plasma T3/T4 ratios for I) or vitamins (liver vitamin B_{12} for Co). Factors such as the specificity and cost of the test, the ease of sample collection, the sensitivity to contamination and the ease of analysis and interpretation determine the method of choice. Some analyses useful for identifying nutrient deficiencies provide an estimate of contemporary nutrient intake (plasma Se), others reflect nutrient reserves (liver Cu concentration) while others indicate changes in normal metabolic processes (plasma methyl malonic acid concentration as an index of Co deficiency). Circulating nutrients are generally accessible but are often under homeostatic control, and concentrations do not reflect nutritional status except in times of crisis. Other nutrients are incorporated into red blood cells during haematopoiesis and concentrations in whole blood reflect intake over an extended period (for example Se). The change from nutrient depletion to deficiency is often characterised by biochemical changes which can sometimes be monitored and usually precede the appearance of clinical disorders. Deficiency does not necessarily lead to such disorders; factors which may be involved include the stage of development and physiological state of the animal, and the challenges to which it may be exposed.

Suttle (1986) developed a framework for classifying indices for detecting trace element deficiencies in grazing livestock which was based on three underlying relationships:

1. the relationship between the index and the intake of nutrients, which is frequently sigmoid; (the index is most sensitive when the slope of the relationship rises steeply, and the location of the points of inflection relative to the animal's requirement for the nutrient determines the suitability of the index for measuring nutrient deficiency or nutrient reserves; an ideal index of nutrient deficiency is not necessarily a suitable index of nutrient reserves);
2. the relationship between the index and time determines the suitability of the index for monitoring acute and chronic deficiencies. Suttle (1986) suggested that indices which fall slowly as the animal is depleted, for example erythrocyte Se concentration, are a better indicator of chronic than of acute deficiencies; indices of acute deficiencies include those with short half lives, for example plasma Se concentration;
3. the relationship between the index and function.

Four stages of depletion were recognised; depletion, deficiency (marginal), dysfunction and disease. During depletion, nutrients are lost from storage organs (for example liver Cu concentration declines) but the animal maintains homeostatic control (plasma Cu concentrations do not decline). Homeostatic control is lost at the deficiency or dysfunction stages and eventually clinical lesions can be identified. Indices of dysfunction typically identify changes in normal metabolic processes.

Another approach to detecting nutrient and other sub-clinical disorders is the `Compton Metabolic Profile Test' (Payne *et al.* 1970), which attempts to identify abnormalities in the composition of blood. Samples are analysed for packed cell volume, haemoglobin, glucose, serum urea, albumin, total protein, and a number of minerals (Ca, Cu, Fe, K, Mg, Na, inorganic P); the test was developed in England for testing dairy herds when equipment for automatic clinical chemistry became available. The profile has been criticized on a number of grounds (Adams *et al.* 1978; Wolff *et al.* 1978) including cost.

Energy

Grazing animals survive in many environments by accumulating reserves in times of plenty, and catabolizing them in times of seasonal or intermittent scarcity. Both liveweight and condition scores can be used to monitor these changes. Langlands and Sutherland (1969) derived relationships between liveweight, age and body composition in fine wool Merino sheep which when differentiated, provide estimates of the quantities of nutrients stored. These estimates ranged from *ca.* 0.8 to 21 MJ /kg liveweight depending on the age and liveweight of the sheep; they must be multiplied by a factor ranging from *ca.* 1.3 to 1.6 (depending on forage quality) to give an estimate of the quantity of dietary metabolizable energy which would provide similar amounts of net energy to a sheep in negative energy balance.

A system for scoring the condition of sheep was first proposed by Jefferies (1961) and has been adopted by others including Russel *et al.* (1969); scores range from 0 to 5 on the basis of the prominence of the spinous processes of the anterior lumbar vertebrae, and the depth of surrounding eye muscle and subcutaneous fat, which is assessed by palpation. A sheep with a score of 0 is severely emaciated while a score of 5 is appropriate if the spinous processes cannot be felt and the eye muscle is well developed with a thick cover of subcutaneous fat. An increase in condition score of 1 unit is associated with an increase in liveweight of 7–12 kg depending on the breed of sheep (J.L. Corbett unpublished). Scoring systems for beef and dairy cattle have been described by Lowman *et al.* (1976) and Earle (1976) respectively; the experience of the operator determines the reliability of the procedure.

Energy reserves in terms of the weight of protein, fat, and energy, and changes in reserves can be estimated from liveweight and the size of the pool of bodywater using tritium or deuterium dilution techniques. The technique relies on the inverse relationship between the weight of bodywater and the weight of fat. The estimation of bodywater is typically undertaken after an overnight fast, and the marker is given intravenously; a period of 4 (for sheep) to 10 (for cattle) hours is allowed for mixing (Searle 1970; Little and Morris 1972) before a blood sample is taken and marker dilution determined. The relationships between composition, total body water and fasted liveweight vary with many factors including age, species and nutritional status and it is prudent to derive relationships appropriate to the animals under study. This involves sacrificing some animals and mincing the whole empty body including the pelage and blood but excluding the gut contents.

Concentrations of some circulating metabolites are affected by both energy intake and requirements, and have been used to determine energy requirements for productive processes such as pregnancy (e.g. Russel *et al.* 1967; Robinson *et al.* 1971). Three metabolites were considered by Russel (1978) for use with sheep:

1. plasma glucose which was the least favoured because of its relative insensitivity to large changes in glucose utilisation due to homeostatic mechanisms, and its sensitivity to sampling stress;

2. plasma free fatty acid concentration which is an indicator of fat mobilization, and was preferred in situations of moderate undernutrition, and
3. plasma ketone concentration, which was preferred when sheep were severely undernourished and were likely to be distressed by sampling procedures.

Particular ketone bodies such as acetoacetate and β-hydroxybutyrate can also be determined. Caple and Halpin (1985) suggested that β-hydroxybutyrate concentrations greater than 1.8 mmol /l in dairy cows 'provide a sensitive indication of sub-clinical ketosis associated with severe underfeeding'. Glucose and free fatty acid concentrations in cattle appear to be less reliable (Rowlands 1980), and may be affected by the nature of the diet and show diurnal variation associated with feeding pattern.

Nitrogen and sulphur

Senescent pastures frequently contain insufficient N and sometimes insufficient S and P. Ruminant microbial function is depressed, intake is reduced and despite an abundance of forage, the grazing animal may exhibit the characteristics of starvation in addition to those specific to N, S or P insufficiencies. Although adult ruminants contain 1 - 1.8 g S and 15 - 30 g N /kg liveweight, these nutrients are not generally available to ameliorate dietary insufficiencies although salivary proteins and catabolites such as sulphate (Kennedy *et al.* 1975) and urea are recycled into the gastro-intestinal tract, and may be incorporated into microbial protein. The animal must therefore rely primarily on dietary sources of N and S. Dietary concentrations can be estimated using oesophageally fistulated animals, or from analysis of ruminal fluid or urine, or from circulating concentrations of sulphate or urea. The oesophageal fistula technique provides satisfactory estimates of both dietary N and S concentrations in cattle (Cohen 1979); dietary concentrations of <10g N and 1g S /kg DM should be regarded as low. Ruminal sulphide and ammonia concentrations reflect dietary intakes, and the extent to which dietary S and N are degraded in the rumen. Bray and Till (1975) suggested that ruminal sulphide-S concentrations of < *ca.* 0.1 mg /l would limit microbial growth; corresponding values for ruminal ammonia are *ca.* 0.50 mg NH_3-N /l (Satter and Roffler 1976). If fistulated animals are not available, samples of rumen fluid can be obtained by passing a tube down the oesophagus. Hunter *et al.* (1979) observed responses to S supplementation in steers grazing pastures with an N:S weight ratio of 25-32. In this experiment serum sulphate-S concentrations ranged from 5-10 mg S /l. Techniques to measure the flow of non-ammonia N of dietary and microbial origin through the gastro-intestinal tract are now available (see above). Their application to determining the quantities of amino acids absorbed by grazing animals may be useful because the productivity of animals grazing senescent pastures frequently responds to high quality proteins such as fishmeal. The response may result from an increase in intake, from stimulating microbial activity in the rumen or from the provision of essential amino acids which escape degradation in the rumen and are absorbed in the small intestine.

Calcium and phosphorus

Characteristics of severe deficiency are similar for both Ca and P and include skeletal and dental abnormalities, lameness and depressed growth, milk production and reproductive performance, with rickets in young and osteomalacia in older animals. Bone chewing (osteophagia) also occurs when cattle are P deficient and death may occur from botulism.

P deficiency is more common than Ca deficiency in grazing animals possibly because P-deficient soils are more widespread and because P content of forage fails more rapidly than Ca content when pastures senesce (Underwood 1981). Ca content of forage is relatively insensitive to senescence, and paddock roughage can constitute a major source of Ca for animals receiving cereals as a supplement in times of drought. Dietary Ca and P concentrations can be estimated using oesophageally fistulated animals as described above, although modifications are necessary to correct for the effects of salivary P contamination. Dietary N content should also be determined where possible to distinguish N-P deficiencies from single element deficiencies. Responses to P supplementation are most likely when the N:P ratio is high. For example, Little (1968) observed a response in intake following phosphate supplementation of steers consuming *Stylosanthes humilis* with a N:P ratio of *ca.* 23:1.

Concentrations of circulating nutrients are less useful as determinants of Ca and P deficiency. Plasma Ca concentrations are regulated homeostatically, are sensitive to changes in albumin concentrations and decline only in severe deficiency, or at parturition when Ca demand for milk production exceeds the rate at which Ca can be supplied, leading to milk fever (parturient paresis). The condition is not attributable to inadequate Ca reserves *per se* but appears to be associated with delays in adjusting the mechanisms for increasing the mobilisation of bone calcium and the availability of dietary Ca to meet the increased demand from the mammary gland. Normal values for plasma Ca range from *ca.* 80 to 120 mg /l and for inorganic P from *ca.* 40 to 70 mg /l. Plasma inorganic P concentrations have been extensively used to assess P status and values of < 20 mg /l suggest deficiency (Gartner *et al.* 1980). However concentrations are sensitive to feeding patterns, age of animal, sampling stress, exercise, storage conditions and sampling site (Teleni *et al.* 1976); consequently bone analysis is preferred.

A 50 kg sheep and a 400 kg cow contain *ca.* 0.5 and 5.6 kg Ca and 0.32 and 3.2 kg P respectively (N.Z. Soc. Anim. Prod. 1983). More than 99% of the Ca and 75% of the P are located in the bone, and in the sheep up to 40% of the P (Benzie *et al.* 1959) can be depleted; Ca and P are preferentially reabsorbed from spongy bone, and the vertebrae and ribs are more sensitive than the long bones to changes in Ca and P status. Bone composition is relatively constant during depletion, the Ca:P ratio is *ca.* 2:1, and the main effect of depletion is to reduce total mineral content, which can be monitored by rib biopsy. Conditions must be standardised, and Little (1984) has described a technique in which a small disc of bone is taken with a trephine from the 12th rib of cattle at the level of the midline. The thickness of compact bone is measured using vernier callipers, and Little (1984) suggested that a thickness of > 3mm indicated P adequacy and of < 2mm indicated deficiency in cattle aged 18 months. Earlier studies by this author were based on the weight of P /unit volume of bone (Little and Shaw 1979), and the criteria then adopted were that < 120 mg /ml was deficient while > 150 mg /ml was adequate. An advantage of the biopsy procedure is that the animal can be resampled, thus eliminating variability between animals from the experimental design; bone analysis may also be applied to samples collected at slaughterhouses and from wildlife. A disadvantage is that such bone disorders are not specific to inadequate intakes of dietary P; cattle receiving a diet containing adequate P but insufficient protein show similar problems (Siebert *et al.* 1975).

Bone quality can also be assessed radiographically (Benzie *et al.* 1959). Sequential measurements can be made on the same animal, and surgical procedures are not required. Suttie *et al.* (1983) described its use to assess Ca and P contents of the metacarpus of red deer. Photon absorptiometry is an alternative procedure in which bone is bombarded with a beam of low energy photons from one or more radionucleide sources such as ^{241}Am or ^{125}I, and the attenuation of the beam is recorded. Jeffcott *et al.* (1986) used the procedure to measure bone mineral content in the leg of the horse and Zetterholm and Dalen (1978) in the coccygeal vertebrae of cattle.

Magnesium

Storage in the adult ruminant is *ca.* 0.3–0.5 g Mg /kg liveweight, *ca.* 70% is located in bone and 25% in skeletal muscle (NZ. Soc. of Anim. Prod. 1983). Some bone Mg is available as a labile reserve particularly in young animals (Blaxter and McGill 1956) and can be assessed by rib biopsy or by the removal of caudal vertebrae (Smith 1959). Insufficient Mg is mobilised to supplement dietary sources when plasma Mg concentration falls from a normal value of 18 – 30 mg /l in response to a decline in Mg intake or availability, sometimes associated with an increased Mg demand at the initiation of lactation. Hypomagnesaemia may lead to grass tetany and death of the animal, and is typically seen in animals grazing heavily fertilized (particularly N and K) grass-dominant pastures in temperate zones in late winter and early spring.

Little (1975) used oesophageally fistulated cattle to compare the Mg content of the diet with that of extrusa; he found analysis of extrusa yielded an unbiased estimate of dietary Mg concentration with a precision of *ca* ± 9%. Unpublished studies by the author using sheep are less encouraging. Eleven forages ranging in Mg content from 1.5 to 3.7 g /kg DM were given to fistulated sheep, and extrusa were collected, freeze dried and analysed for Mg without loss of any fraction. The relationship between the Mg contents of feed and extrusa was significantly quadratic with an RSD expressed as a percentage of the mean dietary Mg content of ± 14%; the Mg content of extrusa was 0.2 to 0.8 g /kg DM less than that of the feed offered.

The urine is the principal means by which absorbed Mg in excess of the animal's requirement is excreted. Kemp *et al.* (1961) regarded Mg content of urine as a better indicator of Mg status than serum Mg concentration because it responds more rapidly to a change in available Mg consumption. They did not observe hypomagnesaemia in cows when urinary Mg excretion exceeded 1 g /d; Cale and Halpin (1985) also regarded a daily excretion of 1 g /d as low while >3 g /d was considered to be adequate.

Sodium

Pasture species vary in their ability to translocate Na (Smith *et al.* 1978) into the aerial portion of the plant, and concentrations are reduced by potassic fertilizers; many tropical species are particularly low in Na, and when other sources such as soil and drinking water are also low, responses to Na supplementation may be observed (Murphy and Plasto 1973; Leche 1977; Davison *et al.* 1980) particularly when the demand for Na is increased by lactation. Na deficiency depresses intake, growth rate and milk production, and deficient animals will search avidly for sources of Na to satisfy their need.

The adult ruminant contains about 1 - 1.4 g Na /kg liveweight and most is located in the extracellular fluids (N.Z. Soc. Anim. Prod. 1983). Concentrations in plasma are homeostatically regulated and do not indicate the likelihood of a response to Na supplementation. Morris (1980) suggested that the Na:K ratio in the parotid saliva and adrenal histology were the most sensitive indices of Na insufficiency; he noted that the Na:K ratio in replete cattle was > 20 (on an equivalent basis) and that animals with such a ratio would not respond to Na supplementation. When Na intake falls, aldosterone production increases which results in a fall in parotid Na concentration and a rise in K concentration; Na:K ratios of < 10 indicate potential responsiveness to Na supplementation. A suction device for collecting saliva under field conditions has been described by Murphy and Connell (1970). Alternatively, appropriately washed foam rubber sponges can be used.

Chronic Na deficiency is associated with an increase in the depth of the zona glomerulosa of the adrenal, and Morris (1980) expressed this value as a percentage of the depth of the cortex, and quoted values of 12.8 - 13.4% for replete cattle and 24 - 31% for deficient animals. Urinary Na excretion is an alternative index of contemporary Na intake which was advocated by Caple and Halpin (1985); they suggested that urinary excretion of < 3 g Na /day in dairy cattle was indicative of marginal insufficiency. Difficulties may arise in collecting representative samples of urine since Kemp *et al.* (1972) reported that the excretion of urinary Na was so irregular that there was no correlation between urinary Na and creatinine concentrations. Differences in urinary Na concentration, peripheral aldosterone and the area of the zona glomerulosa were observed by Blair-West *et al.* (1968) in their study of eastern grey kangaroos (*Macropus giganteus*) and rabbits in Na-adequate and deficient environments.

Copper and molybdenum

Manifestations of Cu deficiency are described by Underwood (1981) and include skeletal, collagen and cardiovascular abnormalities, impaired growth, enzootic ataxia (swayback), depigmentation of hair and pigmented wool and 'steely wool'. Deficiencies can arise from insufficient Cu in the forage and/or from excessive concentrations of elements which interfere with Cu absorption and utilisation; in particular Mo and S. Molybdenum and ruminal sulphide appear to form thiomolybdates in the rumeno-reticulum which complex with Cu and are largely excreted in the faeces (Dick *et al.* 1975). The significance of dietary Cu concentrations must therefore be considered in relation to Mo and S concentrations. The New Zealand Society of Animal Production (1983) presented a table showing the relationship between forage Cu and Mo concentrations and the health of sheep and cattle. Forage containing < 3 mg Cu /kg DM and no Mo was described as 'indicative of simple Cu deficiency', while forage containing 5 mg Mo /kg DM, and 5 (sheep) or 9 (cattle) mg Cu /kg DM as described as 'Mo in excess; Cu inadequate'.

The liver is the main storage tissue for Cu. Liver Cu concentrations vary widely and are a major determinant of total body Cu. Langlands *et al.* (1984) derived a relationship between total body Cu (C, μg), liveweight (W, kg) and liver Cu concentration (H, mg Cu /kg DM) with Merino sheep:

$$C = 989\ W + 6.592\ H.W \qquad \text{equation (4)}$$

and a second relationship between the quantity of Cu stored in the liver as a proportion of total body Cu (E) and liver Cu concentration:

$$E = 0.320 + 0.000542\ H \qquad \text{equation (5)}$$

Equation (4) predicts that a 45 kg sheep with a liver Cu concentration of 200 mg Cu /kg DM would contain 104 mg Cu and equation (5) predicts that 43% of the Cu would be located in the liver.

Liver Cu concentration is the preferred index of Cu status, but samples can only be collected by liver biopsy or by sacrificing the animal. Variability between animals in liver Cu concentration is high, particularly in sheep, and an experimental design incorporating repeated biopsies of the same animal is an effective means of reducing such variability but requires the use of a biopsy technique which is safe and causes minimal distress. Donald *et al.* (1984) described a biopsy procedure for sheep restrained in a sitting position in a laparotomy cradle in which the peritoneal cavity is inflated with oxygen to impart rigidity to the liver, and the liver is entered through the mid-paracostal region. A biopsy technique for cattle has been described by Loosmore and Allcroft (1951). There is no generally accepted liver Cu concentration at which responses to supplementation can be assumed, and different authorities propose different concentrations as indicating deficiency. Concentrations < 5 and < 10 mg Cu /kg DM in cattle and sheep respectively should be regarded as low.

While liver is the preferred tissue for assessing Cu status, samples cannot always be collected, and other tissues are frequently analysed. Most plasma Cu is associated with the enzyme caeruloplasmin, and depressed concentrations of plasma Cu or caeruloplasmin indicate depleted Cu reserves. Caeruloplasmin is less sensitive to contamination, but there is no suitable reference standard and relationships between plasma Cu concentration and caeruloplasmin activity are usually derived (Lorentz and Gibb 1975). Plasma concentrations fall from normal values of 0.6 – 1.2 mg /l when liver Cu concentration in cattle is < *ca.* 40 mg Cu /kg DM (Claypool *et al.* 1975). Plasma concentrations of < 0.5 mg Cu /l are indicative of a low Cu status. Induced Cu deficiencies in sheep may be associated with elevated plasma Cu concentrations at high intakes of Mo. The increase is associated with a fraction which is not caeruloplasmin, contains both Cu and Mo, and is largely insoluble in trichloracetic acid (TCA) (Smith and Wright 1975). Most erythrocyte Cu is associated with the enzyme superoxide dismutase (EC 1.15.1.1), which is incorporated at erythropoiesis, and is retained until the erythrocyte is turned over at *ca.* 3 months. Erythrocyte Cu reflects the quantity of Cu absorbed over an extended period and falls more slowly than plasma Cu as the animal becomes deficient. Other tissues which can be assayed to indicate Cu status include hair in cattle and wool in sheep; care and standardised procedures are needed when cleaning the fibre to remove contaminants but not fibre constituents.

Cobalt and vitamin B_{12}

The only known function of Co in ruminants is for the synthesis of vitamin B_{12} by rumen microorganisms. Other herbivores (for example horses) have a lower dietary requirement for vitamin B_{12} and will survive on pastures on which sheep and cattle die (Underwood 1981). The Co concentrations of some tissues reflect Co intakes but unlike vitamin B_{12} reserves confer little physiological benefit, and are a less sensitive indicator of the ruminants' ability to survive on Co-deficient pastures. Underwood (1981) suggested that the livers of Co-deficient sheep and cattle generally contained < 0.08 mg Co /kg DM whereas concentrations were usually 0.2 to 0.3 mg Co /kg DM in healthy animals. Co deficiency is associated with specific soil types and is observed in all climatic zones. Forage concentrations of < 0.7 mg Co /kg DM are inadequate for sheep and somewhat lower concentrations are inadequate for cattle, but care is necessary when collecting forage samples for Co analysis because they are very sensitive to soil contamination. Soil can be a major source of Co to the grazing animal.

Co (or more properly vitamin B_{12}) deficiency is associated with anaemia, anorexia, and hence depressed productivity, unthriftiness, and emaciation; young animals are particularly susceptible and symptoms appear when vitamin B_{12} reserves are exhausted and within a herd or flock are observed in some animals while others appear unaffected. Vitamin B_{12} is stored primarily in the liver and is required for the utilisation of propionic acid via methylmalonic acid (MMA). MMA and propionic acid accumulate in the blood and excessive amounts of MMA are excreted in urine when animals are B_{12} deficient (Judson *et al.* 1981). Vitamin B_{12} deficiency also impairs the metabolism of folic acid, and formiminoglutamic acid (FIGLU) excretion increases by up to 30 times in vitamin B_{12} deficient sheep (Gawthorne 1968). Both MMA and FIGLU have been suggested as indicators of Co deficiency. Vitamin B_{12} concentrations in plasma/serum and liver are frequently determined, and are believed to reflect dietary Co intake and vitamin B_{12} reserves respectively (Sutherland 1980). Vitamin B_{12} concentrations can be assayed microbiologically but radiodilution procedures are more appropriate for routine analyses. The principle of the dilution procedure is that radioactively labelled cobalamin and the test sample are incubated with a fixed quantity of a binding protein (intrinsic factor) specific for vitamin B_{12}; the greater the quantity of radioactive material lodged on the binding protein, the lower the quantity of cobalamin in the test sample. Binding proteins used in some assays are not specific for cobalamin and may bind physiologically inactive cobalamin analogues. Caple and Halpin (1985) suggest 'plasma levels are of little help in assessing marginal deficiencies in cattle', although hepatic B_{12} concentrations < 0.1 mg /kg fresh weight are indicative of a low Co status. Mean sera concentrations derived from at least six sheep were regarded by Sutherland (1980) as being a useful indicator of dietary Co intake, and Andrews and Stephenson (1966) described concentrations < 0.30 ug /l as 'probably deficient' and ovine liver concentrations < 0.19 mg /kg freshweight as 'borderline' and < 0.10 mg /kg as `low' (Andrews *et al.* 1960). In some situations it may be more appropriate to supplement some animals with vitamin B_{12} and to record changes in liveweight in supplemented and unsupplemented animals. Responses to supplementation are rapid and readily detectable when animals are severely deficient (e.g. Hannam *et al.* 1980).

Selenium and vitamin E

Selenium is an essential component of the enzyme glutathione peroxidase (GSH-px; EC 1.11.1.9) and both GSH-px and vitamin E protect cell membranes from damage by peroxide and other active radicals. Some manifestations of Se insufficiency (for example white muscle disease in sheep and cattle) are moderated by vitamin E, while dietary components such as polyunsaturated fatty acids are readily metabolized to peroxides and aggravate the condition if not first hydrogenated in the rumen. Selenium insufficiencies are typically observed on fertilized pastures on acid soils in high rainfall regions.

Concentrations in forage < 0.05 mg Se /kg DM should be regarded as low and may result in muscular dystrophy, stiffness in the limbs, inability to stand, sudden deaths, depressed growth and wool production, and infertility in domestic ruminants. The muscle lesions are characteristically bilaterally symmetrical. Dietary vitamin E activity is provided by a number of tocopherols and tocotrienols. Concentrations vary widely within plants and decline when pastures mature and senesce, or if forages are conserved and stored. Sheep and cattle grazing green pastures for most of the year as in New Zealand, are more likely to be Se than vitamin E deficient, but cattle in northern Europe are frequently wintered on conserved feeds thus depressing vitamin E status, and are turned out in spring onto pastures which may contain high concentrations of unsaturated fatty acids. If such animals are also of low Se status, they are particularly vulnerable to nutritional myopathy (McMurray *et al.* 1983). Some species are dependent on vitamin E to prevent myopathy, and include the rabbit, quokka (*Setonix brachyurus*) and tammar wallaby (*Macropus eugenii*) (Hume 1982).

Selenium status is often acropus from the Se content of whole blood, plasma or erythrocytes, or GSH-px activity in whole blood or erythrocytes. Most blood Se in sheep and cattle is located in the erythrocytes, and Se concentrations and GSH-px activities are closely correlated, particularly when Se concentrations are low. The enzyme has the advantage as a diagnostic aid that it is the biologically active form of Se, and activity can be determined spectrophotometrically or by a simple spot test (Langlands *et al.* 1980). Another advantage is that perchloric acid digestion is not required. The main disadvantages are that GSH-px is unstable in plasma. Activities in plasma are low and sensitive to haemolysis, and assay procedures are not standardised and may therefore lack comparability between laboratories. Plasma Se concentrations respond more rapidly to changes in contemporary Se intake than whole blood or erythrocyte Se, or GSH-px activities. Erythrocyte Se is incorporated at erythropoiesis and like erythrocyte Cu, reflects dietary intake over an extended period. Selenium concentrations in many other tissues and fluids also reflect Se status including liver, kidney, hair and wool. Total body Se (TSe, µg) varies with the size of animal and its Se status. Langlands *et al.* (1984) presented a relationship for Merino sheep with liveweight (W, kg) and whole blood Se concentration (BSe, µg /ml) as independent variables:

$$TSe = 21.6\ W + 321.0\ BSe.W \qquad \text{equation (6)}$$

Se deposition /kg liveweight change was predicted to be 25 and 86 µg when S was 0.01 (Se-deficient) and 0.2 (Se-adequate) µg /ml respectively.

Blood Se concentrations at which responses to Se supplementation can be expected are likely to vary with vitamin E status, with the animal's demand for Se, which will vary with age, species and physiological state, and with the composition of the diet. Sheppard *et al.* (1984) reported that in New Zealand where vitamin E status is likely to be high, Se-responsive conditions occurred in lambs when blood Se concentrations fell below 0.01 µg /ml. Lamb deaths occurred when blood concentrations were < 0.005 µg /ml. In the author's experience, unthriftiness and depressed wool production occur when concentrations are < 0.02 ug /ml and other manifestations of Se insufficiency may not be evident. Cattle appear to be more tolerant of a low Se status than sheep. Serum and liver are sometimes analysed for α-tocopherol to monitor vitamin E status (Rammell and Cunliffe 1983); values for sheep are generally less than those for cattle. Serum concentrations are closely correlated with total serum lipids, and concentrations < 1 and < 2 mg /l in sheep and cattle respectively should be regarded as low. A number of blood enzymes and particularly creatine kinase (EC 2.7.3.2) may be assayed to indicate the onset of myopathies induced by vitamin E-Se insufficiency.

Iodine and zinc

Iodine is required for the synthesis of the thyroid hormones, tri(T3)- and tetra(T4)-iodothyronine, and is primarily stored in the thyroid gland as thyroglobulin. Clinical signs of insufficiency include enlarged thyroids, neonatal mortality, depressed milk production, and reproductive performance (Underwood 1981); similar conditions are associated with some forms of thyroid malfunction and there is a genetic disposition to such conditions. The demand for I may be increased by some dietary goitrogens; these include cyanogenetic glycosides which are metabolized to thiocyanate in the rumen, but not goitrogens based on thiouracil.

Low concentrations of I in soils are often associated with recent glaciation and remoteness from marine sources of I (Underwood 1981); fertilizer application may also depress I content and pasture species differ in their ability to extract soil I (Hartmans 1971). Soil can also be a significant source of dietary I (Statham and Bray 1975). The existence of enlarged thyroids (> 3 g in lambs), decreased thyroid I concentrations and abnormal thyroid histology (Mason 1976) are indicative of thyroid malfunction which may be attributable to insufficient dietary I. Changes in the I intake of grazing animals can be monitored in the milk and urine. Concentrations < 25 µg I /l in bovine milk are associated with low I intakes (Caple and Halpin 1985). Plasma thyroxine (T4) and other thyroid hormone concentrations can be determined by radioimmunoassay, and have been used to survey the I status of lactating ewes. Concentrations < 40 µg /l were regarded as 'being possibly I deficient' (Wallace *et al.* 1978). Andrewartha *et al.* (1980) concluded from their study of this technique that 'single determinations of thyroxine concentration in ewes in a flock provide very little information on their I nutrition'. They observed seasonal variability in thyroxine concentration in ewes, irregular responses to I supplementation, and suggested that thyroxine concentrations in newly born lambs were more sensitive indicators of thyroid malfunction than concentrations in their dams. Assays of a number of thyroid hormones are used for diagnostic purposes in human medicine but may not be appropriate for animal studies because of the costs involved.

Zinc deficiency is relatively uncommon in grazing ruminants but has been reported in grazing sheep in the Sudan (Mahmoud *et al.* 1983) and in Australia (Masters and Fels 1980), and in cattle in South America and elsewhere (McDowell *et al.* 1983). Deficiencies are difficult to diagnose and may be manifested by depressed intake, retarded growth and reproductive disorders, loss of hair and wool, the development of dry scaly skin which cracks readily, and delayed healing of wounds (Underwood 1981). Deficiency develops rapidly in animals given synthetic diets (Mills *et al.* 1967) and Zn status is usually monitored in plasma which falls from a normal value of 0.6 - 1.2 mg /l; unfortunately similar changes are associated with other factors unrelated to Zn intake including parturition. Mills *et al.* (1967) suggested resampling on two occasions at intervals of at least a week, and that plasma concentrations should be consistently < 0.3 mg Zn /l if animals are to be regarded as deficient. Zinc content of available forage is not a reliable means of identifying responsive situations. Concentrations reported in field studies, in which responses have been observed, are similar to those reported in situations where responses are not recorded (Mills *et al.* 1967). A possible explanation is that Zn released by tissue catabolism (associated for example with insufficient intakes of energy or protein) is available to supplement dietary Zn (Masters 1984). Such interactions highlight one of the problems in assessing nutrient status from tissue analysis, and emphasize the need for caution particularly when applying results from studies with penned animals possibly receiving semi-synthetic diets, to the grazing situation.

Nutrient excesses

Procedures for identifying nutrient excesses are similar to those for nutrient deficiencies, and it is initially desirable to review environmental characteristics including possible sources of contamination, husbandry practices and evidence of clinical symptoms. Grazing herbivores are exposed to environmental contaminants (for example F, Pb), and harmful plant constituents (McDonald 1981) including excessive and possibly toxic concentrations of essential dietary nutrients. The effects of excessive mineral intakes have been reviewed in detail by the National Academy of Sciences (1980) and only a selection of problems are considered here.

Both Mo and to a lesser extent Cu toxicities are observed in grazing animals. Cattle are particularly sensitive to high concentrations of dietary Mo (Underwood 1977), and may show severe scouring within days of grazing pastures high in Mo (teart pastures). Scouring can be prevented by dietary Cu supplementation perhaps because Mo is rendered unavailable as Cu-thiomolybdate. Scouring occurs irrespective of the hepatic Cu reserves. Tissue Mo concentrations respond to elevated intakes of dietary Mo, but the response is reduced in the presence of high concentrations of dietary S presumably because Mo is complexed as the thiomolybdate and excreted. Tissue concentrations which respond to Mo intake when dietary S is low include blood, liver, wool, hair and muscle. Copper toxicities in grazing animals are usually associated with environmental contamination (for example grazing of pastures recently treated with Cu-containing fertilizers or in orchards treated with Cu fungicides), with the ingestion of plants such as Heliotropium which contain hepatotoxic alkaloids which result in the accumulation of hepatic Cu, or from the inappropriate administration of Cu supplements. Sheep are more sensitive to toxicity than cattle and breeds differ in their susceptibility (Wiener *et al.* 1977). Copper can accumulate for many months in the liver and kidney of sheep exposed to chronic copper poisoning without obvious clinical symptoms. A haemolytic crisis then occurs, blood Cu concentrations rise with haemoglobinaemia, haemoglobinuria and extensive jaundice. Intravenous thiomolybdate is effective in preventing the crisis.

Selenium toxicity is associated with seleniferous soils, the consumption of plant species which accumulate Se, and most commonly with the inappropriate administration of Se supplements. The problem was first reported in horses in the United States (alkali disease), and was associated with the loss of hair from the mane and tail, and with lameness and malformation of the hoof (Rosenfeld and Beath 1964). Toxicity also occurs in ruminants in acute and chronic forms. Seleniferous soils are found in many countries including Australia (northwest Queensland). Tissue and blood Se concentrations are very sensitive to an increased Se intake, and hepatic and renal Se concentrations may reach 20-30 mg /kg DM in chronic toxicity (Underwood 1977). In arid and semi-arid zones, vegetation may also contain excessive Na concentrations (*Atriplex* (saltbush) species contain 33-82 g Na /kg DM; Wilson 1966) and drinking water derived from bores may contain high concentrations of Na, Mg, Ca, sulphates, and chlorides. The combined salt load may depress intake, wool production (Wilson 1975) and reproductive performance. The extent of the problem varies with many factors including the composition of the diet selected, the physiological state of the animal, season and hence ambient temperature, water consumption and its chemical composition. The composition of the drinking water should be determined when such problems are suspected, while urinary losses can be used to monitor Na and Mg excesses provided representative samples of urine can be collected as described above.

Most nutrients are toxic when consumed in excess and not all excesses are of mineral origin. Excesses of some nutrients may reduce the availability of other nutrients to the animal (excessive K intakes can reduce the availability of Mg). Inappropriate supplementation procedures are a common source of problems (for example urea toxicity) and the problems are exacerbated when animals are supplemented when they are incorrectly believed to be deficient (for example Cu). Most of the techniques described earlier can be applied to identifying nutrient excesses.

Subclinical and transient deficiencies

The classical symptoms of specific nutrient deficiencies or excesses are relatively uncommon among farm animals today. One reason is that their manifestations are recognized by the farming community in areas with a history of a particular problem. In contrast, sub-clinical deficiencies appear to be widespread. They are frequently transient and reduce productivity with few if any specific symptoms. For example, transient Se deficiencies occur in sheep and cattle grazing on the Great Dividing Range of southeast Australia. The problem occurs in years of above average rainfall when pasture growth is increased and forage Se concentrations are low. It is more serious after several years of above average rainfall when animals' Se reserves are depleted, and it does not occur uniformly across the region. Thus some farms show transient Se deficiency, others are never deficient and some are consistently deficient. Losses in productivity therefore vary between farms, are difficult to predict and are manifested by occasional outbreaks of nutritional myopathy, depressed growth of lambs, a reduction in wool production of up to 15% and reduced reproductive performance. Selenium status can be monitored by blood analysis but it is necessary to assume a relationship between blood Se concentration and the responsiveness to Se supplementation, and then to equate costs and benefits of supplementation. Procedures for deriving such relationships for Co, Se and perhaps Cu have been described by Clark *et al.* (1985) and the approach is to be preferred to subjective assessments of nutrient status such as "probably deficient, adequate, borderline, etc." but there must be a close relationship between tissue concentration and responsiveness which is not always observed in practice. The mechanisms associated with subclinical deficiencies are not clearly established, and explanations are necessarily speculative.

A manifestation of a low dietary concentration of many nutrients is a decline in intake which reduces productivity and hence nutrient demand. If the reduced demand is sufficient to maintain the animal's homeostatic processes, more severe manifestations are unlikely to appear and the deficiency will only be expressed in reduced productivity. Clinical manifestations may be anticipated if equilibrium is further disturbed by falls in nutrient concentration or by an increased demand imposed by stresses such as lactation or parasitism.

STRATEGIES FOR THE FUTURE

Studies with penned animals have contributed greatly to understanding the nutrition of grazing herbivores, but transient, multiple and subclinical deficiencies are difficult to reproduce under pen conditions and appear to be a source of considerable economic loss. Gawthorne (1984) has provocatively suggested that "animal disease resulting from a naturally occurring deficiency or toxicity of one trace element alone is a thing of the past, if indeed it existed at all", and future studies will need to recognize that nutritional insufficiencies encountered in commercial practice are likely to be complex. Sub-clinical deficiencies of mineral origin are particularly difficult to identify at the farm level, and the use of supplementation techniques which are both low cost and effective over long periods, may be more cost-effective than attempting to identify deficiencies which may be transient in their incidence. Alternatively increased emphasis may be placed on the development of "dynamic" indices of nutrient status which involve supplementation with the nutrient under study, and observing the effect on various physiological variables. Measurements which monitor functional performance and hence the specific outcomes of nutritional insufficiency may be particularly useful (Solomons and Allen 1983). They have been classified into seven categories according to function:

1. Structural integrity (e.g. erythrocyte fragility, vitamin E and Se insufficiency).
2. Transport (e.g. thyroid radioiodine uptake, I).
3. Host defence (e.g. leucocyte bactericidal capacity, Se).
4. Haemostasis (e.g. platelet aggregation, vitamin E and Zn).
5. Reproduction (e.g. sperm count, Zn).
6. Nerve function (e.g. dark adaptation, vitamin A and Zn).
7. Work capacity and haemodynamics (e.g. heart rate, protein/energy and Fe).

Some of these tests may be more easily applied to man than to grazing herbivores, and are additional to, rather than substitutes for those described earlier. The range of tests which may be applied to assessing nutritional status is clearly large, and only a small proportion has been discussed in this presentation. The number available is likely to increase with greater knowledge of the nutritional biology of the grazing herbivore.

Adams, R.S., Stout, W.L., Kradel, D.C., Guss, S.B., Moser, B.L., and June, G.A. (1978). *J. Dairy Sci.* **61**, 1671-9.

Andrewartha, K.A., Caple, I.W., Davies, W.D., and McDonald, J.W. (1980). *Aust. Vet. J.* **56**, 18-21.

Andrews, E.D., and Stephenson, B.J. (1966). *N.Z. J. Agric. Res.* **9**, 491-507.

Andrews, E.D., Hart, L.I., and Stephenson, B.J. (1960). *N.Z. J. Agric. Res.* **3**, 364-76.

Beeston, J.W.U., and Hogan, J.P. (1960). *Proc. Aust. Soc. Anim. Prod.* **3**, 79-82.

Benzie, D., Boyne, A.W., Dalgarno, A.C., Duckworth, J., and Hill, R. (1959). *J. Agric. Sci.* **52**, 1-12.

Betteridge, K., and Andrewes, W.G.K. (1986). *J. Agric. Sci.* **106**, 389-92.

Bird, P.R., Watson, M.J., Cayley, J.W.D., and Flinn, P.C. (1984). In "Ruminant Physiology. Concepts and Consequences. A Tribute to R.J. Moir." (Eds S.K. Baker, J.M. Gawthorne, J.B. Mackintosh and D.B. Purser.) pp. 205–16. (University of Western Australia: Perth.)

Blair-West, J.R., Coghlan, J.P., Denton, D.A., Nelson, J.F., Orchard, E., Scoggins, B.A., Wright, R.D., Myers, K., and Junqueira, C.L. (1968). *Nature*, London. **217**, 922–8.

Blaxter, K.L., and McGill, R.F. (1956). *Vet. Revs. Annot.* **2**, 35–55.

Bray, A.C., and Till, A.R. (1975). In "Digestion and Metabolism in the Ruminant." (Eds I.W. McDonald and A.C.I. Warner.) pp. 243–60. 4th Int. Symp. on Ruminant Physiol. (University of New England Publ. Unit: Armidale.)

Caple, I.W., and Halpin, C.G. (1985). In "Refresher Course for Veterinarians." Proceedings No. 76 pp. 307–37. (Post-Graduate Committee in Veterinary Science: University of Sydney.)

Caple, I.W., Doake, P.A., and Ellis, P.G. (1982). *Aust. Vet. J.* **58**, 125–31.

Clark, R.G., Wright, D.F., and Millar, K.R. (1985). N.Z. Vet. J. **33**, 1–5.

Claypool, D.W., Adams, F.W., Pendell, H.W., Hartmann, N.A., and Bone, J.F. (1975). *J. Anim. Sci.* **41**, 911–4.

Cohen, R.D.H. (1979). *J. Agric. Sci. Camb.* **93**, 607–18.

Corbett, J.L. (1968). *Aust. J. Agric. Res.* 19, 283–94.

Corbett, J.L. (1978). In "Measurement of Grassland Vegetation and Animal Production." (Ed. L. 't Mannetje.) pp. 163–231. (Commonw. Bur. of Pasture and Field Crops: Hurley, UK.)

Corbett, J.L. (1981a). In "Forage Evaluation: Concepts and Techniques." (Ed. J.L. Wheeler and R.D. Mochrie.) pp. 287–97. (CSIRO: Canberra.)

Corbett, J.L. (1981b). In "Forage Evaluation: Concepts and Techniques." (Eds J.L. Wheeler and R.D. Mochrie.) pp. 383–99. (CSIRO: Canberra.)

Davison, T.M., Murphy, G.M., Maroske, M.M., and Arnold, G. (1980). *Aust. J. Exp. Agric. Anim. Husb.* **20**, 543–6.

Dick, A.T., Dewey, D.W., and Gawthorne, J.M. (1975). *J. Agric. Sci.* **85**, 567–8.

Donald, G.E., Paull, D.R., and Langlands, J.P. (1984). Aust. Vet. J. **61**, 121–2.

Earle, D. (1976). J. Agric. Vic. **74**, 228–31.

Faichney, G.J. (1975). In "Digestion and Metabolism in the Ruminant." (Eds I.W. McDonald and A.C.I. Warner.) pp. 277–91. 4th Int. Symp. on Ruminant Physiol. (University of New England: Armidale.)

Field, A.C. (1964). Brit. J. Nutr. **18**, 357–68.

Gartner, R.J.W., McLean, R.W., Little, D.A., and Winks, L. (1980). *Trop. Grassl. J.* **14**, 266–72.

Gawthorne, J.M. (1968.) *Aust. J. Biol. Sci.* **21**, 789–94.

Gawthorne, J.M. (1984.) In "Ruminant Physiology. Concepts and Consequences. A Tribute to R.J. Moir." (Eds S.K. Baker, J.M. Gawthorne, J.B. Mackintosh and D.B. Purser.) pp. 357–71. (University of Western Australia: Perth.)

Hacker, R.B. (1980). *Trop. Grassl.* **14**, 288–95.

Hamilton, B.A., and Hall, D.G. (1975). *J. Br. Grassld. Soc.* **30**, 229–35.

Hannam, R.J., Judson, G.J., Reuter, D.J., McLaren, L.D., and McFarlane, J.D. (1980). *Aust. J. Agric. Res.* **31**, 347–55.

Hartmans, J. (1974). *Neth. J. Agric. Sci.* **22**, 195–206.

Holechek, J.L., Galyean, M.L., Wallace, J.D., and Wofford, H. (1985). *Grass and Forage Sci.* **40**, 489–92.

Hume, I.D. (1982). "Digestive Physiology and Nutrition of Marsupials." (Cambridge Univ.: Cambridge.)

Hunter, R.A., Siebert, B.D., and Webb, C.D. (1979.) *Aust. J. Exp. Agric. Anim. Husb.* **19**, 517–21.

Jeffcott, L.B., McCartney, R.N. and Speirs, V.C. (1986). *Vet. Rec.* **118**, 499–505.

Jefferies, B.C. (1961). *Tasm. J. Agric.* **32**, 19–21.

Jones, D.I.H., and Hayward, M.V. (1973). *J. Sci. Fd. Agric.* **24**, 1419–26.

Judson, G.J., Hannam, R.J., Benson, T.H., and Reuter, D.J. (1981). In "Trace Element Metabolism in Man and Animals." (Eds J.McC. Howell, J.M. Gawthorne and C.L. White.) pp. 386–9. (Aust. Acad. Sci.: Canberra.)

Kemp,.A. (1964). *Neth. J. Agric. Sci.* **12**, 263–80.

Kemp, A., Deijs, W.B., Hemkes, O.J., and Van Es, A.J.H. (1961). *Neth. J. Agric. Sci.* **9**, 134–49.

Kemp, A., Korzeniowski, A., and Geurink, J.H. (1972). *Z. Tierphysiol. Tierernahrg Futtermittelkde.* **29**, 257–63.

Kennedy, P.M., Williams, E.R., and Siebert, B.D. (1975). *Aust. J. Biol. Sci.* **28**, 31–42.

Kotb, A.R., and Luckey, T.D. (1972). *Nutr. Abstr. Rev.* **42**, 813–45.

Laby, R.H., Graham, C.A., Edwards, S.R., and Kautzner, B. (1984). *Can J. Anim. Sci.* **64**, (Suppl.) 337–8.

Langlands, J.P. (1966). *Aust. J. Agric. Res.* **17**, 757–63.

Langlands, J.P. (1972). *Anim. Prod.* **14**, 317–22.

Langlands, J.P. (1974). *Anim. Prod.* **19**, 249–52.

Langlands, J.P. (1975). In "Digestion and Metabolism in the Ruminant." (Eds I.W. McDonald and A.C.I., Warner.) Proc. 4th Int. Symp. on Ruminant Physiol. pp. 320–32. (University of New England: Armidale.)

Langlands, J.P., and Sutherland, H.A.M. (1969). *Brit. J. Nutr.* **23**, 603–9.

Langlands, J.P., Bowles, J.E., Donald, G.E., and Smith, A.J. (1984). *Aust. J. Agric. Res.* **35**, 701–7.

Langlands, J.P., Corbett, J.L., McDonald, I., and Reid, G.W. (1963). *Brit. J. Nutr.* **17**, 211–26.

Langlands, J.P., Donald, G.E., Bowles, J.E., and Smith, A.J. (1980). *Aust. J. Agric. Res.* **31**, 357–67.

Leche, T.F. (1977). *Papua New Guinea Agric. J.* **28**, 11-7.
Little, D.A. (1968). *Proc. Aust. Soc. Anim. Prod.* **7**, 376-80.
Little, D.A. (1975). *Aust. J. Exp. Agric. Anim. Husb.* **15**, 437-9.
Little, D.A. (1984). *Can. J. Anim. Sci.* **64**, (Suppl.) 229-31.
Little, D.A., and Morris, J.G. (1972). *J. Agric. Sci.* **78**, 505-8.
Little, D.A., and Shaw, N.H. (1979). *Aust. J. Exp. Agric. Anim. Husb.* **19**, 645-51.
Little, D.A., McLean, R.W., and Winter, W.H. (1977). *J. Agric. Sci.* **88**, 533-8.
Loosmore, R.M., and Allcroft, R. (1951). *Vet. Rec.* **63**, 414-6.
Lorentz, P.P., and Gibb, F.M. (1975). *N.Z. Vet. J.* **23**, 1-3.
Lowman, B.G., Scott, N.A., and Somerville, S.H. (1976). East of Scotland College of Agriculture Bull. No. 6.
Mahmoud, O.M., Samani, F.E., Bakheit, A.O., and Hassan, M.A. (1983). *J. Comp. Path.* **93**, 591-5.
Mannetje, L. t' (1978). In "Measurement of Grassland Vegetation and Animal Production." (Ed. L. 't Mannetje.) pp. 63-95. (Commonw. Bur. of Pastures and Field Crops: Hurley, UK.)
Mason, R.W. (1976). *Br. Vet. J.* **132**, 374-9.
Masters, D.G. (1984). In "Ruminant Physiology. Concepts and Consequences. A Tribute to R.J. Moir." (Eds S.K. Baker, J.M. Gawthorne, J.B. Mackintosh and D.B. Purser.) pp. 337-46. (University of Western Australia: Perth.)
Masters, D.G., and Fels, H.E. (1980). *Biol. Trace Elem. Res.* **2**, 281-90.
Mayes, R.W. and Lamb, C.S. (1983). *Proc. Nutr. Soc.* **43**, 39A.
Mayes, R.W., Lamb, C.S. and Colgrove, P.M. (1986). *J. Agric. Sci.* **107**, 161-70.
McDonald, I.W. (1968). *Nutr. Abstr. Rev.* **38**, 381-400.
McDonald, I.W. (1981). In "Grazing Animals. World Animal Science Series." Vol. B1. (Ed. F.H.W. Morley.) pp. 349-60. (Elsevier: Amsterdam.)
McDowell, L.R., Conrad, J.H., Ellis, G.L., and Loosli, J.K. (1983). "Minerals for Grazing Ruminants in Tropical Regions." Bull. University of Florida and U.S. Agency for Intern. Develop.
McMurray, C.H., Rice, D.A., and Kennedy, S. (1983). In "Trace Elements in Animal Production and Veterinary Practice." pp. 61-73. (Occ. Publ. No. 7: Brit. Soc. Anim. Prod.)
Mills, C.F., Dalgarno, A.C., Williams, R.B., and Quarterman, J. (1967). *Brit. J. Nutr.* **21**, 751-68.
Moir, K.W. (1960a). *Qld. J. Agric. Sci.* **17**, 373-83.
Moir, K.W. (1960b). *Qld. J. Agric. Sci.* **17**, 385-9.
Morris, J.G. (1980). *J. Anim. Sci.* **50**, 145-52.
Murphy, G.M., and Connell, J.A. (1970). *Aust. Vet. J.* **46**, 595-8.
Murphy, G.M., and Plasto, A.W. (1973). *Aust. J. Exp. Agric. Anim. Husb.* **13**, 369-74.
Myers, V.I. (1983). In "Manual of Remote Sensing." Vol. II. 2nd Ed. (Eds R.N. Colwell, J.E. Estes and G.A. Thorley.) pp. 2111-228. (American Society of Photogrammetry: Falls Church, Virginia.)

National Academy of Sciences. (1980). "Mineral Tolerance of Domestic Animals." (NRC: Washington.)

New Zealand Society of Animal Production. (1983). "The Mineral Requirements of Grazing Ruminants." (Ed. N.D. Grace.) N.Z. Soc. Anim. Prod. Occas. Publ. 9.

Payne, J.M., Dew, S.M., Manston, R., and Faulks, M. (1970). *Vet. Rec.* **87**, 150–8.

Pickering, F.S., Nolan, J.V., and Corbett, J.L. (1982). *Proc. Nutr. Soc. Aust.* **7**, 127–30.

Playne, M.J., Khumnualthong, W., and Echevarria, M.G. (1978). *J. Agric. Sci.* **90**, 193–204.

Rammell, C.G., and Cunliffe, B. (1983). *N.Z. Vet. J.* **31**, 203–4.

Read, M.V.P., Engels, E.A.N., and Smith, W.A. (1986). *S. Afr. J. Anim. Sci.* **16**, 18–22.

Reid, R.L., and Horvath, D.J. (1980). *Anim. Fd Sci. Technol.* **5**, 95–167.

Robinson, J.J., Fraser, C., and Bennett, C. (1971). *J. Agric. Sci.* **77**, 141–5.

Rosenfeld, I., and Beath, O.A. 1964. "Selenium. Geobotany, Biochemistry, Toxicity, and Nutrition." (Academic: New York.)

Rowlands, G.J. (1980). *World Rev. Nutr. Diet.* **35**, 172–235.

Russel, A.J.F. (1978). In "Blood Profiles in Animal Production." (Ed. D. Lister.) British Society of Animal Production. Occas. Publ. No.1. pp.31–9.

Russel, A.J.F., Doney, J.M., and Gunn, R.G. (1969). J. Agric. Sci. **72**, 451–4.

Russel, A.J.F., Doney, J.M., and Reid, R.L. (1967). *J. Agric. Sci.* **68**, 359–63.

Satter, L.D., and Roffler, R.E. (1976) In "Tracer Studies on Non-protein Nitrogen for Ruminants III." pp. 119–38. (IAEA: Vienna.)

Saul, G.R., Flinn, P.C., and Heard, J.F. (1986). *Proc. Aust. Soc. Anim. Prod.* **16**, 351–4.

Searle, T.W. (1970). *J. Agric. Sci.* **74**, 357–62.

Sheppard, A.D., Blom, L., and Grant, A.B. (1984). *N.Z. Vet. J.* **32**, 91–5.

Siebert, B.D., Newman, D.M.R., Hart, B., and Michell, G.L. (1975). *Aust. J. Exp. Agric. Anim. Husb.* **15**, 321–4.

Smith, R.H. (1959). *Biochem. J.* **71**, 609–14.

Smith, B.S.W., and Wright, H. (1975). *Clin. Chim. Acta.*, **62**, 55–63.

Smith, G.S., Middleton, K.R., and Edmonds, A.S. (1978). *N.Z. J. Exp. Agric.* **6**, 183–8.

Solomons, N.W., and Allen, L.H. (1983). *Nutr. Revs.* **41**, 33–50.

Statham, M., and Bray, A.C. (1975). *Aust. J. Agric. Res.* **26**, 751–68.

Sutherland, R.J. (1980). *N.Z. Vet. J.* **28**, 169–70.

Suttie, J.M., Wenham, G., and Kay R.N.B. (1983). *Vet. Rec.* **113**, 393–4.

Suttle, N.F., Alloway, B.J., and Thornton, I. (1975). *J. Agric. Sci.* **84**, 249–54.

Suttle, N.F. (1986). *Vet. Rec.* **119**, 148–52.

Teleni, E., Dean, H., and Murray, R.M. (1976). *Aust. Vet. J.* **52**, 529-33.

Tilley, J.M.A., and Terry, R.A. (1963). *J. Brit. Grassl. Soc.* **18**, 104-11

Towers, N.R., and Clark, R.G. (1983). In "The Mineral Requirements of Grazing Ruminants." (Ed. N.D. Grace.) pp. 13-21. N.Z. Soc. Anim. Prod. Occas. Publ. 9. pp. 13-21.

Underwood, E.J. (1977). "Trace Elements in Human and Animal Nutrition." 4th Ed. (Academic: New York.)

Underwood, E.J. (1981). "The Mineral Nutrition of Livestock." 2nd Ed. (Commonw. Agric. Bur.: Hurley, UK.)

Vickery, P.J., and Nicol, G.R. (1982). Anim. Res. Lab. Tech. Paper No. 9. (CSIRO: Australia.)

Wallace, A.L.C., Gleeson, A.R., Hopkins, P.S., Mason, R.W., and White, R.R. (1978). *Aust. J. Biol. Sci.* **31**, 39-47.

Webb, J.S., Thornton, I., and Fletcher, K. (1968). *Nature*, London. **217**, 1010-2.

Wiener, G., Field, A.C., and Smith, C. (1977). *Vet. Rec.* **101**, 424-5.

Wilson, A.D. (1966). *Aust. J. Agric. Res.* **17**, 155-63.

Wilson, A.D. (1975). *Aust. J. Exp. Agric. Anim. Husb.* **15**, 760-5.

Wolff, J.E., Bryant, A.M., Cordes, D.O., Ramberg, C.F., Saunders, W.M.H., and Sutherland, R.J. (1978). *N.Z. Vet. J.* **26**, 266-9.

Zetterholm, R. and Dalen, N. (1978). *Acta. Vet. Scand.* **19**, 1-17.

17

ADAPTATION OF HERBIVORES TO SEASONAL CHANGES IN NUTRIENT SUPPLY

by S. J. McNaughton

Wild herbivores are confronted by a variety of unpredictable environmental variables that animal husbandry has, to a greater or lesser extent, rendered more predictable for their domesticated counterparts. A principal aggregate objective of animal husbandry, I believe, has been to increase predictability in a generally variable world. This, it seems to me, has been as assiduously sought as the more explicitly recognized objective of increasing the proportion of ecosystem productivity that is channelled into humans. Pastoral societies have little impact on total solar energy captured by the foods that they utilize indirectly through livestock, but they channel that energy through the food web to increase food stability in inherently variable and harsh environments (Coughenour *et al.* 1985). Where spatial changes are temporally predictable, both wild and domesticated animals will undergo predictable migrations; opportunistic nomadism increases as environmental predictability declines (Coughenour 1986). Because the uncultivated vegetation of grassland–savanna ecosystems is incapable of supporting significant human populations, conversion of that vegetation into animal products is a major objective of agriculture in those ecosystems. Herding, in the beginning, and fencing, later, increased the predictability of animal spatial utilization patterns. Haymaking and other food preservation techniques increased the temporal predictability of food supply to animals and, hence, of food yield to humans. Nutritional supplementation and provision of supplementary water sources extended husbandry into regions previously unexploitable and veterinary practices extended control over pathogens and parasites. The principal concern of this paper is how large, free-ranging herbivores have evolved to exploit inherently stochastic environments in the absence of direct human management. It must be born in mind, however, that no ecosystem is now free of human influence and my consideration is tempered by that fact.

MIGRATION

Large nomadic ungulates were major features of grassland ecosystems throughout the Earth until they were inundated by the tide of European colonisation in the 18th and 19th centuries, except in South America where they were eliminated by Pleistocene extinctions, and Australia where ecologically similar macropods dominated the fauna. These

NUTRITION OF HERBIVORES
ISBN 0 12 750052 9

ungulates often ranged over wide geographic areas, moving seasonally in response to the temporal abundance and quality of forages in different portions of their total range. Early settlers on the North American Great Plains often depended upon bison (*Bison bison*) as a major meat source and their historical records of hunting documented the unpredictability of bison spatial distributions in different seasons and years (Roe 1951; Dary 1985). Seasonal migrations of varying spatial extent are characteristic of elk (*Cervus canadensis*) in the Rocky Mountains (Knight 1970; Houston 1982), red deer (*Cervus elaphus*) in Scotland (Staines 1974; 1976), African equids and several species of bovids (Watson 1967; Maddock 1979), and caribou (*Rangifer tarandus granti*) in the arctic (Lent 1966; Hemming 1971). Spatial variation of food supply through the year is commonly invoked to explain the major seasonal movements of all of these large, wild grazers. Since many of these species are now confined to national parks and other wildlife reserves that are incomplete seasonal ranges for such wide-ranging animals, seasonal variations in the spatial distribution of food resources are of major management importance. When considering animal movement patterns in relation to food supplies, it should be remembered that food requirements are not fixed but will vary within certain bounds. White-tailed deer (*Odocoileus virginianus*) raised on different diets underwent seasonal hormonal changes that were related to diet and would influence metabolic demands (Bahnak *et al.* 1981). Decreased concentrations of circulating thyroid hormones during periods of acute malnutrition can function as a mechanism to reduce energy requirements, bringing them more into concordance with intake. This endocrine regulation may explain why animals fed *ad libitum* still restrict food intake during periods of normal food shortage (Robbins 1983).

Insect harassment can modify group size, activity budgets and distribution; seasons of severe harassment are characterized by reduced feeding time and large, highly mobile groups (White *et al.* 1975). The groups are the result of congregation in specialized habitats such as wind-swept ridges, where insect attack is reduced, as animals are driven out of otherwise preferred habitats. However, such movements are commonly of short duration and limited in spatial extent compared with seasonal travel associated with variations in food supply.

There also are commonly seasonal variations in group size and social organization in wild ungulates (Underwood 1982). Mean group size increases during period of food abundance during the wet season in the tropics or summer in temperate regions. Group size often declines as these large groups fragment during periods of food shortage. As food supply for impala (*Aepyceros melampus*) diminished during the dry season, herd size declined, individuals became more widely spaced within herds, and both the rate of individual movement and total distance moved daily while feeding increased as animals sought progressively rarer high-quality foods (Jarman and Jarman 1974).

Migration is not a species-specific trait even in such major migratory species as caribou in the arctic (White *et al.* 1975) or wildebeest (*Connochaetes taurinus*), zebra (*Equus burcheli*), and Thomson's gazelle (*Gazella thomsonii*) in the Serengeti region of Tanzania and Kenya (Estes 1974; Maddock 1979). Each of these species form large, migratory herds and smaller, resident herds that are sedentary within comparatively small home ranges. Resident herds in the Serengeti are associated with regions of high habitat diversity and localized climate gradients allowing them to duplicate many features of the annual cycle of food search, characteristic of the much larger migratory herds (McNaughton 1983). As the migrants move from regions of low mean annual rainfall in the wet season to regions of high mean annual rainfall in the dry season (Maddock 1979), residents concentrate on the drier upper catenas during the wet season and move progressively downhill as the dry season progresses until they are exploiting meager food resources on bottom catenas at the end of the dry season (Bell 1970). Similarly, small sedentary caribou herds are able to exploit localized resources during the winter that are insufficient to support the large migratory herd (White *et al.* 1975).

ALLOMETRY, NUTRITION, AND LIFE HISTORY

Recognition of the tendency of African herbivores with varying body sizes to utilize grasses of different statures, with large herbivores feeding on rank vegetation and smaller species exploiting finer canopy structures (Vesey-Fitzgerald 1960), led to the hypothesis that a nutritional quality gradient is partitioned among herbivore species based on their body size (Jarman 1968; Bell 1969, 1970). Organization of the digestive tract might also contribute to dietary resource partitioning since monogastric fermenters, like equids, could be differentiated from ruminants (Bell 1969). The more rapid passage rate with less efficient total digestion in equids allows them to extract equivalent energy from a lower quality diet provided that diet has sufficient readily metabolizable energy to substitute for the greater extractive efficiency, due to greater retention times, in ruminants (Bell 1969, 1970).

There is a general interplay among forage quality, digestive efficiency, and retention time in the digestive tract that applies to both individuals (Baile and Forbes 1974) and species (Demment and Van Soest 1985). In addition, a number of behavioural traits are related to feeding ecology. There is a continuum from small, solitary, more selective concentrate-feeders to larger, gregarious, less selective roughage-feeders (Jarman 1974). It should be noted, however, that the diets of wild animals (Demment and Van Soest 1985) rarely approach the level of low fibre (<100g /kg DM) commonly used to define foodstuffs as concentrates in the diets of domestic livestock (Baile and Forbes 1974).

A major constraint is placed on small animals by their higher specific metabolic costs (Kleiber 1961). Because gut volume scales linearly with body mass (Parra 1978; Demment 1982), while maintenance metabolism scales as a fractional power of body mass (Kleiber 1961), feeding ecology is constrained significantly by allometric balances (Demment and Van Soest 1985). Particularly important ecologically is the ratio metabolic requirement/gut capacity, which is inversely related to body mass. That is, the metabolic requirement per unit of gut capacity declines as body mass increases.

Total digestibility of a diet, of course, is a summation of the fractional digestibilities of a heterogeneous assortment of cell and tissue types. Herbivores ingest a variable mixture of cell contents, primary cell walls, secondary cell walls, and digestion inhibitors. Cell contents are readily digestible and provide the major nutrients of herbivores with simple digestive tracts while more recalcitrant plant components become progressively more available to herbivores with progressively more complicated digestive systems and associated symbiotic gut microbes (Hungate 1966).

Fibre curves relate the abundance of plant material (as mass/area) to the percentage of that mass that is fibre (Demment and Van Soest 1985). Those curves reveal that as the apparent availability of plant material increases, the proportion of that mass in fibre increases dramatically. Small herbivores, therefore, with high specific metabolic rates and high ratios of metabolic requirement to gut capacity, must become increasingly selective as plant standing crop increases until at the extreme, in forests, they are reduced to searching for very rare plant material capable of sustaining them in a general environment with a plentiful supply of unsuitable foodstuffs. Ruminants dominate the African fauna at intermediate (10 – 600 kg) body masses (Demment and van Soest 1985). Below 10 kg, non ruminants with hindgut fermentation of low fibre diets, often coprophagic, are the most common species in the fauna. Above a 600 kg body size, retention time in the gut is so high even in the absence of a rumen that a relatively complete digestion of fibre can be accomplished and ruminants are a comparatively uncommon contributor to faunal diversity.

Allometric constraints on dietary nutritional properties are associated with a wide array of additional evolutionary consequences important to an animal's ecology (McNab 1974) so caution must be exercised in ascribing any single consequence of body size as a primary selective gradient.The population parameter, intrinsic rate of increase, r_m, is related to body mass as W (Hennemann 1983). Similarly, birthrate is a negative exponent of mass of value −0.33 and the ratio of population production to biomass has an identical value (Western 1983). Therefore, the high dietary quality imposed on small herbivores by their specific metabolic costs and gut capacity is also associated with high intrinsic rates of population increase and high rates of flux of individuals (in other words, both birth and death rates are high) through populations. This suggests that smaller animals might be more important in ecosystem process dynamics than their numbers or standing crops, interpreted simply, would reveal (McNaughton and Georgiadis 1986).

Interactions between animals and their food supplies also vary in a fashion suggesting that population densities of large animals may more often be regulated by extrinsic ecological factors while the densities of animals of small body size may be regulated more frequently by intrinsic factors (Caughley and Krebs 1983). Intrinsic regulation (self-regulation) occurs when a population's realized rate of increase (r rather than r_m) is affected principally by behavioural or physiological mechanisms, while extrinsic regulation is due to the direct effects of food supply, predators, diseases, or weather. Small herbivores with high values of r_m have reproductive capacities tending to destabilize the relationship of population density to food supply leading to destruction of the latter with potential extinction of localized populations due to limit cycles and other unstable population trajectories through time. Selection for behavioural or physiological responses within individuals could reduce the extinction probability for such populations (but note the group selectionist element of this argument). Larger herbivores, in contrast, with lower population r_m due to large body mass, may be less subject to unstable relationships with their food supplies, and therefore may be less subject to natural selection for intrinsic regulatory mechanisms. Caughley and Krebs (1983) argued that a body mass near 30 kg, with a resultant annual r_m of 0.45, was a critical level separating small mammalian herbivores with intrinsic regulation from large animals with extrinsic regulation. This is near, but not at, the lower limit of ruminant body size in many ecosystems (Demment and Van Soest 1985). Since food shortage in many ruminants leads to a reduction of female body size and nutritional status to below the level required to reproduce (Thomas 1982; Loudon *et al.* 1983; Skogland 1983), reducing population fecundity, a plausible mechanism exists for translating extrinsic regulation in these animals into population reproductive rate.

Individual growth rates of uniparous ruminants are also systematically related to body mass and sexual dimorphism (Georgiadis 1985). Although the absolute growth rates of larger species are greater than those of smaller bodied species, relative growth rates are lower. In addition, sexually dimorphic species are commonly polygynous and have a female-biased sex ratio. Males of these species attain a higher adult body mass than females primarily due to extended periods of growth, although their growth rates also are slightly higher.

Body size also is related to travel speed (Pyke 1981), which obviously could be important to migratory species searching for food. Three criteria might regulate travel speed:-

minimization of the rate of energy expenditure,
minimization of energy cost per unit distance covered,
or maximization of net rate of energy gain.

Pyke (1981) concluded that the latter criterion was most likely for foraging activities. Moreover, the rate accomplishing maximization of net rate of energy gain was greater than the rates associated with either minimization criterion.

In spite of these major trends relating allometry to foraging and social organization of wild herbivores, selection experiments within laboratory populations indicate that broad phyletic patterns can be modified by selection within species. Selection for body weight at six weeks of age in laboratory mice (*Mus musculus*) indicated that realized heritabilities were 0.24 and 0.35 for higher liveweight and lower liveweight strains, respectively (Fuente and San Primitivo 1985). Associated with weight divergence was a divergence of litter size; selection for greater body weight produced increased litter size due to increased ovulation rates and uterine capacities. Cattle selected for higher performance under stress provoked by infection with bovine infectious keratoconjunctivitis, helminths, heat, and lower planes of nutrition tended to have reduced growth potentials under unstressed conditions (Frisch 1981). Thus, selection for performance under stress, which is likely in free-ranging wild animals, should have resulted in stress resistance and not to increased inherent performance ability. This suggests, as a general principle, that selection for high performance in a certain environment is not transferable across environments and that the genes that control performance in different environments differ (Falconer 1977). Therefore, it is likely that wild animals with fragmented ranges distributed over broad geographic areas consist of genetically distinct subspecific populations differing according to the levels and varieties of stresses that prevail.

FORAGING HIERARCHIES

Forage selection by large herbivores ranges in scale from the level of geographic regions to plant organs (McNaughton 1983) and on time scales from months to seconds (R.L. Senft pers. comm.). Therefore, a hierarchial approach to food selection is required. Decision-making on progressively longer time scales accompanies variability at successively larger spatial scales. At the regional level, animals will select landscapes within regions, through migration, home range placement, or nomadism at behavioural frequencies of a few times a year. The presumed foraging goal is fitness maximization by maintaining intake and avoiding environmental stresses. Such extrinsic, or noninteractive variables as climate, soils, proximity to water or salt licks, and physical barriers to movement can regulate the freedom with which decisions at this level are made. Within landscapes, animals will select plant communities, or large vegetation patches, at a frequency of many times each year to optimize foraging efficiency due to forage quantity (density per unit area) and quality (nutrient density). Within plant communities, animals select feeding stations (areas within which they can feed without moving) at a frequency of perhaps 10^7 times annually, for a momentary maximization of foraging efficiency based on forage density, availability, and quality. At the level of feeding stations, local nutritional variability due to microsite variation can act as an extrinsic constraint on freedom of choice.

At the sward level, where momentary choices of high frequency and resolution are required, both sward and forager properties are important (E. Laca and M.A. Demment pers. comm.). A hierarchical approach is also required at this level since foraging consists of, at least, feeding stations and bites. In modelling the feeding process at this level, Laca and Demment included a number of sward variables: distance between feeding stations, vegetation height, standing crop, bulk density (forage per unit volume), mean and standard deviation of forage quality (nutrient density per unit plant tissue), and sward structural type. The latter was categorized as temperate or regrowth type where high quality forage was at the sward top and quality declined with depth in the canopy, and the tropical or mature type with parts of different quality intermixed throughout the canopy. Animal variables included bite depth and area, feeding station area, rate of movement between feeding stations, maximum chewing and ingestion rate, handling time for a bite, a low quality limit of acceptable food, and minimum bite size as a proportion of potential bite size. The animal, therefore, can balance both rate of intake (a quantity criterion) and forage quality as it feeds. Model output included digestible dry matter intake rate, mean dietary quality, mean bite size, mean biting rate, and time spent at each feeding station. Running the model over a wide range of sward and animal characteristics indicated that mastication and formation of a swallowable bolus were important limitations on forage intake when feeding stations have high bulk densities. As has frequently been observed (Dunham 1980; Hudson and Nietfeld 1985), there was a negative relationship between bite size and biting rate as sward structure and forage availability changed that allowed animals to maintain similar intake rates over a wide range of sward properties. However distinct limits were evident in the interplay between these two foraging traits in cattle; below a bite size of about 0.7 g dry matter, intake dropped in spite of increased biting rate.

At low selectivities, which would be characteristic of wild animals described as roughage-feeders, most bites were large and biting rate was low so intake was limited by mastication and bolus formation, but intake was constant over a wide range of sward conditions (E. Laca and M.A. Demment pers. comm.) When selectivity was high, which would be characteristic of wild animals characterized as concentrate-feeders, bite size and travel time per bite became important constraints. Whereas moderate changes in selectivity had little effect on the intake rates of roughage-feeders, slight increases in the selectivity of concentrate-feeders reduced intake rates drastically.

FORAGE QUALITY

Although there may be a general positive relationship between forage quantity and quality on a seasonal basis, with both quality and quantity high during periods of rapid plant growth, there is evidence that these two food properties may not be concordant geographically. In arid, sub-Sahelian rangelands, forage quality, as indexed by nitrogen content, was inversely related to the rainfall gradient with arid areas producing higher quality forage than more mesic regions (Breman and de Wit 1983). Bell (1982) suggested that there was a broad nutritional dichotomy in African savanna-grasslands between arid ecosystems on high nutrient soils supporting high densities of medium-sized ungulates to mesic, low fertility soils supporting typical miombo woodland with low densities of larger-bodied herbivores. This conundrum relating forage quality and quantity, if it has general utility, complicates both the development of forage management criteria and evaluations of forage quality for wild herbivores.

While nitrogen has been the forage quality criterion most often emphasized in wild herbivore studies (Bell 1982; Owen-Smith 1982; McNaughton and Georgiadis 1986), a wide variety of other factors also can be important (McDonald *et al.* 1981; Robbins 1983; McDowell 1985). Vitamin deficiencies are common in captive animals fed artifical diets, but only vitamin E deficiency has commonly been reported in wild herbivores (Robbins 1983). However, tropical regions with extended dry seasons might subject grazers to vitamin A deficiencies if liver stores became depleted (McDowell 1985).

Mineral deficiencies, in contrast, may be a common occurrence in wild herbivores. In addition to their own mineral requirements, ruminants must meet the mineral requirements of their microbial symbionts to maintain adequate levels of nutrition (Butler and Jones 1973). Tropical and subtropical soils, in particular, are often deficient in a wide variety of mineral elements with the exact deficiencies varying from locality to locality (McDowell 1985). An extensive survey of Latin American forages revealed major deficiencies compared to feeding standards for domestic livestock; percentage of all samples containing deficient or near deficient levels for various minerals are given in Table 1.

Similarly, a survey of forages in the African country of Malawi indicated deficiencies, according to livestock feeding standards, of a variety of elements and found that percentages of pastures deficient for the various elements differed in the wet and dry seasons (Table 1). These data suggest that either the mineral requirements of Africa's indigenous fauna must differ substantially from those of domestic livestock, which seems highly unlikely, or that fauna must be highly selective of habitats and forage species in different seasons, which is well known (McNaughton and Georgiadis 1986). It also is worth remarking that wet-season forages, which are generally considered to be of consistently higher quality than dry season forages,

are more likely to lead to deficiencies of Ca and Mg. This is undoubtedly due to their low retranslocation rates from senescing plant tissue, resulting in their enrichment in above-ground plant tissues as other elements are exported at the beginning of dry season dormancy.

Table 1. Percentage of pastures showing evidence of mineral deficiency in Latin America and Malawi. Values based on feeding standards for domestic livestock.

Mineral	Latin America	Malawi(wet season)	Malawi(dry season)
Calcium	31	81	13
Phosphorus	73	56	96
Magnesium	35	77	31
Potassium	15	3	57
Sodium		96	97
Cobalt	43	14	13
Copper	47	47	91
Iron	24	6	3
Manganese	21	6	3
Molybdenum	86		
Selenium		55	96
Zinc	75	96	94

In addition to the absolute nutrient content of plant tissue, the balance of various nutrients can influence food quality. Lactating wildebeest in the Serengeti ecosystem can meet their Ca requirements only in the south east Serengeti Plains where high Ca contents and Ca/P ratios favourable to Ca assimilation occur in forages (Kruelen 1975). There also is an interaction between Se concentration and assimilation of vitamin E by herbivores (McDowell 1985). On bison ranges in NW Canada, about 75% of all forage samples were Se deficient by beef cattle standards; excessive atmospheric sulphur levels and resultant S-enriched rain, a growing environmental problem, could further exacerbate Se deficiency (Shaw and Reynolds 1985).

SECONDARY PLANT CHEMICALS

Compared to the massive interest in the role of plant secondary chemicals in plant-insect interactions due to the impact of Ehrlich and Raven's (1964) classic paper on coevolution of butterflies and plants, little attention has been directed toward investigating the role of such chemicals in the relationships between mammalian herbivores and their food sources (McNaughton and Georgiadis 1986), except for some excellent work on small mammals (Negus and Berger 1977; Bryant 1981) and on the effects of conifer oils on deer (Oh *et al.* 1967, 1968). Just as mineral balances are important to the rumen microbial community, so also secondary compounds can have important effects on ruminants through their impact on that community (Nagy and Tengerdy 1968). Slow acclimation to a diet of increasing concentrations of secondary chemicals can render the animal much less susceptible to deleterious effects than is evident when there is a rapid diet switch to foods containing those chemicals (Hegarty *et al.* 1964). There also is some evidence of complimentarity since deer will eat more of a mixture of tissues of plant species containing a variety of different secondary chemicals than they will eat of any plant species separately (Smith 1959).

There are three broad classes of effects that plant secondary chemicals can have upon herbivores, acting as digestion inhibitors, toxins, and feeding deterrents. Any given chemical may function in all three ways simultaneously against an array of herbivores or sequentially against the same herbivore at different times; it also might have functions in addition to its impact upon herbivores. For example, gossypol, a terpenoid produced by cotton (*Gossypium barbadense*), acts as a feeding deterrent and digestion inhibitor against some herbivores, a feeding stimulant for yet others, and a phytoalexin protecting the plants against pathogens (Hedin *et al.* 1968; Shaver *et al.* 1970). The frequency of alkaloids in floras varies geographically, being most common of tropical plants and least frequent in Australia and New Zealand, which may reflect different selection regimes associated with the abundance of browsers in the faunas of different regions (Levin 1976). Local plant populations also can be polymorphic for the concentration or presence of secondary chemicals known to be deleterious to herbivores (Cooper-Driver and Swain 1976). Jones and Ramnani (1985) argued that phenotypic shifts of the expression of cyanogenesis in plant populations could act to disrupt the learning patterns of herbivores and that some plants could, therefore, be catagorized as altruists.

The palatability of 14 species of woody plants in southern African savannas to three species of browsers was most closely related to concentrations of condensed tannins, with this effect tending to override nutritional quality as indexed by nitrogen, phosphorus, sodium and fibre content (Cooper and Owen-Smith 1985). There was a threshold of tannin content, about 5%, below which there was no effect. Since tannins are strong digestion inhibitors (van Hoven 1984) reducing the intake of plant tissues containing tannins would be advantageous and their bitter taste provides browsers with an evident gustatory clue to their presence and concentration. In addition to differences between plant species or tissues, developmental stage of a plant tissue may be an important determinant of its secondary chemistry. Young birch twigs were decidedly more inhibitory to digestion than older twigs due primarily to different concentrations of water-soluble phenolics (Palo 1985).

Isolation from willow (*Salix* sp.) catkins of an oestrogenically active compound similar in physical properties to oestriol revealed that plants could be sources of hormone-mimic chemicals affecting vertebrates in fundamental ways (Skarzynski 1933). This finding led to isolation of plant chemicals founding the birth control pill industry. One of the best known of such hormonal influences, because of its economic importance, is infertility in sheep and other livestock feeding on subterranian clover (*Trifolium subterraneum*), which contains phytoestrogens (Bennetts *et al.* 1946; Bradbury and White 1951). Concentrations of such chemicals were affected by rainfall but were relatively insensitive to many other factors (Gil *et al.* 1984a, 1984b). In spite of the potential major effects of hormone-mimics upon animals and their broad distribution in plants (Slama 1980), I know of no study of wild animals that adduces such secondary chemicals as an ecological factor. Similarly, ryegrass (*Lolium perenne*) staggers due to a fungal endophyte has major effects upon domestic livestock (Fletcher 1982), but similar effects of plants on wild animals have not been reported, to my knowledge. A fruitful area of research on large wild herbivores could be investigating plant secondary chemistry and its effects in more detail, cautioned by the fact that wild animals are capable of eating some plants that are fatal to domestic livestock (Owen-Smith 1982).

SPECIFICS OF FORAGING BEHAVIOUR

Herbivores spend most of their active time either looking for food or digesting it. Both herd scans and focal animal observation of wildebeest in Etosha National Park, Namibia, recorded that, on an annual basis, 32% of their time was spent actively grazing and 53% "resting", which presumably includes rumination (Berry *et al.* 1982). Mixed feeders that depend upon both grass and non-graminoids in their diet commonly concentrated on grasses during periods of high rainfall and high rates of grass growth, switching to other foods during dry periods (Dunham 1980; Spinage *et al.* 1980). Diet quality, as indexed by the crude protein percentage of rumen contents, was highest during periods when grass made up the largest proportion of rumen fill (Dunham 1980). Greater kudu (*Tragelaphus strepsiceros*) in Kruger National Park adjusted feeding time per unit distance travelled more than the proportion of foraging time devoted to feeding when moving between habitats differing in forage abundance and as seasonal food abundance changed (Owen-Smith 1979). Habitat differences were compensated for to some extent seasonally by habitat selection, modification of the range of foods eaten, and diversion of time from other activities to feeding. Similar diversion of behaviour from non-feeding to feeding activities can result in almost complete energetic compensation in deer fawns fed a restricted diet (Muller-Schwarze *et al.* 1982). Spinescent foods of higher nutrient content were sought by kudu during the growing season since nutrient content was higher (Owen-Smith *et al.* 1983). When food availability declined as the dry season progressed, preference shifted to plants allowing highest intake rates.

The presence of recurved thorns on *Acacia tortilis* reduced the bite rate of impala for a given bite size compared to non-spinescent plants because the spines caught the animal's tongue and lips as it fed (Dunham 1980). There may be a general tendency for physically protected plants to produce forage of higher quality (Cooper and Owen-Smith 1986). This forces animals to make a choice between quality and intake rate. Both biting rate and bite size are reduced by thorns and this effect is greater for large animals than for smaller species. Therefore, an availability gradient could be superimposed on quality gradients to influence resource partitioning in a community of diverse browsers.

Studies of domestic livestock indicate that taste and odour can influence food preferences (Arnold *et al.* 1980). Glycine, wintergreen oil, peppermint oil, aniseed oil, and 5-hexane-1-ol all acted as feeding deterrents when applied to sheep diets There also was evidence of acclimation to odours and taste over time since the animals lost an initial dislike for cedarwood oil, tannic acid, propionic acid and amyl alcohol; they also developed a liking for butyric acid. Several compounds applied to the diet affected intake of anosmic and agustatory individuals as well as controls, so sensory stimuli were not necessary for some diet adjustments Dietary preferences of sheep in other experiments could be shifted by spraying juice from preferred species on avoided species, and *vice versa* (Scehovic *et al* 1985). These studies indicate that immediate food choice is under a complicated combination of sensory and nonsensory cues.

Comparative studies of the foraging behaviours of impala, a mixed feeder of higher selectivity, and blesbok (*Damaliscus dorcas phillipsi*), a grazer of lower selectivity, indicated that the latter spent less of its activity budget actively feeding (Klein and Fairall 1986). Considering the general trends of body size and feeding strategy discussed above in relation to the diets is instructive since these species differ in diet but have similar body mass. Many of the dietary traits related to allometry also applied to these species in spite of that body size similarity. Blesbok has a greater rumen capacity and intestine length, reflecting a lower quality diet and the requirement for efficient extraction of nutrients from that diet. Blesbok spent less time each day feeding and also were less sensitive to high temperatures. The higher crude protein of the impala diet was associated with a rate of water turnover, as indicated by isotopic dilution, twice as high as blesbok (Fairall and Klein 1984), suggesting that the more selective, mixed feeder has a higher metabolic rate even though body size is similar. Therefore, the grazer has an energetically conservative life history involving unselective consumption of poorer quality food while the mixed feeder has an energetically more profligate life history requiring selective consumption of higher quality, less common foodstuffs.

DIGESTIBILITY

In addition to all the factors discussed above which can influence forage consumption and quality of the forage consumed, yield to the animal is finally determined by the interaction between a food and its digestive system; in ruminants, in particular, the gut microbe community is of major importance (Hungate 1966; Blaxter 1962). Ruminal volatile fatty acid (VFA) and osmolarity levels in a wildebeest grazing natural dry season forages reached a daily peak about three hours after feeding commenced and then dropped off somewhat as the day progressed (Kreulen and Hoppe 1979). Levels of VFA ranged from 65 to 194 mequiv /l, about the levels found in domestic ungulates. The yield of VFA per unit forage, calculated using certain explicit assumptions, also was comparable to domesticated animals of comparable body size.
A comparative study of *in vitro* dry matter digestibility (IVDMD) of 13 shrubs, 8 forbs, 4 grasses and a cactus using steer, sheep, goat and white-tailed deer (*Odocoileus virgianus*) rumen liquor obtained at four different seasons revealed consistent seasonal trends of IVDMD (Blankenship *et al.* 1982). IVDMD was highest in spring and winter, lowest in summer and fall. Over all forages and seasons, IVDMD averaged 0.525 for deer, 0.492 for goat, 0.475 for sheep, and 0.466 for steer. In addition, certain plant species were consistently more digestible and others consistently of low digestibility for all animal species. Comparing sheep with eland (*Taurotragus oryx*), bush duiker (*Sylvicapra grimmia*), Coke's hartebeest (*Alcelaphus buselaphus cokei*), and Thomson's gazelle fed pelleted diets ranging in crude protein content from 65–135 g /kg DM revealed distinctive digestibility patterns in the five species (Arman and Hopcraft 1975). DMD was highest in duiker and hartebeest, lower in sheep, lower yet in gazelle, and lowest in eland. Crude protein digestibility, in contrast, was higher in eland, duiker and hartebeest, lower in sheep and gazelles. Although sex, age, and body mass were confounding variables in this study, the results did not indicate that the wild animals were either inherently more or less efficient than domesticated livestock.

Digestibilities *in vitro* of summer diet mixtures incubated in "summer" rumen innoculum of Peary caribou (*Rangifer tarandus pearyi*) ranged from 0.50–0.80 for vascular plants, 0.18–0.85 for lichens, with mosses from 0.11–0.35 (Thomas and Kroeger 1980). When winter dietary components were used with "winter" rumen liquor, green sedge tissues had a digestibility of 0.65–0.74, dry sedge tissues were 0.25–0.43 digestible, and values were 0.54–0.83 for lichens, 0.03–0.11 for mosses. There were clear seasonal differences in the digestive capacities of rumen fluids and in the digestibilities of different forage components in different seasons. For composite diets, digestibilities were similar to those calculated from the weighted content of samples digested separately. Digestive capacity of kudu rumen innoculum acting on three identical diets also varied seasonally (Boomker 1984), generally being highest in the dry, cool months and lower in the wet, warm months. For wild animals in variable environments, then, there is evidence that seasonal changes in digestive efficiency modify digestive yield to increase yield during periods of poor food supply.

When elk were fed diets with crude protein content varying from 38 to 293g /kg, apparent digestible protein increased in a curvilinear fashion so that the amount of digestible protein was related linearly to dietary crude protein (Mould and Robbins 1981). Faecal N content was linearly related to dietary N, but diets of plant tissues rich in phenolic compounds resulted in much greater faecal N than predicted from dietary N and reduced digestible N significantly. The proportion of urea recycled ranged from 0.18–0.85 of the urea produced, and this proportion was inversely related to plasma urea concentrations. This recycling could be a major element of an animal's N budget when dietary N was restricted during poor seasons. It seems likely that elk could easily meet their N requirements for maintenance except during winter when both low N foods and browsing upon plant tissues high in phenolics could force them onto negative N budgets, excreting more than is taken in.

CONCLUSION: UNDERNUTRITION AND ITS CONSEQUENCES

It is unfortunate that nutritional studies have concentrated much more upon weight gain than weight loss (Robbins 1983), particularly in wild animals where a vague, qualitative state of undernutrition is often adduced as a major cause of death and, therefore, of population regulation in natural communities (McNaughton and Georgiadis 1986). Moreover, reliance upon the many indirect estimates of nutritional state, for example various "condition indexes," is of dubious value at best (Kirkpatrick 1980).

Not only can nutritional stress place an animal at risk of disease, disease also can increase nutritional demands upon the individual (Arman and Hopcraft 1971). Nutritional imbalances leading to such forage diseases as grass tetany due to K, Ca, and Mg imbalance in forages also would predispose herbivores to greater risk of predation due to mobility disruption long before the disease reached an acute phase of obvious locomotory dysfunction. Moreover, although fat status and catabolism have been emphasized in most studies of animal condition, this is impossible to justify because simultaneous carbohydrate metabolism is required for complete oxidation of fatty acids and glucose is required by the brain (Robbins 1983). Wild animals lacking the dietary benefits derived from husbandry, reserve forage stores and nutritional supplement, must store nutrients and energy during favourable periods for utilization during unfavourable periods. Alternating periods of weight gain, perhaps even compensatory growth of the type known for domesticated livestock (Drennen 1979; Imaizumi 1981), and of weight loss involving a good deal of protein catabolism and water loss, are probably general features of the nutrition of wild, free-ranging herbivores. Yet, little is known of either the extent or magnitude of such changes.

A temporal run of seasons or years of low primary productivity or of low forage quality, could gradually drag wild herds into, first, subreproductive physical condition, then undernutrition predisposing animals to disease and predation, and, finally starvation. Much more research emphasis on weight loss is required for a fundamental understanding to emerge of the physiological and ecological consequences of wild animal nutrition. Migration, social organization, endocrine regulation of intake, allometric constraints and consequences, activity costs, foraging hierarchies and selective feeding, forage quality, plant secondary chemistry, time budgets, and food digestibility all contribute not only to weight gain, but also to weight loss. A balanced approach to wild animal nutrition must consider both.

Preparation of this chapter was facilitated by US National Science Foundation Grant BSR 8508562.

Arman, P., and Hopcraft, D. (1971). *Proc. Nutr. Soc.* **30**, 65–6.
Arman, P., and Hopcraft, D. (1975). *Br. J. Nutr.* **33**, 255–64.
Arnold, G.W., de Boer, E.S., and Boundy, C.A.P. (1980). *Aust. J. Agric. Res.* **31**, 571–87.
Bahnak, B.R., Holland, J.C., Verme, L.J., and Ozoga, J.J. (1981). *J. Wildl. Manage.* **45**, 140–7.
Baile, C.A., and Forbes, J.M. (1974). *Physiol. Rev.* **54**, 160–214.
Bell, R.H.V. (1969). Ph.D. thesis. Univ. Manchester, England.
Bell, R.H.V. (1970). In "Animal Populations in Relation to Their Food Resources." (Ed. A. Watson.) pp.111–24. (Blackwell: Oxford.)
Bell, R.H.V. (1982). In "Ecology of Tropical Savannas." (Eds B.J. Huntley and B.H. Walker.) pp.193–216. (Springer-Verlag: Berlin.)
Bennetts, H.W., Underwood, E.J., and Shier, F.L. (1946). *Vet. J.* **22**, 2–12.
Berry, H.H., Siegfried, W.R., and Crowe, T.M. (1982). *Z. Tierpsychol.* **59**, 229–46.
Blankenship, L.H., Varner, L.W., and Lynch, G.W. (1982). *J. Range Manage.* **35**, 664–6.
Blaxter, K.L. (1962). "The Energy Metabolism of Ruminants." (Hutchinson: London.)
Boomker, E.A. (1984). *Can. J. Anim. Sci.* **64** (Suppl.), 208–9.
Bradbury, R.B., and White, D.E. (1951). *J. Chem. Soc.* **1951**, 3447–9.
Breman, H., and de Wit, C.T. (1983). *Science* **221**, 1341–7.
Bryant, J.P. (1981). *Science* **213**, 889–90.

Butler, G.W., and Jones, D.I.H. (1973). In "Chemistry and Biochemistry of Herbage." (Eds G.W. Butler and R.W. Bailey.) Vol. 2. pp.127–62. (Academic: New York.)

Caughley, G., and Krebs, C.J. (1983). *Oecologia* (Berl.) **59**, 7–17.

Cooper, S.M., and Owen-Smith, N. (1985). *Oecologia* (Berl.) **67**, 142–6.

Cooper, S.M., and Owen-Smith, N. (1986). *Oecologia* (Berl.) **68**, 446–55.

Cooper-Driver, G.A., and Swain, T. (1976). *Nature* (Lond.) **260**, 604.

Coughenour, M.B. (1986). Proc. IV Int. Congr. Ecol., p.121.

Coughenour, M.B., Ellis, J.E., Swift, D.M., Coppock, D.L., Galvin, K., McCabe, J.T., and Hart, T.C. (1985). *Science* **230**, 619–25.

Dary, D.A. (1985). "The Buffalo Book." (Avon: New York.)

Demment, M.W. (1982). *Afr. J. Ecol.* **20**, 43–7.

Demment, M.W., and Van Soest, P. J. (1985). *Am. Nat.* **125**, 641–72.

Drennan, M.J. (1979). *Ir. J. Agric. Res.* **18**, 131–43.

Dunham, K.M. (1980a). *J. Zool.* (Lond.) **192**, 41–57.

Dunham, K.M. (1980b). *Afr. J. Ecol.* **18**, 253–7.

Dunham, K.M. (1982). *S. Afr. J. Wildl. Res.* **12**, 36–40.

Ehrlich, P.R., and Raven, P.H. (1964). *Evolution* **18**, 586–608.

Estes, R.D. (1974). In "The Behaviour of Ungulates and its Relation to Management." (Eds V. Geist and R. Walther.) pp.166–205. (IUCN: Morges, Switz.)

Fairall, N., and Klein, D.R. (1984). *Can. J. Anim. Sci.* **64** (Suppl.), 212–4.

Falconer, D.S. (1977). *Proc. Nutr. Soc.* **36**, 47–51.

Fletcher, L.R. (1982). *N.Z. J. Exp. Agric.* **10**, 203–7.

Frisch, J.E. (1981). *J. Agric. Sci.* **96**, 23–38.

Fuente, L.F. de la, and San Primitivo, R. (1985). *Z. Tierz. Zuechtungsbiol.* **102**, 221–9.

Georgiadis, N. (1985). *Afr. J. Ecol.* **23**, 75–87.

Gil, L.A., Diaz, M.C., Ramirez, J., and Mayorga, M. (1984a). *Turrialba* **34**, 147–56.

Gil, L.A., Ramirez, J., and Diaz, J.C. (1984b). *Turrialba* **34**, 437–44.

Hedin, P.A., Miles, L.R., Thompson, A.C., and Minyard, J.P. (1968). *J. Agric. Food Chem.* **16**, 505–13.

Hegarty, M.P., Schinckel, P.G., and Court, R.D. (1964). *Aust. J. Agric. Res.* **15**, 153–67.

Hemming, J.E. (1971). Alaska Dept. Fish Game Tech. Bull. No. 1.

Hennemann, W.W. (1983). *Oecologia* (Berl.) **56**, 104–8.

Houston, D.B. (1982). "The Northern Yellowstone Elk." (Macmillan: New York.)

Hudson, R.J., and Nietfeld, M.T. (1985). *J. Range Manage.* **38**, 80–2.

Hungate, R.E. (1966.) "The Rumen and its Microbes." (Academic: New York.)

Imaizumi, E. (1981). *Japan Agric. Res. Quart. 15*, 50–7.

Jarman, P.J. (1968). Ph.D. Thesis. Univ. Manchester: England.

Jarman, P.J. (1974). *Behaviour* **48**, 215–67.

Jarman, P.J., and Jarman, M.V. (1974). In "The Behaviour of Ungulates and its Relation to Management." (Eds V. Geist and F. Walther.) pp.871–81. (IUCN: Morges, Switz.)

Jones, D.A., and Ramnani, A.D. (1985). *Evol. Theory* **7**, 143–8.
Kirkpatrick, R.L. (1980). In "Wildlife Management Techniques Manual." (Ed. S.C. Schemnitz.) (Wildl. Soc.: Washington, D.C.)
Kleiber, M. (1961). "The Fire of Life; an Introduction to Animal Energetics." (Wiley: New York.)
Klein, D.R., and Fairall, N. (1986). *J. Appl. Ecol.* **23**, 489–502.
Knight, R.R. (1970). Wildl. Monogr. No. 23.
Kreulen, D. (1975). *E. Afr. Wildl. J.* **13**, 297–304.
Kreulen, D.A., and Hoppe, P.P. (1979). *Afr. J. Ecol.* **17**, 53–63.
Lent, P.C. (1966). In "Environment of the Cape Thompson Region, Alaska." (Eds N.J. Wilimovsky and J.N. Wolfe.) (US AEC Div. Tech. Inf. Ext., Oak Ridge, Tenn.)
Levin, D.A. (1976). *Am. Nat.* **110**, 261–84.
Loudon, A.S.I., McNeilly, A.S., and Milne, J.A. (1983). *Nature* (Lond.) **302**, 145–7.
Maddock, L. (1979). In "Serengeti, Dynamics of an Ecosystem." (Eds A.R.E. Sinclair and M. Norton-Griffiths.) (Univ. Chicago: Chicago.)
McDonald, P., Edwards, R.A., and Greenhalgh, J.F.D. (1981). "Animal Nutrition." (Longman: London.)
McDowell, L.R. (1985). "Nutrition of Grazing Ruminants in Warm Climates." (Academic: New York.)
McNab, B.K. (1974). *Ohio J. Sci.* **74**, 370–80.
McNaughton, S.J. (1983). *Ecol. Monogr.* **53**, 291–320.
McNaughton, S.J., and Georgiadis, N.J. (1986). *Ann. Rev. Ecol. Syst.* **17**, 39–65.
Mould, E.D., and Robbins, C.T. (1981). *J. Wildl. Manage.* **45**, 323–34.
Muller-Schwarze, D., Stagge, B., and Muller-Schwarze, C. (1982). *Science* **215**, 85–7.
Nagy, J.G., and Tengerdy, R.P. (1968). *Appl. Microbiol.* **16**, 441–4.
Negus, N.C., and Berger, P.J. (1977). *Science* **196**, 1230–1.
Oh, H.K., Jones, M.B., and Longhurst, W.M. (1968). *Appl. Microbiol.* **16**, 39–44.
Oh, H.K., Sakai, T., Jones, M.B., and Longhurst, W.M. (1967). *Appl. Microbiol.* **15**, 777–84.
Owen-Smith, N. (1979). *S. Afr. J. Wildl. Res.* **9**, 102–10.
Owen-Smith, N. (1982). In "Ecology of Tropical Savannas." (Eds B.J. Huntley and B.H. Walker.) pp.359–404. (Springer-Verlag: Berlin.)
Owen-Smith, N., Cooper, S.M., and Novellie, P.A. (1983). *S. Afr. J. Anim. Sci.* **13**, 35–8.
Palo, R.T. (1985). *Oecologia* (Berl.) **68**, 10–4.
Parra, R. (1978). In "The Ecology of Arboreal Folivores." (Ed. G.G. Montgomery) (Smithsonian Inst.: Washington, D.C.)
Pyke, G.H. (1981). *Am. Nat.* **118**, 475–87.
Robbins, C.T. (1983). "Wildlife Feeding and Nutrition." (Academic: New York.)
Roe, F.G. (1951). "The North American Buffalo." (Univ. Toronto: Toronto.)

Scehovic, J., Poisson, C., and Gillet, M. (1985). *Agronomie* **5**, 347–54.
Shaver, T.N., Lukefahr, M.J., and Garcia, J.A. (1970). *J. Econ. Entomol.* **63**, 1544–6.
Shaw, G.G., and Reynolds, H.W. (1985). *Arctic* **38**, 61–4.
Skarzynski, B. (1933). *Nature* (Lond.) **131**, 766.
Skogland, T. (1983). *Oecologia* (Berl.) **60**, 156–68.
Slama, K. (1980). *Biochem. Physiol. Pflanzen* **175**, 177–93.
Smith, A.D. (1959). *J. Range Manage.* **12**, 8-13.
Spinage, C.A., Ryan, C., and Shedd, M. (1980). *Afr. J. Ecol.* **18**, 19–25.
Staines, B.W. (1974). *Mammal Rev.* **4**, 79-91.
Staines, B.W. (1976). *J. Zool.* (Lond.) **180**, 1-8.
Thomas, D.C. (1982). *Can. J. Zool.* **60**, 597-602.
Thomas, D.C., and Kroeger, P. (1980). *Arctic* **33**, 757–67.
Underwood, R. (1982). *J. Zool.* (Lond.) **196**, 191-205.
Van Hoven, W. (1984). *Can. J. Anim. Sci.* **64** (Suppl.), 177–8.
Vesey-Fitzgerald, D.F. (1960). *J. Mammal.* **41**, 161–72.
Watson, R.M. (1967). Ph.D. Thesis. Cambridge Univ., Cambridge.
Western, D. (1983). *Oecologia* (Berl.) **59**, 269–71.
White, R.G., Thomson, B.R., Skogland, T., Person, S.J., Russell, D.E., Holleman, D.F., and Luick, J.R. (1975). In "Ecological Investigations of the Tundra Biome in the Prudhoe Bay Region, Alaska." (Ed. J. Brown.) (Univ. Alaska: Fairbanks.)

18

SUPPLEMENTATION OF RUMINANT DIETS WITH FORAGE

by R. Elliott and N.P. McMeniman

For the purposes of this paper, supplementation is defined as the addition, on a daily basis, of a proportion of, in this case forage, to a basal diet. The supplement may be obtained by grazing or it may be a cut forage or a by-product from another industry.

Supplements, by definition, provide nutrients that are deficient in or not supplied in sufficient quantities by the basal diet and the beneficial effect of a supplement may be exerted by promoting increased intake and/or digestion and/or utilization of the basal diet. Alternatively, a basal diet may contain all of the essential nutrients for the animal but not in the quantities necessary to achieve the desired level of animal production. Under these circumstances a supplement would provide additional balanced nutrients, preferably without reducing intake of the basal diet.

In the following discussion attention will be focussed on the use of forages as supplements to various basal diets that are fed to cattle, sheep and goats.

SUPPLEMENTING GRASS DIETS

Nutrient limitations in pasture

Nutrients limiting ruminant production from tropical and subtropical pastures have been discussed by, amongst others, Mannetje (1982, 1984), Wilson (1982), Minson (1982), Norton (1982) and McDowell *et al.* (1984). As pastures mature, protein concentration falls and this can lead to deficiencies of rumen-fermentable nitrogen and intestinally digestible protein. Mature pastures usually have a low organic matter digestibility (OMD) due to increased lignification associated with increased stem to leaf ratios and are commonly deficient in a number of minerals, namely sulphur, phosphorus, sodium and in certain locations, calcium, zinc, iodine, cobalt and copper. Under tropical conditions many green pastures have low OMD as well as low concentrations of protein and minerals.

Grass supplements. During the dry season or during periods of drought, supplements of grass with a relatively higher nutrient content than the low quality native pasture can be fed in several ways.

Native pastures can be oversown with introduced species that, for a number of reasons, have a higher nutritive value than the native species. The introduced grasses could have longer growing periods, produce more DM /ha, have higher nutrient contents or retain their

NUTRITION OF HERBIVORES
ISBN 0127500529

nutrients at higher concentrations after growth has ceased. The usual aim of introducing a new grass species into a sward, however, is eventually to replace the native pasture.

Studies on a number of introduced grass species in tropical north Queensland (McIvor 1982) have shown that, when compared to native grasses growing in the region, the introduced species in general have a quicker growth response to early wet season and dry season rain. The introduced species also remain green for longer into the dry season and as a consequence their nutritive value stays higher. The stimulating effect that small quantities of green grass in a predominantly dry grass pasture have on liveweight and wool production of grazing sheep has been noted by Willoughby (1959). Mannetje (1974) showed that the amount of green pasture on offer and animal productivity are positively related.

The potential of introduced grass species may not be exhibited however, unless they are fertilised or grown in fertile soil (Ng 1972) and if grown on soils with sub-optimal fertility animal production may be no better than that obtained from native pastures grown on the same soils (Chen *et al*. 1981). Integration of native savanna grasses with small areas of improved pastures to which grazing cattle had continuous or restricted access has been evaluated in the Llanos Orientales region of Colombia. Such semi-intensive systems of pasture management, however, did not provide sufficient nutrients to improve the growth rate and fertility of breeding cattle (CIAT 1984).

One problem with improved grass species is that they may be more palatable than the native pasture and as a consequence be preferentially grazed and disappear from the pasture within a short period.

Under arid and semi-arid conditions pasture growth is frequently limited by lack of soil moisture rather than daily temperature, light or soil fertility. Strategies to improve soil moisture status could result in the provision of additional dry matter yields and in some instances, production of green forage outside the normal growing season.

Frequently, rainfall in arid and semi-arid tropical regions is intense and as a result a large proportion of the precipitated moisture runs off the soil surface. By spreading the run-off water from watercourses back onto areas of flat ground and holding it there by means of banks, the moisture status of the soil may be improved to enable growth of additional native or introduced pasture or in some instances crops. Investigations into such a waterspreading scheme in south-west Queensland have been reported by Pressland (1977). This study showed that waterspreading increased the yield of native pasture on a flooded area with a change in the composition of the sward towards species with a higher level of production (and probably moisture use). Satisfactory production of winter cereal crops was also possible. Batianoff and Burrows (1973), however, noted that waterspreading favoured the growth of useless woody species. Condon

(1985) stated that in western New South Wales waterspreading systems have significantly improved carrying capacity and financial returns on a number of properties.

An alternative approach that has been used in tropical north Queensland (Wilson *et al.* 1980) is to collect runoff water in a shallow catchment dam and then sow forage sorghum into the exposed soil as the water recedes. A one kilogram per head supplement of fresh sorghum to weaner sheep consuming a basal diet of dry Flinders grass (*Iseilema* spp.) pasture hay resulted in a daily weight gain of 10 g /head compared with a weight loss of 26 g /head in the unsupplemented group. The authors considered that as the yield of forage sorghum was 5 tonnes /ha, an area of 4–5 ha should be sufficient to provide a flock of 1000 weaners with a daily supplement of 0.5 kg /head for a 6–8 week period in winter when pasture quality is low.

Shallow water storage areas can also be used to extend the growth into the dry season of productive wetland grasses. Wilden (1985) reported that in central Queensland, Paragrass (*Brachiaria mutica*) has been successfully grown in ponded areas on a number of properties with resultant increases in cattle production.

In higher rainfall areas forage crops can be grown in most years and the crops used as supplements. In coastal tropical and subtropical environments pasture quality, especially during winter and spring, can limit milk production of dairy cows. Royal and Hughes (1976) supplemented dairy cows grazing subtropical Kikuyu pastures during winter with forage oats (*Avena sativa*), Italian ryegrass (*Lolium multiflorum*), additional Kikuyu grass (*Pennisetum clandestinum*), chopped sugar cane or forage lupins (*Lupinus luteus*). The sugar cane was green chopped and fed in troughs whereas the forages were strip grazed. All supplements were fed at the daily rate of 7 kg DM /cow. The results showed that milk production was strongly correlated with digestibility of the supplements, the temperate species oats, ryegrass and lupins giving the highest production.

Conservation of forage as hay or silage is common practice in temperate and sub-tropical areas. The processes involved and nutritive value of the products have been reviewed by Jarrige *et al.* (1982). Although silage or hay may constitute the principal component of ruminant diets during critical periods of the year, these products have also been used as supplements to grazing animals. The benefits accrued by feeding hay in this manner are subject to many variables including weather conditions, stocking rates, quality of hay. Early work by Hamilton *et al.* (1972) showed that at high stocking rates, livestock productivity could be penalised by excessive confinement of animals during the period of fodder conservation. Birrell and Bishop (1980) found that, at low stocking rates, conservation of excess spring pasture as hay increased wool growth simply by improved pasture conditioning following hay removal (the hay made was never fed to the animals). Similarly Scattini (1984) and Stephenson *et al.* (1985) have observed that animal production from pastures that have previously been harvested for the production of pasture hay can be superior to that obtained from pastures that have not been harvested. Presumably the higher quality forage produced from the limited areas subsequent to hay-making acted as supplement to the hay diet. Recently Birrell (1984) found that hay consumption by animals at pasture was to some extent dictated by animal liveweight; heavy animals in good condition preferentially consumed hay while lighter animals consumed the higher quality pasture. Saul (1984) reported that weight gain in heifers at pasture was significantly improved by *ad libitum* access to hay, but emphasized that although this was common practice throughout Victoria, the results depend on the quality of the hay being fed (Flinn and Saul 1983). Pasture hay offered to grazing dairy heifers has been reported to compensate adequately for reduced pasture availability during autumn resulting in both improved milk production and body condition during late lactation. In a study of dairy cows, hay supplementation was found not to influence selection of pasture although the degree of pasture substitution caused by hay supplementation was found to depend on the quality of pasture available (Eldridge and Kat 1980).

With the recent availability of machinery capable of producing large (500 kg) round bales, there is now considerable interest in subtropical and tropical Australia in conserving excess native pasture to be used as a supplementary feed source during ensueing periods when standing dry pasture has a lower nutritive value or is in short supply. The harvesting of native pasture hay usually has to be delayed until the pasture is in an advanced stage of maturity. Earlier harvesting may be impractical due to the difficulty of moving machinery over wet ground or to low yields. As a result of late harvesting the nutrient content of native pasture hay is relatively low. Stephenson *et al.* (1986) reported that the average nitrogen and cell wall content of hays obtained from three different sites in western Queensland were 10 g and 750 g /kg DM, respectively; weaner sheep fed these hays as the sole component of the diet barely maintained or lost liveweight. However, the nutrient content of hay can be slightly higher than that of the standing feed available later in the dry season. Sturtz and Parker (1974) found that, although the crude protein content of a previously harvested Townsville stylo/native pasture hay was not different from that of dry pasture on an area adjacent to where the hay had been harvested, the digestibility of the hay was superior. This hay grown as a supplement to the native pasture promoted increased DM intake and liveweight gain in cattle.

Stephenson and Hopkins (1985) reported that the nitrogen content, dry matter digestibility, and intake of native Flinders grass hay conserved in 500 kg round bales can be improved by adding a solution containing urea and molasses (0.02 urea and 0.06 molasses w/w of hay). The solution was injected into the hay bales by means of a spear. When fed alone, hay treated in this way promoted slight liveweight gain in weaner sheep compared to significant liveweight loss in animals eating the untreated hay.

Little work has been conducted on the use of silages as supplements except for dairy production. Silages made from tropical grass and legume pastures cannot in general support high milk yields because of their low digestible energy content (Esperance and Guerra 1978; Hamilton *et al.* 1978). Davison *et al.* (1982) evaluated the effect of maize silage supplements on milk yield in dairy cows grazing basically green panic/glycine pastures in North Queensland and found that although significant increases in milk production were obtained, yields were further increased by protein supplementation. A significant interaction between maize silage supplementation and protein intake on milk production was also reported by Moran *et al.* (1986).

Legume supplements. The effects of oversowing native tropical pastures with legumes or of providing areas of pure legume adjacent to native pasture on liveweight gain and reproductive performance of cattle have been reviewed by Paladines (1984). In summary, oversowing with legumes and applying a phosphorus fertiliser results in increases in animal production in the order of 200 to 400%. If the legumes are not fertilised, however, the level of production is at best 150% of that obtained with native pastures.

Browse supplements. The nutrient content of a large number of browse species has been reported although the list is far from being exhaustive. These data are mainly presented as proximate analyses of browse leaves or pods. Survey of this information indicates that, compared with dry grass, browse species have higher concentrations of crude protein and minerals and lower crude fibre concentrations.

In vivo dry matter, organic matter and nitrogen digestibility values that are available for browse are presented in Table 1. Except for the chenopodiaceous species, the majority of browse types that have been examined have low dry matter and organic matter digestibilities. Wilson (1977) has shown that the low organic matter digestibilities of the leaves of the four Australian browse trees, *Acacia pendula*, *Casuarina cristata*, *Heterodendrum oleifolium* and *Geijera parviflora* are associated with the depressed digestion of the fibre fraction of their leaves, in particular the acid detergent fibre.

Nitrogen digestibility values are available for a smaller number of species; again, apart from the chenopodiaceous species, apparent digestibility is low. True nitrogen digestibilities of the species examined by Wilson (1977), while still being considerably lower than values recorded for green forages, were higher than the corresponding apparent digestibility figures. For example, while the apparent nitrogen digestibility of *H. oleifolium* leaves fed to sheep was 0.14 the true digestibility was 0.66 and Wilson ascribed this difference to the low organic matter digestibilities of the browse leaves. It is known that the low nitrogen digestibility of the leaves of one browse species, mulga (*Acacia aneura*), is due, at least in part, to its tannin content (Gartner and Hurwood 1976). Daily dosing of mulga-fed sheep with olyethylene glycol (PEG) at the rate of 8 g /head to displace tannins from protein-tannin complexes improved nitrogen digestibility (Pritchard *et al.* 1985); subsequent experiments have shown that dosing with PEG at higher levels improves both nitrogen and organic matter digestibilities (Pritchard unpublished). The presence of tannin, however, does not always mean that protein digestibility will be low. Although leucaena (*Leucaena leucocephala*) is reported to contain 20 g tannin /kg DM (NAS 1977) it has reasonably high apparent nitrogen digestibility (Table 1). It is possible that tannins in low concentrations enhance the flow of undigested food protein to the small intestines where the amino acids then become available to the animal.

Table 1. The apparent *in vivo* digestibility of browse.

Browse species	Digestibility % DM	OM	N	Source
Acacia aneura	47-64	47-54	36-63	2,3,5,9
Acacia pendula	43	63	4	
Anthyllis vulneraria		57	36	7
Arbutus unedo	75	7		
Artemisia herba alba		69		7
Atalaya hemiglauca	36			8
Atriplex glauca		68		7
Atriplex nummularia	68-74	65-67	78-83	1,4
Atriplex semibaccata		60		7
Atriplex vesicaria	52-54	44-53	71-74	1,4
Brachychiton populneum	62	63	76	2
Casuarina cristata		29	22	4
Ceratonia siliqua		90		7
Geijera parviflora		60		4
Heterodendrum oleifolium		40	14	4
Leucaena leucocephala		71	78	6
Maireana pyramidata		58	55	4
Olea europaea oleaster		44		7
Opuntia ficus indica		60		7
Phillyrea angustiflora		51		7
Pistacia lentiscus		37		7
Prosopis juliflora		76		7
Sclerolaena diacantha		41	49	4
Ventilago viminalis	50			8

Sources: 1. Wilson (1966); 2. Norton *et al.* (1972); 3. McMeniman (1976); 4. Wilson (1977) 5. Gartner and Niven (1978); 6. Jones (1979); 7. Le Houerou (1980b); 8. Wilson and Harrington (1980) 9. McMeniman *et al.* (1981).

In vitro digestibility (OM and DM) of a range of browse species has been reported by McLeod (1973), Wilson (1977), and McDonald and Ternouth (1979). This information must be viewed with a certain amount of caution however, as Wilson (1977) has shown that *in vivo* and *in vitro* estimates of DM digestibility are not always well correlated. Indirect estimates of digestibility based on chemical analyses for a number of species are available. Except for values derived from chemical analyses, present *in vitro* techniques provide no measures of protein digestibility.

Bamualim *et al.* (1980) determined the nylon bag dry matter digestibility (NBDMD) of the leaves and stems of 24 tropical browse species and found that NBDMD was most closely correlated to the lignin content of the samples ($r = -0.924$).

There is little information available on the quantities of browse eaten. Under pen feeding conditions it has been demonstrated that voluntary intake of mulga leaves by sheep can be increased by supplementing the diet with molasses (or sulphur), phosphorus and protein; there is no response to phosphorus supplementation unless molasses (or sulphur) is also provided (McMeniman and Little 1974; Entwistle and Baird 1976; McMeniman 1976; Gartner and Niven 1978; McMeniman *et al.* 1981). These findings show that this browse itself has nutrient deficiencies that limit its intake. In another series of pen studies, Wilson (1977) showed that neither organic matter nor nitrogen digestibility of four browse shrub and four browse tree species was associated with their voluntary intake. These results indicate that for browse species, digestibility is not necessarily associated with voluntary intake; however, factors that influence palatability of browse have not been extensively studied.

Various authors have recorded observational evidence detailing browse species that are consumed (Everist 1969; Lamprey *et al.* 1980; Le Houerou 1980a,b). In general, it can be stated that, if browse is available, goats will usually consume significant quantities but when pasture conditions are good, browse intake by cattle and sheep is not extensive. Deteriorating pasture conditions stimulate all three species to consume more browse if it is available.

With the exceptions of leucaena and mulga, it is not known whether browse species fulfil the criteria for a supplement as defined previously. Bamualim *et al.* (1984b) have shown that leucaena provides supplementary protein when fed with a basal ration of black spear grass (*Heteropogon contortus*). However, preliminary results (Goodchild and McMeniman 1986) show that including mulga in graded amounts in a low quality sorghum stubble diet, while increasing total DM intake up to an inclusion level of 0.27, had no influence on stubble intake; dry matter digestibilities of the diets were inversely related to the level of inclusion of mulga. Presumably these responses to the mulga supplement were associated with its tannin content.

Comparisons of the production responses of grazing animals to the inclusion of browse in the diet are few. Jones (1979), in reviewing the use of leucaena for ruminants in the tropics, has shown that given the right conditions of soil and climate, leucaena supplementation can improve liveweight gain in grazing cattle. In Australia, however, intensive grazing of leucaena has previously produced adverse effects on reproductive performance and weight gain in beef animals. These effects were attributed to the presence of the non-protein amino acid mimosine and its goitrogenic breakdown product dihydroxypyridone (DHP). Recently, the introduction into Australia of rumen bacteria capable of totally degrading DHP has alleviated this problem (Jones and Lowry 1984). In a large scale grazing trial in southeast Queensland, control cattle (without introduced bacteria) maintained on a sole diet of leucaena gained 0.83 kg /d whereas inoculated steers gained 1.11 kg /d. When leucaena comprised less than 0.30 of the diet performance of uninoculated animals was equal to that of inoculated animals (Anon 1985). In general, if cattle are allowed access to leucaena planted in rows 2-4 metres apart or allowed restricted access to blocks of the legume there are no adverse effects on reproductive performance.

In view of the range of browse and tree species available as animal feed and the dependence of animals in many regions on such materials during certain critical times of the year it is important to maintain screening procedures to identify possible toxins. The insidious nature of subclinical toxicity, especially on reproductive performance, is particularly damaging to long-term animal productivity.

The effect of including *Gliricidia maculata* in the diet of rams (Kantharaju and Chadhokar 1981) and ewes (Chadhokar and Kantharaju 1980) fed a poor quality grass diet was to increase weight gains, wool growth and reproductive performance; the authors concluded that the optimum rate of inclusion of *Gliricidia* to rectify the protein deficiency in the grass was 0.20-0.25. At inclusion levels above this further substitution of *Gliricidia* for the grass was evident. When *Gliricidia* was included in mature grass diets at a level of 0.25 or 0.50, milk production of dairy cows was improved (Chadhokar and Lecamwasam 1982).

The value of the browse shrubs *Kochia aphylla, Atriplex vesicaria* and *A. nummullaria* as supplements to sheep grazing pastures on the Riverine plain in southern Australia has been examined by Wilson (1966) and Leigh *et al.* (1970). The presence of the shrubs in the sward had no effect on reducing seasonal fluctuations in wool growth and the value of the shrubs lay in their ability to provide nutrients during drought years. A preliminary report from an American study, however, (Otsyina *et al.* 1982) suggested that a crested wheatgrass (*Agropyron cristatum*)/saltbush (*Atriplex canescens*) association maintained the liveweight of sheep better than a similar association without the saltbush. Clearly much more work is required to determine whether the numerous browse species that are available have any beneficial effect when fed with low quality forage diets.

SUPPLEMENTING OTHER FORAGE DIETS

The supplementation of sugar cane

The concept of using sugar cane as an animal feed has changed in recent years from that of an emergency measure in times of drought to its use as a feed lot ration for growing/fattening steers (Preston 1977).

One major constraint in the feeding of sugar cane is that it must first be chopped. Although chopping of the stem fraction to particle sizes of 5-30 mm does not significantly affect animal performance (Silvestre *et al.* 1976; Montpellier *et al.* 1977), there is a need to include a second fibre type in the diet to stimulate voluntary intake of cane stalk and hence liveweight gain. Ferreiro and Preston (1977) found that optimal intake of cane stalk was achieved by including cane tops in the diet in the ratio 25:75 or 30:70, a ratio which encompasses that normally observed in the plant. Not only was the proportion of tops to stalk important in maintaining adequate intake but the particle size of the tops was critical. Large particles (5-20 cm in length) promoted good intake whereas grinding the tops to less than 10 mm depressed digestibility and dry matter intake (Salais *et al.* 1977).

When cane is grown under extreme dry season conditions the proportion of tops in the plant can be low and in this situation restricted pasture grazing (2-3 hours /d) would maximise use of the chopped cane (Preston 1977).

The need for an extra fibre source in diets of chopped cane stalk, coupled with a high requirement for rumen-available nitrogen on such a diet (Ferreiro *et al.* 1977) led to the use of restricted grazing of protein-rich legumes; the use of leucaena, particularly in beef breeding operations, has been particularly successful (Alvarez and Preston 1976). Subsequent studies have evaluated the production responses to other protein-rich forage supplements such as cassava tops (Meyreles *et al.* 1977a,b), banana foliage (Ffoulkes and Preston 1978a, 1979) and sweet potato tops (Meyreles *et al.* 1979). Agronomic studies on these crops indicate that with careful management, removal of material for animal feed does not seriously affect production of tubers or fruit for human consumption (Ruiz *et al.* 1980; Pound and Fernandez 1981).

Attempts have been made to store excess high protein forages such as cassava or sweet potato tops as silage (Ruiz *et al.* 1981); however such materials have high buffering capacity and low sugar contents, characteristics which are not inducive to consistent silage manufacture (Waldo 1984). Sugar cane tops rapidly become unpalatable when dry. However, they can easily be ensiled, thus extending the availability of forage (Deville *et al.* 1979).

In most of the feeding experiments where high protein forages were fed as supplements to sugar cane diets, partial substitution of the supplements for the basal diets was observed. Total feed intakes however were greater in supplemented animals and consequently animal performance was improved.

The supplementation of sisal wastes

There is potential for ruminant production from the pulp and bagasse produced as by-products of the sisal (*Agave fourcroides*, *Agave sisalana*) industries. The wastes contain 800-900 g moisture and 8-10 g N /kg DM and have a pH of 3.8-4.1. Sisal production is well established in many countries which have a pronounced dry season (Central America, East Africa, Indonesia etc.) but the by-products arising from sisal fibre production are not extensively utilised.

Sisal wastes are very unstable and ferment rapidly to produce unpalatable material but, by utilising silage-making techniques, an acceptable conserved forage can be obtained (Godoy *et al.* 1979). However voluntary feed intakes (VFI) of both fresh and ensiled by-products are usually low (less than 20 g DM /kg liveweight) and supplementation with urea and protein meals is required to increase VFI.

Studies with cattle (Ferreiro *et al.* 1978) showed that VFI of sisal pulp was significantly increased when animals had access to pasture. This effect was confirmed in a pen study (Gutierrez *et al.* 1981) in which the influence of low quality African star grass (*Cynodon plectostachyus*) supplements on rumen fibre digestion and VFI was measured in sheep. Inclusion of African star grass at a level of 0.25 of the total diet increased rumen cellulose digestibility from 0.40 to 0.75 and VFI increased by 0.80.

Increased intakes of sisal pulp have not been reported in cattle fed supplements of the high protein browse forages leucaena (Priego *et al.* 1979a; Herrera *et al.* 1980; Dixon *et al.* 1981) or *Brosimum alicastrum* (Priego *et al.* 1979b; Ferreiro *et al.* 1979). Recently, however, Rodriguez (1983) reported that chopped *Brosimum* foliage fed at approximately 0.25 of DMI to sheep offered *ad libitum* sisal waste significantly increased intake of the basal diet. It appears though that the basal diet was not supplemented with rumen-available nitrogen and the *Brosimum* foliage (250g /kg CP) may have stimulated intake by providing this limiting nutrient.

It is possible that on high moisture feeds such as sisal wastes, the presence of dry grass or straw fulfils roles other than that of providing extra nitrogen or minerals. In this regard, supplements of hay and straw have significantly increased digestible energy intake of root crops (Drew 1967; Valderrabano *et al.* 1986). However, Mader *et al.* (1983) and Mader and Horn (1986) failed to obtain improvements in animal performance when steers were fed wheat straw supplements to a basal diet of high moisture wheat pasture.

The supplementation of straws

Cereal crop residues comprise an abundant feedstock for ruminants throughout the world. As with low-quality mature pasture grasses, utilisation of such materials is restricted by their high content of lignocellulosic material and low level of nitrogen. Although treatment of such materials with alkalis and/or acids to improve digestibility and intake has proved successful (Sundstol and Owen 1984) such chemicals are expensive and not readily available in many rural communities. It is in these areas, however, that improved utilisation of straws would be most beneficial. In South East Asia, for example, the imbalance between increasing animal populations and the availability of grazing lands has focussed attention on better utilisation of rice straw and a recent publication has summarised the results of considerable research into the use of this material as a ruminant feed (Doyle *et al.* 1986). These authors suggest that future investigations into rice straw utilisation be focussed on the potential of green forages as supplements.

In this regard, attention is immediately focussed on the use of forage and browse legumes and examples of possible legume production systems have been described by Shelton (1980) and Gutteridge (1981). However, subsistence farmers will not grow forage legumes in place of food crops simply to supply food for livestock. One way of increasing the availability of legume residues could be to intensify cropping using such dual purpose crops as mung bean (*Vigna radiata*), cowpea (*V. unguiculata*), lablab bean (*Lablab purpureus*), broad bean (*Vicia faba*), etc. Not only would increased production yield more human food and straws for animal feeding, but the nitrogen fixed by rhizobia would be available for subsequent cereal crops. Recently McMeniman *et al.* (1987) reported on the influence of legume straw supplementation on the intake and digestion of rice straw by sheep. The legume straws evaluated were mung bean, cowpea, peanut (*Arachis hypogaea*) and pigeonpea (*Cajanus cajan*); lucerne was also fed as a supplement. In this experiment all forms of legume supplementation increased total dry matter intake but partial substitution of rice straw was recorded in sheep fed pigeonpea, cowpea and peanut straws; intake of rice straw was not influenced by mung bean straw or lucerne supplements. Nylon bag studies showed that the rate of rice straw digestion was increased by all of the legume sources and that the extent of digestion of the leaf and stem fractions of the legumes was superior for the mung bean and lucerne; it has been known for some time that extracts or ash fractions of lucerne stimulate fibre digestion (Shrewsbury *et al.* 1943; Swift *et al.* 1951). It could be argued, however, that DM intakes recorded by McMeniman *et al.* (1987) were highest in sheep fed the lucerne and mung bean straw simply because these supplements not only provided essential nutrients to optimise ruminal activity but also were themselves very degradable in the rumen. The more slowly degradable straws with a high proportion of ADF, although contributing certain factors to assist rumen activity, took considerably longer to break down and hence reduced rice straw intake.

Supplementing molasses diets

Molasses is widely used both as a drought supplement and as the basal energy source for feedlot cattle production. Since the introduction to Cuba of intensive beef production systems based on molasses/urea diets during the late sixties the importance of providing forage supplements to these diets has been widely appreciated (Munoz *et al.* 1970). Adequate forage intakes are important not only for stimulating molasses intake (Clark *et al.* 1972) but also for preventing molasses toxicity (Losada *et al.* 1971; Preston 1972).

Geerkin and Sutherland (1969) observed low rumen turnover rates in cattle fed molasses diets and fibre supplements were required to stimulate rumen outflow. At high levels of forage supplementation, however, total DM intakes were reduced, probably due to a decreased digestibility of the entire diet (Clark *et al.* 1972). Optimal levels of forage supplementation (fresh basis) of molasses diets range from 0.015–0.02 of animals liveweight (equivalent to 0.005–0.008 of liveweight when dry straws were fed) (Peron and Preston 1971; Losada and Preston 1973).

Martin (1971) fed bulls a molasses/urea diet supplemented with either elephant grass (*Pennisetum purpureum*), sugar cane bagasse, rice straw or rice husks at the low level of 0.4 kg DM /100 kg liveweight; molasses intakes (kg /d) were 7.5, 7.3, 6.8 and 6.1 and liveweight gains were 0.94, 0.79, 0.73 and 0.58 kg /d, respectively. There were five cases of molasses toxicity, one in animals fed the sugar cane bagasse and four in those fed rice husks. Both of these supplements are unpalatable to cattle, particularly the rice husks and any reduction in intake of the supplement would predispose the animal to molasses toxicity.

Complete withdrawal of forage from a molasses based diet fed to cattle usually results in molasses toxicity within seven days (Losada *et al.* 1971; Rowe *et al.* 1980). However, no toxic signs appeared in cattle fed up to 6 kg molasses /day without forage for 6 months (Gulbransen 1984).

Edwin *et al.* (1979) found that the symptoms of molasses toxicity were similar to those associated with the disease known in Europe as cerebrocorticonecrosis and in the USA as polioencephalomalacia, both of which are due to a deficiency of thiamine. Mella *et al.* (1976) suggested that in animals fed molasses/urea diets, low rumen turnover rates could stimulate excessive production of thiaminase-producing bacteria and that low rates of digesta flow from the rumen would limit the quantity of thiamine available for absorption in the small intestine. However, Calderon Cortes (1987) found that rumen thiaminase activity was not related to dietary fibre levels in cattle fed *ad libitum* molasses/urea diets.

The need for feed proteins to escape degradation in the rumen of cattle fed molasses diets is well documented (Preston and Willis 1974). In this context there is considerable interest in the use of high protein forages to provide both fibre and protein on such diets, in particular, cassava foliage, sweet potato tops or leucaena forage (Ffoulkes and Preston 1978b; Meyreles *et al.* 1982). Apart from the nutritional advantages of using these materials, the highly palatable nature of such forages should ensure the absence of molasses toxicity.

THE ROLE OF FORAGE SUPPLEMENTS

The net effect of adding forage supplements to many basal diets is to increase the metabolisable energy intake by the animal. The legume straws fed as supplements to a basal diet of rice straw by McMeniman *et al.* (1987) doubled ME intakes. Similar large increases in ME intakes were reported by Mosi and Butterworth (1985) who fed combinations of cereal crop residues with *Trifolium tembense* hay. The increases in ME intake were similar, if not superior, to those expected following chemical treatment of straws with strong alkalis. In this regard one overlooked attribute of green forages is their relatively high level of lipid material contained in the chloroplasts.

Preston and Leng (1984) have listed rumen-available nitrogen as first priority in a supplement in order to stimulate digestion of low quality roughages but there is still much debate concerning the optimum concentrations of rumen ammonia for stimulating microbial activity. Many studies have shown that adequate rumen ammonia concentrations should not be less than 70–80 mg /l (Satter and Slyter 1974; Elliott *et al.* 1984). However, Krebs and Leng (1984) have suggested that levels above 150 mg /l should be maintained throughout the day. This latter recommendation, however, arose from studies involving the use of molasses feed blocks as supplements to straw diets and it is probable that the presence of molasses increases the requirement for rumen-available nitrogen.

It is generally believed that a sustained degradation of protein in the rumen leads to a more efficient utilisation of the ammonia produced; the slow continuous ammonia production matching the slow degradation of the structural polysaccharides in the plant fibres. Aii and Stobbs (1980) showed that there is considerable variation in the rate at which the nitrogen in legumes and grasses is digested in the rumen. In this context condensed tannins released from plant materials during chewing may bind feed proteins (Jones *et al.* 1973) and limit the digestion and assimilation of these nutrients (Barry and Duncan 1984). McLeod (1974), concluded that plant tannins appeared not to be detrimental to the health of the ruminant although Hegarty *et al.* (1985) have pointed out that phenolic compounds in plants have been associated with animal deaths both in Australia and the USA. High levels of tannins in some browse species (e.g. *Calliandra*) may also act as a feeding deterrent (Brewbaker 1985).

Sulphur adequacy is also important in maintaining rumen digestion of plant fibre (Kennedy and Siebert 1972; Bird 1974) and factors affecting the availability of this element parallel those affecting nitrogen metabolism. Bray and Till (1975) recommended that rumen sulphide concentrations should be maintained above 10 mg /l but the results of McMeniman *et al.* (1987) would suggest that concentrations as low as 0.5 mg /l were sufficient for optimal digestion of rice straw. Although rumen sulphide levels are important for adequate microbial protein production in the rumen (Elliott and Armstrong 1982) sulphur also stimulates the growth and development of rumen fungi (Akin and Hogan 1982). The vegetative growth of rumen fungi decreases the tensile strength of plant fibre (Akin *et al.* 1983) and although they are not capable of degrading lignin and pectins they do possess active glycosides which degrade cellulose and hemicellulose to simple sugars (Bauchop 1984). Supplementation of low-quality diets with protein-rich forages will provide extra sulphur to the rumen environment and this may stimulate greater fungal colonisation leading to improved fibre digestion. Supplementation of forage diets with inorganic sulphur sources may not be as effective as organic sources in promoting the growth of fungi (Gordon and Ashes 1984). The leaves of the browse legumes leucaena and gliricidia support large populations of rumen fungi whereas mulga leaves, which contain substantial quantities of protein-binding tannins, support low fungal populations (V. Rudwick, B.W. Norton and R. Elliott, unpublished).

Enhanced feed intake in animals fed low-quality roughages has also been induced by increasing the supply of amino acids to the tissues (Weston 1971; Egan 1977). In contrast to the considerable information available on the extent to which feed proteins escape rumen degradation in animals fed temperate forages there are only few reports in which this parameter has been measured in animals fed tropical feeds. Bamualim *et al.* (1984b) reported substantial quantities of feed protein leaving the rumen of sheep when supplements of dried leucaena was added to a basal diet of black spear grass (*Heteropogon contortus*). These increases in protein flows were associated with an 11% increase in spear grass intake. Feeding fresh leucaena as a supplement to goats also resulted in significant quantities of leucaena protein leaving the rumen and improved intake of spear grass (Bamualim *et al.* 1984a). In general, however, leucaena supplementation of rice straw diets (Moran *et al.* 1983; Devendra 1983), chopped sugar cane (Alvarez and Preston 1976), sisal wastes (Priego *et al.* 1979a; Herrera *et al.* 1980) or native grass diets (Wahyuni *et al.* 1982) has failed to increase intake of the basal diet. Interestingly, in the study of Moran *et al.* (1983), a comparison was made between untreated and alkali-treated rice straw, supplemented with leucaena. Intake in cattle fed the treated straw increased by 20% but no increases in DMI were observed in cattle offered untreated material.

With the exception of leucaena and mulga the extent to which proteins in other legumes and legume straws (and other high protein forages) escape rumen fermentation is not known. Sun drying of cut legume forage has been shown to increase the quantity of protein escaping rumen fermentation (Nolan and Leng 1972) and this process may be of some use in "cut and carry" management practices to decrease protein solubility in the rumen.

Forage supplements which contain significant quantities of soluble nutrients could also influence the rate of fibre digestion in the rumen by stimulating colonisation of plant materials by rumen microflora (Orpin and Bountiff 1978). Bauchop (1981) noted that the extent of colonisation of wheat straw by microfungi varied in sheep fed different diets and early work by McManus *et al.* (1972) showed that the digestibility of test substances in nylon bags was affected by the type of roughage fed to the animal. The stimulatory effect of a digestible fibre type on straw degradation has also been shown by Silva and Ørskov (1985). This area of research warrants greater attention.

Aii, T., and Stobbs, T.H. (1980). *Anim. Feed Sci. Technol.* **5**, 183-92.

Akin, D.E., and Hogan, J.P. (1982). *Proc. Aust. Soc. Anim. Prod.* **14**, 680.

Akin, D.E., Gordon, G.L.R., and Hogan, J.P. (1983). *Appl. Environ. Microbiol.* **46**, 738-48.

Alvarez, F.J., and Preston, T.R. (1976). *Trop. Anim. Prod.* **1**, 112-9.

Anon, (1985). In "Brian Pastures Annual Report." Qld. Dept. Prim. Ind. p.15.

Bamualim, A., Jones, R.J., and Murray, R.M. (1980). *Proc. Aust. Soc. Anim. Prod.* **13**, 229-32.

Bamualim, A., Stachiw, S., Jones, R.J., and Murray, R.M. (1984a). *Proc. Aust. Soc. Anim. Prod.* **15**, 259-62.

Bamualim, A., Weston, R.H., Hogan, J.P., and Murray, R.M. (1984b). *Proc. Aust. Soc. Anim. Prod.* **15**, 255-8.

Barry, T.W., and Duncan, S.J. (1984). *Br. J. Nutr.* **51**, 485-91.

Batianoff, G.N., and Burrows, W.H. (1973). *Qld. J. Agric. Anim. Sci.* **30**, 65-71.

Bauchop, T. (1981). *Agric. Environ.* **6**, 333-48.

Bauchop, T. (1984). In "Reviews in Rural Science 6. Biotechnology and Recombinant DNA Technology in the Animal Production Industries." (Eds R.A. Leng, J.S.F. Barker, D.B. Adams and K.J. Hutchinson.) pp. 118-23. (Univ. New England: Armidale, NSW.)

Bird, P.R. (1974). *Aust. J. Agric. Res.* **25**, 631-42.

Birrell, H.A. (1984). *Aust. J. Exp. Agric. Anim. Husb.* **24**, 26-33.

Birrell, H.A., and Bishop, A.H. (1980). *Aust. J. Exp. Agric. Anim. Husb.* **20**, 406-12.

Bray, A.C., and Till, A.R. (1975). In "Digestion and Metabolism in the Ruminant." (Eds I.W. McDonald and A.C.I. Warner.) pp. 243-60. (Univ. New England: Armidale, NSW.)

Brewbaker, J.L. (1985). In "Forages in Southeast Asia and South Pacific Agriculture." (Eds G.J. Blair, D.A. Ivory and T.R. Evans.) pp. 43-50. ACIAR Proceedings Series No. 12, Canberra.

Calderon Cortes, F.J. (1987). PhD Thesis, University of Queensland.

Centro Internacional de Agricultura Tropical - CIAT. (1984). Tropical Pastures, Annual Report 1984. pp. 221-5.

Chadhokar, P.A., and Kantharaju, H.R. (1980). *Trop. Grassl.* **14**, 78-82.

Chadhokar, P.A., and Lecamwasam, A. (1982). *Trop. Grassl.* **16**, 46-8.

Chen, C.P., Ajit, S.S., and Evans, T.T. (1981). *MARDI Res. Bull.* **9**, 207–24.

Clark, J., Preston, T.R., and Zamora, A. (1972). *Cuban J. Agric. Sci.* **6**, 19–25.

Condon, R.W. (1985). Proc. 15th Int. Grassl. Congr., pp. 1305–6. Kyoto, Japan.

Davison, T.M., Marschke, R.J., and Brown, G.W. (1982). *Aust. J. Exp. Agric. Anim. Husb.* **22**, 147–54.

Devendra, C. (1983). *MARDI Res. Bull.* **11**, 272–90.

Deville, J., Wong You Cheong, Y., Leclezio, P., and Duvivier, P. (1979). *Trop. Anim. Prod.* **4**, 134–7.

Dixon, R.M., Priego, A., Wyllie, D., and Preston, T.R. (1981). *Trop. Anim. Prod.* **6**, 37–43.

Doyle, P.T., Devendra, C., and Pearce, G.R. (1986). "Rice Straw as a Feed for Ruminants." (IDP: Canberra.)

Drew, K.R. (1967). *Proc. N.Z. Soc. Anim. Prod.* **27**, 210–22.

Edwin, E.E., Markson, L.M., Shreeve, J., Jackman, R., and Carroll, P.J. (1979). *Vet. Rec.* **104**, 4–8.

Egan, A.R. (1977). *Aust. J. Agric. Res.* **28**, 907–15.

Eldridge, G.A., and Kat, C. (1980). *Proc. Aust. Soc. Anim. Prod.* **13**, 245–8.

Elliott, R., and Armstrong, D.G. (1982). *J. Agric. Sci..* **99**, 51–61.

Elliott, R., McMeniman, N.P., Norton, B.W., and Calderon Cortes, F.J. (1984). *Proc. Aust. Soc. Anim. Prod.* **15**, 337–40.

Entwistle, K.W., and Baird, D.A. (1976). *Aust. J. Exp. Agric. Anim. Husb.* **16**, 174–80.

Esperance, M., and Guerra, A. (1978). *Cuban J. Agric. Sci.* **12**, 131–41.

Everist, S.L. (1969). "Use of Fodder Trees and Shrubs." Qld. Dept. Prim. Ind. Div. Plant Ind. Adv. Leaflet No. 1024.

Ferreiro, H.M., and Preston, T.R. (1977). *Trop. Anim. Prod.* **2**, 90–100.

Ferreiro, H.M., Elliott, R., Rios, V., and Preston, T.R. (1978). *Trop. Anim. Prod.* **3**, 69–73.

Ferreiro, H.M., Preston, T.R., and Herrera, F. (1979). *Trop. Anim. Prod.* **4**, 73–7.

Ferreiro, H.M., Sutherland, T.M., and Preston, T.R. (1977). *Trop. Anim. Prod.* **2**, 213–8.

Ffoulkes, D., and Preston, T.R. (1978a). *Trop. Anim. Prod.* **3**, 125–9.

Ffoulkes, D., and Preston, T.R. (1978b). *Trop. Anim. Prod.* **3**, 186–92.

Ffoulkes, D., and Preston, T.R. (1979). *Trop. Anim. Prod.* **4**, 37–41.

Flinn, P.C., and Saul, G.R. (1983). In "Feed Information and Animal Production." (Eds G.E. Robards and R.G. Packham.) pp. 259–63. (CAB: Farnham Royal, UK.)

Gartner, R.J.W., and Hurwood, I.S. (1976). *Aust. Vet. J.* **52**, 194–6.

Gartner, R.J.W., and Niven, D.R. (1978). *Aust. J. Exp. Agric. Anim. Husb.* **18**, 768–72.

Geerken, C.M., and Sutherland, T.M. (1969). *Cuban J. Agric. Sci.* **3**, 217–22.

Godoy, R., Herrera, F., Carcano, R., Elliott, R., and Preston, T.R. (1979). *Trop. Anim. Prod.* **4**, 268–75.
Goodchild, A.V., and McMeniman, N.P. (1986). In "Australian Acacias in Tropical Countries." ACIAR (In press).
Gordon, G.L.R., and Ashes, J.R. (1984). *Can. J. Anim. Sci.* **64**, (Supplement) 156–7.
Gulbransen, B. (1984). *Proc. Aust. Soc. Anim. Prod.* **15**, 213–6.
Gutierrez, E., Elliott, R. and Harrison, D. (1981). *Trop. Anim. Prod.* **6**, 361.
Gutteridge, R.C. (1981). Proc. 14th Int. Grassl. Congr. (Eds J.A. Smith and V.W. Hays.) pp. 226–9. (Westview: Colorado, USA.)
Hamilton, B.A., Hutchinson, K.J., and Swain, F.G. (1972). *Proc. Aust. Soc. Anim. Prod.* **9**, 214–20.
Hamilton, R.I., Catchpoole, V.R., Lambourne, L.J., and Kerr, J.D. (1978). *Aust. J. Exp. Agric. Anim. Husb.* **18**, 16–24.
Hegarty, M.P., Lowry, J.B., and Tangendjaja, B. (1985). In "Forages in Southeast Asia and South Pacific Agriculture." (Eds G.J. Blair, D.A. Ivory and T.R. Evans.) pp. 129–32. ACIAR Proceedings Series No. 12, Canberra.
Herrera, F., Wyllie, D., and Preston, T.R. (1980). *Trop. Anim. Prod.* **5**, 18–24.
Jarrige, R., Demarquilly, C., and Dulphy, J.P. (1982). In "Nutritional Limits to Animal Production from Pastures." (Ed. J.B. Hacker.) pp. 363–87. (CAB: Farnham Royal, UK.)
Jones, R.J. (1979). *Wld Anim. Rev.* **31**, 13–23.
Jones, R.J., and Lowry, J.B. (1984). *Experientia* **40**, 1435–6.
Jones, W.T., Anderson, L.B., and Ross, M.D. (1973). *N.Z. J. Agric. Res.* **16**, 441–6.
Kantharaju, H.R. and Chadhokar, P.A. (1981). *Indian Vet. J.* **58**, 157–61.
Kennedy, P.M., and Siebert, B.D. (1972). *Aust. J. Agric. Res.* **24**, 143–52.
Krebs, G., and Leng, R.A. (1984). *Proc. Aust. Soc. Anim. Prod.* **15**, 704.
Lamprey, H.F., Herlocker, D.J., and Field, C.R. (1980). In "Browse in Africa – The Current State of Knowledge." (Ed. H.N. Le Houerou.) pp. 33–54. (ILCA: Addis Ababa.)
Le Houerou, H.N. (1980a). In "Browse in Africa – The Current State of Knowledge." (Ed. H.N. Le Houerou.) pp. 55–82. (ILCA: Addis Ababa.)
Le Houerou, H.N. (1980b). In "Browse in Africa – The Current State of Knowledge." (Ed. H.N. Le Houerou.) pp. 261–89. (ILCA: Addis Ababa.)
Leigh, J.H., Wilson, A.D. and Williams, O.B. (1970). Proc. 11th Int. Grassl. Congr. (Ed. M.J.T. Norman.) pp. 55–9. (Univ. Qld: Brisbane, Australia.)
Losada, H., and Preston, T.R. (1973). *Cuban J. Agric. Sci.* **7**, 185–90.
Losada, H., Dixon, F., and Preston, T.R. (1971). *Cuban J. Agric. Sci.* **5**, 369–78.
Mader, T.L., and Horn, G.W. (1986). *J. Anim. Sci.* **62**, 1113–9.

Mader, T.L., Horn, G.W., Phillips, W.A., and McNew, R.W. (1983). *J. Anim. Sci.* **56**, 1021–8.

Mannetje, L. 't (1982). In "Nutritional Limits to Animal Production from Pastures." (Ed. J.B. Hacker.) pp. 67–85. (CAB: Farnham Royal, UK.)

Mannetje, L. 't (1974). Proc. 12th Int. Grassl. Congr. (Eds V.G. Iglovikov and A.P. Movsissyants.) V.3 pp. 299–304. (Moscow.)

Mannetje, L. 't (1984). In "Herbivore Nutrition in the Subtropics and Tropics." (Eds F.M.C. Gilchrist and R.I. Mackie.) pp. 51–66. (Science Press: Craighall, South Africa.)

Martin, J.L. (1971). *Cuban J. Agric. Sci.* **5**, 131.

McDonald, W.J.F., and Ternouth, J.H. (1979). *Aust. J. Exp. Agric. Anim. Husb.* **19**, 344–9.

McDowell, L.R., Conrad, J.H. and Ellis, G.L. (1984). In "Herbivore Nutrition in the Subtropics and Tropics." (Eds F.M.C. Gilchrist and R.I. Mackie.) pp. 67–88. (Science Press: Craighall, South Africa.)

McIvor, J.G. (1982). CSIRO Div. Trop. Crops Past. Tech. Memo. No. 29.

McLeod, M.N. (1973). *Aust. J. Exp. Agric. Anim. Husb.* **13**, 245–50.

McLeod, M.N. (1974). *Nutr. Abstr. Rev.* **44**, 803–15.

McManus, W.R., Manta, L., McFarlane, J.D., and Gray, A.C. (1972). *J. Agric. Sci.* **79**, 27–40.

McMeniman, N.P. (1976). *Aust. J. Exp. Agric. Anim. Husb.* **16**, 818–22.

McMeniman, N.P., and Little, D.A. (1974). *Aust. J. Exp. Agric. Anim. Husb.* **14**, 316–21.

McMeniman, N.P., Elliott, R., and Ash, A.J. (1987). *Anim. Feed Sci. Technol.* (In press).

McMeniman, N.P., Niven, D.R., Crowther, D., Gartner, R.J.W., and Murphy, G.M. (1981). *Aust. J. Exp. Agric. Anim. Husb.* **21**, 404–9.

Mella, C.M., Perez-Oliva, O. and Loew, F.M. (1976). *Can. J. Comp. Med.* **40**, 104–10.

Meyreles, L., MacLeod, N.A., and Preston, T.R. (1977a). *Trop. Anim. Prod.* **2**, 73–80.

Meyreles, L., MacLeod, N.A., and Preston, T.R. (1977b). *Trop. Anim. Prod.* **2**, 200–5.

Meyreles, L., Pound, B., and Preston, T.R. (1982). *Trop. Anim. Prod.* **7**, 92–7.

Meyreles, L., Rowe, J.B., and Preston, T.R. (1979). *Trop. Anim. Prod.* **4**, 255–62.

Minson, D.J. (1982). In "Nutritional Limits to Animal Production from Pastures." (Ed. J.B. Hacker.) pp. 167–82. (CAB: Farnham Royal, UK.)

Montpellier, F.A., Valdez, R.E., and Preston, T.R. (1977). *Trop. Anim. Prod.* **2**, 206–13.

Moran, J.B., Satoto, K.B., and Dawson, J.E. (1983). *Aust. J. Agric. Res.* **34**, 73–84.

Moran, J.B., Stockdale, C.R., and Trigg, T.E. (1986). *Proc. Aust. Soc. Anim. Prod.* **16**, 283–6.

Mosi, A.K., and Butterworth, M.H. (1985). *Anim. Feed Sci. Technol.* **12**, 241–51.

Munoz, F., Morciego, F., and Preston, T.R. (1970). *Cuban J. Agric. Sci.* **4**, 91–6.

NAS (1977). "Leucaena – Promising Forage and Tree Crop for the Tropics." (National Academy of Sciences: Washington, DC.)

Ng, T.T. (1972). *Trop. Grassl.* **6**, 229–36.

Nolan, J.V., and Leng, R.A. (1972). *Br. J. Nutr.* **27**, 177–94.

Norton, B.W. (1982). In "Nutritional Limits to Animal Production from Pastures." (Ed. J.B. Hacker.) pp. 89–110. (CAB: Farnham Royal, UK.)

Norton, B.W., Rohan-Jones, W., Ball, F., Leng, R.A., and Murray, R.M. (1972). *Proc. Aust. Soc. Anim. Prod.* **9**, 346–51.

Orpin, C.G., and Bountiff, L. (1978). *J. Gen. Microbiol.* **104**, 113–22.

Otsyina, R., McKell, C.M., and van Epps, G. (1982). *J. Range Mngt* **35**, 751–3.

Paladines, O. (1984). In "Herbivore Nutrition in the Subtropics and Tropics." (Eds F.M.C. Gilchrist and R.I. Mackie.) pp. 145–66. (Science Press: Craighall, South Africa.)

Peron, N., and Preston, T.R. (1971). *Cuban J. Agric. Sci.* **5**, 49–57.

Pound, B., and Fernandez, A. (1981). *Trop. Anim. Prod.* **6**, 261–6.

Pressland, A.J. (1977). *Qld. J. Agric. Anim. Sci.* **34**, 213–23.

Preston, T.R. (1972). *Wld Rev. Nutr. Diet.* **17**, 250–311.

Preston, T.R. (1977). *Wld Anim. Rev.* **21**, 11–7.

Preston, T.R., and Leng, R.A. (1984). In "Straw and Other Fibrous Byproducts as Feed." (Eds F. Sundstol and E. Owen.) pp. 373–413. (Elsevier: Amsterdam.)

Preston, T.R., and Willis, M.B. (1974). "Intensive Beef Production." (Pergamon: Oxford, UK.)

Priego, A., Dixon, R.M., Elliott, R., and Preston, T.R. (1979a). *Trop. Anim. Prod.* **4**, 281–6.

Priego, A., Elliott, R., and Preston, T.R. (1979b). *Trop. Anim. Prod.* **4**, 287–91.

Pritchard, D.A., Weider, G., and Hopkins, P.S. (1985). "Recent Advances in Animal Nutrition in Australia." (Univ. New England: Armidale.)

Rodriguez, A. (1983). *Trop. Anim. Prod.* **8**, 74–5.

Rowe, J.B., Bobadilla, M., Preston, T.R., and Leng, R.A. (1980). *Trop. Anim. Prod.* **5**, 63–70.

Royal, A.J.E., and Hughes, R.M. (1976). *Proc. Aust. Soc. Anim. Prod.* **11**, 513–6.

Ruiz, M.E., Lozano, E., and Ruiz, A. (1981). *Trop. Anim. Prod.* **6**, 234–44.

Ruiz, M.E., Pezo, D., and Martinez, L. (1980). *Trop. Anim. Prod.* **5**, 144–51.

Salias, F.J., Wilson, A., and Elliott, R. (1977). *Trop. Anim. Prod.* **2**, 306–9.
Satter, L.D., and Slyter, L.L. (1974). *Br. J. Nutr.* **32**, 199–208.
Saul, G.R. (1984). *Proc. Aust. Soc. Anim. Prod.* **15**, 569–72.
Scattini, W.J. (1984). *Aust. J. Exp. Agric. Anim. Husb.* **24**, 20–5.
Shelton, H.M. (1980). *Exp. Agric.* **16**, 57–66.
Shrewsbury, C.L., Andrews, F.N., Harper, C., and Zelle, M.R. (1943). *J. Anim. Sci.* **2**, 209–20.
Silva, A., and Ørskov, E.R. (1985). *Proc. Nutr. Soc.* **44**, 50A.
Silvestre, R., MacLeod, N.A., and Preston, T.R. (1976). *Trop. Anim. Prod.* **2**, 86–94.
Stephenson, R.G.A., and Hopkins, P.S. (1985). Proc. Int. Conf. Anim. Prod. Arid Zones, Damascus. (In press).
Stephenson, R.G.A., Martin, P., Pritchard, D.A., Cobon, D.H., and Sullivan, M. (1986). *Proc. Aust. Soc. Anim. Prod.* **16**, 363–6.
Stephenson, R.G.A., Pritchard, D.A., Pepper, P.M., and Connelly, P.T. (1985). *Aust. Rangel. J.* **7**, 75–9.
Sturtz, J.D., and Parker, G.V. (1974). *Proc. Aust. Soc. Anim. Prod.* **10**, 344–8.
Sundstol, F., and Owen, E. (1984). "Straw and Other Fibrous Byproducts as Feeds." (Elsevier: Amsterdam.)
Swift, R.W., Cowan, R.L., Barron, G.P., Maddy, K.H., and Grose, E.C. (1951). *J. Anim. Sci.* **10**, 434–8.
Valderrabano, J., Uriarte, J., and Munoz, F. (1986). *Anim. Prod.* **42**, 241–6.
Wahyuni, S., Yulianti, E.S., Komara, W., Yates, N.G., Obst, J.M., and Lowry, J.B. (1982). *Trop. Anim. Prod.* **7**, 275–83.
Waldo, D.R. (1984). In "Forage Legumes for Energy Efficient Animal Production." (Eds F. Barnes, R. Bull, R.W. Brougham, G.C. Morten and D.J. Minson.) pp. 220–5. (USDA Agricultural Research Services: USA.)
Walker, B.H. (1980). In "Browse in Africa – The Current State of Knowledge." (Ed. H.N. Le Houerou.) pp. 7–24. (ILCA: Addis Ababa.)
Weston, R.H. (1971). *Aust. J. Agric. Res.* **22**, 469–79.
Wilden, J.H. (1985). Proc. 15th Int. Grassl. Congr. pp. 1303–4. Kyoto, Japan.
Willoughby, W.M. (1959). *Aust. J. Agric. Res.* **10**, 248–68.
Wilson, A.D. (1966). *Aust. J. Agric. Res.* **17**, 147–53.
Wilson, A.D. (1977). *Aust. J. Agric. Res.* **28**, 501–8.
Wilson, A.D., and Harrington, G.N. (1980). In "Browse in Africa – The Current State of Knowledge." (Ed. H.N. Le Houerou.) pp. 291–7. (ILCA: Addis Ababa.)
Wilson, J.R. (1982). In "Nutritional Limits to Animal Production from Pastures." (Ed. J.B. Hacker.) pp. 111–31. (CAB: Farnham Royal, UK.)
Wilson, K.J., Edwards, J.C., and Scanlan, J.C. (1980). *Qld. Agric. J.* **106**, 2–5.

19

SUPPLEMENTS OTHER THAN FORAGES

by P.T. Doyle

A supplement is something added to remedy a deficiency. However, many "supplements" confer other advantages or disadvantages in the processes of utilisation of forage components. Positive effects may occur through increased forage intake or increased digestion of forage constituents, while negative associative effects lead to depressed intake and/or decreased digestion of major constituents of the forage.

Herbivores inhabit many diverse environments and the domesticated species are kept under a wide range of husbandry conditions, ranging from pastoral nomadism to extensive grazing on native or improved pastures and to more intensive systems involving confinement. The effects of different environments and management conditions on forage availability and quality have been considered in detail earlier in this symposium and in previous conferences for example, the Nutritional Limit to Animal Production Conference (Hacker 1982). When forage quality is low, then high quality forages can be used as supplementary feeds for grazing herbivores, as discussed in the previous chapter (Elliott and McMeniman *ibid*. pp. 409–28). In these circumstances, other supplementary feeds, such as grains or meals, may be given to enable domesticated herbivores to better utilize the available forage or to improve their production. Where forage availability is also low, grains or even lower quality roughages may be used as supplements to enable survival. In contrast to the domesticated animals, free-ranging or wild herbivores are generally left to fend for themselves and supplementary feeding is rare. These animals generally expand the range of forage species eaten or migrate when quantity or quality of preferred forage is low.

This review focuses primarily on the provision of supplementary nutrients in concentrate feeds or in discrete forms to herbivores consuming low quality forages. It does not consider strategic, short-term provision of supplements, a practice which is currently being investigated as a means of improving reproductive performance by providing surges in nutrient supply at critical times. However, some consideration is given to the use of concentrates when only limited amounts of forage are available and to the extreme situation of zero grazing or feed lot systems. By necessity, most of the information discussed relates to work with domesticated herbivores.

DIGESTIVE ANATOMY AND SUPPLEMENTATION

There are many differences in the digestive anatomy of herbivores which have implications for selection and subsequent ingestion and

NUTRITION OF HERBIVORES
ISBN 0 12 750052 9

digestion of forages. The herbivores can be simply classified into those which ferment ingested feed prior to gastric digestion (foregut fermenters) and those which ferment digesta in the hindgut (hindgut fermenters) (Hume and Warner 1980). Important differences occur between these two groups as regards the digestion of forages and of supplementary feeds. With foregut fermenters the components of ingested feed become involved in fermentation interactions and the amounts which pass unaltered to the sites at which mammalian enzymes are secreted vary with the nature of the digestive system, the type of feed and the level of feeding. In contrast, feeds are exposed to "true" digestion prior to fermentation in the hindgut fermenters.

Within the foregut fermenters, the ruminant digestive system is characterised by selective retention of larger digesta particles in the reticulo-rumen, a phenomenon controlled by the reticulo-omasal orifice. Structures or practices which result in long retention times of feed in the fermentative regions of the stomach lead to more complete fermentation of feed components. Other foregut fermenters, such as the macropod marsupials, have no constrictions which impede passage of digesta from the fermentative to the acidic region of the stomach. Retention times in the fermentative region are, therefore, generally shorter and may result in less complete fermentation. Opposed to this argument are the observations that some non-ruminant foregut fermenters in the Macropodidae masticate their food more thoroughly (Hume and Warner 1980), and may have higher fermentation rates than ruminants (Hume 1982).

Hindgut fermenters have also been classified into two groups: the colon fermenters in which the proximal colon and caecum function as a single fermentation unit (e.g. equids), and caecum fermenters in which there is selective retention of small particles in the caecum (for example, rabbits) (Hume and Warner 1980). The consequences of these differences as regards supplementary feeding are less important than those occurring within the foregut fermenters, as supplements of energy-yielding substrates, protein, minerals and vitamins will be largely digested in the stomach and small intestines and absorbed for tissue utilization.

Herbivores with different feeding habits and different digestive anatomies are affected to differing degrees by fluctuations in availability and quality of forage. As the cell wall content of forage increases, the equids seem better able to increase feed intake relative to ruminants (Janis 1976; Hume and Warner 1980). Another hindgut fermenter, the capybara (*Hydrochoerus hydrochaeris*) has been found to harvest short forages of high nutritive value when cattle cannot (Gonzalez-Jimenez 1977). Within the ruminants, Langlands and Bowles (1976) reported that sheep select forage of considerably higher nitrogen content than cattle grazing the same pasture and it is likely that their diet would also be more digestible. It has also been observed that the more specialized grazers and browsers in Africa tend to continue to select among a narrow range of feed components, even through dry seasons when choice is limited (Giesecke and van Gylswyk 1975; Hoppe *et al.* 1977a, b). These few examples illustrate that the need for supplementary nutrients can vary greatly when different herbivores are grazing under similar conditions.

It is well known that small ruminants have proportionally higher energy requirements per unit liveweight than do larger ones, and that energy needs for growth, pregnancy and lactation are considerably greater than for maintenance. The need for small animals to consume more digestible energy relative to body weight can be met not only by eating more, but also by selecting a more digestible diet (Kay *et al.* 1980). Conversely, larger non-productive ruminants may obtain sufficient energy for maintenance from poor quality forages when smaller ruminants cannot; for example, some rice straws given alone can provide sufficient digestible energy for maintenance of mature cattle and water buffaloes, but not for sheep and goats (Doyle *et al.* 1986). This also has obvious implications as regards supplementary feeding.

The types and proportions of energy-yielding substrates available to the tissues of foregut and hindgut fermenters vary considerably. All fermentation systems yield volatile fatty acids (VFA), microbial protein, B-vitamins and vitamin K. VFA are readily absorbed from the foregut and hindgut, but the microbial protein and vitamins synthesized are only of use where they are subsequently exposed to intestinal digestion and absorption. Some hindgut fermenters reingest faeces and in this way make use of these microbial products, but this is not universally true and some lose a substantial source of specific nutrients in their faeces (Hornicke and Bojrnhag 1980). When grazing low quality forages, which are characteristically high in cell wall constituents and low in nitrogen, foregut fermenters gain advantages from more complete digestion of structural carbohydrates and from the utilization of recycled nitrogen for microbial protein synthesis. However, when ingesting better quality forages they suffer the disadvantages that the fermentation of soluble carbohydrates is energetically inefficient and VFA are less efficiently utilised than hexoses by mammalian tissues. As a consequence of the fermentation of soluble carbohydrates, blood glucose levels are generally lower in ruminant and non-ruminant foregut fermenters than in other herbivores (Moir 1965). Thus, the amounts of gluconeogenic substrates, propionate and amino acids, absorbed are important to efficient tissue metabolism in these animals.

In both foregut and hindgut fermenters, the efficiency of tissue utilization of absorbed nutrients is dependent upon the amounts and proportions of energy-yielding substrates and the balance of essential nutrients supplied. In the foregut fermenters, the efficiency of the dominant fermentation process is also dependent upon the nature and balance of dietary components as they contribute to the needs of the microbial population.

READILY DIGESTIBLE CARBOHYDRATE SUPPLEMENTS

Cereal grains and crop by-products, such as brans or molasses, which are high in soluble carbohydrates, are used as supplementary feeds when forage availability is low during droughts. They may also be used as supplements where grazing animals are losing liveweight due to the low digestibility and essential nutrient content of forages. It may be desirable to feed concentrates containing readily fermentable carbohydrates to reduce liveweight losses, to maintain liveweight or even to support growth. However, because compensatory liveweight gains usually occur when forage quality increases after rains, it is often desirable to provide these supplements only at levels which will optimize the utilization, intake and digestion, of available dry forage. Supplementation with such concentrates is not always the best option for this purpose for the reasons discussed below.

Theoretical considerations

Supplementary feeds providing readily digestible carbohydrates have associative effects on forage utilization by herbivores. These effects can be mediated at two levels: the supplement may reduce forage intake (substitution) or it may depress the digestion of forage components.

Substitution effects. When included at low levels in the diet of ruminants, readily fermentable carbohydrate supplements have been reported to depress (Hennessy *et al.* 1983; Smith and Warren 1986b), to not affect (Hennessy and Murison 1982) or to increase (Cohen 1974; Ernst *et al.* 1975; Lee *et al.* 1987) the intake of low quality forages. When sufficient of a readily fermentable carbohydrate feed is given to ruminants substitution invariably occurs. However, the level of supplementation at which substitution begins is not always the same. In some instances the decrease in forage intake with increasing amounts of supplement given is apparently linear (Mulholland *et al.* 1976b; Hodge and Bogdanovic 1983), while in other cases forage intake may be maintained or increased at low levels of supplementation and reduced as the amount of supplement given is increased (Crabtree and Williams 1971a, b).

Several factors contribute to the variable substitution effects. Supplements contain different amounts of various essential nutrients and, in the case of some brans, may provide significant amounts of fat. Consequently, their effects on the intake of forages may be mediated, in part, through nutrients or energy sources other than readily fermentable carbohydrates. For example, supplementation of low quality feeds with molasses does not generally increase forage intake, but when given in combination with urea forage intake may be substantially increased (Coombe and Tribe 1962). Responses to cereal grain supplements have also at times been improved by the addition of urea or other non-protein nitrogen (NPN) sources (Kondos and Mutch 1975) although this is not always the case (Smith and Warren 1986b).

Crabtree and Williams (1971b) found that increasing amounts of mixtures of barley and soyabean meal depressed hay intake, but that higher protein content in the supplement tended to increase forage intake. Clearly, the presence of nutrients other than readily fermentable carbohydrates can influence substitution effects.

Within the rumen, most of the readily digestible carbohydrates of concentrates are fermented, but with some feeds, such as maize (Waldo 1973), sorghum (Sutton 1980) and rice polishings (Elliott *et al.* 1978) appreciable amounts of starch may escape to the intestines. These supplements may have beneficial effects on tissue metabolism by increasing glucose supply and this may influence the degree of substitution.

The quality or characteristics of the forage being consumed also influence the degree of substitution. Gulbransen (1974) and Leibholz and Kellaway (1984) provide evidence that the reduction in intake is greater with high compared to low quality forages. In other instances, however, the substitution rate has not differed greatly with forage quality, particularly at higher levels of supplementation (Crabtree and Williams 1971a). The reduced intake of forage by ruminants in response to these supplements has often been attributed to their effects on cell wall digestion in the rumen, but if this were the sole cause of substitution then the effects might be expected to be greater with more fibrous feeds.

Decreases in forage consumption by ruminants, as levels of energy-yielding concentrates in the diet are increased, are mediated by many complex factors operating within the fermentation system and within the animal's tissues. The effects of these supplements in depressing fermentation of cell wall constituents is undoubtedly linked to the reduced intake. However, the efficiency of use of energy in the tissues, which is influenced by the types of energy substrates absorbed, by the amounts of other essential nutrients available and by the physiological needs of the animal, is also likely to be crucial in determining the degree of substitution.

Information on the effects of readily fermentable carbohydrate supplements on the intake of forage by foregut fermenters which do not selectively retain large particles in the stomach is not available. In hindgut fermenters, readily digestible carbohydrates are primarily digested in the small intestine yielding hexoses for absorption. Also in these animals there is not the same physical restriction to passage of larger digesta particles through the gastro-intestinal tract (GIT) that occurs in ruminants. It might therefore be expected that such animals would eat to meet their energy requirements and that concentrate supplements would only replace forage in the diet when they are readily available and they are the preferred feed component.

Depressed cell wall digestion. In ruminants, rapid ingestion of large amounts of readily-fermentable carbohydrates can lead to adverse effects on the fermentation system and on the animal. However, even at low levels of intake these supplements have negative associative effects on the fermentation of cell wall constituents (McDonnell *et al.* 1979). The level of supplementation at which reductions in cell wall digestibility occur is influenced by several inter-related factors, but can occur when concentrates comprise as little as 0.15 to 0.20 of the diet (Mulholland *et al.* 1976b).

It has been shown in *in vitro* fermentation systems that, as the pH declines from 6.5 to 6.0 or less, there is a reduction in cellulose digestion (Terry *et al.* 1969). A similar situation exists *in vivo*, with cellulase activity and, hence, cell wall digestion decreasing with reducing pH, until negligible digestion occurs below pH 6.0 (Mould and Ørskov 1983). Readily fermentable supplements lead to rapid pH reduction in the rumen due to production of VFA; the degree of reduction being dependent upon the nature of the readily fermented carbohydrates ingested, the amount ingested, the frequency of ingestion and in some instances, on the processing of the supplements (Dixon 1986a). Considering the nature of the supplement, rolled barley supplements have been found to depress cellulose digestion more than whole barley (Ørskov and Fraser 1975) or sugar beet pulp (Fahmy *et al.* 1984) presumably due to development of a lower pH. However, at low levels in the diet some concentrate supplements do not drastically reduce pH. For example, Silva and Ørskov (1985) found that unmolassed sugar beet pulp, which provides β-glucans, only depressed pH from 7.0 to 6.9 and improved the rate of digestion of straw.

Feeding concentrates in small meals throughout the day minimizes their effects on rumen pH and limits the wild fluctuations and instability in the microbial environment associated with once daily feeding (Moir and Somers 1957; Kaufmann 1976). However, this frequency of feeding is difficult to achieve under grazing conditions. Restricting the frequency of feeding of concentrates to less than once daily, for example twice weekly, could also conceivably result in smaller effects on cell wall digestion as rumen pH is depressed less frequently. Fredericks *et al.* (1986) found that liveweight loss by grazing lambs was reduced and wool growth increased when oat or triticale grain supplements were given on every third day compared to daily. However, the mechanisms associated with these responses were not elucidated.

When buffers have been used to maintain rumen pH, the depression in cell wall digestion due to concentrate supplements is reduced, but not completely removed (Mould *et al.* 1983). This indicates that factors other than pH are involved. Some workers (El-Shazly *et al.* 1961; Henning *et al.* 1980) have found that concentrate supplements reduce the numbers and/or activity of cell wall digesting bacteria. It is conceivable that competition for essential nutrients, such as nitrogen and sulphur, occurs between populations which ferment cell walls and those which ferment readily digestible carbohydrates resulting in detrimental effects on fermentation of structural carbohydrates. However, at low levels of supplementation rapid uptake of nutrients by amylolytic organisms may be advantageous through subsequent recycling to cellulolytic types.

Forage characteristics also influence the extent to which cell wall digestion is depressed. Burroughs *et al.* (1949) found that maize cob digestibility was depressed by starch supplements, while lucerne hay digestibility was not. Dixon (1986b) found that concentrate supplementation reduced cell wall digestion more in graminaceous than non-graminaceous forages and that the reduction was positively related to the cellulose content of the forage. Thus, the effects are greater with more fibrous feeds.

Depressions in cell wall digestion are also likely to occur when energy concentrates are given to foregut fermenters which do not selectively retain large particles in the stomach. However, the degree to which this occurs in particular species and the subsequent effects on forage intake are unknown.

In hindgut fermenters, depressions in cell wall digestion attributable to fermentable carbohydrate supplements do not occur to the same extent as in ruminants. Baldizan *et al.* (1982) found that while 0.30 cassava flour in the diet depressed cell wall digestion in sheep there was no effect in capybara. In ponies, Hintz *et al.* (1971) reported no depression in forage digestibility when the diet was changed from hay to 0.50 hay and 0.50 shelled corn or 0.20 hay and 0.80 shelled corn. However, in horses, replacing hay in the diet with 0.80 oats has reduced cell wall digestibility (Thompson *et al.* 1981). Pre-caecal digestibility of starch in these animals is generally high, but may range from 0.60 (Hinkle *et al.* 1983) to 0.90 (Meyer *et al.* 1983) and the proportion digested is apparently independent of the ratio of forage to concentrate when feed intake is restricted (Hinkle *et al.* 1983). This indicates that the amounts of starch reaching the hindgut undigested can be significant when high proportions of concentrate are given. Under these conditions depressed cell wall digestion may occur and it is likely the effects will be exacerbated at high levels of intake when rate of passage to the hindgut would be increased.

Practical considerations

The efficacy of supplementation of grazing herbivores is dependent upon the time taken to adapt to or accept the supplement and on the achievement of the desired level of intake by individual animals. Acceptance of cereal grain supplements has been studied in ruminants, but to my knowledge has not been researched for other herbivores. With sheep, non-feeders which make no attempt to consume cereal grain are often observed (Arnold and Bush 1968; Squires and Daws 1975). As a consequence these animals may not survive in drought conditions, while others in the flock may ingest too much supplement resulting in acidosis. Effective use of cereal grains at low levels in the diet is dependent upon their rapid acceptance by a majority of grazing animals with little variation in intake between individuals within a flock or herd.

It has been shown with trained sheep that wheat and barley are preferred foods (Hutson and van Mourik 1981; Hutson and Wilson 1984), indicating that these supplements may be readily accepted by grazing animals. However, with untrained sheep, acceptability is much poorer and Chapple (1985) found they were neophobic to containers, such as troughs holding supplements and to grain itself. In this regard, sheep that have previously been exposed to wheat adapt more rapidly to it, and to other cereals, than those with no previous experience of grain (Mottershead *et al.* 1985). Thus, they seem to remember from season to season.

Lynch *et al.* (1983) and Green *et al.* (1984) have found that lambs exposed to supplements prior to weaning in the presence of their mothers subsequently accept these supplements more readily than lambs without this experience. It appears that young ruminants are strongly influenced by their mothers acceptance of supplements, although Lynch (1985) has indicated that they may also be influenced by other adults.

Under extensive grazing conditions it is apparent that adaptation to supplements by individuals is also influenced by the placement of troughs and the amount of trough space available (Arnold and Maller 1974). As well as variable adaptation times, problems exist with variation in the amounts of cereal grains consumed by sheep within a flock. Lobato *et al.* (1980a) found in a group of sheep offered oat grain at a daily rate of 300 g /head (2100 g /week) that individual intakes ranged from 1086 to 2679 g /week. Similar variation has been reported by Green *et al.* (1984) for sheep which had been previously adapted to wheat supplements. It is often suggested that variation in grain intake within a flock can be reduced by offering the supplement every two or three days instead of daily.

In summary, supplementary feeds containing significant amounts of readily fermentable carbohydrates can only optimize the utilization of low quality forages by foregut fermenters when the supplementary feed comprises a small proportion of the diet and is given in combination with other limiting essential nutrients. Outside these conditions, these supplements cause perturbations in the fermentation environment leading to depressed digestion of major forage components. They also substitute for forage in the diet of ruminants, an effect which is linked to their effects on fermentation and also to the balance of nutrients yielded to the tissues. Under grazing conditions variable acceptance and intake of the supplements means that the desired level of supplement intake by individual animals is seldom achieved. Thus, from a practical viewpoint, it would seem that while research continues to focus on understanding the complex interactions which lead to substitution and depressed forage digestion, a better understanding of the behavioural aspects of supplement acceptance, a problem which by comparison has received little attention, would lead to more immediate gains. It should also be remembered that, despite their negative associative effects, these supplements generally result in increased digestible energy intake by ruminants, unless extreme instability in intake or frank acidosis are encountered. The problem of depressed forage digestion which occurs in foregut fermenters, does not appear to be as serious in hindgut fermenters. While little is known about the variability in supplement acceptance and intake by hindgut fermenters it is likely that the problems will be similar to those encountered with ruminants.

NITROGEN SUPPLEMENTS

Micro-organisms inhabiting the gastro-intestinal tract utilize ammonia, amino acid and peptide nitrogen for protein synthesis, while mammalian tissues require nitrogen principally in the form of amino acids, but also in nucleotides and vitamins. Supplementary nitrogen is often required to optimize the utilization of mature forages. It is generally agreed that when forages contain less than 70g/kg crude protein that intake and digestion by ruminants will be limited by nitrogen supply. Obviously with hindgut fermenters, protein supplements which are digested in the stomach and intestines will be the only effective means of providing additional essential amino acids. However, with foregut fermenters it is necessary to consider nitrogen requirements of the micro-organisms as well as those of the tissues.

Theoretical considerations

In ruminants most of the ingested forage protein is fermented in the reticulo-rumen to yield peptides, amino acids and eventually ammonia. This generally accounts for about 75% of forage protein (Leibholz and Kellaway 1984). Ammonia, which is the major nitrogen source for microbial protein synthesis in the reticulo-rumen (Nolan and Leng 1972) and in the forestomach of the Macropodidae (Hume 1982) is derived from forage protein and NPN, from recycled protein and NPN and may be provided by NPN or protein supplements.

The utilization of ammonia by micro-organisms is influenced by carbohydrate availability; maximum utilization of ammonia occurring when the rate of ammonia production is synchronized with the rate of carbohydrate fermentation (Johnson 1976). Other nutrients such as amino acids or peptides (Hume 1970b; Maeng and Baldwin 1976; Cotta and Russell 1982), essential minerals, in particular sulphur (Bird and Hume 1971; Elliott and Armstrong 1982; Lindsay *et al*. 1983), and higher VFA (Hume 1970a) are also necessary for efficient uptake of ammonia. While the ability of micro-organisms to utilize NPN sources to synthesize protein is potentially of enormous benefit in foregut ferments grazing low quality forages, it is often necessary to provide other supplements with the nitrogen to ensure efficient utilization.

The eating pattern of grazing domesticated herbivores generally involves periods of forage ingestion early in the day and during the late afternoon and early evening, leading to relative continuity of fermentative activity throughout each 24 h. Exceptions to this do occur as in lactation when ruminants spend more time eating (Campling 1966; Arnold 1975; Dulphy *et al.* 1980). Consequently, it is desirable to give NPN supplements either frequently or in a form which is hydrolyzed slowly by the microbial population. Urea is hydrolyzed rapidly by rumen populations, the rate of release of ammonia being as much as four times the rate at which it can be assimilated (Bloomfield *et al.* 1960). It also appears that little or no adaptation of the micro-organisms is required for urea breakdown (Nikolic *et al.* 1980).

Under experimental conditions, frequent administration of urea, achieved by spraying it onto low quality forage or by continuous infusion into the rumen, tends to consistently increase intake of such feeds and increase nitrogen retention by ruminants relative to less frequent administration (Tudor and Morris 1971; Romero *et al.* 1976; Umunna 1982). In cattle fed a straw and concentrate diet, Meggison *et al.* (1979a) found that the efficiency of microbial protein synthesis was higher when urea was given continuously than when it was given twice daily. However, the benefits from continuous administration of urea have not always been as great as might be expected (e.g. Knight and Owens 1973) and the possibilities that other nutrients limit the magnitude of the response or that recycled nitrogen contributes significantly to the response when urea has been given infrequently need to be considered. While frequent administration of urea is desirable, it is not always possible to achieve in practical situations as is discussed later.

Rapid release of ammonia from urea has led to a continuing interest in NPN sources which break down more slowly in the rumen. Biuret (Oltjen *et al.* 1968; Fonnesbeck *et al.* 1975; Bartley and Deyoe, 1977), isobutylidene diurea (IBDU) (Meggison *et al.* 1979b), lactosyl urea (LU) (Merry *et al.* 1982a, b), and uric acid, which occurs in poultry manure (Jacobs and Leibholz 1977), all release ammonia at slower rates than does urea. They would seem to be the most acceptable of the NPN sources tested although others do exist (Milligan *et al.* 1969; Martin *et al.* 1982; Bergner 1984).

Ruminal micro-organisms may take from 3 to 10 weeks to adapt to biuret and in order to maintain biuretase activity the supplement must be provided regularly (Bartley and Deyoe 1977; Nikolic *et al.* 1980). While some work has indicated biuret is used more efficiently than urea (Kondos and Mutch 1975) evidence presented by Fonnesbeck *et al.* (1975) and Nikolic *et al.* (1980) indicates this is not always the case.

With IBDU no adaptation period seems to be required (Nikolic *et al.* 1980) and it has been found in some instances to give superior results to urea (Meggison *et al.* 1979b) although these occurred when diets containing concentrates were fed. In contrast, Smith and Warren (1986b) found no advantage in roughage intake or liveweight gains by lambs given a supplement of oats/IBDU compared to oats/urea. As pointed out for LU supplements by Merry *et al.* (1982b), incomplete release of ammonia to the fermentation system can be a problem with slow-release NPN sources.

The use of IBDU, LU and other NPN sources in practice may be limited by their cost. However, in some circumstances uric acid, which forms the major source of NPN in poultry litter, may be a useful alternative to urea. Nickolic *et al.* (1980) indicate that microbial degradation of uric acid is slower than urea, but faster than that of biuret, and present evidence that uric acid can be successfully incorporated into diets for ruminants.

At present, urea is still the most commonly used NPN source and will probably remain so until it is conclusively demonstrated that slow-release ammonia products are used more efficiently for microbial protein synthesis and that subsequent production benefits warrant the additional costs.

The nature of responses to NPN supplementation have not always been similar. The effects are generally attributed to increased microbial activity resulting in increased rates of digestion within the rumen and consequently increased feed intake (Campling *et al.* 1962). However, responses in intake have also occurred due to increased microbial protein synthesis without concurrent increases in the rate or extent of digestion (Egan and Doyle 1985). Obviously, the characteristics of low quality forages (potential digestibility, nitrogen and mineral contents) affect the efficiency of NPN utilization. For example, the rumen ammonia-nitrogen levels in sheep fed cereal straws of similar nitrogen content (ca. 4.5 g /kg DM) but different *in vitro* organic matter digestibilities (0.30, 0.47 and 0.55) were 34, 20 and 14 mg /l, respectively (P.T. Doyle unpublished). The high rumen ammonia level in sheep fed the least digestible straw indicates that utilization of an NPN supplement would be poor, unless a readily fermentable carbohydrate supplement was also provided.

The contents of nitrogen and essential minerals in low quality forages also varies widely. For example, rice straws can contain from 3.5 to 12.0 g N /kg DM (Doyle *et al.* 1986). As the nitrogen in low quality feeds is extensively degraded (Egan *et al.* 1975), it is likely that the fermentation of those containing reasonable levels of nitrogen would not increase greatly with additions of NPN.

In foregut fermenters without constrictions between the fermentitive and gastric regions of the stomach, cell wall digestion may be lower than in ruminants because of shorter retention times and in spite of more rapid fermentation rates (Hume 1982). However, in the euro (*Macropus robustus erubescens*) (Brown 1969) and tammar wallaby (*Macropus eugenii*) (Lintern-Moore) 1973) the retention of nitrogen from dietary urea can be as efficient as that from casein. This indicates NPN can be utilized by foregut micro-organisms in such animals, and the principles that apply are likely to be similar to those in ruminants. As urea appears to be absorbed from the small intestine (Egan *et al.* 1985), and the primary requirement of the tissues is for amino acids, it is unlikely that NPN supplements would contribute significantly to the nitrogen status of hindgut fermenters. In this respect, most of the nitrogen from urea supplements given to rats is excreted in urine (see Moir 1965). However, nitrogen recycled to the hindgut may improve the energy supply to tissues of hindgut fermenters by enabling greater microbial fermentation of cell wall constituents.

Protein nitrogen supplements, such as protein meals or oil seeds, may be fermented in part by microbes in the foregut, but can also contribute undegraded dietary protein (UDP) for gastric digestion. In sheep fed protein-free diets, Hume (1974) reported that only 0.29 of a fishmeal supplement and 0.39 of soyabean meal were degraded in the reticulo-rumen compared to about 0.65 for peanut and lupin meals. The degradability of protein meals in the rumen is undoubtedly dependent upon retention time which will be influenced by the nature and level of intake of the base forage. The extent to which shorter retention times in other foregut fermenters may lead to more UDP reaching the intestines is unknown. In addition, the susceptibility of plant and by-product proteins to microbial fermentation can be altered by heat and chemical treatments (Beever and Thomson 1977).

Thus, protein supplements given to foregut fermenters may improve low quality forage utilization through increased microbial activity resulting in increased digestion and intake and through increased microbial and UDP supply to the intestines. Increased protein digestion in the intestines may improve animal performance by providing limiting essential amino acids, by providing amino acids that are catabolized to improve supplies of energy or gluconeogenic substrates or by resulting in increased urea nitrogen recycling to the forestomach.

It is generally agreed that, providing rumen microbial requirements for nitrogen are satisfied, proteins which are less readily degraded in the rumen may elicit greater responses in low quality forage utilization than those which are rapidly degraded. Considerable conjecture exists as to the mechanisms of responses to protein supplements which are slowly fermented. The effects have been attributed to additional amino acids acting at the tissue level (Egan, 1965; Kempton *et al.* 1979) or to prolonged availability of ammonia and amino acids in the reticulo-rumen (Leibholz and Kellaway 1984). It would seem that the supplements may lead to responses through a combination of these effects.

Readily fermentable pure protein supplements, such as casein, have been shown to have little advantage when compared to urea where the supplements are ingested with the feed and other essential nutrients are provided (Kempton and Leng 1979; Kempton *et al.* 1979; Redman *et al.* 1980). However, treatment of casein with formaldehyde has in some instances resulted in increased intake and liveweight gain (Kempton and Leng 1979), while in others where ruminal requirements were satisfied no such response occurred (Redman *et al.* 1980). No doubt protein in seeds and meals will be degraded more slowly in the rumen than urea leading to more efficient utilization of the peptides and ammonia released. In this regard, Hunter and Siebert (1980) and Coombe (1985) report that supplementation of mature forages with protein meals gives a more consistent increase in feed intake and digestibility than urea supplements. This is not always the case, but even when intake responses are not superior, protein meals may result in greater increases in growth rate (Lee *et al.* 1987).

Different protein feeds have varying effects on feed intake and animal performance. Smith and Warren (1986a, b) found a pelleted cottonseed meal supplement was superior to soybean meal or lupins as regards feed intake or liveweight gain by ruminants. Cottonseed meal supplements have been reported to increase liveweight gain by animals by about 0.45 kg /kg supplement (Smith and Warren 1986a), 0.62 kg /kg (Smith and Warren 1986b) and 0.5 kg /kg (Hennessy *et al.* 1981). The variable responses were due in part to differences in the quality of forage being grazed.

Practical considerations

Non-protein nitrogen supplements. Grazing herbivores generally select a diet of higher quality (digestibility and nitrogen content) than the average of the available forage. This has been found to preclude the need for NPN supplementation of sheep grazing some cereal grain stubbles (Mulholland *et al.* 1976a; Coombe and Mulholland 1983) and may explain many of the poor responses of grazing ruminants to NPN supplements.

Urea supplements have been provided in several different ways. Spraying urea solutions onto forages has the advantage of ensuring relatively continuous supply of the supplement, but is not practical in grazing situations as nitrogen which does not contact the plant material is lost and further losses occur due to microbial breakdown. Where crop residues are harvested and given to confined ruminants as a major component of the diet, as occurs in small farm systems in Asia, this means of supplementation would seem to be attractive. However, the technology has, in general, not been adopted, presumably due to experiences with toxicity, poor responses due to palatability problems resulting from inadequate mixing or from limitations imposed by the reluctance to provide other relatively expensive energy-providing feeds and minerals with the NPN. More recently, urea-sprayed forages have been ensiled prior to feeding which leads to more uniform distribution of NPN as well as increased susceptibility of cell walls to digestion (Doyle *et al.* 1986). This means of providing NPN may prove successful where the economic benefits justify its use, as in dairy production systems on small farms.

Urea and mineral supplements can be supplied in drinking water using dispensers capable of metering their release (Stephenson 1983), a practice which has been used under grazing conditions (Stephenson *et al.* 1981, 1984). The efficacy of this practice is dependent on the supplemented water being the only drinking water available, and is influenced by the relatively infrequent, once daily, ingestion of water. While some workers (Stephenson and Hopkins 1985) have found good relationships between supplemented water intake and liveweight, others (Wright *et al.* 1978) have indicated that the efficiency of this means of supplementation is limited by the variability in water intake between individual animals. Variations in temperature and humidity will also affect the amount of solution ingested. Nonetheless, this means of supplementation has been used with some degree of success in northern Australia.

Urea supplements can also be provided with carrier feeds, cereal grains or crop by-products such as cassava chips or molasses. The practical difficulties encountered with cereal grain supplements have already been discussed, and will apply when urea-fortified grains or specific products, such as starea, a processed mixture of finely ground grains and urea (Bartley and Deyoe 1977), are offered to grazing animals.

Molasses has also been used as a carrier for urea as it makes the supplement more palatable, provides readily fermentable carbohydrates and some minerals, such as sulphur. Liquid urea/molasses supplements are provided in roller drum or float lick systems in which the molasses is diluted, while solid supplements are offered in block form. The blocks can be formulated to contain minerals or energy or protein concentrates. The acceptance of block licks (Lobato and Pearce 1980a; Pollard 1981) and liquid supplements (Nolan *et al.* 1974) by sheep and cattle is variable. At the extremes, Lobato and Pearce (1980a) found that no sheep in one flock licked a block supplement while 0.80 of another flock did; Coombe and Mulholland (1983) reported that supplement acceptability increased to include all sheep after six weeks.

Confinement of sheep (Lobato and Pearce 1980a) and cattle (Pollard 1981) and various practices such as (a) providing a grain supplement on the block, (b) including animals known to readily consume blocks within a flock or (c) providing a range of block types (Lobato and Pearce 1980b) have been used to induce ingestion of urea/molasses blocks. However, such practices by no means guarantee these supplements will be ingested. Lambs exposed to blocks pre-weaning subsequently ingested this type of supplement more readily than those without previous experience Lobato *et al.* 1980b). While the reasons for this response are not clear, it indicates that taste acquisition via the ewe's milk, or observed learning from the mother may be learning practices which can be exploited to ensure greater acceptability of these supplements. Other factors, such as the quality of the available forage and ease of access to the supplements, will also influence acceptance.

The intake of urea/molasses supplements by grazing ruminants, whether in the form of liquid (Entwistle and Knights 1974; Nolan *et al.* 1974, 1975; Langlands and Donald 1978; Mulholland and Coombe 1979; Coombe and Mulholland 1983) or as a block (Mulholland and Coombe 1979; Lobato and Pearce 1980a; Lobato *et al.* 1980a; Coombe and Mulholland 1983) is extremely variable. Low intakes of NPN may be insufficient to produce benefits in forage intake and digestibility, while higher than desired consumption may lead to substantial nitrogen losses in the urine. Recently, production responses by housed ruminants to urea/molasses blocks have been reported (Kunju 1986) and Leng and Preston (1983) have suggested these supplements may be more efficiently used by housed or tethered ruminants than by grazing ruminants.

Protein supplements. Protein feeds may be given alone or with other supplements in troughs or licks to avoid wastage. When given at moderate levels to grazing ruminants they can improve performance through increased intake of low quality forages and through increased supply of essential nutrients (Hennessy *et al.* 1981; Lindsay *et al.* 1982; Smith and Warren 1986a, b). However, at higher levels in the diet they may still sustain increased liveweight gains but will substitute for forage (Crabtree and Williams 1971a).

As with cereal grain and NPN supplements, variable acceptance and intake of free-choice protein supplements has been reported (Langlands and Bowles 1976; Llewelyn *et al.* 1978; Dove 1984; Dove *et al.* 1984). For example, sunflower meal supplements appear to be more acceptable to sheep than rapeseed meal (Dove 1984). Variability in intake between animals can be large, as evidenced by a range of 20 to 1030 g /day of sunflower meal in a small group of lactating ewes (Dove *et al.* 1984).

Intake and digestibility of mature forages, which are low in protein, by ruminants can undoubtedly be increased by supplementary nitrogen; particularly if the opportunity to select more nutritious feed components is limited. In the first instance, supplementation should provide nitrogen to ensure efficient fermentation, as this process contributes to the removal of digesta from the rumen. Efficient use of NPN for this purpose is usually dependent upon the provision of other supplementary nutrients essential to ammonia uptake by the micro-organisms and is limited in grazing situations by variable acceptance and intake of the supplement. Protein in seeds and meals is fermented more slowly and these feeds also provide additional energy and other essential nutrients. In addition, protein supplements which are resistant to microbial fermentation increase UDP supply to the intestines and offer a means of providing specific limiting amino acids to the tissues. Thus, protein supplements may elicit production responses in ruminants by stimulating fermentation, through the increased intake of forage, through the additional energy supplied by the supplement and by altering the proportion of nutrients absorbed as essential amino acids. Because of this, supplements of protein meals result in greater improvements in forage intake and animal performance than supplements which primarily supply readily-fermentable carbohydrates (Hennessy *et al.* 1981; Hennessy and Murison 1982; Smith and Warren 1986b).

In other foregut fermenters, supplementary protein will have similar effects. However the degree to which responses are elicited by increased microbial activity or by altering nutrient balance at the tissue level will depend upon retention times of supplements in the fermentation system. Protein supplements given to hindgut fermenters will act primarily at the tissue level, although increases in the amounts of nitrogen flowing to or recycled to the fermentative system may lead to increased VFA absorption. With all herbivores the problems of variable supplement acceptance and intake represent major constraints to efficient utilization in extensive grazing situations.

SUPPLEMENTATION WITH MINERALS

Essential macro- and trace minerals are required by the micro-organisms inhabiting the gastro-intestinal tract and for metabolic functions within the tissues. Supplements of minerals can have positive associative effects on forage utilization when given at appropriate levels, but adverse interactions or toxicities occur when the level or balance of added minerals is inappropriate. In this review, attention is focussed on the deficiencies that occur, on the sites at which mineral supplements might mediate positive effects and on the practical problems associated with appropriate supplementation.

Macro-minerals

Of the essential macro-minerals, deficiencies in grazing herbivores have only been reported with calcium (McKenzie *et al.* 1981), magnesium (Rook and Storry 1962) phosphorus (Cohen 1975, 1980), sodium (Ealey 1967a, b) and sulphur (Hunter *et al.* 1979). It is not always clear whether macro-mineral deficiencies manifest themselves through effects on the micro-organisms or on the animal itself. For example, sulphate or amino acid sulphur supplements may increase intake (Kennedy and Siebert 1972), digestibility (Guardiola *et al.* 1980) and nitrogen retention (Kennedy and Siebert 1973) in ruminants on low quality forages, responses which are indicative of improved microbial growth and activity. This is supported by measurements of increased microbial protein flow to the intestines when sulphur supplements have been given (Bird and Hume 1971; Lindsay *et al.* 1983). Responses may vary in different species due to differences in sulphur recycling to the rumen. In this regard, Bird (1974) found cattle were less responsive to supplementary sulphur than were sheep. In addition to effects within the rumen, the amounts of sulphur-containing amino acids absorbed in forage-fed ruminants often limits nitrogen retention and some productive functions like wool growth (Doyle 1981; Egan *et al.* 1984). Other foregut fermenters might be expected to respond in a similar manner to sulphur supplements.

Other macro-minerals may not elicit their effects via improved microbial activity. For example, ruminal levels of inorganic phosphorus are unlikely to limit microbial activity as copious quantities of this element are recycled in saliva (Hogan 1984). Thus, the major effect of phosphorus supplementation in deficient sheep was found to occur at the tissue level (Milton and Ternouth 1985).

In addition to the sites at which macro-mineral deficiencies might manifest themselves, there is evidence of genetically controlled differences in the metabolism of minerals. Field and Wooliams (1984) found plasma concentrations, urinary and endogenous faecal excretion and efficiencies of absorption of phosphorus were similar within, but different between four sets of triplet sheep.

Deficiencies also occur in herbivores other than ruminants. Sodium deficiency occurs in the eastern grey kangaroo (*Macropus giganteus*) and the common wombat (*Vombatus ursinus*) in the Snowy Mountains in south-eastern New South Wales (Blair-West *et al.* 1968) and in the euro and red kangaroo (*Macropus rufus*) in parts of north-western Australia (Ealey 1967a, b). As the saliva secreted by the red kangaroo, like that in sheep and cattle, contains large amounts of phosphorus, sodium and chlorine (Forbes and Tribe 1969) deficiencies of these minerals in the fermentative region of the stomach are unlikely. The concentration of potassium in the saliva of the kangaroo may be low. Nevertheless, it is likely that the sodium deficiency which occurs in Macropods is of greater consequence to the tissues than to microbial fermentation. Deficiencies also occur in hindgut fermenters, as with calcium and phosphorus in horses grazing tropical pastures in Queensland (McKenzie *et al.* 1981). It is likely that macro-mineral supplements act primarily at the tissue level in hindgut fermenters. However, recycling of some minerals to hindgut fermentation systems may be substantially less than that which occurs through saliva secretions in the foregut fermenters. Hence, the possibility exists that mineral supplements which are not absorbed pre-caecum may stimulate microbial activity resulting in increased VFA supply to the tissues.

Importantly, responses to supplementation with macro-minerals can only be anticipated where energy supplies or nitrogen-containing compounds are not limiting. For example, the response phosphorus supplementation of ruminants has been found to be dependent on the supply of sufficient energy (McMeniman and Little 1974) or protein and energy (van Niekerk and Jacobs 1985). Supplements of macro-minerals can be provided by fertilization, by giving feeds rich in macro-minerals or by giving supplements in discrete forms. However, the forms in which elements are present in feeds or in discrete supplements can markedly affect their biological availability and hence their nutritional worth (Peeler 1972). In addition, minerals may interact in the gastro-intestinal tract with other feed components such as oxalates, making them relatively unavailable.

Fertilization. Application of minerals to pastures can affect plant growth and composition and can also influence the amount of a mineral consumed. Supplements of phosphorus (Ozanne *et al.* 1976) and sulphur (Rees *et al.* 1974, 1982) have not always resulted in equivalent improvements in intake and digestibility to those obtained from fertilization. This occurs because fertilizer can alter the proportions of plant parts and may lead to greater availability of more digestible leaf. With calcium, Rees and Minson (1976) found no effect of direct supplementation on the intake of pangola grass (*Digitaria decumbens*) by sheep, while fertilizer application increased consumption, apparently due to a decrease in acid detergent fibre content of the grass. Hogan (1984) has discussed the reasons for differences in response between fertilizer application and supplementation. He indicated the need for improved knowledge of the effects of fertilizers on cell wall structure and of the susceptability of cell walls of different plant species to microbial degradation.

Fertilizer application does not guarantee that minerals will be taken up by the plant, and even if they are they may be relatively unavailable to the animal. For example, Playne *et al*. (1978) found that in spear grass (*Heteropogon contortus*) hay the ruminal availability (release) of calcium, phosphorus, sodium and sulphur was less than half of that present. Availability of these elements was generally higher from lucerne (*Medicago sativa*), while the availability from Townsville stylo (*Stylosanthes humilis*) and chloris (*Chloris barbata*) varied with the different elements. With phosphorus, while the availability was high, less than 60% was absorbed from hays other than lucerne. It was suggested that chemical binding of elements to indigestible molecules, physical enclosure in an undigested fibre bundle, or the lack of a concentration gradient between the residues and the rumen fluid may limit availability.

Provision of other feeds. Providing minerals in feeds which contain high concentrations of specific elements or by incorporating them into concentrate supplements is only likely to be practised where the supplements are also intended to provide additional energy-yielding substrates or protein. The problems discussed previously in relation to efficient provision of supplements will also limit the efficacy of providing mineral supplements to grazing herbivores by these means.

Provision of discrete supplements. Salt licks are commonly used to provide mineral supplements to grazing animals, as they are convenient and relatively inexpensive. It is believed that the salt regulates the intake of supplement. Wheeler *et al*. (1980) found the intake of a coarsely-powdered mixture of sodium and potassium chlorides varied considerably between sheep which had no previous experience of licks, with 35% of the flock consuming no supplement. Inclusions of molasses led to all sheep consuming some supplement, but intakes remained variable. Rocks *et al*. (1980) confirmed that additions of 0.04 to 0.08 liquid molasses enhanced acceptance of salt licks; lower levels, 0.01 or 0.02, reduced acceptance. Other additives, such as dried molasses, did not improve lick intake while bentonite reduced intake, and additions of aniseed improved intake over time, indicating an acquired taste.

In Queensland, horses grazing improved tropical pastures have been found to suffer from nutritional secondary hyperparathyroidism, a problem attributed to dietary oxalate resulting in low absorption of calcium and phosphorus (Blaney *et al.* 1981). In pen feeding studies, Gartner *et al.* (1981), found that horses would not ingest ground limestone, rock phosphate or dicalcium phosphate when offered alone. However, additions of liquid molasses markedly improved acceptance and led to hourly intakes up to 200 g calcium and 50 g phosphorus with 1 kg of the mineral supplements being consumed within 1 or 2 days. McKenzie *et al.* (1981) found that grazing horses consumed the required amounts of these supplements within two days, but that behavioural traits led to excessive consumption by some animals at the expense of others. Inclusions of salt, which decreased consumption rate, assisted in overcoming this problem.

Other attempts to provide supplements of macro-minerals in discrete forms have generally not been successful for practical or economic reasons. For example, a high-density magnesium pellet was not successful due to low probability of retention in the rumen resulting from the low density of the bullet (Laby 1980). The progress made in the development of variable geometry, continuous release devices, which is discussed later, may lead to renewed interest and successful developments in this area.

Trace minerals

While trace minerals are required by micro-organisms, deficiencies are unlikely to occur in the fore-gut fermenters as dietary supply is supplemented by recycling of these nutrients. An exception to this is cobalt which is required for vitamin B_{12} synthesis. While the situation in hindgut fermenters is not clear, it would seem that deficiencies are more likely to occur in the tissues of all herbivores. The supply of trace minerals to the tissues is related to plant composition, soil conditions, diet selection, and to the availability of ingested elements for absorption. The biological availability of trace minerals in different feeds or in different chemicals is variable (Ammerman and Miller 1972). In foregut fermenters, availability can be affected by uptake by micro-organisms or by mineral interactions such as that with copper-molybdenum-sulphur (Gawthorne 1984). Competition for absorptive sites in the small intestine, as occurs with chemically similar elements (copper, zinc, iron and cadmium), can also affect availability.

As with other nutrients, requirements for trace minerals might be expected to increase with increased metabolic activity, as in growth, pregnancy and lactation, and are thus related to energy and protein intake. For example, Masters (1984) suggests that zinc requirements during pregnancy may be greatly reduced if ewes are losing weight due to deficiencies of major nutrients, partly because it can be mobilized from body reserves. This situation may also occur with other trace minerals. Conversely, the provision of protein and energy supplements may induce trace mineral deficiencies by increasing dietary requirements. In this regard, many of the production responses to trace mineral supplements have occurred in herbivores grazing green pastures (see Hosking *et al.* 1986).

In addition to this, there are apparent differences between herbivores in requirements for trace minerals. Underwood (1977) reported that cobalt requirements of sheep are 0.08–0.11 μg/g feed, while captive quokkas (*Sentonix brachyurus*) have been found to survive on diets containing as little as 0.01–0.03 μg/g for 12 weeks (Barker 1960). While liver cobalt levels, and vitamin B_{12} concentrations in plasma and liver declined markedly in these quokkas, no symptoms typical of cobalt deficiency in sheep appeared. In addition, quokkas maintained on high-protein commercial sheep pellets (Kakulas 1961) and the tammar wallaby and red-necked pademelon (*Thylogale thetis*) given lucerne hay (Hume 1982) have suffered from vitamin E deficiency. While addition of selenium to vitamin E-deficient diets can generally prevent 'white muscle disease' in lambs (Underwood 1977), in these marsupials nutritional muscular dystrophy was only responsive to vitamin E (Kakulas 1961, Hume 1982). Within species differences are also apparent. These differences indicate the need for additional information on requirements for trace minerals and vitamins, and on how they interact with other nutrients in different herbivores, and illustrate that extrapolations between species may be dangerous. Wiener (1979), in reviewing the evidence relating to copper, concluded that differences in nutritional requirements for trace elements may occur between breeds of sheep, the differences being manifested in genetic influences on the control of uptake and utilization of minerals.

In ruminants under grazing conditions, deficiencies of cobalt (Judson *et al.* 1981b; Duncan *et al.* 1986), copper (Hunter 1977; Richards *et al.* 1985), iodine (Azuolas and Caple 1984), manganese (Egan 1972), selenium (McDonald 1975; Johnson *et al.* 1981) and zinc (Egan 1972; Spais and Papasteriadis 1974; Masters and Fels 1980) have been reported. Trace mineral supplements can be provided in many ways, the delivery method being determined by economics, by the type of animal, and by whether the mineral is stored to any extent in body tissues.

Application in fertilizers. Mixing trace minerals in fertilizer when top- dressing pastures has been used to provide cobalt, copper, iodine and selenium (Underwood 1977). It has advantages in that some elements are required for both plant and animal growth. However, problems are encountered which relate to the availability of trace element fertilizers to plants and subsequently to animals. Consequently, it is only likely to be used where domestic livestock graze improved pastures and may only be economical when these pastures are able to support high stocking rates (Siebert and Hunter 1982).

Other drawbacks include the frequency at which trace element applications are required, a factor influenced by soil type and the characteristics of the mineral. For example, iodized fertilizers may only maintain satisfactory levels in forage with frequent application (Hartmans 1974), whereas copper and zinc need only be applied infrequently (Gartrell and Glencross 1969; Siebert and Hunter 1982). The use of selenium-fortified fertilizers has not been accepted because of difficulties in supplying appropriate concentrations which will supplement stock for a reasonable period of time without causing toxicity (Halpin *et al.* 1985). However, recently an encapsulated selenium fertilizer granule (Selcote prill) has been tested in New Zealand (Watkinson 1983) and Australia (Halpin *et al.* 1985) and may overcome safety and efficiency problems.

Provision in salt licks. Provision of trace minerals in salt licks has been found to be effective in overcoming clinical deficiencies. For example, selenium has been provided by this method in the prevention of nutritional muscular dystrophy in lambs (Paulson *et al.* 1968) and calves (Jenkins *et al.* 1974). However, salt licks have many disadvantages in that exposure to hot or humid conditions or to rain can lead to losses of minerals by volatilization or leaching (see Underwood, 1977) and there is variable acceptance and intake.

Pulse dosing by drenching or injection. The major advantage of pulse dosing is that individual animals receive the desired amount of supplement. It is generally only used for trace minerals which are stored in body tissues and not for those which require frequent administration. It has major disadvantages in that it is time and labour consuming and consequently expensive.

Oral drenches may be used to provide supplemental selenium and cobalt, while subcutaneous and intramuscular injections are used to provide vitamin B_{12} (as a replacement for cobalt), copper and selenium (Underwood 1977). Pulse dosing may still be used to treat clinical deficiencies, but other delivery systems are likely to be cheaper in overcoming subclinical deficiencies.

Recently, MacPherson (1981) reported that provision of copper and cobalt through the drinking water, where the amounts added were controlled by a metering device, was as effective as copper injections or cobalt bullets in overcoming limitations imposed by pasture deficiencies of those minerals. However, this method of supplementation has drawbacks which were discussed earlier.

Heavy intraruminal pellets. These pellets lodge in the reticulo-rumen and have been used to provide cobalt (Dewey *et al.* 1958), selenium (Kuchel and Buckley 1969) and more recently zinc (Masters and Moir 1980). The composition of pellets used for sheep is generally as follows: cobalt pellets, 0.05 Co and 0.95 Fe; selenium pellets, 0.05 elemental Se and 0.95 Fe; and zinc pellets, 0.50 Zn shot and 0.50 Fe. Release is believed to be by both abrasion, aided by the inclusion of another heavy object or second pellet, and dissolution.

Release rates into digesta from cobalt and selenium pellets are greatest in the first weeks after administration and thereafter decline (Dewey *et al.* 1969; Hunter *et al.* 1981). Cobalt pellets are as effective as regular injections of vitamin B_{12} in calves for up to 20 weeks, after which liveweight gains by the injected group tend to be greater (Judson *et al.* 1981b). Coating of these pellets with calcium phosphate (e.g. Dewey *et al.* 1958) apparently contributes to the slowing of release, while surface coating has only minor effects on selenium pellet efficacy (Hunter *et al.* 1981). Selenium grain size influences release rates and the efficacy of these pellets as judged by plasma selenium levels (Hudson *et al.* 1981a, b; Peter *et al.* 1981). Coarser grained pellets have a slower release rate due initially to a lower surface area at which reduction of elemental selenium can occur, but maintain elevated plasma selenium concentrations for considerably longer.

With zinc pellets, Masters and Moir (1980) reported release rates of about 15 mg Zn /d in the first week after administration but only 9 mg in the third week, and as little as 3 mg /d after 7 weeks. Analysis of pellets recovered from the rumen indicated that, with time, a coat probably of zinc phosphate forms, and this slows release. The problems with release rates from heavy intra-ruminal pellets have recently led to more intensive investigation of the factors involved and also to the development of controlled release devices.

Controlled release devices. These include controlled release glasses (Knott *et al.* 1985) and variable geometry devices, the principles of which have been described by Laby (1970). Controlled release glasses are phosphate based glasses into which essential elements can be incorporated. The rate at which they dissolve and release required nutrients is achieved by varying their composition. Hence, it is possible to develop different glasses to provide different rates of release or different effective delivery times. Soluble glass boluses are administered to the rear of the throat of sheep and cattle using a balling gun; they generally lodge in the reticulum with retention being about 0.95 (Allen *et al.* 1985). A glass containing cobalt, copper and selenium produced by Telfer *et al.* (1983) and manufactured commercially has been found to successfully provide cows, calves and adult sheep with adequate amount of these minerals for 12 months and to protect lambs for up to 6 months (Carlos *et al.* 1985; Care *et al.* 1985; Knott *et al.* 1985). Glasses which provide single trace elements such as cobalt or copper (Allen *et al.* 1985; Judson *et al.* 1985a, b) have also been tested. However, Petterson *et al.* (1985) have indicated that the physical integrity of some glass prototypes needs to be improved.

The rate of dissolution of a glass providing copper is too slow to provide an immediate boost to liver reserves (Judson *et al.* 1985a), indicating that for the treatment of clinical deficiencies glasses would have to be augmented with other means of administration. These are still technical difficulties with the release rates from soluble glasses, as evidenced by slow release rates of selenium when compared to that from heavy intraruminal pellets (Judson *et al.* 1985b). However, it would seem that problems with release rate and integrity of soluble glasses will be overcome by altering their composition.

Variable geometry devices, which are retained in the reticulo-rumen by the opening of pairs of fingers on one or both ends of a cylinder or by the opening of two half-cylinders hinged along one edge (Laby 1980), are another method of achieving controlled release of trace elements. The capsule portion of these devices consists of a plastic cylinder containing the mineral homogeneously dispersed throughout a core matrix. The carrier matrix dissolves at the open end of the cylinder. A constant area of dissolution can be maintained by moving the core towards the opening with a spring, which also leads to fairly constant release of the supplement. Release rate can be varied by altering the spring strength, the orifice diameter and the matrix composition (Laby *et al.* 1984). Some capsules, such as the iodine capsule, have been constructed differently, with the mineral surrounded by a plastic envelope and the resistance to release being related to solubility of iodine in the plastic envelope and the vapour pressure of iodine (Laby 1980).

Recently, Panggabean *et al.* (1984) have used a general purpose variable geometry device to provide supplementary cobalt, copper, molybdenum, manganese, selenium and zinc to sheep under experimental conditions. Besides their use in ruminant animals, these devices have been used to administer a digesta marker to deer (Kelly *et al.* 1985) and it has been suggested that, by varying their geometry, they may prove useful in some foregut fermenters which do not selectively retain large particles in the stomach (Ellis *et al.* 1985). Laby (1980) points out that planned failure of the retention mechanism at the end of the active life of the devices is possible.

Copper oxide particles. Supplementary copper has been given to ruminants by oral administration of particles which lodge in the abomasum (Dewey 1977). The high specific gravity and low mass of the particles ensures transfer from the rumen and that a proportion of them is retained in the abomasum, where trace amounts of copper are solubilized. The proportion of particles and the time for which they are retained in the abomasum determine the efficacy of supplementation. If the particles lodge in the rumen, then copper supplementation may be less effective due to complexing of released copper with other minerals. Particles not retained in either the abomasum or reticulo-rumen will be excreted in faeces and of no value. However, when particles of appropriate size and form are used, sustained release of copper appears to occur in most animals, but some do not receive sufficient of the supplement (Costigan and Ellis 1980).

Two forms of copper oxide suitable for use are copper oxide needles, as a laboratory chemical, or oxidized copper wire. The retention time of copper oxide needles is variable, evidence indicating retention times of one month (Dewey 1977), to one year (Judson *et al.*

1985a) in sheep and from 40 days in calves (Suttle 1979) to between 100 (Costigan and Ellis 1980) and 180 days (MacPherson 1985) in mature cattle. Oxidized copper wire has been reported to be as effective as subcutaneous injection of copper glycinate (Judson *et al.* 1981a) and more effective than a glass copper/cobalt pellet (Judson *et al.* 1985a) in elevating tissue copper concentrations.

Subcutaneous or intramuscular implants. Controlled release glasses containing trace minerals have been implanted subcutaneously behind the ear or intramuscularly into the neck of ruminants (Allen *et al.* 1979; 1981). While there was an initial rapid release of selenium during the first few days after implantation of a selenium-containing glass, the rate decreased rapidly before stabilizing at a low level. A surface deposit was thought to have contributed to the decline in release rate, but it was concluded that new formulations of selenium controlled-release glasses could be developed to give almost linear release rates.

Subsequently Allen *et al.* (1985) have indicated that parenteral administration (in other words not via the gastro-intestinal tract) is generally unsatisfactory due to adverse tissue reactions at the site of administration. In addition, Laby (1980) points out that a non-zero concentration of the substance being delivered in tissues surrounding the implant would lead to an animal-based source of variability in release rate and a consequent uncertainty of the expiry time. Despite these problems, it would seem that improvements in the technology of such devices and in the delivery process have the potential to make them a useful means of providing trace elements and supplements of other nutrients in the future.

In summary, it would seem that heavy pellets, controlled release devices and copper oxide particles can be used effectively to improve the trace mineral status of foregut fermenters which selectively retain particles in the stomach compartments. Further improvements in these technologies will undoubtedly occur. However, a major challenge exists in defining where these supplements can be used economically as production responses may be the over-riding factor in determining whether trace mineral supplements are given. This necessitates a knowledge of the requirements for these minerals in relation to the likely intake of other major nutrients. Some of these developments such as the variable geometry devices are also likely to be used in the future to provide medicaments, such as anti-bloat agents or anthelmintics, substances which alter fermentation patterns, and perhaps macro-minerals. Other means of providing trace-minerals to herbivores have obvious disadvantages in the efficacy and/or cost of supplementation.

SUPPLEMENTATION WITH VITAMINS

Micro-organisms within the gastro-intestinal tract synthesize B-vitamins and vitamin K and these subsequently become available for absorption from the small intestine in foregut fermenters and in hindgut fermenters which ingest faeces. Other hindgut fermenters obtain these vitamins in the diet. However, as there appears to be evidence of primary deficiencies in herbivores such as the horse, it may be that they sythesize and absorb vitamins in the hindgut. The question of whether or not vitamins are absorbed from the hindgut of herbivores warrants further investigation. Importantly most B-vitamins, an exception being vitamin B_{12}, are not stored in appreciable quantities in body tissues, but as they are only required in minute amounts deficiencies are rare in grazing animals. Vitamin B_{12} deficiency does occur in ruminants in specific environments and is generally rectified by supplying cobalt to the ruminal micro-organisms.

Vitamins A, D and E are obtained from the diet. Carotenoids, the precursors of vitamin A, are present in high concentrations in green plant tissue, but are generally low in dry forage and seeds. Hence deficiencies may occur when herbivores deplete tissue stores during prolonged periods on dry forage or grain diets. Such deficiencies are more likely in young animals with limited body stores or in high-producing animals fed grain-based diets where requirements are high.

Vitamin D_3 is generally obtained from ergosterol in plant materials and from exposure of skin sterols to sunlight. It is important in calcium and phosphorus metabolism and deficiencies primarily affect bone development in young growing animals. Deficiencies are more likely to be a problem with housed animals which are not exposed to direct sunlight.

Vitamin E is present in forage as tocopherols and is generally in high concentrations in green pastures. Like selenium, this vitamin acts as an antioxidant and prevents muscle damage by oxidants produced during tissue metabolism. `White muscle disease', which may successfully be treated with selenium, can occur in lambs, calves, foals and kids and may occur during spring on green pastures or in summer and autumn on dry pastures (Hosking *et al.* 1986). However, in some cases nutritional myopathy of weaners fed on dry forages is only responsive to α-tocopherol (Steele *et al.* 1980; McDonald and Caple 1983), and in captive marsupials is also only vitamin E responsive (see Hume 1982).

Supplements of vitamins for grazing animals are given by pulse dosing techniques using the oral, subcutaneous or intramuscular routes.

SUPPLEMENTATION OF CONFINED HERBIVORES

The dietary requirements of some essential minerals and vitamins, notably iron and vitamin D, may be greater in confined than free-ranging herbivores. It is also inevitable that dietary requirements for essential nutrients will differ in housed and grazing herbivores due to differences in the amounts and nature of the energy-yielding substrates ingested and consequent differences in levels of production. Herbivores are kept confined for many reasons, two examples are given here.

In the small farm systems of Asia, where farmers only possess at most a few animals, large ruminants are routinely confined overnight to prevent theft and may be continuously confined or tethered during cropping seasons. They are fed primarily on fibrous crop residues and hand-harvested forages. While the circumstances indicate that the intake and utilization of these feeds could be improved greatly by provision of appropriate amounts of supplementary nutrients, this is seldom practised. Cereal grains and concentrate by-products are preferentially used for human consumption or for feeding pigs and poultry, while supplements of NPN or minerals are seldom used due to cost and lack of effective responses. The probable reason for the lack of response is the need for balanced nutrient supply to promote microbial activity in the forestomach or more efficient tissue utilization.

In Europe and North America, herbivores are kept confined to avoid losses during winter and/or for feeding in production systems. Production feeding systems may be based on conserved forage, such as maize silage, or on concentrate diets. In these systems it is possible to formulate diets which achieve balanced nutrient supplies for both fermentation and tissue metabolism. Obviously, the difficulties of providing supplements to individual animals are greatly reduced when compared to grazing animals. However, production systems based on intensive feeding of confined animals may only exist while surplusses of concentrates prevail, and the continuing need to increase cropping to meet human requirements may eventually lead to a reduction in concentrate-based feeding. It is likely that, when this occurs, herbivores will be raised in confinement, but feeding will be based more on the utilization of crop residues and by-products. It is worth mentioning that at present in Australia, where supplements are required for animal survival during droughts, it is generally cheaper on the basis of digestible energy supplied to feed cereal grains than conserved forages.

CONCLUSIONS

Effective supplementation of herbivores is dependent upon the achievement of balances in the supply of different energy-yielding substrates and essential nutrients to the tissues. In foregut fermenters, balances in the supply of these substrates also determine the efficiency of fermentation. Thus, specific nutrients will only improve forage utilization where the supply of other nutrients is satisfactory. Examples of this are the general lack of response to provision of mineral supplements when energy or protein supplies are limiting and the poor utilization of NPN supplements by micro-organisms when energy or essential nutrients are poorly available.

While substantial advances have been made towards effective provision of some specific nutrients, such as trace element provision by controlled release devices, an understanding of how much of particular forages different herbivores will consume is lacking. As this ultimately determines the supply of energy-yielding substrates, it is a prerequisite to effective and cost efficient supplementation. This suggests a need for a better understanding of the effects of the physical and chemical characteristics of forages on feed selection which ultimately determine their intake.

Diet selection not only presents problems as regards the prediction of forage intake. It is obvious that the efficiency of supplementation of grazing herbivores is severely limited by variable acceptance and intake of the supplement. Further research designed to overcome these problems is warranted. In the first instance, the use of attractants which make supplements more acceptable may be successful. However, it has been observed that early pre-weaning exposure of animals to supplements improves their subsequent acceptance. This suggests that techniques, such as exposing young animals to flavours, which may later be associated with supplements, may be the key to achieving quicker and more uniform acceptance later in life.

Allen, W.M., Drake, C.F., Sansom, B.F., and Taylor, R.J. (1979). *Ann. Rech. Vet.* **10**, 356-8.

Allen, W.M., Drake, C.F., and Tripp, M. (1985). In "Trace Elements in Man and Animals - TEMA 5." (Eds C.F. Mills, I. Bremner and J.K. Chesters.) (CAB: Farnham Royal, UK.)

Allen, W.M., Moore, P.R., and Sansom, B.F. (1981). In "Trace Element Metabolism in Man and Animals - TEMA-4." (Eds J. McC. Howell, J.M. Gawthorne and C.L. White.) (Aust. Acad. Sci.: Canberra.)

Ammerman, C.B., and Miller, S.M. (1972). *J. Anim. Sci.* **35**, 681-94.

Arnold, G.W. (1975). *Aust. J. Agric. Res.* **26**, 1017-24.

Arnold, G.W., and Bush, I.G. (1968). CSIRO, Division of Plant Industry, Field Station Record, **7**, 47-58.

Arnold, G.W., and Maller, R.A. (1974). *Anim. Prod.* **19**, 309-19.

Azuolas, J.K., and Caple, I.W. (1984). *Aust. Vet. J.* **61**, 223-7.

Baldizan, A., Dixon, R., Parra O, and Parra, R. (1982). In "Universidad Central De Venezuela, Facultad De Agronomia, Instituto De Produccion Animal, Informe Anual '82." (Instituto De Produccion Animal, Universidad Central De Venezuela.)

Barker, S. (1960). Ph.D. thesis, Univ. Western Australia (cited by Hume, 1982).

Bartley, E.E., and Deyoe, C.W. (1977). In "Recent Advances in Animal Nutrition - 1977." (Eds W. Haresign and D. Lewis.) (Butterworths: London.)

Beever, D.E., and Thomson, D.J. (1977). In "Recent Advances in Animal Nutrition - 1977." (Eds W. Haresign and D. Lewis.) (Butterworths: London.)

Bergner, H. (1984). *Can. J. Anim. Sci.* **64** (Suppl.), 37-8.

Bird, P.R. (1974). *Aust. J. Agric. Res.* **25**, 631-42.

Bird, P.R., and Hume, I.D. (1971). *Aust. J. Agric. Res.* **22**, 443-52.

Blair-West, J.R., Coghlan, J.P., Denton, D.A., Nelson, J.F., Orchard, E., Scoggins, B.A., Wright, R.D., Myers, K., and Junqueira, C.L. (1968). *Nature*, Lond. **217**, 922-8.

Blaney, B.J., Gartner, R.J.W., and McKenzie, R.A. (1981). *J. Agric. Sci.* **97**, 507-14.

Bloomfield, R.A., Garner, G.B., and Muhrer, M.E. (1960). *J. Anim. Sci.* **19**, 1248.

Brown, G.D. (1969). *Aust. J. Zool.* **17**, 187-94.

Burroughs, W., Gerlaugh, P., Edgington, B.H., and Bethke, R.M. (1949). *J. Anim. Sci.* **8**, 271-8.

Campling, R.C. (1966). *Br. J. Nutr.* **20**, 25-39.

Campling, R.C., Freer, M., and Balch, C.C. (1962). *Br. J. Nutr.* **16**, 115-24.

Care, A.D., Anderson, P.J.B., Illingworth, D.V., Zervas, G., and Telfer, S.B. (1985). In "Trace Elements in Man and Animals - TEMA 5." (Eds C.F. Mills, I. Bremner and J.K. Chesters.) (CAB: Farnham Royal, UK.)

Carlos, G., Zervas, G., Driver, P.M., Anderson, P.J.B., Illingworth, D.V., Al-Tekrity, S.A.A., and Telfer, S.B. (1985). In "Trace Elements in Man and Animals - TEMA 5" (Eds C.F. Mills, I. Bremner and J.K. Chesters.) (CAB: Farnham Royal, UK.)

Chapple, R.S. (1985). B. Sc. (Hons.) Thesis, Univ. New England, Armidale, NSW. (cited by Lynch, 1985).

Cohen, R.D.H. (1974). *Aust. J. Exp. Agric. Anim. Husb.* **14**, 589-92.

Cohen, R.D.H. (1975). *World Rev. Anim. Prod.* **11** (2), 27-43.

Cohen, R.D.H. (1980). *Livestock Prod. Sci.* **7**, 25-37.

Coombe, J.B. (1985). *Aust. J. Agric. Res.* **36**, 717-28.

Coombe, J.B., and Mulholland, J.G. (1983). *Aust. J. Agric. Res.* **34**, 767-80.

Coombe, J.B., and Tribe, D.E. (1962). *J. Agric. Sci.* **59**, 125-41.

Costigan, P., and Ellis, K.J. (1980). *Proc. Aust. Soc. Anim. Prod.* **13**, 451.

Cotta, M.A., and Russell, J.B. (1982). *J. Dairy Sci.* **65**, 226-34.

Crabtree, J.R., and Williams, G.L. (1971a). *Anim. Prod.* **13**, 71-82.

Crabtree, J.R., and Williams, G.L. (1971b). *Anim. Prod.* **13**, 83-92.

Dewey, D.W. (1977). *Search* **8**, 326-7.

Dewey, D.W., Lee, H.J., and Marston, H.R. (1958). *Nature*, Lond. **181**, 1367–71.

Dewey, D.W., Lee, H.J., and Marston, H.R. (1969). *Aust. J. Agric. Res.* **20**, 1109–16.

Dixon, R.M. (1986a). In "Ruminant Feeding Systems Utilizing Fibrous Agricultural Residues – 1985." (Ed. R.M. Dixon.) (IDP: Canberra.)

Dixon, R.M. (1986b). *Anim. Feed Sci. Technol.* **14**, 193–202.

Doyle, P.T. (1981). *Aust. J. Biol. Sci.* **34**, 47–59.

Doyle. P.T., Devendra, C., and Pearce, G.R. (1986). "Rice Straw as a Feed for Ruminants." (IDP: Canberra.)

Dove, H. (1984). *Aust. J. Exp. Agric. Anim. Husb.* **24**, 484–93.

Dove, H., Freer, M., Egan, A.R., and Doyle, P.T. (1984). *Proc. Nutr. Soc. Aust.* **9**, 166.

Dulphy, J.P., Remond, B., and Theriey, M. (1980). In "Digestive Physiology and Metabolism in Ruminants." (Eds Y. Ruckebusch and P. Thivend.) pp. 103–122. (AVI.: Westport, Connecticut).

Duncan, I.F., Greentree, P.L., and Ellis, K.J. (1986). *Aust. Vet. J.* **63**, 127–8.

Ealey, E.H.M. (1967a). *CSIRO Wildl. Res.* **12**, 9–25.

Ealey, E.H.M. (1967b). *CSIRO Wildl. Res.* **12**, 27–51.

Egan, A.R. (1965). *Aust. J. Agric. Res.* **16**, 451–62.

Egan, A.R. (1972). *Aust. J. Exp. Agric. Anim. Husb.* **12**, 131–5.

Egan, A.R., Benevenga, N.J., and Radcliffe, B.C. (1984). In "Ruminant Physiology, Concepts and Consequences." (Eds S.K. Baker, J.M. Gawthorne, J.B. MacKintosh and D.B. Purser.) (Univ. W. Aust.: Nedlands.)

Egan, A.R., Boda, K., and Varady, J. (1985). In "Control of Digestion and Metabolism in Ruminants." (Eds L.P. Milligan, W.L. Grovum and A. Dobson.) pp. 386–402. (Prentice-Hall: Englewood Cliffs, New Jersey.)

Egan, A.R., Walker, D.J., Nader, C.J., and Storer, G. (1975). *Aust. J. Agric. Res.* **26**, 909–22.

Egan, J.K., and Doyle, P.T. (1985). *Aust. J. Agric. Res.* **36**, 483–95.

Elliott, R., and Armstrong, D.G. (1982). *J. Agric. Sci.*, **99**, 51–60.

Elliott, R., Ferreiro, H.M., Priego, A., and Preston, T.R. (1978). *Trop. Anim. Prod.* **3**, 30–5.

Ellis, K.J., Costigan, P., and Schlink, A.C. (1985). Seminar on Research and Development of Controlled-release Technology for Agrochemicals using Isotopes, Vienna, Austria, 1–5 July, 1985.

El-Shazly, K., Dehority, B.A., and Johnson, R.R. (1961). *J. Anim. Sci.* **20**, 268–73.

Entwistle, K.W., and Knights, G. (1974). *Aust. J. Exp. Agric. Anim. Husb.* **14**, 17–22.

Ernst, A.J., Limpus, J.F., and O'Rourke, P.K. (1975). *Aust. J. Exp. Agric. Anim. Husb.* **15**, 451–5.

Fahmy, S.T.M., Lee, N.H., and Ørskov, E.R. (1984). *Anim. Prod.* **38**, 75-81.
Field, A.C., and Wooliams, J.A. (1984). *Can. J. Anim. Sci.* **64** (Suppl.), 232-3.
Fonnesbeck, P.V., Kearl, L.C., and Harris, L.E. (1975). *J. Anim. Sci.* **40**, 1150-84.
Forbes, D.K., and Tribe, D.E. (1969). *Aust. J. Zool.* **17**, 765-75.
Fredericks, F., Dixon, R.M., Holmes, J.H.G., and Egan, A.R. (1986). *Proc. Aust. Soc. Anim. Prod.* **16**, 207-10.
Gartner, R.J.W., Blaney, B.J., and McKenzie, R.A. (1981). *J. Agric. Sci.* **97**, 581-9.
Gartrell, J.W., and Glencross, R.N. (1969). W.A. Dept. Agric. Bulletin No. 3614.
Gawthorne, J.M. (1984). In "Ruminant Physiology, Concepts and Consequences." (Eds S.K. Baker, J.M. Gawthorne, J.B. Mackintosh and D.B. Purser.) (School of Agriculture, Univ. W. Aust.: Nedlands.)
Giesecke, D., and van Gylswyk, N.O. (1975). *J. Agric. Sci.* **85**, 75-83.
Gonzalez-Jimenez, E. (1977). *World Anim. Rev.* **21**, 24-30.
Green, G.C., Elwin, R.L., Mottershead, B.E., Keogh, R.G., and Lynch, J.J. (1984). *Proc. Aust. Soc. Anim. Prod.* **15**, 373-5.
Guardiola, C.M., Fahey, G.C., Spears, J.W., and Garrigus, U.S. (1980). *Can. J. Anim. Sci.* **60**, 337-44.
Gulbransen, B. (1974). *Proc. Aust. Soc. Anim. Prod.* **10**, 74-7.
Hacker, J.B. (1982). (Ed.) "Nutritional Limits to Animal Production from Pastures." (CAB: Farnham Royal, UK.)
Halpin, C., McDonald, J., Hanrahan, P., and Caple, I. (1985). In "Trace Elements in Man and Animals - TEMA 5." (Eds C.F. Mills, I. Bremner and J.K. Chesters.) (CAB: Farnham Royal, UK.)
Hartmans, J. (1974). *Neth. J. Agric. Sci.* **22**, 195-206.
Hennessy, D.W., and Murison, R.D. (1982). *Aust. J. Exp. Agric. Anim. Husb.* **22**, 140-6.
Hennessy, D.W., Williamson, P.J., Lowe, R.F., and Baigent, D.R. (1981). *J. Agric. Sci.* **96**, 205-12.
Hennessy, D.W., Williamson, P.J., Nolan, J.V., Kempton, T.J., and Leng, R.A. (1983). *J. Agric. Sci.* **100**, 657-66.
Henning, P.A., Linden, Y.V.D., Mattheyse, M.E., Nauhaus, W.K., Schwartz, H.M., and Gilchrist, F.M. (1980). *J. Agric. Sci.* **94**, 565-73.
Hinkle, D.K., Potter, G.D., and Kreider, J.L. (1983). In "Proceedings of the Eighth Equine Nutrition and Physiology Symposium." pp. 227-30.
Hintz, H.F., Hogue, D.E., Walker, E.F., Lowe, J.E., and Schryver, H.F. (1971). *J. Anim. Sci.* **32**, 245-8.
Hodge, R.W., and Bogdanovic, B. (1983). *Aust. J. Exp. Agric. Anim. Husb.* **23**, 19-23.

Hogan, J.P. (1984). In "Ruminant Physiology, Concepts and Consequences." (Eds S.K. Baker, J.M. Gawthorne, J.B. Mackintosh and D.B. Purser.) (Univ. W. Aust.: Nedlands.)

Hoppe, P.P., Qvortrup, S.A., and Woodford, M.H. (1977a). *J. Agric. Sci.*, **89**, 129-35.

Hoppe, P.P., Qvortrup, S.A., and Woodford, M.H. (1977b). *J. Zool.* Lond. **181**, 1-9.

Hornicke, H., and Bjornhag, G. (1980). In "Digestive Physiology and Metabolism in Ruminants." (Eds Y. Ruckebusch and P. Thivend.) pp. 707-30. (AVI: Westport, Connecticut.)

Hosking, W.J., Caple, I.W., Haplin, C.G., Brown, A.J., Paynter, D.I., Conley, D.N., and North-Coombes, P.L. (1986). "Trace Elements for Pastures and Animals in Victoria." (Vict. Gov. Printing Office: Melbourne.)

Hudson, D.R., Hunter, R.A., and Peter, D.W. (1981a). In "Trace Element Metabolism in Man and Animals - TEMA 4." (Eds J. McC. Howell, J.M. Gawthorne and C.L. White.) (Aust. Acad. Sci.: Canberra.)

Hudson, D.R., Hunter, R.A., and Peter, D.W. (1981b). *Aust. J. Agric. Res.* **32**, 935-45.

Hume, I.D. (1970a). *Aust. J. Agric. Res.* **21**, 297-304.

Hume, I.D. (1970b). *Aust. J. Agric. Res.* **21**, 305-14.

Hume, I.D. (1974). *Aust. J. Agric. Res.* **25**, 155-65.

Hume, I.D. (1982). "Digestive Physiology and Nutrition of Marsupials." (Cambridge Univ.: Cambridge.)

Hume, I.D., and Warner, A.C.I. (1980). In "Digestive Physiology and Metabolism in Ruminants." (Eds Y. Ruckebusch and P. Thivend.) pp. 665-84. (AVI: Westport, Connecticut.)

Hunter, A.P. (1977). *New Zealand Vet. J.* **25**, 305-7.

Hunter, R.A., Peter, D.W., Hudson, D.R., and Chandler, B.S. (1981). *Aust. J. Agric. Res.* **32**, 927-33.

Hunter, R.A., and Siebert, B.D. (1980). *Aust. J. Agric. Res.* **31**, 1037-47.

Hunter, R.A., Siebert, B.D., and Webb, C.D. (1979). *Aust. J. Exp. Agric. Anim. Husb.* **19**, 517-21.

Hutson, G.D., and van Mourik, S.C. (1981). *Aust. J. Exp. Agric. Anim. Husb.* **21**, 575-82.

Hutson, G.D., and Wilson, P.N. (1984). *Anim. Prod.* **38**, 145-6.

Jacobs, G.J.L., and Leibholz, J. (1977). *J. Agric. Sci.*, **89**, 699-710.

Janis, C. (1976). *Evolution* **30**, 757-74.

Jenkins, K.J., Hidiroglou, M., Wauthy, J.M., and Proulx, J.E. (1974). *Can. J. Anim. Sci.* **54**, 49-60.

Johnson, R.R. (1976). *J. Anim. Sci.* **43**, 184-91.

Johnson, W.H., Norman, B.B., and Dunbar, J.R. (1981). In "Trace Element Metabolism in Man and Animals - TEMA-4." (Eds J. McC. Howell, J.M. Gawthorne and C.L. White.) (Aust. Acad Sci.: Canberra).

Judson, G.J., Brown, T.H., and Dewey, D.W. (1985a). In "Trace Elements in Man and Animals - TEMA 5." (Eds C.F. Mills, I. Bremner and J.K. Chesters.) (CAB: Farnham Royal, UK.).

Judson, G.J., Dewey, D.W., McFarlane, J.D., and Riley, M.J. (1981a). In "Trace Elements in Man and Animals - TEMA - 4." (Eds J. McC. Howell, J.M. Gawthorne and C.L. White.) (Aust. Acad. Sci.: Canberra.)

Judson, G.J., Koh, Tee-Siaw, McFarlane, J.D., Turnbull, R.K., and Kempe, B.R. (1985b). In "Trace Elements in Man and Animals - TEMA 5." (Eds C.F. Mills, I. Bremner and J.K. Chesters.) (CAB: Farnham Royal, UK.)

Judson, G.J., McFarlane, J.D., Riley, M.J., Milne, M.L., and Horne, A.C. (1981b). In "Trace Elements in Man and Animals - TEMA-4." (Eds J. McC. Howell, J.M. Gawthorne and C.L. White.) (Aust. Acad. Sci.: Canberra).

Kakulas, B.A. (1961). *Nature*, Lond. **191**, 402-3.

Kaufmann, W. (1976). *Livestock Prod. Sci.* **3**, 103-14.

Kay, R.N.B., Engelhardt, W.V., and White, R.G. (1980). In "Digestive Physiology and Metabolism in Ruminants." (Eds Y. Ruckebusch and P. Thivend.) (AVI: Westport, Connecticut.)

Kelly, T.J., Thompson, J.M., Ellis, K.J., and Rowe, M.D. (1985). Paper presented at 55th ANZAAS Congress, 26-30 August, 1985.

Kempton, T.J., and Leng, R.A. (1979). *Br. J. Nutr.* **42**, 289-302 .

Kempton, T.J., Nolan, J.V., and Leng, R.A. (1979). *Br. J. Nutr.* **42**, 303-15.

Kennedy, P.M., and Siebert, B.D. (1972). *Aust. J. Agric. Res.* **23**, 45-56.

Kennedy, P.M., and Siebert, B.D. (1973). *Aust. J. Agric. Res.* **24**, 143-52.

Knight, W.M., and Owens, F.N. (1973). *J. Anim. Sci.* **36**, 145-9.

Knott, P., Algar, B. Zervas, G., and Telfer, S.B. (1985). In "Trace Elements in Man and Animals - TEMA 5." (Eds C.F. Mills, I. Bremner and J.K. Chesters.) (CAB: Farnham Royal, UK.)

Kondos, A.C. and Mutch, B. (1975). *J. Agric. Sci.*, **85**, 359-68.

Kuchel, R.E., and Buckley, R.A. (1969). *Aust. J. Agric. Res.* **20**, 1099-107.

Kunju, P.J.G. (1986). In "Rice Straw and Related Feeds in Ruminant Rations." (Eds M.N.M. Ibrahim and J.B. Schiere.) (Straw Utilization Project, Publication No. 2: Kandy, Sri Lanka.)

Laby, R.H. (1970). Australian Patent No. 449,029 (cited by Laby 1980).

Laby, R.H. (1980). *Proc. Aust. Soc. Anim. Prod.* **13**, 6-10.

Laby, R.H., Graham, C.A., Edwards, S.R., and Kautzner, B. (1984). *Can. J. Anim. Sci.* **64** (Suppl.), 337-8.

Langlands, J.P., and Bowles, J.E. (1976). *Aust. J. Exp. Agric. Anim. Husb.* **16**, 630-5.

Langlands, J.P., and Donald, G.E. (1978). *Aust. J. Agric. Res.* **29**, 875-83.

Lee, G.J., Hennessy, D.W., Nolan, J.V., and Leng, R.A. (1987). *Aust. J. Agric. Res.* **38**, (in press).

Leibholz, J., and Kellaway, R.C. (1984). AMRC Review No. 48. (Australian Meat Research Committee, Sydney, NSW).

Leng, R.A., and Preston, T.R. (1983). *Proc. 5th World Conference on Animal Production* **1**, pp. 310–8.

Lindsay, J.A., Mason, G.W.J., and Toleman, M.A. (1982). *Proc. Aust. Soc. Anim. Prod.* **14**, 67–72.

Lindsay, J.R., Weston, R.H., Purser, D.B., and Davis, P. (1983). *Proc. Nutr. Soc. Aust.* **8**, 148.

Lintern-Moore, S. (1973). *Comp. Biochem. Physiol.* **46A**, 345–51.

Llewelyn, D., Kempton, T.J., and Nolan, J.V. (1978). *Proc. Aust. Soc. Anim. Prod.* **12**, 174.

Lobato, J.F.P., and Pearce, G.R. (1980a). *Aust. J. Exp. Agric. Anim. Husb.* **20**, 417–21.

Lobato, J.F.P., and Pearce, G.R. (1980b). *Aust. J. Exp. Agric. Anim. Husb.* **20**, 422–6.

Lobato, J.F.P., Pearce, G.R., and Beilharz, R.G. (1980b). *Appl. Anim. Ethol.* **6**, 149–61.

Lobato, J.F.P., Pearce, G.R., and Tribe, D.E. (1980a). *Aust. J. Exp. Agric. Anim. Husb.* **20**, 413–6.

Lynch, J.J. (1985). In "Recent Advances in Animal Production in Australia 1985." (Ed. R.B. Cumming.) (Univ. New England: Armidale, NSW.)

Lynch, J.J., Keogh, R.G., Elwin, R.L., Green, G.C., and Mottershead, B.E. (1983). *Anim. Prod.* **36**, 175–83.

MacPherson, A. (1981). In "Trace Element Metabolism in Man and Animals – TEMA-4." (Eds J.McC. Howell, J.M. Gawthorne and C.L. White.) (Aust. Acad. Sci.: Canberra.)

MacPherson, A. (1985). In "Trace Elements in Man and Animals – TEMA-5." (Eds C.F. Mills, I. Bremner and J.K. Chesters.) (CAB: Farnham Royal, UK.)

Maeng, W.J., and Baldwin, R.L. (1976). *J. Dairy Sci.* **59**, 643–7.

Martin, P.J., Milligan, L.P., and Mathison, G.W. (1982). *Can. J. Anim. Sci.* **62**, 1129–34.

Masters. D.G. (1984). In "Ruminant Physiology, Concepts and Consequences." (Eds S.K. Baker, J.M. Gawthorne, J.B. Mackintosh and D.B. Purser.) (School of Agriculture, Univ. W. Aust.: Nedlands.)

Masters, D.G., and Fels, H.E. (1980). *Biol. Trace Elem. Res.* **2**, 281–90.

Masters. D.G., and Moir, R.J. (1980). *Aust. J. Exp. Agric. Anim. Husb.* **20**, 547–51.

McDonald, J.W. (1975). *Aust. Vet. J.* **51**, 433–5.

McDonald, J.W., and Caple, I.W. (1983). In "Proceedings No. 67, Sheep Production and Preventive Medicine." (Postgraduate Committee Vet. Sci., Univ. Sydney: Sydney.)

McDonnell, M.L., Klopfenstein, T.J., and Britton, R.A. (1979). *J. Anim. Sci.* **49**, Suppl. 1, 267.

McKenzie, R.A., Gartner, R.J.W., Blaney, B.J., and Glanville, R.J. (1981). *Aust. Vet. J.* **57**, 554–7.

McMeniman, N.P., and Little, D.A. (1974). *Aust. J. Exp. Agric. Anim. Husb.* **14**, 316–21.

Meggison, P.A., McMeniman, N.P., and Armstrong, D.G. (1979a). *Proc. Nutr. Soc.* **38**, 146A.

Meggison, P.A., McMeniman, N.P., and Armstrong, D.G. (1979b). *Proc. Nutr. Soc.* **38**, 147A.

Merry, R.J., Smith, R.H., and McAllan, A.B. (1982a). *Br. J. Nutr.* **48**, 287–304.

Merry, R.J., Smith, R.H., and McAllan, A.B. (1982b). *Br. J. Nutr.* **48**, 305–18.

Meyer, H., Schmidt, H., and Lindemann, G. (1983). In "Proceedings of the Eighth Equine Nutrition and Physiology Symposium." pp. 231–3.

Milligan, L.P., Robblee, A.R., Wood, J.C., Kay, W.C., and Chakrabartty, S.K. (1969). *Can. J. Anim. Sci.* **49**, 135–41.

Milton, J.T.B., and Ternouth, J.H. (1985). *Aust. J. Agric. Res.* **36**, 647–54.

Moir, R.J. (1965). In "Physiology of Digestion in the Ruminant." (Eds R.W. Dougherty, R.S. Allen, W. Burroughs, N.L. Jacobson and A.D. McGilliard.) (Butterworths: London.)

Moir, R.J., and Somers, M. (1957). *Aust. J. Agric. Res.* **8**, 253–65.

Mottershead, B.E., Lynch, J.J., Elwin, R.L., and Green, G.C. (1985). *Anim. Prod.* **41**, 257–9.

Mould, F.L., and Ørskov, E.R. (1983). *Anim. Feed Sci. Technol.* **10**, 1–14.

Mould, F.L., Ørskov, E.R., and Mann, S.O. (1983). *Anim. Feed Sci. Technol.* **10**, 15–30.

Mulholland, J.G., and Coombe, J.B. (1979). *Aust. J. Exp. Agric. Anim. Husb.* **19**, 23–31.

Mulholland, J.G., Coombe, J.B., Freer, M., and McManus, W.R. (1976a). *Aust. J. Agric. Res.* **27**, 881–93.

Mulholland, J.G., Coombe, J.B., and McManus, W.R. (1976b). *Aust. J. Agric. Res.* **27**, 139–53.

Nikolic, J.A., Pavlicevic, A., Zeremski, D., and Negovanovic, D. (1980). In "Digestive Physiology and Metabolism in Ruminants." (Eds Y. Ruckebusch and P. Thivend.) pp. 603–20. (AVI: Westport, Connecticut.)

Nolan, J.V., Ball, F.M., Murray, R.M., Norton, B.W., and Leng, R.A. (1974). *Proc. Aust. Soc. Anim. Prod.* **10**, 91–4.

Nolan, J.V., and Leng, R.A. (1972). *Br. J. Nutr.* **27**, 177–94.

Nolan, J.V., Norton, B.W., Murray, R.M., Ball, F.M., Roseby, F.B., Rohan-Jones, W., Hill, M.K., and Leng, R.A. (1975). *J. Agric. Sci.*, **84**, 39–48.

Oltjen, R.R., Slyter, L.L., Kozak, A.S., and Williams, E.E. (1968). *J. Nutr.* **94**, 193–202.

Ørskov, E.R., and Fraser, C. (1975). *Br. J. Nutr.* **34**, 493–500.

Ozanne, P.G., Purser, D.B., Howes, K.M.W., and Southey, I. (1976). *Aust. J. Exp. Agric. Anim. Husb.* **16**, 353-60.

Panggabean, T., Moir, R.J., Purser, D.B., and Laby, R. (1984). In "Ruminant Physiology, Concepts and Consequences." (Eds S.K. Baker, J.M. Gawthorne, J.B. Mackintosh and D.B. Purser.) (Univ. W. Aust.: Nedlands.)

Paulson, G.D., Broderick, G.A., Baumann, C.A., and Pope, A.L. (1968). *J. Anim. Sci.* **27**, 195-202.

Peeler, H.T. (1972). *J. Anim. Sci.* **35**, 695-712.

Peter, D.W., Hunter, R.A., and Hudson, D.R. (1981). In "Trace Element Metabolism in Man and Animals - TEMA 4." (Eds J. McC. Howell, J.M. Gawthorne and C.L. White.) (Aust. Acad. Sci.: Canberra.)

Petterson, D.S., Casey, R.H., Masters, H.G., and Wilson, P.E. (1985). In "Trace Elements in Man and Animals - TEMA 5." (Eds C.F. Mills, I. Bremner and J.K. Chesters.) (CAB: Farnham Royal, UK.)

Playne, M.J., Echevarria, M.G., and Megarrity, R.G. (1978). *J. Sci. Fd. Agric.* **29**, 520-6.

Pollard, T.G. (1981). M.Agr.Sc. Thesis, Univ. Melbourne.

Redman, R.G., Kellaway, R.C., and Leibholz, J. (1980). *Br. J. Nutr.* **44**, 343-54.

Rees, M.C., Davis, P., and Hogan, J.P. (1982). *Proc. Aust. Soc. Anim. Prod.* **14**, 681.

Rees, M.C., and Minson, D.J. (1976). *Br. J. Nutr.* **36**, 179-87.

Rees, M.C., Minson, D.J., and Smith, F.W. (1974). *J. Agric. Sci.*, **82**, 419-22.

Richards, D.H., Hewett, G.R., and Parry, J.M. (1985). In "Trace Elements in Man and Animals - TEMA 5." (Eds C.F. Mills, I. Bremner and J.K. Chesters.) (CAB: Farnham Royal, UK.)

Rocks, R.L., Wheeler, J.L., and Hedges, D.A. (1980). *Proc. Aust. Soc. Anim. Prod.* **13**, 293-6.

Romero, A., Siebert, B.D., and Murray, R.M. (1976). *Aust. J. Exp. Agric. Anim. Husb.* **16**, 308-14.

Rook, J.A.F., and Storry, J.E. (1962). *Nutr. Abstr. Rev.* **32**, 1055-77.

Siebert, B.D., and Hunter, R.A. (1982). Supplementary feeding of grazing animals. In "Nutritional Limits to Animal Production from Pastures." (Ed. J.B. Hacker.) pp. 409-26 (CAB: Farnham Royal, UK.)

Silva, A.T., and Ørskov, E.R. (1985). *Proc. Nutr. Soc.* **44**, 50A.

Smith, G.H., and Warren, B. (1986a). *Aust. J. Exp. Agric.* **26**, 1-6.

Smith, G.H., and Warren, B. (1986b). *Aust. J. Exp. Agric.* **26**, 7-12.

Spais. A.G., and Papasteriadis, A.A. (1974). In "Trace Element Metabolism in Animals - 2." (Eds W.G. Hoekstra, J.W. Suttie, H.E. Ganther and W. Mertz.) pp. 628-31. (Univ. Park: Baltimore.)

Squires, V.R., and Daws, G.T. (1975). *Appl. Anim. Ethol.* **1**, 263-74.

Steele, P., Peet, R.L., Skirrow, S., Hopkinson, W., and Masters, H.G. (1980). *Aust. Vet. J.* **56**, 529–32.

Stephenson, R.G.A. (1983). *Aust. Rangl. J.* **5**, 83–8.

Stephenson, R.G.A., Cobon, D., McGuigan, K.R., and Hopkins, P.S. (1984). *Proc. Aust. Soc. Anim. Prod.* **15**, 601–3.

Stephenson, R.G.A., Edwards, J.C., and Hopkins, P.S. (1981). *Aust. J. Agric. Res.* **32**, 497–509.

Stephenson, R.G.A., and Hopkins, P.S. (1985). In "Recent Advances in Animal Nutrition in Australia 1985." (Ed. R.B. Cumming.) (Univ. New England: Armidale, NSW.)

Suttle, N.F. (1979). *Proc. Nutr. Soc.* **38**, 135A.

Sutton, J.D. (1980). In "Digestive Physiology and Metabolism in Ruminants." (Eds Y. Ruckebusch and P. Thivend.) pp. 271–90. (AVI: Westport, Connecticut.)

Telfer, S.B., Zervas, G., and Knott, P. (1983). UK. Patent G.B. 21164249, pp. 7–12. (Cited by Knott *et al.* 1985.)

Terry, R.A., Tilley, J.M.A., and Outen, G.E. (1969). *J. Sci. Food Agric.* **20**, 317–20.

Thompson, K.N., Baker, J.P., Lew, J.P., and Baruc, C.J. (1981). In "Proceedings of the Seventh Equine Nutrition and Physiology Symposium." pp. 3–7.

Tudor, G.D., and Morris, J.G. (1971). *Aust. J. Exp. Agric. Anim. Husb.* **11**, 483–7.

Umunna, N.N. (1982). *J. Agric. Sci.* **98**, 343–6.

Underwood, E.J. (1977). "Trace Elements in Human and Animal Nutrition." (Academic: New York.)

van Niekerk, B.D.H., and Jacobs, G.A. (1985). *S. Afr. J. Anim. Sci.* **15**, 133–6.

Waldo, D.R. (1973). *J. Anim. Sci.* **37**, 1062–74.

Watkinson, J.H. (1983). *N.Z. Vet. J.* **31**, 78–85.

Wheeler, J.L. Rocks, R.L., and Hedges, D.A. (1980). *Proc. Aust. Soc. Animl Prod.* **13**, 297–300.

Wiener, G. (1979). *Livestock Prod. Sci.* **6**, 223–32.

Wright, D.E., Towers, N.R., and Sinclair, D.P. (1978). *N.Z. J. Agric. Res.* **21**, 215–21.

20

HERBIVORES IN COLD CLIMATES

by R.G. White, D.F. Holleman, M.E. Hubbert and H. Staaland

The circumpolar arctic and subarctic land masses are inhabited by animals that have evolved mechanisms that allow them to survive in a highly seasonal environment dominated by cold, darkness and reduced accessibility of food in the winter and by a variable but warm, long-daylight and food-rich summer period. In this paper a review is presented of the nutritional adaptations made by terrestrial herbivorous mammals indigenous to these extreme cold-dominated environments. The adaptations of these indigenous animals have a markedly seasonal basis (Irving 1972) which allow optimization in timing of events such as matching the breeding cycle to the seasonal food supply. Thus exploitation of the food resource can be partitioned seasonally. In winter, competition between species for the food resource is apparently minimized by preferential and specialized foraging that involves a combination of chemo-taxis plus alimentary functions that maximize fibre processing in relation to body size constraints and energy requirements. A continuum of fore-gut and hind-gut fermenters are included in the herbivorous mammalian array, but missing are small fore-gut fermenters and large hind-gut fermenters, species that typically require an abundant year-round food supply.

From an animal production standpoint, these seasonally cold environments present challenges that must be overcome by man when introducing animals that have not evolved to maximize productivity in highly seasonal cold environments. With the possible exception of reindeer (*Rangifer tarandus*), common domestic species have not been artificially selected in these cold environments. Cold affects the food requirements, forage digestion and metabolic rate in the domesticated species in an almost contrary fashion to the indigenous herbivores (Young *ibid*. pp. 163–190). Most of these differences are attributable to domestic animals being less cold tolerant, in that they infrequently experience temperatures below their thermoneutral range, and have inappropriate timing of breeding relative to the food supply. Thus, in winter, acute and chronic cold stress in sheep and cattle increases metabolic rate and food intake which can lead to a lowered food digestibility and lowered production efficiency. Under similar environmental conditions large indigenous ruminants, such as reindeer (*R. tarandus tarandus*), caribou (*R. tarandus* ssp.), moose (*Alces alces*), muskox (*Ovibos moschatus*) and deer (Family: Cervidae) are usually within their thermoneutral range (Parker and Robbins 1985) and show a high degree of nutrient economy in winter (White 1979).

This paper will concentrate on aspects of the influence of both temperature and nutrition on the year-round residents. These are

NUTRITION OF HERBIVORES
ISBN 0 12 750052 9

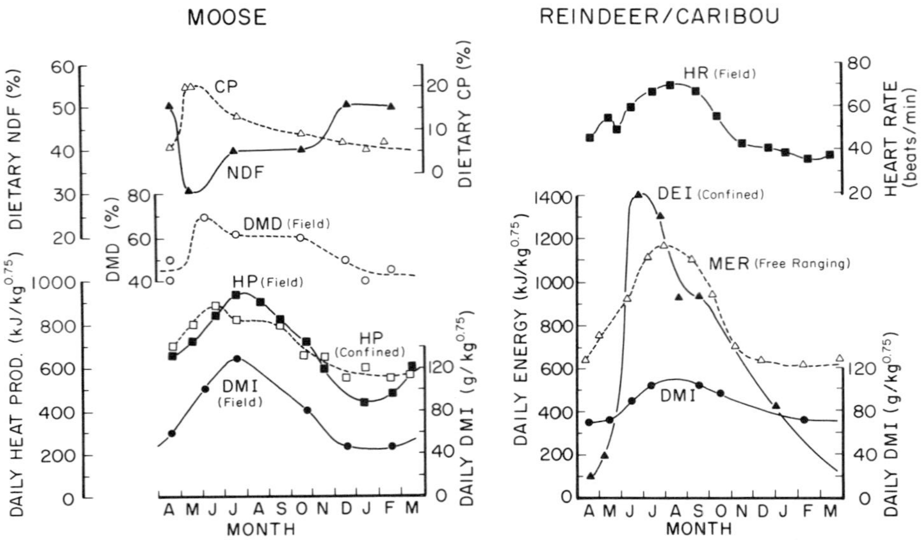

Figure 1. Seasonal changes in diet quality, heat production (HP) and dry matter intake (DMI) in comparison with an index of resting HP (heart rate, HR, Fancy 1986; Fancy and White 1986), digestible energy intake (DEI, Fancy and White 1986), maintenance energy requirements (MER, Boertje 1985) and DMI (Boertje 1985) in reindeer/caribou and moose. Except for the HP of confined moose (Regelin *et al.* 1985) all other estimates are from unrestrained animals (moose, reindeer) or are based on summation models (Boertje 1985). A common feature in these two species is that field HP or HR peaks in August, well after the peak in food quality (crude protein (CP), and dry matter digestibility (DMD) as measured with nylon bags, Reneker and Hudson 1985).

considered fully acclimatized to cold-dominated environments and since this book emphasizes herbivore nutrition, more emphasis is placed on seasonal effects and food handling dynamics than on cold acclimatization *per se*. For detailed analyses of energetic and thermoregulatory adaptations of these species the reader is directed to reviews by Scholander *et al.* (1950a, b, c), Whittow (1971), Irving (1972), Moen (1973), Wang and Hudson (1978), Christopherson *et al.* (1979), Parker and Robbins (1985) and Feist and White (1987).

SEASONAL INFLUENCES AND ENDOGENOUS RHYTHMS

Adaptation to extreme seasonal environments include regulatory mechanisms that prepare animals for winter maintenance and survival and maximize reproduction success by timing growth and birth to the food supply (Irving 1972). These effects are very clear for the large animals that must live above the snow. All are highly seasonal breeders that display a voluntary reduction in appetite and resting-fasting metabolism in winter (Fig. 1). Studies with deer (Pollock 1975; Kay and Ryder 1978; Abbott *et al.* 1984) and reindeer (Ryg 1983) show that the appetite is controlled by photoperiod but no studies to date have shown whether the appetite and metabolic responses are truly circa-annual. Although winter-induced reduction in appetite and metabolism is well documented, recent studies have shown that in winter compensatory or catch-up body growth can be made in young reindeer (Ryg and Jacobsen 1982a, b; Ryg 1986) and relatively mature moose (Hubbert 1987). These studies suggest an important feedback role of appetite in relation to target body weights or body condition levels that the animal strives to achieve at a particular time of the year if the food resource is, or becomes, abundant (Hubbert 1987).

Controversy exists over whether the appetite or metabolism is the pace-setter of these seasonal changes. The fundamental question as to which has higher control has not been fully answered. On the one hand it has been shown that restriction of food intake in summer or winter results in equivalent reductions in fasting metabolic rate in both reindeer and moose (Nilssen *et al.* 1984a, b; Hubbert 1987), suggesting that appetite is pace-setting. Alternately, concurrent measurement of voluntary food intake and metabolic rate, estimated from heart rate, shows that a seasonal spring increase in metabolism precedes the increase in food intake in caribou (Fancy 1986). This suggests that metabolic rate increases in anticipation of the spring flush and therefore metabolic regulation precedes the seasonal change in appetite (Fancy 1986).

Little is known of the seasonal appetite and metabolic rates of the indigenous hind-gut fermenters. However, given that their food supply is limiting and of low quality in winter and that breeding is highly seasonal and confined to spring and early summer, we suggest that a seasonal preparedness through appetite and metabolism might be expected (Holleman *et al.* 1982).

In addition to seasonal appetite, the body condition of northern ruminants has a remarkably high seasonal pattern (Dauphine 1976; Robbins 1983; Reimers and Ringberg 1983; Huot and Goudreault 1985). Although seasonal pattern in body weight has been documented for hares and other rodents, their change in body condition is not very marked. Whether nutrition is at the basis of this lack of body condition change is not known. However, it cannot be a body mass constraint as other small species such as the arctic ground squirrel (*Citelus undulatus*) store extreme quantities of body fat towards late summer, in preparation for their hibernation through winter. The question as to why hares and rodents do not fatten is perhaps linked to their requirement to maintain high winter metabolism. Allometry may be important, however, as their high surface area to body weight ratio means they cannot store sufficient fat to last more than a few days without eating in winter (Morrison 1960). Therefore selection for small body reserves while maintaining a high metabolic rate is a highly successful strategy when there is a sufficient food base, as it provides for good predator avoidance. Highly obese animals may be less agile, and unable to avoid predators and would be highly sought after as a food resource.

Many arctic and subarctic species undergo cyclical fluctuations in numbers. A good series of years during which reproduction rates are high may result in large populations of arctic hares (*Lepus arcticus*). Such population outbreaks have been documented on the high arctic islands. In the boreal forest of North America, population outbreaks of snowshoe hares (*Lepus americanus*) are characteristic and occur on approximately 10 year cycles. Other smaller-bodied rodents are subnivean. Usually they breed during the summer months; however, a number of species, including lemmings and voles, display occasional reproductive activity beneath the snow that is apparently a result of a high-quality food base. In populations of some species which show cyclical population fluctuations, a population outbreak phase is based on subnivean reproduction (Batzli *et al.* 1980; Batzli 1981). These animals are capable of high rates of food ingestion in winter, and reproduction is based on utilization of good quality frozen green forage supplemented by mosses. Clearly the role of photoperiod in controlling the breeding cycle in these subnivean species differs from the full photoperiodic effect as it affects the larger herbivores.

FOOD RESOURCES OF NORTH-TEMPERATE HERBIVORES

The major nutritional or ecological factors to which these species have evolved include a highly seasonal pattern of food supply. Phenological gradients and the level of plant productivity are created by seasonal snow ablation patterns, aspect, altitude, permafrost and soil substrate.

The large northern herbivores respond to these patterns (Klein 1964, 1968, 1970; Skogland 1978; Reimers *et al.* 1983; Kuropat 1984) while the same patterns are responsible for the spatial distribution of the small mammals (Batzli *et al.* 1980). In general, both plant species diversity and primary production, including the production of leaf and floral components, decreases with latitude (Klein 1986) and these biomass changes are associated with changes in the grazing density of reindeer and caribou (Skogland 1978, 1983, 1985, 1986; White *et al.* 1981). This cline in productivity was studied during the IBP era and is detailed in the Arctic Biome synthesis volume (Bliss *et al.* 1981). More recent studies have detailed latitudinal clines in productivity for grasses (Riley and Skjelvag 1984) and willows (Klein 1986). Plant biomass below 100 g DM /m^2 may limit feeding rate in northern ruminants (White *et al.* 1975; Trudell and White 1981; Kuropat 1984; Skogland 1985; Wickstrom *et al.* 1984; Arnold 1985; Renecker and Hudson 1986b) and therefore food availability may limit daily intake on extreme northern ranges.

In contrast to plant biomass and productivity, the quality of this vegetation expressed as crude protein, phosphorus and non-fibrous cell components, increases with latitude (Batzli *et al.* 1980; Van Soest 1982). The high food quality supports rapid body growth of herbivorous mammals in summer and compensatory growth can occur in response to winter food restriction if plant biomass is not limiting daily intake (Suttie *et al.* 1983, 1984; Ryg 1986). A further major limitation of food intake on an annual basis is the duration of high food quality and its availability (Klein 1986). Recent findings for geographic clines in moose and red deer body weight in Norway have been explained in terms of these plant responses, rather than by the classical explanation based on Bergmann's Rule for adaptation to cold (Saether 1985; Langvatn and Albon 1986).

Winter food supply and classification of northern herbivores

The winter period is characteristically a time of low food availability for all herbivores except perhaps for some of the small rodents which can forage on frozen green vegetation and on buried plant parts, stolons and roots. When combined with severe cold and low light availability, a variable snow cover becomes a strong selective influence on the resident herbivores (Klein 1970) as it influences both food availability and the energetics associated with finding food (Parker and Robbins 1985; Fancy and White 1985a, b), escaping predators and ameliorating the effects of cold stress (Fancy 1986). These nutritional-environmental factors have presented very different selective pressures on small and large herbivores. Although a wide array of large herbivores, representing both fore- and hind-gut fermenters, was present in the most northern refugia during the Pleistocene, only the fore-gut fermenters survived to present times. In contrast, the small hind-gut species such as the lagomorphs are extremely successful, e.g. arctic, tundra and snowshoe hares.

The winter food supply consists of a highly variable array of non-vascular (lichens, mosses and fungi) and vascular (graminoids and woody browse) plants. Independent of body size, indigenous herbivores with both the fore-gut and hind-gut types of digestive systems show winter dietary adaptations favouring either a browse, monocot or non-vascular plant specialization. The extreme grazer (monocot) specialists include bison (*Bison bison*), arctic hare and brown lemming (*Lemmus trimucronatus*), while the extreme woody browse (dicot) specialists include the moose, snowshoe hare and collared lemming (*Dicroxtonyx groenlandicus*). The only herbivores that can be considered to be concentrate selectors in winter are the reindeer and caribou which specialize on terrestrial lichens which are of high energy and organic matter digestibility (White *et al.* 1975; Jacobsen and Skjenneberg 1976, 1979; Trudell *et al.* 1980).

Woody browse is usually associated with high concentrations of secondary chemicals that modulate food selection patterns (Bryant and Kuropat 1980; Jung and Batzli 1981) and intake (Klein 1977; Bryant 1981a, b; Pehrson 1983; Bryant *et al.* 1983; Fox and Bryant 1984). Thus adaptations associated with food handling (Schwartz and Hobbs 1985) and with the avoidance or metabolism of secondary compounds dominate the winter adaptations of the herbivores which specialise in woody browse. The adaptations of this group of ruminants has been summarized by Kay *et al.* (1980) but we emphasize that this basic alimentary/morphological classification that is usually associated with concentrate selectors, as espoused by Hofmann (1973, 1983), refers to their summer rather than winter dietary preferences.

The interaction of seasonal food resources and body size constraints

Hypotheses based on African grazer-browse systems would suggest that there should be a gradient in food selectivity patterns based on body size, with the smaller animals utilizing plants highest in nutrients, particularly in energy and protein digestibility (Bell 1970, Demment and Van Soest 1985). We can find less evidence for strict adherence to this generalization in northern systems; the nutrient content of winter food is not the only factor determining intake and limited studies suggest that the smallest herbivores, such as brown lemming, are capable of processing more food than the larger animals (Batzli *et al.* 1981), thereby compensating for the higher relative requirements that are associated with small body size. Extreme alimentary adaptation in the form of caecotrophy and coprophagy increases the total mean retention time of the liquid and potentially digestible dietary components (Bjornhag 1972; Uden *et al.* 1982; Van Soest *et al.* 1983a,b). In addition, arguments based on relationships of metabolism and alimentary fill to body weight do not take into account seasonal adjustments. In particular, the possibility that nutrient requirements of the small herbivores may be lower in winter than in summer (White 1979) has not been seriously tested, perhaps because a lowering in nutrient and/or metabolic requirements, other than by torpor or hibernation, has not been considered a viable survival strategy in mammals of small body size. We (Holleman *et al.* 1982) have good evidence to the contrary as field studies on the red-backed vole (*Clethrionomys rutilis*) show they undergo a significant decline in water metabolism in winter compared with summer, even though average daily energy metabolism is as high or higher in winter than in summer (Table 1). Holleman *et al.* (1982) interpret the reduced water turnover as indicative of a reduced intake of protein and minerals (Cameron 1972; Cameron *et al.* 1982), an adaptation paralleling the large species, and pointing to an uncoupling of protein/mineral and energy metabolism in this small herbivore (Table 1).

Table 1. Comparison of seasonal uncoupling of energy and water metabolism in the red-backed vole and reindeer. Energy metabolism is expressed as CO_2 production. For both species water transfer rate was significantly higher in summer than in winter.

	Red-backed vole[1]		Reindeer	
	Winter	Summer	Winter	Summer
Ambient temp. (^oC)	-23	8	-21	7
Body wt. (g)	13.6	15.9	80,000	96,000
CO_2 production				
l /d	2.50	2.54	693[2]	1,337[4]
ml /$g^{0.73}$d	396	345	182[2]	308[4]
Water transfer rate				
ml /d	3.6	5.7	2,320[3]	5,490[3]
ml /$g^{0.8}$d	0.42	0.63	0.28[3]	0.57[3]
CO_2 /water	0.69	0.45	0.30	0.24

1, Holleman *et al.* (1982); 2, Fancy (1986); 3, Cameron and Luick (1972); 4, Boertje (1985).

Hibernation

The phenomenon of hibernation is an extreme adaptation to reduce food requirements. In cold-dominated environments the phenomenon must be accompanied by habitat selection that maintains an ambient temperature between -5 and 5^oC, although ambient temperatures may decline to less than -50^oC. Ground squirrels are the most successful hibernaters in the arctic. In addition, bears undergo a seasonal lowering of metabolism that approaches hibernation, and bats undergo short torpor bouts. Hibernation in the arctic ground squirrel involves reducing metabolism to a minimum by lowering deep body temperature to that in the hibernaculum. This phenomenon lasts about 14 days and is followed by a 1-2 day bout when intense metabolism raises body temperature to 38^oC, following which it again declines. The arousal bout is responsible for most of the winter energy and protein metabolism and is supported energetically by the animal's fat reserves. Arctic ground squirrels must fatten sufficiently in summer to live through the winter, and also to support early summer homeothermy, as animals may leave the hibernaculum to set up breeding territories before food becomes available the following spring. The storing of winter food resources for use during short-term arousal bouts is common for many hibernaters but is not performed by arctic ground squirrels. The thermogenic system that allows the animal to arouse periodically is driven by brown adipose tissue fuelled by lipid reserves. The ability of this species to convert summer food to body lipid is of interest to the nutritional ecologist.

BODY SIZE CONSTRAINTS

Current theories that explain dietary specialization in herbivorous mammals involve nutritional constraints arising from body size differences that either limit animals to niches and to dietary specialization or allow wide exploitation of lower quality food resources (Short 1964; Bell 1970; Janis 1976; Demment and Van Soest 1985; Hudson 1985).

Historically body size differences have been considered to be the most important influence on metabolic rate, food requirements and resistance to cold (Kleiber 1961; Irving 1972). These and earlier avian studies (Calhoun 1947, Van Balen 1967) report that intraspecific body size is inversely related to environmental temperature, supporting Bergmann's Rule (Bergmann 1847) that large body size is advantageous in cold-dominated climates. The rule was based on the physics of heat flow in relation to surface area, the reduction in relative surface area reducing heat loss. Therefore, at a given ambient temperature, as animal body size increases the surface area to volume ratio declines, thus lowering energy requirements per unit body weight. Small individuals would be highly constrained in their ability to exploit cold environments due to increasing heat loss per unit body weight. These heat balance relations should have implications for foraging in cold-dominated systems.

On the other hand, some species have adapted to overcome body size constraints and as a result do not adhere to Bergmann's rule. The physiological, biochemical, behavioural and social processes that allow relaxation of the body size constraint in relation to cold are shaped by individual selection. The adaptations themselves are often related to body size. For example, deposition of and dependency on thermogenic review of brown adipose tissue is a thermogenic process restricted to small mammals, and is the basis of a behavioural adaptation that allows these small mammals to exploit the subnivean space in which the environment is considerably warmer than at the snow surface. In contrast, seasonal pelage changes may drastically increase whole body insulation and as a result dramatically widen the thermoneutral zone. This classical winter adaptation has maximum effect in large animals (Scholander *et al.* 1950c; Irving 1972; Parker and Robbins 1985, Renecker and Hudson 1985). These general cold adaptations have a very marked seasonal basis and daylength is the dominant proximal cue for timing.

Figures 1 and 2 summarize some changes in energetic and insulative relationships of larger herbivores. The thick winter coat of outer hair and underwool (Bohl and Nikolajewski 1931; Ryder 1977; Timisjarvi *et al.* 1984) is shed annually, driven proximately by photoperiod but with ultimate control over timing controlled by temperature (Fig. 2, Jackes and Watson 1975), sex-specific processes and nutrition (Ryder and Stephenson 1968; Ryder 1977). Activity of the primary and secondary follicles precedes coat regrowth (Fig. 2) and the combined fibre types greatly influence the insulative value of the pelt which drastically reduces the heat loss of the animal (Irving 1972; Nilssen *et al.* 1984 a, b; White and Fancy 1986). Insulation brought about by the winter pelt increases with depth of pelt and body weight, scaling as $W^{0.65}$ (or conductance scales as $W^{-0.35}$, Parker and Robbins 1985) indicating that insulation per unit body weight is greater in small animals. However, pelt insulation must be balanced against heat loss per unit area which scales as $W^{0.66}$ suggesting that the extra insulative value of pelt per unit body weight in small animals counteracts heat loss per unit area. A seasonal change in pelt or whole body insulation of 3-fold affects both the thermoneutral range and the increase in thermogenesis below the lower critical temperature (Scholander *et al.* 1950a, b, c; Irving 1972; Nilssen *et al.* 1984a, b; Parker and Robbins 1985; White and Fancy 1986).

In this paper emphasis is placed on the implications for herbivores of nutrition in cold-dominated environments. However, most recent emphasis relating nutrition, diet and niche specialization and social organization in grazing systems has taken place in more tropical systems such as the savannas. Therefore, theories relating food requirements to body size do not take into account the influence of highly repeatable seasonal changes which add a different constraint to the less time-dependent temperature, food supply and water (drought) influences of the tropics and sub-tropics.

An explosion of ideas and ramification of the way body size influences energy and nutrient requirements of animals (Robbins 1983; Hudson 1985) has led many nutritional ecologists to speculate that nutrition is the overwhelming environmental factor which controls body size (Demment and Van Soest 1985), reproductive strategies (Clutton-Brock *et al.* 1982) and speciation (Geist 1971, 1986). With respect to herbivorous mammals, realization of the way in which herbivores handle dietary fibre relative to their energy requirements has led to a better understanding of the role of alimentary differentiation and habitat use (Bunnell and Gillingham 1985).

Energy expenditure, and therefore energy requirements, scale to body weight (W; kg) as $W^{0.75}$, whereas the wet weight of alimentary contents scales as $W^{1.04}$ (Demment and Van Soest 1985). Therefore

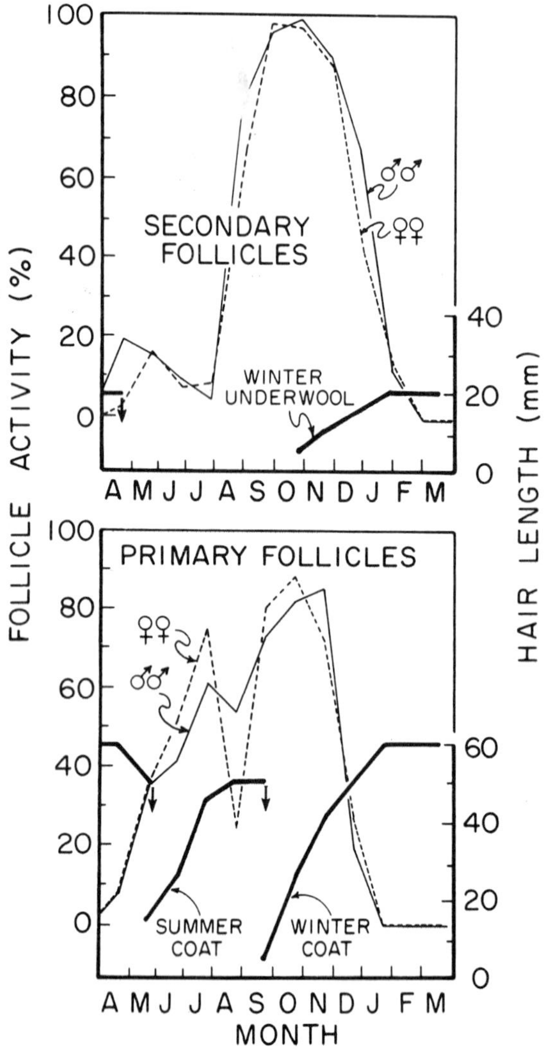

Figure 2. Seasonal pattern of skin follicle activity and fibre depth in red deer (redrawn from Ryder 1977). Whereas two outer hair coats are produced annually, and are accompanied by a double cycle in primary follicle activity, underwool is only grown in fall by secondary follicles which increase in activity in the fall. Not shown in the figure are details on moulting dates which are cohort specific for under-coat growth and moulting (downward-pointing arrow).

body weight has a profound effect on the retention time of digesta in the alimentary tract and, in consequence, on the digestibility of the food. Based on this hypothesis, the expectation is that as body size increases animals are no longer committed to the fast digesta flow, low digestibility constraint of small body size. As food retention time increases, digestibility is maximized such that the acquisition and absorption of energy exceeds the animals requirements because energy requirements increase at a relatively slower rate (i.e. $W^{0.75}$) than intake ($W^{1.09}$). Alternatively, the energy requirements can be met by a lower quality food resource of the larger herbivores. In contrast, the constraint of small size and reduced digesta retention time lowers diet digestibility and when combined with high relative energy requirements, small animals must select high quality food in order to meet their energy demands (Bunnell and Gillingham 1985). In an extreme example from the sub-tropics, selective foraging by small ruminants results in extremely high rates of ruminal fermentation (Hoppe 1977, 1984).

We would also argue that some animals have evolved modified digestive system that release them from body size constraints, or at least the adaptations ease the effect of the constraint. In small herbivores these adaptations can take two extreme forms. Hind-gut (caecal and colon) modification may improve the retention time of potentially digestible forage when combined with caecotrophy and or coprophagy. An alternate modification involves fore-gut simplification in small ruminants (for example, the African grey duiker (*Syloicapra grimmia*), Hofmann 1973; Hoppe 1977, 1984) to allow an increase in food passage rate; an advantageous strategy when food resources are not limiting. Additionally, small macropods (2-3 kg) selectively digest food particles by either fermentation or gastric processes and their fore-gut morphology ensures a separation of forage components directing them to these separate digestive processes. In large ruminants the rumeno-reticulum delays food particles and maximizes digestion which is advantageous when food resources are limited. Again a contrasting adaptation in large herbivores such as equids is hind-gut simplification to allow a high food intake.

Are these same theories applicable to northern herbivorous mammals? A major feature that differentiates the northern species from those in the tropics is the strict seasonality of the environment. How then has seasonality shaped the body scaling factors and adaptations to overcome these constraints in northern species?

ALIMENTARY FUNCTION IN NORTHERN HERBIVORES

Given that the food supply is both most limiting and at its lowest quality in winter, it is to be expected that the alimentary tract may also change seasonally. For ruminants, seasonal comparison of the wet weight of rumeno-reticulum contents in field-shot animals shows a lower relative fill in winter than in summer for reindeer and caribou,

Table 2. Seasonal changes in rumen contents fill (kg FW /100 kg BW) in field-shot animals.

		Summer	Winter	Reference
Reindeer	N	15–20	9–12	Staaland *et al.* (1979) Reimers *et al.* (1983)
	S	15	13	Staaland *et al.* (1979)
Muskox	14	11		H. Thing and H. Staaland (unpublished)
Moose	10	9		Gasaway and Coady (1974)

N, Norway; S, Svalbard (Spitzbergen).

Table 3. Seasonal change in digesta passage rate and mean alimentary fill in muskoxen given *ad libitum* a pellet plus hay diet.

	Winter	Summer	Sig. of diff.†
Number	8	8	
Body wt. (kg)	152	115	
Daily dry matter intake ($g /kg^{0.75}$)	38	102	***
Transit time (h)	19.7	12.1	*
Mean rentention time	53.8	22.0	*
Rumen turnover time	24.6	22.0	NS
Alimentary fill (g DM /kg)			
Indigestible	4.3	6.0	*
Total	13.5	17.9	*
DM digestibility	0.81	0.74	**

† Significance of difference between seasons:
NS, not significant; *, $P<0.05$; **, $P<0.01$; ***, $P<0.001$.

and muskoxen but to a lesser extent in moose (Table 2). These data refute the hypothesis that ruminants increase rumen fill in winter to compensate for a declining plane of nutrition. The results raise questions as to whether food availability, or quality, controls rumen fill in the field. Under pen-feeding conditions, rumen fill in winter can be maintained at summer levels when reindeer or moose are given *ad libitum* good quality food (Jacobsen and Skjenneberg 1979; Schwartz *et al.* 1985). These observations support the hypothesis that food availability may be limiting rumen fill in winter. Recent *in vivo* studies with tame muskoxen and moose suggest that total alimentary fill estimated using non-absorbed digestive markers is lower (Table 3, Holleman *et al.* 1984) and less adaptive to plane of nutrition in winter than in early spring and summer (Hubbert 1987). Estimates of total and indigestible alimentary fill in north temperate ruminants agrees with field estimates (Fig. 3) and with predictions from Hofmann's classification (Hofmann 1983) which shows an increasing cline from browser/concentrate selectors (moose, red deer) to mixed foragers (reindeer), and are compared with estimations for domestic sheep in Table 4.

These observations contrast to simple stomached animals and those with hind-gut adaptations, for in the snowshoe hare an increase in caecal size has been found in winter (Smith *et al.* 1980). This observation supports the idea that hares increase food intake as food quality declines in order to meet metabolic requirements, as has also been found with equids (Janis 1976).

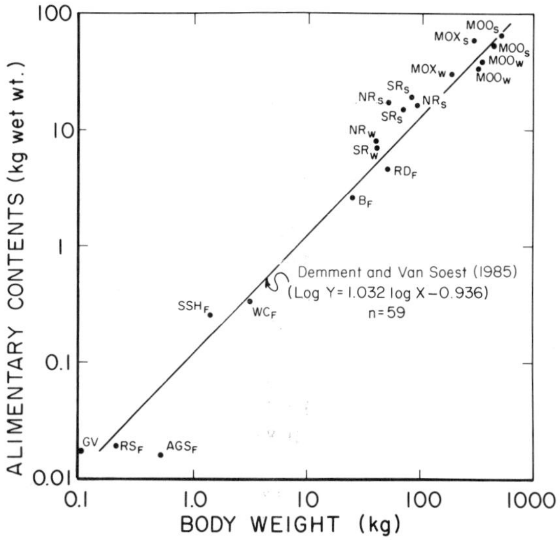

Figure 3. Alimentary contents fill in relation to body weight in some arctic herbivores compared with the interspecies line composed of 16 ruminant and 35 non-ruminant species. MOO, moose (Gasaway and Coady 1974; Demment and Van Soest 1985); MOX, muskoxen (H. Thing and H. Staaland unpublished); NR, Norwegian reindeer; SR, Svalbard reindeer (Staaland *et al.* 1979); RD, red deer; B, beaver; WC, woodchuck; SSH, snowshoe hare; AGS, arctic ground squirrel; RS, red squirrel (Staaland, White and Holleman unpublished).

Table 4. Alimentary fill and digesta rate of passage in some arctic ruminants compared with domestic sheep. All estimates were made in winter (January – March) and ^{141}Ce or ^{103}Ru was used as a particle marker.

Species	Red deer	Reindeer		Muskox	Moose		Sheep
	B + C	L +	H	H + C	B	P	H + C
No.	3	2	4	8	4	7	5
Body wt. (kg)	50	101	53	152	432	420	72
Rumen contents (g FW /kg)	68	169	158	119	–	48	29
Total alimentary fill (g /kg)							
Fresh wt.	92	268	–	170	–	(76)	171
Dry wt.	17	30	–	–	–	(12)	21
Marker, DM	11	16	23	13	11	16	20
Indigestible fill (g DM /kg)	7.6	6.5	12.6	4.3	8.5	9.8	10.5
Transit time (h)	20.0	6.7	9.3	19.7	16.8	–	18.6
Mean retention time (h)	54	51	44	54	60	49	53
Rumen turnover time (h)	25.7	33.6	26.0	24.6	35.0	24.8	23.8
Reference	1	2	2	3	4	5	6

B = browse; C = concentrate; L = lichen; H = hay; P = pellets. Rumen contents and alimentary fill are estimates for animals at slaughter. Estimates of total alimentary fill and indigestible fill were made according to Holleman and White (unpublished) where indigestible DM fill (IDF) = faeces output x total mean retention time; total DM fill = IDF + (IDF.D)/(2.1–D), where D = DM digestibility. 1. Staaland, White and Langvatn (unpublished). 2. White and Gau (1975); White, Gaare and Staaland, (unpublished). 3. Holleman *et al.* (1984). 4. Schwartz *et al.* (unpublished). 5. Hubbert (1987). 6. White, Holleman and Nielsen (unpublished).

A summary of estimates of the food–handling capacity of a number of northern ruminants is shown in Table 5. In spite of relatively large differences in body size between young red deer (50 kg) and adult moose (430 kg), these winter estimates, including the transit time, total mean retention time and rumen turnover time, are remarkably similar. Except for the browse–based diets given red deer and moose, the diets were similar in being chopped to a small particle size and were of moderate to high digestibility. In muskoxen, a marked increase in digesta passage rate accompanies the high voluntary intake in summer. Dry matter intake was inversely correlated with total mean retention time in muskoxen (Holleman *et al.* 1984). This same effect is seen in moose (Fig. 3); however, the relationship is more closely associated with a digestibility change (Fig. 4). In our studies to date, turnover time of the liquid component of rumen digesta in northern ruminants (reindeer/caribou, muskoxen, moose) is highly inversely correlated with food intake (White *et al.* 1984a) and the species differences are not significant. These data do not support predictions that rate of passage in browser species, such as moose, should be higher than in mixed foragers, such as reindeer and muskoxen.

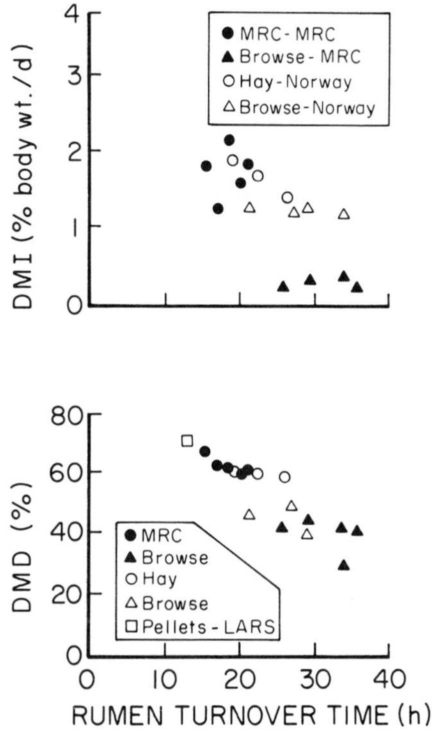

Figure 4. Comparison of the effect of rumen turnover time on dry matter intake and diet digestibility in moose. MRC, Moose Research Center (Schwartz, Regelin, Franzman, White and Holleman unpublished; White *et al.* 1984a), Norway, Hjeljord *et al.* (1982); LARS, Large Animal Research Station (White and Holleman, unpublished).

These studies indicate that an increase in rumen and total alimentary fill is a major component of the enhanced ability of the ruminant to maximize food intake in summer. When combined with the seasonal change in digestibility (Fig. 1), these two factors account for the extremely high voluntary intake potential of reindeer/caribou, moose and muskoxen in summer (White and Fancy 1986). The significance of these results for digestion processes is considerable. Details on the separate effects of liquid and particle flow are necessary to determine the separate effects on the plant cell contents, cell wall, and protein, and on microbial protein production. An important general implication of the summer alimentary hypertrophy is that it mediates against an extreme reduction in the residence time of digesta, as intake increases, thereby modifying rumen fermentation and maintaining an optimal, or near optimal, digestibility of the diet.

Table 5. Digesta rate of passage and alimentary fill of some small arctic herbivores. Studies were made in early winter (October–November) and alimentary fill was measured in animals from the laboratory (L) or taken from the field (F).

Species	Red squirrel[F]	Arctic ground squirrel[L]	Snowshoe hare[F]	Woodchuck[L]	Beaver[F]
Diet	Natural+ grain	Rabbit pellets, lettuce	Natural browse	Carrots, cabbage, rabbit cubes	Access to logs - willow, alder
No.	6	3	1	2	6
Body wt. (kg)	0.23	0.53	1.40	2.91	24.9
Transit time (h)	-	3.7	-	18.6	9.8
Mean retention time (h)	-	11.0	63[1]	33.2	30.4
Total alimentary fill at slaughter					
g FW	19.2	16.2	253	323	225
g FW /kg	84	29	181	110	105
Distribution of contents (% total contents)					
stomach	19	18	11	57	15
caecum	35	28	56	23	38

[1]Estimated using equation in footnote of Table 4; total fill = 253g, intake = 85 g /kg daily intake (Reichardt *et al.* 1984), DM digestibility = 0.38 (Reichart *et al.* 1984), and indigestible dry matter fill = 194 g.

Less information is available on the alimentary fill and rate of passage in the smaller herbivores. However, Parra (1978) and Demment and Van Soest (1985) have shown that across a wide species range a common exponent accounts for the relation between alimentary contents and body weight ($W^{1.04}$) for 16 ruminants and 35 non-ruminant herbivorous mammals. Estimates for northern herbivores fit this relationship (Fig. 3). Therefore, alimentary capacity scales independent of resting metabolism ($W^{0.75}$, Kleiber 1961) and this difference has been stressed as the main evolutionary factor explaining between-species differences in diet composition (Short 1964; Bell 1970; Janis 1976; Parra 1978; Hanley 1980; Hanley and Hanley 1982; Van Soest 1982; Robbins 1983; Demment and Van Soest 1985).

A summary of our recent estimates of alimentary contents for local Alaskan non-ruminant herbivores is shown in Table 5. Variation in relative fill at 29 to 181 g /kg body weight is large and, except for the very high fill in reindeer fed lichens, the range for northern ruminants (76-171 g /kg body weight) was within that of the non-ruminants. The higher alimentary fill in the non-ruminants was for those species processing high fibre diets (snowshoe hare, woodchuck (*Marmota monax*), beaver (*Castor canadensis*) and the lowest fill was for the concentrate selectors (arctic ground squirrel, red squirrel) (Fig. 3). The distribution of digesta within the alimentary tract of these non-ruminants varied depending on the development of the hind-gut. Thus, over 50% of the digesta was in the caecum of the snowshoe hare and only 23 to 28% in the caeca of the woodchuck and arctic ground squirrel (Table 5). An obvious hypertrophy of the stomach was noted in the woodchuck, suggesting that it serves a relatively greater role as a storage organ in the woodchuck than in the other four species.

These data on alimentary fill support the findings of Demment and Van Soest (1985), and therefore it would be predicted that the rate of food handling should be highly related to body size, digesta transit time and total mean retention time, being considerably less in the small animals than in the large ruminants, a generalization shown by Robbins (1983). Demment (1983) reports that digesta retention time scales at $W^{0.28}$, which is not different to what would be expected for metabolic time (Kleiber 1961) that scales as $W^{0.25}$, if food intake were to scale at $W^{0.75}$ (White 1975). However, Demment and Van Soest (1985) show that digestibility is an important covariate accounting for spread about the regression line in such retention time-body weight relations.

The total mean retention time for the arctic ground squirrel (11 h) was very short compared with the values found for the woodchuck and beaver (*ca* 33h), and is in line with its small alimentary fill per unit body weight (Table 5). Values for the woodchuck and beaver are approximately 35% lower than those obtained for the ruminants (Table 4), a surprising finding given the huge difference in body weight.

The enlarged stomach capacity of the woodchuck was associated with an extreme delay in digesta transit time, compared with the other non-ruminants and was close to that found for the ruminants (Table 4). The total mean retention time estimate for snowshoe hare was made indirectly, but indicates a value approximating moose feeding on woody browse of similar digestibility. The basis of the delayed digesta passage rate in the snowshoe hare could be caecotrophy. Van Soest *et al.* (1983a, b) report that in the domestic rabbit, retention time of particles excreted in the dry faeces is 9 h but for the liquid phase it is 193 h and the latter reflects the passage of very small food particles as well as soluble components of the liquid phase. Clearly the rate of passage of liquid and particle phases of digesta warrant further study in the northern herbivores to determine how these species have adapted to body size constraints, and how the seasonal nature of the food resource has influenced food handling-digestion adaptations.

The work presented was supported by grants to the Large Animal Research Station, Institute of Arctic Biology from the State of Alaska. The muskox studies were supported by a major grant from the NSF, Division of Polar Programs (DPP-77-18384) and the moose research by Alaska State funding of the Moose Research Center, Soldotna, Alaska.

Abbott, M.J., Ullrey, D.E., Ku, P.K., Schmitt, S.M., Romsos, D.R., and Tucker, H.A. (1984). *J. Wildl. Manage.* **48**, 776-87.

Arnold, G.W. (1985). In "Bioenergetics of Wild Herbivores." (Eds R.J. Hudson and R.G. White.) p. 81. (CRC: Boca Raton, Florida.)

Batzli, G.O. (1981). In "Tundra Ecosystems: A Comparative Analysis." (Eds L.C. Bliss, W.O. Heal and J.J. Moore.) p. 377. (Cambridge Univ.: Cambridge.)

Batzli, G.O. (1983). *Oikos* **40**, 396-406.

Batzli, G.O., Jung, H.-J., and Guntenspergen, G,. (1981). *Oikos* **37**, 112-6.

Batzli, G.O., White, R.G., MacLean, S.F., Pitelka, F.A., and Collier, B.D. (1980). In "An Arctic Ecosystem: The Coastal Tundra at Barrow, Alaska." (Eds P.C. Miller, L.L. Tieszen, and F.L. Bunnell.) p. 335. (Dowden, Hutchinson and Ross: Stroudsburg, Pennsylvania.)

Bell, R.H.V. (1970). In "Animal Populations in Relation to their Food Resources." (Ed. A. Watson.) pp. 111-24. (Blackwell: Oxford.)

Bergmann, C. (1847). *Gottinger Studien* **3**, 595.

Bjornhag, G. (1972). *Swedish J. Agr. Res.* **2**, 125.

Bliss, L.C., Heal. O.W., and Moore, J.J. (Eds) (1981). "Tundra Ecosystems: A Comparative Analysis." (Cambridge Univ.: Cambridge.)

Bohl, B.K., and Nikolajewski, L.D. (1931). *Zeitschrift Mikr.-anat. Forschung.* **25**, 1.

Boertje, R.D. (1985). *J. Range Manage.* **38**, 468.

Bryant, J.P. (1981a). Proc. World Lagomorph Conf. (Eds K. Myers and C.D. McIness.) p. 720.

Bryant, J.P. (1981b). *Science* **213**, 889-90.

Bryant, J.P., and Kuropat, P.J. (1980). *Ann. Rev. Ecol. Syst.* **11**, 261–85.
Bryant, J.P., Chapin, F.S., and Klein, D.R. (1983). *Oikos* **40**, 357–68.
Bunnell, F.L., and Gillingham, M.P. (1985). In "Bioenergetics of Wild Herbivores." (Eds R.J. Hudson and R.G. White.) p. 53. (CRC: Boca Raton, Florida.)
Calhoun, J.B. (1947). *Am. Nat.* **8**, 208.
Cameron, R.D. (1972). Ph.D. Thesis., Univ. Alaska, Fairbanks, AK, USA.
Cameron, R.D., and Luick, J.R. (1972). *Can. J. Zool.* **50**, 107–16.
Cameron, R.D., Luick, J.R., and White, R.G. (1982). In "Use of Tritiated Water in Studies of Production and Adaptation in Ruminants." p. 189. (International Atomic Energy Agency: Vienna.)
Christopherson, R.J., Hudson, R.J., and Christophersen, M.K. (1979). *Can. J. Anim. Sci.* **59**, 611–7.
Clutton-Brock, T.H., Guinness, F.E., and Albon, S.D. (1982). "Red Deer: The Behaviour and Ecology of Two Sexes." (Univ. Chicago: Chicago, USA.)
Dauphine, T.C. (1976). "Biology of the Kaminuriak Population of Barren-Ground Caribou: Part 4; Growth, Reproduction and Energy Reserves." Canadian Wildl. Serv. Report Ser. No. 38.
Demment, M.W. (1983). *Afr. J. Ecol.* **21**, 219–33.
Demment, M.W., and Van Soest, P.J. (1985). *Am. Nat.* **125**, 641–72.
Fancy, S.G. (1986). Ph.D. Thesis., Univ. Alaska, Fairbanks, AK, USA.
Fancy, S.G., and White, R.G. (1985a). *J. Wildl. Manage.* **49**, 987–93.
Fancy, S.G., and White, R.G. (1985b). In "Bioenergetics of Wild Herbivores." (Eds R.J. Hudson and R.G. White.) p. 143. (CRC: Boca Raton, Florida.)
Feist, D.D., and White, R.G. (1987). In "Animal Adaptation to Cold." (Ed. L.C.H. Wang.) (Springer-Verlag: Heidelberg.) (In press).
Fox, J.F., and Bryant J.P. (1984). *Oecologia* **63**, 128–35.
Gasaway, W.C., and Coady, J.W. (1974). *Naturaliste Can.* **101**, 227.
Geist, V. (1971). "Mountain Sheep - A Study in Behaviour and Evolution." (Univ. Chicago: Chicago, USA.)
Geist, V. (1986). *Natural History* **3**, 54.
Hanley, T.A. (1982). *J. Range Manage.* **35**, 146–51.
Hanley, T.A., and Hanley, K.A. (1982). *J. Range Manage.* **35**, 152–8.
Hjeljord, O., Sundstol, F., and Haagenrud, H. (1982). *J. Wildl. Manage.* **46**, 333–43.
Hofmann, R.R. (1973). "The Ruminant Stomach: Stomach Structure and Feeding Habits of East African Game Ruminants." (East African Literature Bureau: Nairobi.)
Hofmann, R.R. (1983). *R. Soc. N.Z. Bull.* **20**, 51–8.
Holleman, D.F., White, R.G., and Feist, D.D. (1982). *J. Mamm.* **63**, 293–6.
Holleman, D.F., White, R.G., Frisby, K., Jourdan, M., Henrichsen, P., and Tallas, P.G. (1984). Proc. First Int. Muskox Symp. (Eds D.R. Klein, R.G. White and S. Keller.) Biol. Papers Univ. Alaska, Spec. Report No. 4. p. 188.
Hoppe, P.P. (1977). Proc. 13th Int. Congr. Game Biologists. (Ed. T.J. Peterle.) p. 141. (Wildlife Society: Washington, DC.)

Hoppe, P.P. (1984). In "Herbivore Nutrition in the Subtropics and Tropics." (Eds F.M.C. Gilchrist and R.I. Mackie.) pp. 222–43. (Science Press: Craighall, South Africa.)
Hubbert. M.E. (1987). Ph.D. Thesis., Univ. Alaska, Fairbanks, AK, USA.
Hudson, R.J. (1985). In "Bioenergetics of Wild Herbivores." (Eds R.J. Hudson and R.G. White.) p. 1. (CRC: Boca Raton, Florida.)
Huot, J., and Goudreault, F. (1985). In "Caribou Management, Census Techniques, Status in Eastern Canada." (Eds T.C. Meredith and A.M. Martell.) p. 157. McGill Subarctic Research Paper No. 4.
Irving, L. (1972). "Arctic Life of Birds and Mammals Including Man." (Springer-Verlag: Berlin.)
Jackes, A.D., and Watson, A. (1975). *J. Zool.* **176**, 403–9.
Jacobsen, E., and Skjenneberg, S. (1976). *Forsk. Fors. Landbrukt* (Aas, Norway) **27**, 287.
Jacobsen, E., and Skjenneberg, S. (1979). *Meld. Nor. Landbrukshogsk.* **58** (No.34), 1–11.
Janis, C. (1976). *Evolution* **30**, 757–74.
Jung, H-J., and Batzli, G.O. (1981). *J. Mamm.* **62**, 286–92.
Kay, R.N.B., and Ryder, M.L. (1978). *J. Zool.* **185**, 505–10.
Kay, R.N.B., Engelhardt, W.V., and White, R.G. (1980). In "Digestive Physiology and Metabolism in Ruminants." (Eds Y. Ruckebusch and P. Thivend.) pp. 743–61. (MTP: Lancaster, UK.)
Kleiber, M. (1961). "The Fire of Life; An Introduction to Animal Energetics." (John Wiley: New York.)
Klein, D.R. (1964). *J. Mamm.* **45**, 226–35.
Klein, D.R. (1968). *J. Wildl. Manage.* **32**, 350–67.
Klein, D.R. (1970). *J. Range Manage.* **23**, 8–14.
Klein, D.R. (1977). *Proc. Int. Congr. Game Biol.* **13**, 266.
Klein, D.R. (1986). In "Grazing Research at Northern Latitudes." (Ed. O. Gudmundsson.) p. 237. (Plenum: New York.)
Kuropat, P. (1984). M.S. Thesis. Univ. Alaska, Fairbanks, AK, USA.
Langvatn, R., and Albon, S.D. (1986). *Holarctic Ecol.* **9**, (In press.)
Moen, A.N. (1973). "Wildlife Ecology." (W.H. Freeman: San Francisco.)
Morrison, P.R.M. (1960). *Bull. Mus. Comp. Zool.* **124**, 75–92.
Nilssen, K.J., Johnsen, H.K., Rognmo, A., and Blix, A.S. (1984a). *Am. J. Physiol.* **246**, R963–7.
Nilssen, K.J., Sundsfjord, J.A., and Blix, A.S. (1984b). *Am. J. Physiol.* **247**, R837–41.
Parker, K.L., and Robbins, C.T. (1985). In "Bioenergetics of Wild Herbivores." (Eds R.J. Hudson and R.G. White.) p. 161. (CRC: Boca Raton, Florida.)
Parra, R. (1978). In "The Ecology of Arboreal Folivores." (Ed. G.G. Montgomery.) p. 205. (Smithsonian Inst.: Washington, DC.)
Pehrson, A. (1983). *Holarctic Ecol.* **6**, 395.
Pollock, A.M. (1975). *J. Physiol.* **244**, 95–6P.
Regelin, W.L., Schwartz, C.C., and Franzmann, A.W. (1985). *J. Wildl. Manage.* **49**, 388–93.

Reichardt, P.B., Bryant, J.P., Clausen, T.P., and Wieland, G.D. (1984). *Oecologia* **65**, 58–69.

Reimers, E., Klein, D.R., and Sorumgaard, R. (1983). *Arctic Alp. Res.* **15**, 107.

Reimers, E., and Ringberg, T. (1983). *Acta Zool. Fennica* **175**, 69–72.

Renecker, L.A., and Hudson, R.J. (1985). *J. Wildl. Manage.* **49**, 785–92.

Renecker, L.A., and Hudson, R.J. (1986a). *Can. J. Zool.* **64**, 322–7.

Renecker, L.A., and Hudson, R.J. (1986b). *J. Wildl. Manage.* **50**, 143–7.

Riley, H., and Skjelvag, A.O. (1984). (Eds) Proc. 10th General Meeting of the European Grassld Federation. As, Norway.

Robbins, C.T. (1983). "Wildlife Feeding and Nutrition." (Academic: New York.)

Ryder, M.L., and Stephenson, S.K. (1968). "Wool Growth." (Academic: London.)

Ryder, M.L. (1977). *J. Zool.* **181**, 137–43.

Ryg, M. (1983). *Polar Research* **1**, 249.

Ryg, M., and Jacobsen, E. (1982a). *Can. J. Zool.* **60**, 15–23.

Ryg, M., and Jacobsen, E. (1982b). *Can. J. Zool.* **60**, 1562–7.

Ryg, M. (1986). *Rangifer* Special Issue No. 1, p. 261.

Saether, B.E. (1985). *J. Wildl. Manage.* **49**, 977–83.

Scholander, P.F., Hock, R., Walters, V., and Irving, L. (1950a). *Biol. Bull.* **99**, 259–71.

Scholander, P.F., Hock, R., Walters, V., Johnson, F., and Irving, L. (1950b). *Biol. Bull.* **99**, 237–71.

Scholander, P.F., Walters, V., Hock, R., and Irving, L. (1950c). *Biol. Bull.* **99**, 225–36.

Schwartz, C.C., and Hobbs, N.T. (1985). In "Bioenergetics of Wild Herbivores." (Eds R.J. Hudson and R.G. White.) p. 25. (CRC: Boca Raton, Florida.)

Schwartz, C.C., Regelin, W.L., and Franzmann, A.W. (1985). *J. Wildl. Manage.* **49**, 137–41.

Short, H.L. (1964). *J. Wildl. Manage.* **28**, 445–58.

Skogland, T. (1978). *Arctic Alp. Res.* **10**, 569.

Skogland, T. (1983). *Oecologia* **60**, 156–68.

Skogland, T. (1985). *J. Anim. Ecol.* **54**, 359–74.

Skogland, T. (1986). *J. Wildl. Manage.* **50**, 314–9.

Smith, R.L., Hubartt, D.J., and Shoemaker, R.L. (1980). *J. Wildl. Manage.* **44**, 719–24.

Staaland, H., Jacobsen, E., and White, R.G. (1979). *Arctic Alp. Res.* **11**, 457.

Suttie, J.M., Goodall, E.D., Pennie, K., and Kay R.N.B. (1983). *Br. J. Nutr.*, **50** 737.

Suttie, J.M., Lincoln, G.A., and Kay R.N.B. (1984). *J. Repro. Fert.* **71**, 7.

Timisjarvi, J., Nieminen, M., and Sippola, A-L. (1984). *Comp. Biochem. Physiol.* **79A**, 601–9.

Trudell, J., and White, R.G. (1981). *J. Appl. Ecol.* **18**, 63–81.

Trudell, J., White, R.G., Jacobsen, E., Staaland, H., Ekern, E., and Kildemo, K. (1980). Proc. 2nd Int. Reindeer and Caribou Symp. (Eds E. Reimers, E. Gaare and S. Skjenneberg.) p. 262. (Directoratet for vilt og ferskvannsfisk: Trondheim, Norway.)

Uden, P., Rounsaville, T.R., Wiggans, G.R., and Van Soest, P.J. (1982). *Br. J. Nutr.* **48**, 329–39.

Van Balen, J.H. (1967). *Ardea* 55, 1.

Van Soest, P.J. (1982). "Nutritional Ecology of the Ruminant." (O and B Books: Corvallis, Oregon.)

Van Soest, P.J., Jeraci, J., Foose, T., Wrick, K., and Ehle, F. (1983a). *R. Soc. N.Z. Bull.* **20**, 75–80.

Van Soest, P.J., Uden, P., and Wrick, K.F. (1983b). *Nutr. Rep. Int.* **27**, 17.

Wang, L.C.H., and Hudson, J.W. (Eds.) (1978). "Strategies in Cold: Natural Torpidity and Thermogenesis." (Academic: New York.)

White, R.G. (1975). In "Physiological Adaptation to the Environment." (Ed. F. J. Vernberg.) pp. 239–68. (Intext Educ.: New York.)

White, R.G. (1979). In "Comparative Mechanisms of Cold Adaptation." (Eds L.S. Underwood, L.L. Tieszen, A.B. Callahan and G.E. Folk.) pp. 13–50. (Academic: New York.)

White, R.G., and Fancy, S.G. (1986). In "Grazing Research at Northern Latitudes." (Ed. O. Gudmundsson.) p. 259. (Plenum: New York.)

White, R.G., and Gau, A. (1975). Proc. First Int. Reindeer and Caribou Symp. (Eds J.R. Luick, P.C. Lent, D.R. Klein and R.G. White.) Biol. Papers, Univ. Alaska, Spec. Report No. 1, p. 284.

White, R.G., Bunnell, F.L., Gaare, E., Skogland, T., and Hubert, B. (1981). In "Tundra Ecosystems: A Comparative Analysis." (Eds L.C. Bliss, J.B. Cragg, D.W. Heal and J.J. Moore.) p. 397. (Cambridge Univ.: Cambridge.)

White, R.G., Holleman, D.F., Schwartz, C.C., Regelin, W.L., and Franzman, A.W. (1984a). *Can. J. Anim. Sci.* **64** (Suppl.), 349–50.

White, R.G., Holleman, D.F., Wheat, P., Tallas, P., Jourdan, M., and Henrichsen, P. (1984b). (Abstract) Biol. Pap. Univ. Alaska Spec. Rep. No. 4, p. 193.

White, R.G., Thomson, B.R., Skogland, T., Person, S.J., Russell, D.E., Holleman, D.F., and Luick, J.R. (1975). In "Ecological Investigations of the Tundra Biome in the Prudhoe Bay Region, Alaska." (Ed. J. Brown.) Biol. Papers, Univ. Alaska, Spec. Report No. 2, p. 151.

Whittow, G.C. (1971). In "Comparative Physiology of Thermoregulation." Vol. II (Ed. G.C. Whittow.) pp. 191–281. (Academic: New York.)

Wickstrom, M.L., Robbins, C.T., Hanley, T.A., Spalinger, D.E., and Parish, S.M. (1984). *J. Wildl. Manage.* **48**, 1285–301.

21

GENETIC ASPECTS OF MAMMALIAN SURVIVAL AND GROWTH IN RELATION TO BODY SIZE

by St C.S. Taylor and J.I. Murray

Presented at the 2nd International Symposium on the Nutrition of Herbivores at the University of Queensland as the Douglas Ormonde Butler Memorial Lecture for 1987.

There are animals of every conceivable size living in a somewhat one-sided symbiosis with man. All sizes appear to function quite happily, except perhaps the extremes of any order, species or strain (Lerner 1954). All sizes of mammal have virtually the same biochemistry, physiology and anatomy. Yet they can look so different! What could be superficially less alike than a seal, a tiger, a bat, a man?

The traits of interest to ecologists and nutritionists, and of considerable relevance to animal production, are the amount of food ingested and metabolised, the growth of muscle, fat and bone, the yield of milk and wool, reproductive rate, survival rate, and whether the generation interval is short or long. All these traits are quantitative and most of them are related to body size. This is a review of how body size influences food intake, growth rate, body composition, production and reproduction. The relationship of these traits to body size is surprisingly uniform over the whole mammalian range, so much so that most of these associations can be encapsulated in two formal genetic size-scaling rules. These can then be used to furnish a quantitative description of the growth of a typical mammal (Taylor 1985).

Genetic size-scaling is thus the body of theory that stems from the underlying physical, physiological and biochemical similarity of mammalian growth. Genetic size-scaling depends on each genotype having its own genetic size factor operating throughout growth from embryo to adult. This genetic size factor expresses itself in many traits at every stage of growth.

Defining adult body size

Body weight, changing continuously during growth, eventually becomes stationary at an average level intimately related to the genotype of the individual. For this reason, adult body weight plays a dominant role in studies of mammalian growth. The main non-genetic variable influencing adult body weight in healthy, moderately active mammals in a thermoneutral environment is nutrition. Provided they have not been

NUTRITION OF HERBIVORES
ISBN 0 12 750052 9

severely undernourished during growth and thereby permanently stunted, mammals will always grow to the same adult body weight. Adult body weight changes as the level of food intake changes, varying from the lower extreme of an emaciated adult, all skin and bone, to an excessively fat adult. Body fat represents reserves of nutrition stored internally. How much reserve should a mammal have? It all depends on the circumstances. Let us therefore agree upon a conventional standard for fat reserves where lipid constitutes 15% of adult body weight. The associated adult body weight, A, can then be adopted as the quantified expression of an animal's genetic size-scaling factor.

As an example, consider the transgenic mice of Hammer *et al.* (1984). Normal controls had an adult body weight of 26 g. A line of lit/lit homozygotes, presumably deficient in growth hormone, had an adult body weight of 12 g. The new transgenic genotype carried both the lit/lit and rat growth hormone genes. It had its own new genetic size factor and grew to an adult body weight of 36 kg. The growth pattern of all three genotypes was largely dictated by the different genetic size factors. Observed growth was reasonably well described in each case by rescaling the same growth pattern with a different genetic size factor, A.

Adult size, A, always in kilograms, will permeate the whole of this review.

Genetic variation

Genetic variation and covariation with body size will be examined primarily at the level of mammalian species, but also to some extent at the level of breeds within species.

BODY SIZE AND BIOLOGICAL TIME

Genetic size-scaling for time variables had its genesis in 1927, when A.J. Clark published his extraordinary collection of data on heart pulse rate in which he gave an inter-species regression coefficient of -0.27 for both mammals and birds. Since then the length of most cyclical biological periods have been found to increase systematically with increasing body size of species. The relationship was well documented by Günter and Guerra (1955) whose average coefficient for many traits was 0.27. More recently Lindstedt and Calder (1981) extended the list of traits and gave inter-species regression coefficients ranging from 0.18 to 0.39. Peters (1983) further extended the number of traits and gave coefficients ranging from 0.08 to 0.49. The average coefficient was close to 0.27 in both cases. In particular, Peters (1983) lists coefficients from 27 different sources for speed of locomotion ranging from 0.12 to 0.38. Their average value was 0.23 with a large standard error of ± 0.06, calculated on the assumption that the observed range was entirely due to sampling. An interesting trait not included in these lists was the retention time (T_r) of a food particle in the gut, for which Demment (1983) gave the relationship $T_r = 0.589\ D\ A^{0.28} \approx 0.60\ D\ A^{0.27}$ where D is the digestibility of the diet.

Taylor (1986) gave coefficients for seven traits and the average was again 0.27 (range 0.21 to 0.31); he also illustrated the wide range of coefficients obtained by different authors for the same trait, with apparent significant differences. Taylor (1986) dismissed such differences as due to non-random sampling of species and therefore unimportant. He had already suggested (Taylor 1980a) that virtually all regression coefficients are consistent with a single common value of 0.27 which, if adopted, gave a uniform scale for genetic comparisons.

These genetic relationships between biological time variables and body size are now well authenticated and well established in the literature. What is not clear is whether all coefficients are uniformly 0.27 or whether there are variations about 0.27 for different traits and in different subgroups of species. Peters (1983) and Taylor (1980a, 1985) both believe that all the empirical relations reported are compatible with 0.25 or 0.27. Even if they are not, it is preferable to postulate that they are, and treat the variations about 0.27 as deviations.

Genetic deviations

Genetic deviations are an all-important part of any study of regression relationships. In the last few decades, the main emphasis has been on establishing the systematic relationships. The study of genetic deviations - so valuable to geneticists - has not yet received the same amount of attention. Deviations have to be partitioned into components that are genetic and non-genetic. This is often a major difficulty because, once the predominant influence of body size has been removed, the remaining deviation usually contains so much non-genetic variation that even the best estimate of the genetic deviation may be unreliable. Nevertheless it is clear that there are many species deviations, some of considerable magnitude. Species deviations are illustrated for gestation length, which can usually be determined with high accuracy (Fig. 1). The kangaroo rat (a marsupial) is a glaring deviant with a gestation length only one fifth of that expected. Whales and rabbits have unusually short gestations, even shorter than the pig. The majority of species are close to the mean. The bat has an exceptionally long gestation, about three times longer than expected.

Metabolic age

In order to make full use of the clear relationship between biological time variables and adult body weight, Taylor (1965) proposed replacing ordinary age by a metabolic age scale. This metabolic age scale combines the properties of Brody's (1945) physiological age and Kleiber's (1961) metabolic turnover time. Unlike Brody's physiological age scale, which is based on three parameters, metabolic age is based on only one, namely mature body weight. Metabolic age, θ, is defined as

$$\theta = \gamma\ (t-3.5)/A^{0.27} \qquad \text{Equation 1}$$

where age in days from conception, t, is empirically adjusted to an age origin 3.5 days after conception, where A is a standard adult weight always measured in kilograms, and where γ is a dimensioned unit that allows metabolic age to be expressed in metabolic days, (which can readily be converted to metabolic seconds, minutes, hours, weeks or years as seems appropriate). It should be stressed that the metabolic age scale is always based on adult body weight, A. A quantity expressed per kilogram mature body weight per metabolic day equals the same quantity expressed per unit mature metabolic weight ($A^{0.73}$) per day, but this does not hold for immature body weights. When all time variables associated with mammalian growth and metabolism are transformed to this metabolic age scale there emerges a pattern of developmental events common to all species.

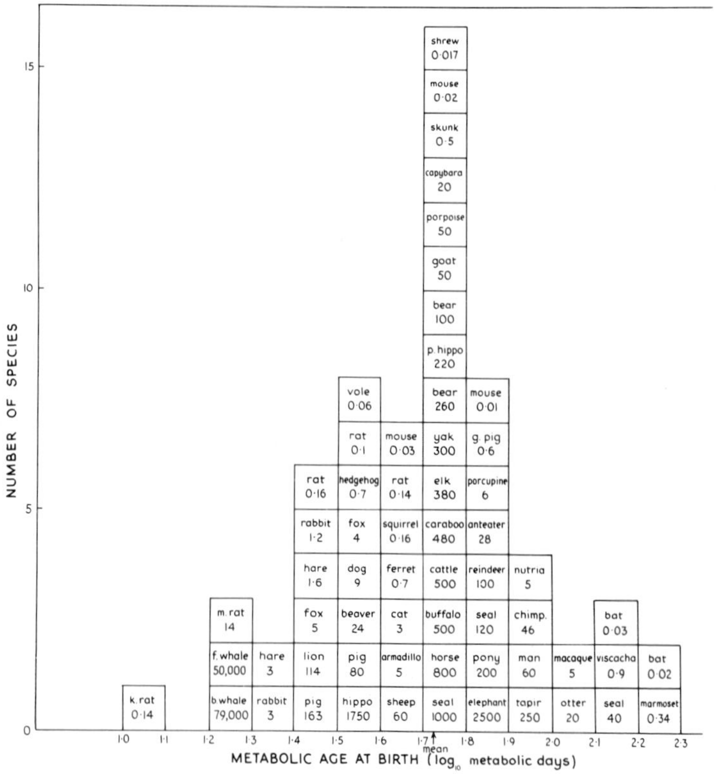

Figure 1. Frequency distribution of length of gestation in mammals when measured as metabolic age at birth.

Table 1. Expected temporal life-plan of a mammal

Metabolic age	Developmental event
0	egg enters uterus (3.5 days after ovulation)
4	implants
7	first somites appear
10	mammal becomes 0.00001 mature in body weight
14	end of metamorphosis
17	foetal eyelids close
40	is 0.04 mature in body weight
50	is born 0.05 mature in body weight
90	is weaned
100	is 0.4 mature in body weight
110	becomes sexually mature
115	is 0.5 mature in body weight
150	starts lactating
200	ends first lactation
270	tibial epiphyses fuse
440	0.98 mature in body weight
500	is fully mature in body weight
700	end of prime of life
1000	first onset of cancer
2000	dies

The temporal life plan of a typical mammal was given by Taylor (1985, 1986) (Table 1). At a metabolic age of zero (which is 3.5 days after conception), the typical mammalian fertilised egg enters the uterus. Four metabolic days later, that is, at a metabolic age of 4 days, the egg implants. At 7 metabolic days of age the first somite appears, and so on, until at 2000 metabolic days of age, the typical mammal dies.

Duration of metabolic processes

In the same way as for developmental events, formulae can be given for the duration of short-term metabolic processes in adult mammals. Thus the heart beat of a typical mammal lasts 0.3 metabolic seconds, it breathes every 1.2 metabolic seconds, and its gut pulses every 3 metabolic seconds. A typical mammals blood circulates every 0.3 metabolic minutes, or every 60 heart beats. The half-life of γ-globulin is 6 metabolic days. These expected values given in terms of metabolic age implicitly involve the 0.27 power of adult weight. Formulae for many such traits are listed by Peters (1983), Calder (1984) and elsewhere. Being based on variable coefficients, however, the constants of proportionality cannot be compared unreservedly and all require separate adjustments. The advantages of accepting a uniform coefficient for all biological time variables is therefore enormous. For example, the number of heart beats in an adult life time has an expected value given by the number of times 0.3 metabolic seconds goes into 1500 metabolic days or about 400 million, with about another 100 million during the growing period. Different species come reasonably close to this expected value (Table 2).

Table 2. Mammalian lifetime metabolism means and species deviations.

Species	Heat production ($\log_e$ kJ /kg)	Heart beats ($\log_e$ units)
Mouse	13.1	20.3
Rat	12.7	20.0
Guinea Pig	13.1	20.3
Rabbit	13.1	20.0
Cat	13.8	20.3
Dog	12.9	19.8
Goat	12.4	19.8
Sheep	12.7	20.0
Cattle	12.9	20.3
Horse	13.1	20.0
Elephant	13.6	20.5
Mean mammalian lifetime total	13.03 ± 0.32 (500 MJ /kg)	20.11 ± 0.25 (500 million beats)

The main conclusion is that expected ages for any species can be calculated from the formulae given and from the temporal life plan of a typical mammal: all that need be known is the adult body weight of the species. The size of experiment needed to detect whether a genotype deviates from its *expected performance* is often surprisingly large. Yet if a genetic experiment cannot demonstrate deviants, then it is merely demonstrating in a highly complex indirect way, that one genotype is bigger than another, which would be obvious before the experiment began.

Body size and metabolic time within species

If metabolic age is to be used in breed comparisons in the same way as for species comparisons, then it must be shown that the biological time-scale of different breeds increases systematically with their adult body weight raised to the 0.27 power. Taylor (1985) reviewed the few available results. He concluded that there was reasonably good evidence for using metabolic age for time taken to mature in live weight and to reach constant backfat thickness in cattle and some for age of attaining sexual maturity. Gestation length and age at emergence of permanent incisors were clearly unrelated to breed size. Hence there appeared to be two distinct types of time control.

A subsequent updating (Taylor 1987a) gave a statistically significant regression coefficient of 0.31 ± 0.13 for time taken to mature body weight by sheep although Kirkwood's (1985) results cast doubt on the use of metabolic age at early stages of maturity but not later stages. There was some moderately convincing evidence that age at puberty occurred at about the same metabolic age in both large and small breeds of cattle, while age at which a constant subcutaneous fat concentration was attained was much greater in large breeds than small in both cattle and sheep. The fact that two of the four coefficients for age at constant backfat were significantly greater than 0.27 merely emphasised the need to apply some age adjustment.

Kirkwood's (1985) coefficient of 0.19 ± 0.04 for heart beat duration in 50 breeds of dogs suggests that metabolic age can be used for metabolic processes. His coefficient of 0.24 ± 0.05 for age of onset of osteosarcoma appeared convincing but was based on only 4 breeds. The use of metabolic age is therefore recommended until more accurate information has accumulated, but not for gestation length nor for age at which teeth erupt, and possibly not for early growth.

When using metabolic age to study genetic differences within a species, a major difficulty is how to make an appropriate allowance for the effect of a constant gestation length. One possibility is discussed in the section on maternal performance. The current doubt about the general applicability of genetic size-scaling for time variables within a species can be avoided by using degree of maturity in body weight instead of metabolic age in growing animals, and cumulate food intake instead of metabolic age in mature animals.

Within-breed genetic selection

Almost no information is available for strains produced by genetic selection. An exception is the half life of tissue protein which was studied in two lines of mice selected for large and small body size together with a control line (Priestly and Robertson 1973). Assuming that tissue protein half life is 10.7 metabolic days, observed and expected half lives in days were 5.0 (4.8 expected) for the large selected line, 3.7 (4.4 expected) for the control line and 4.2 (4.1 expected) for the small selected line. The agreement is much better in terms of metabolic age than ordinary age.

THE MEAN MAMMALIAN GROWTH CURVE

Inter-species relationships

Time taken to mature to a given proportion of adult body weight was shown to be proportional to $A^{0.27}$ by Brody (1945) and Günter and Guerra (1955). This genetic relationship was generalised by Taylor (1965) to all degrees of maturity from embryo to adult, and he proposed that mammalian growth be expressed in terms of metabolic age rather than age. In consequence, he produced a mean mammalian growth curve (Taylor 1980), around which curves for individual species cluster. Moore (1985) produced a growth equation that described this mean curve from which he derived mean growth rate and acceleration curves.

Taylor (1985) pointed out that an animal's stage of development is more fully represented by its degree of maturity in body weight than by its metabolic age. If an animal is nutritionally restricted and obliged to grow slowly, it will be less well developed in other aspects also. Thus when degree of maturity is affected by various treatments, intended or unintended, its relationship to similarly affected variables may be only slightly altered. Comparison at the same metabolic age would then show noticeable differences in both degree of maturity and associated traits. Conversely, comparisons made at the same degree of maturity in body weight might not be greatly affected by these nutritional differences. For this reason traits will often be presented in terms of degree of maturity instead of metabolic age.

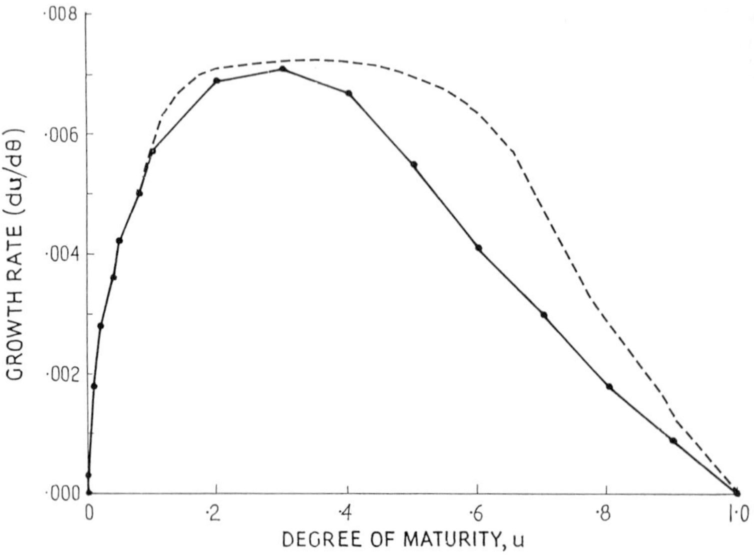

Figure 2. Genetically size-scaled mean mammalian growth curve (——) together with probably curve for improved breeds on good modern diet (----).

The mean growth-rate curve is shown in terms of degree of maturity in Figure 2. It was based on data published 30 or more years ago and appears to have a less extended period of maximum growth than that found for modern domesticated breeds on good modern diets (Fig. 2). A feature of mammalian growth is that between 0.15 and 0.50 mature growth rate does not change much (Fig. 2) and the mean growth curve is almost a linear function of metabolic age. Mammalian growth rates being the ratio of weight to a time, are expected to be proportional to $A/A^{0.27} = A^{0.73}$. Such proportionality has been amply established in Peters' (1983) review of the published evidence.

Inter-breed relationships

Inter-breed genetic correlations for body weights at different ages are extremely high in cattle being 0.99 for ages separated by 6 months and 0.97 for ages separated by 1 year (Thiessen 1985). The inference is that breed curves cross only rarely. Animals with the gene for double-muscling are an exception : they grow faster than expected at early ages, presumably laying down 1.3 to 1.6 times the normal amount of muscle, and slow down more rapidly at later ages, presumably laying down no body fat (Hansett 1986). As part of a more comprehensive study of inter-breed inter-age correlations, Thiessen (1985) also showed that the growth rate of a breed up to 1 year of age was genetically highly correlated with its body weight, most correlations being above 0.85.

Analysis of results reported by Cundiff *et al.* (1986) showed that the highly inter-correlated early weights and growth rates of different breeds were also highly correlated with mature body weight, with correlations between 0.80 and 0.90 (Taylor 1987a).

As also discussed earlier the evidence generally favours the use of metabolic age in breed comparisons although perhaps not prenatally and with some reservations about early postnatal ages. Growth rates of different breeds are strongly related to breed size and some scaling is essential. It is therefore a reasonable procedure to use metabolic age and see whether the growth curves for different breeds are similar, and, if not, the way in which their shape has been altered in early life.

Genetic covariation within breeds

Genetic correlations between body size at one age and a later age are high within breeds. Taylor and Craig (1965) found an average genetic correlations of 0.95 between body size in first and second years for 12 body measurements in twin cattle. High inter-age genetic correlations were also found for degree of maturity in body weight by Fitzhugh and Taylor (1971) in cattle and by Stobart *et al.* (1986) in sheep, animals more mature at one age generally being more mature at all other ages. The overall pattern is one of strong genetic control of growth at all immature ages, primarily in relation to mature body weight but not exclusively so.

Within-breed selection. The consequence of high inter-age genetic correlations is most simply seen in selection experiments. Selection for body weight at any one age has resulted in growth curves that differ at all ages and it is difficult to detect the age at which genetic selection occurred (for example, Land and Legates 1969). In fact, a general overall increase (or decrease) in body size at all ages has proved to be probably the easiest mammalian characteristic to alter by genetic selection and it has been widely practiced in mice, rats, dogs, sheep and cattle.

Maximum expected growth rates were calculated as $7A^{0.73}$ g /day for lines of mice selected by Falconer (1973). The agreement between observed and expected values is acceptable (Table 3). Nevertheless selection appears to have altered growth rate by slightly more than the amount expected from the change in mature body weight.

Table 3. Observed and expected mean growth rates from 3 to 6 weeks of age in unselected mice and in lines selected for large and small body weight (Falconer 1973).

	Strain of selected mice		
	large	control	small
Observed	1.01	0.62	0.33
Expected	0.72	0.60	0.43

BODY SIZE AND BODY COMPOSITION IN ADULT MAMMALS

In general, body tissues and organs comprise the same proportion of adult body weight in all species with certain notable exceptions. Thus on average blood is 6.0% of adult body weight, lungs are 1.1%, the heart is 0.6%, the gut is 6.5% and its contents 12.3%. Muscle comprises about 35% to 40% of body weight and fat 10% to 15%. It is claimed (Pitts and Bullard 1968) that the lion has almost 60% muscle, with the cat family in general in the range 50 to 60%. Surprisingly, domestic mammals seem to have the least muscle, about 30% or half the maximum.

Bone. On average, bone is about 6% of adult body weight. It has been suggested that heavier bones with a wider cross section are needed to support large mammals. This has reinforced the belief that the proportion of bone is less in small than in large animals. Published data range from 5% in small mammals up to 27% in elephants. The inter-species regression of 1.14 given by Keyser and Heusner (1964) has been superceded by a lesser trend of 1.09 given by Prange *et al.* (1979).

Pelt. The pelt shows an opposite trend from the skeleton, being a smaller proportion in large mammals. Calder (1984) suggests that pelt plus skeleton together comprise about 17% of adult body weight both in large and small species.

Liver. The liver is clearly relatively smaller in large mammals with an expected value $0.033A^{-0.15}$ (Schmidt-Nielsen 1984), being 5.6% of a mouse but only 1.3% of a cow. The same is true for kidney, with corresponding values 10.2% and 0.3% and an expected value of $0.0073A^{-0.15}$. Their inter-species regression coefficients of about 0.85 (Brody 1945), are intermediate between the coefficients of 1.0 for most parts of the body and 0.73 for the ingested food that the liver has to process.

Brain. The most extremely deviant organ is the brain, with an expected proportion given by $0.01A^{-0.30}$ (Brody 1945), thus comprising almost 3% of a 30g mammal and 0.4% in a 500kg mammal.

Metabolic rate. The liver, kidney and brain are metabolically extremely active tissues. In small mammals especially, they are responsible for a large proportion of the total metabolism, both energy and protein (Lobley *et al.* 1980). Liver may be regarded as providing the power for growth, and there is a dramatic change in liver and kidney proportions as mammals decrease genetically in body size and time to maturity. Thus these organs comprise 1.7, 3.2 and 6.4% of body weight respectively in the cow, dog and mouse.

Another interesting characteristic of the mammalian liver is the number of mitochondria, given by Smith (1956) as proportional to $A^{-0.10}$. The more mitochondria per gram of liver (and therefore per cell), the shorter the life of the liver and the animal. The data of Krebs (1950) was used to regress the oxygen consumption of liver slices of different species on adult body weight and gave O_2 consumption per g liver as $13.9A^{-0.12}$, a significant result. Combining this equation with that for liver proportion ($0.033A^{-0.13}$; Schmidt-Nielsen 1984) gives O_2 consumption by the entire liver as proportional to A to the power $1.0-0.12-0.13 = 0.75$, and hence directly proportional to O_2 consumption by the whole body. An even more interesting result is obtained by combining the equation for O_2 consumption by the liver with that for number of mitochondria to give *oxygen consumption per mitochondrion as independent of body size.* Hence number of mitochondria in the liver could be regarded as determining the metabolic rate of a mammal. The same is probably true for number of mitochondria in other tissues. For example, the only feature for distinguishing easily between muscles from large or small mammals is the number of mitochondria per gram (Schmidt-Nielsen 1984).

Table 4. **Nutritional association between daily fasting heat production (MJ /kg $^{0.75}$LW) and weight of organs (as a % of liveweight) in lambs fed for 6 weeks on an initial high, medium or low feeding level and for 6 weeks on a subsequent level (from Ferrell and Jenkins 1985).**

Feeding regime	Fasting heat production	Stomach	Small intestine	Large intestine	Liver	Kidney	Spleen	Heart
Low–High	0.41	2.21	1.21	0.88	1.52	0.28	0.15	0.35
Medium–Medium	0.29	1.71	0.83	0.64	1.08	0.24	0.18	0.30
(adjusted to 0.35% heart)	(0.34)	(2.00)	(0.97)	(0.75)	(1.26)	(0.28)	(0.21)	(0.35)
High–Low	0.29	1.71	0.78	0.73	1.03	0.23	0.17	0.35
(Low–High)/(High–Low)	1.41	1.29	1.55	1.21	1.48	1.22	0.88	1.00

Non-genetic differences. The main difficulty in making genetic comparisons of liver and small intestine is the marked effect of level of nutrition. The difference between lambs kept on a high-low *versus* a low-high nutritional regime was a 55% increase in size of small intestine, 48% increase in liver size and a 42% increase in fasting heat production (Table 4).

Chemical composition

Mature animals of different species have, on average, a body composition of 18.4% fat, 58.6% water, 18.7% protein and 4.3% ash, which might be taken as a standard mammalian body composition (Mitchell 1962). Taking the heat of combustion of protein as 23.8 kJ /g and lipid as 39.6 kJ /g gives an average heat of combustion for mature body tissue of 11.7 kJ /g. Species variation is greatly reduced by taking fat-free chemical composition.

The average chemical composition of muscle, fat and bone given in Table 5 applies to well-fed animals; it changes as fat reserves are depleted on low levels of nutrition. There were no obvious trends with body size. Cattle and mice can be equally fat. The proportion of ash in larger mammals may increase to the extent that the proportion of skeleton apparently increases, but such an increase in ash has not been demonstrated.

Table 5. Average chemical composition (%) of muscle, fat and bone in well-fed mammals.

	Muscle	Fat	Bone	Whole body	Fat-free whole body
Water	74	20	28	58.6	71.8
Protein	20	8	24	18.7	22.9
Ash	1	0	34	4.3	5.3
Lipid	5	72	14	18.4	-

Body size and adult breed differences

Muscle, fat and bone. Breed difference in carcass bone, muscle and fat in relation to body size in cattle and sheep were reviewed by Taylor (1987a). For bone proportion, there were clear breed deviations but no systematic trend with breed size except when domestic and feral breeds were compared (Fig. 3). Expected mature proportion of dry bone from the inter-species relationship $0.06A^{1.08}$ gave 7.7% for the small feral Soay breed and 8.6% for the large Oxford Down compared with mature values of 9.2% and 8.2% for undried bone estimated on the assumption

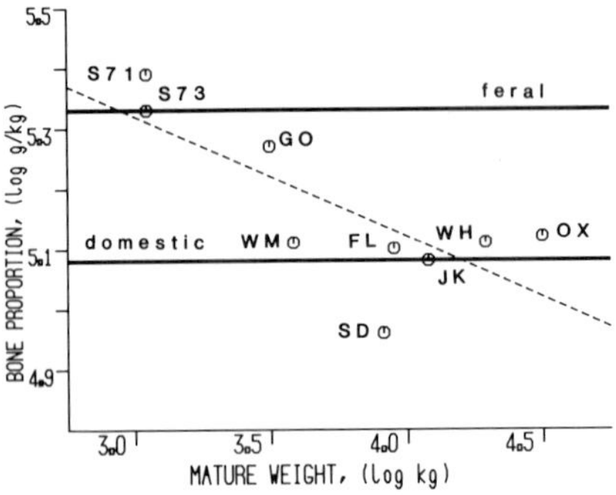

Figure 3. Inter-bred regression of the natural logarithm of proportion of the carcass that is bone (averaged over four degrees of maturity) on the natural logarithm of mature female weight (kg). Observed (---), expected (——). Breeds were Soay-71 (S 71), Soay-73 (S 73), feral goat (GO), Welsh Mountain (WM), Southdown (SD), Finnish Landrace (FL), Wiltshire Horn (WH) and Oxford Down (OX).

that carcass bone was 0.74 of total bone. Muscle proportion was clearly independent of breed size. Similarly the proportion of body fat appeared to be independent of breed size when the breeds compared were either domestic or feral and had the same proportion of subcutaneous fat or were equally mature in liveweight (Fig. 4). The large Oxford Down was fatter than average but the even larger Wensleydale was leaner than average (D. Croston, A.J. Kemster, D.R. Guy and D.W. Jones unpublished). Large breeds of cattle may have less fat than average because they were originally selected for draught purposes. Similarly, small feral breeds of sheep like the Soay are clearly leaner than average.

In particular situations, a genetic association between body size and body composition can be generated. A number of breeds put together on fairly poor pasture will show a negative correlation between mean body size and percentage body fat. Another example, somewhat artificial biologically, yet one of the most widely studied, is the genetic correlation between body size and body composition found when animals of different mature size are all observed at a fixed immature body weight. There is then a high negative genetic correlation between percentage body fat and mature weight, artificially introduced because fat percentage increases as animals become more mature and because at a fixed weight animals with a large adult weight are always less mature.

As an overall conclusion, body composition shows a great deal of variation but, for breeds equally mature in body weight, there is in general no genetic association with mature body size.

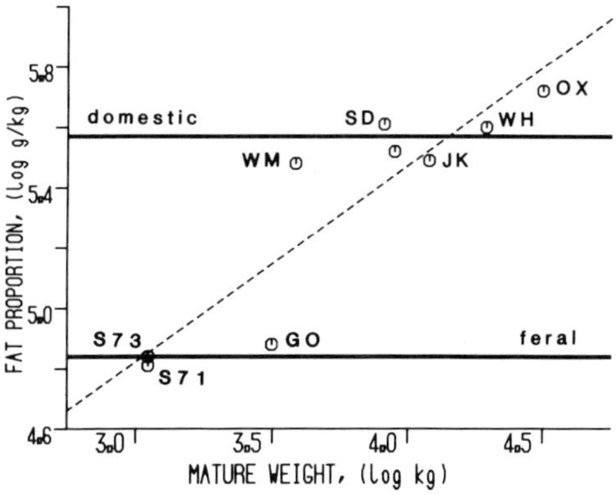

Figure 4. Inter-breed regression of the natural logarithm of the proportion of the carcass that is fat (averaged over four degrees of maturity) on the natural logarithm of mature female weight. Observed (----), expected (——). See legend to Figure 3 for breed symbols.

Within-breeds genetic selection

A correlated response to genetic selection for body weight at an early age is nearly always an increase in proportion of body fat at all subsequent ages McCarthy (1980). Selection for large and small Merino sheep resulted in an exceptionally heavy skeleton in the large line compared with the small line (Thompson *et al.* 1985). Lines of mice selected for large body size had larger cells and more cells in their internal organs than had mice selected for small body size (Falconer *et al.* 1978). Moreover, during growth, the large selected mice had a larger proportion of liver and kidney than the small line (Table 6). This genetic trend is the opposite of the inter-species decrease in liver and kidney size with increasing body size.

Table 6. Correlated response in organ size to selection for body size in mice (Falconer *et al.* 1978).

Lines	6-week weight(g)	Lungs	Spleen	Liver	Kidney
		(Organ weight as % of body weight)			
Large	32.34	0.68(0.68)	0.42	7.54(6.60)*	1.70(1.58)
Control	24.48	0.68	0.42	6.88	1.65
Small	16.51	0.74(0.68)	0.35	6.29(7.29)	1.53(1.75)
(large-small)/small		-0.08	0.20	0.20(-0.09)	0.11(-0.10)

*expected value in brackets

CHANGE IN BODY COMPOSITION DURING NORMAL GROWTH OF A TYPICAL MAMMAL

Cleavage of the mammalian fertilised egg, differentiation and organogenesis, if studied on a metabolic age scale, are found to proceed in a remarkably similar fashion in all mammals (Table 1). From the foetal stage onwards, studies have usually concentrated on the relative growth rates of different body organs and tissues, changes in chemical composition and in cell size and number.

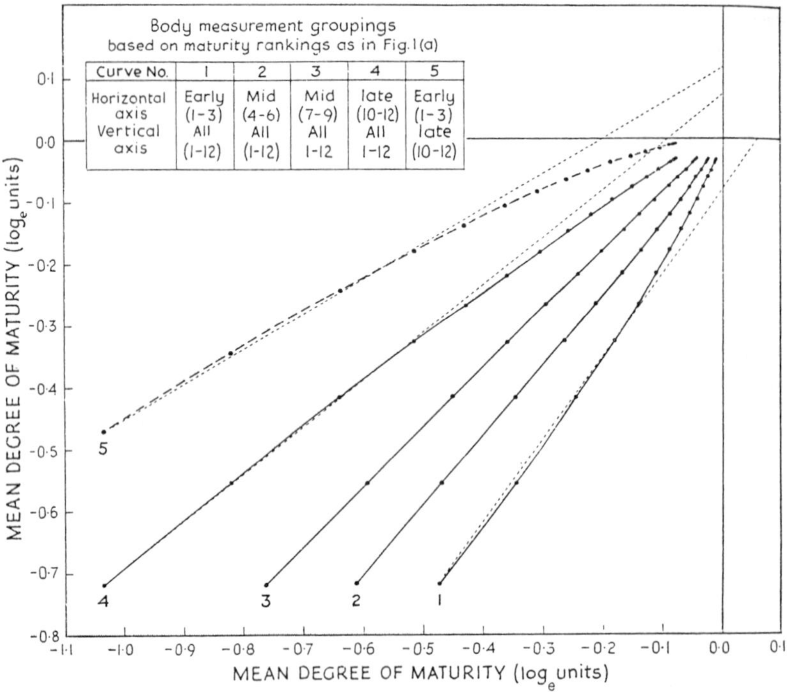

Figure 5. Allometric plots of mean degree of maturity of linear body measurements in different maturing groups (Russell 1975).

Allometry of body components

The allometric equation (Huxley 1932) has been and still is by far the most widely used method of describing relative growth. Within the range 0.25 to 0.75 mature in body weight, allometry seems to describe a wide range of traits with reasonable success. Foetal and early postnatal growth exhibit many rapid and complex changes that simple allometry cannot deal with (Kidwell *et al*. 1952; Berg and Butterfield

1976). Similarly, the relative growth rates maintained during the main period of postnatal growth can change as full maturity is approached and the allometry lines can become curved (Fig. 5). Prediction of adult values from allometric relationships estimated during the main period of growth may be misleading for body shape but appear to be reasonably acceptable for many organ and tissue proportions.

The genetically-standardised allometry square

Every body component can be partitioned into an adult value and a degree of maturity in the same way as for body weight. The allometry equation for degree of maturity in a body component, u_y, in relation to degree of maturity in the whole body, u, is simply:

$$u_y = u^b \qquad \text{Equation 2}$$

where b is the allometric coefficient that describes the curvature or relative maturing or relative growth rate (Fig. 6). The degrees of maturity for both part and whole are dimensionless units : the allometric coefficient b is therefore dimensionless and independent of adult values. Alternatively the genetic size-scaling rules of Taylor (1980a) can be applied and lead to the same equation with the coefficient b unaffected. This is an extraordinary property of the allometric coefficient. It allows direct comparison of species of widely different body size and shape. Genetically size-scaled values of b can therefore be estimated without knowing adult values for component traits, only the adult value for body weight. When body weight is expressed as degree of maturity in body weight, but the component trait y, has a mature value y_A, the allometric equation becomes:

$$y = y_A\, u^b \qquad \text{Equation 3}$$

and the constant of proportionality, y_A, is the mature value of the y trait, that is, the value of y when u = 1.

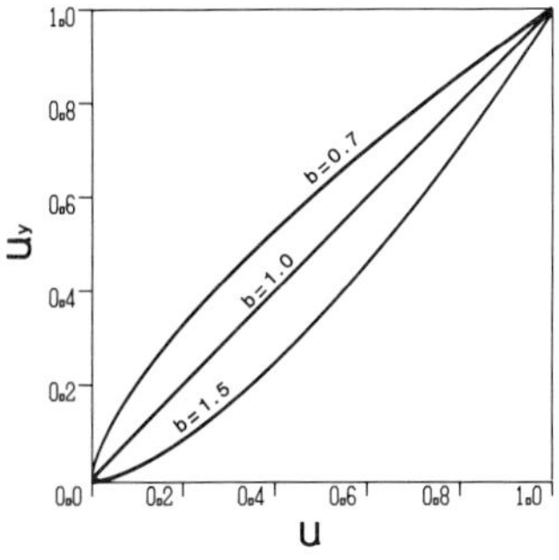

Figure 6. The standardised allometric square illustrating the equation $u_y = u^b$ which relates degree of maturity, u_y in component part y, to degree of maturity, u, in the whole. Early maturing parts have b <1 and late maturing parts have b >1.

Mean mammalian relationship for compositional maturity

Tulloh (1964) summarised allometric coefficients for relative growth of carcass muscle, fat and bone in cattle, pigs and sheep. When averaged over the three species, mean values of the allometric coefficients (for proportions) were 0.65 for fat, −0.02 for muscle and −0.26 for bone. These mean coefficients were used by Taylor (1985) to produce mean curves for relative compositional maturity in body composition. Early maturing organs and tissues (e.g. bone), which have b <1.0, are a declining proportion of the whole body as animals mature. Late maturing parts with b >1.0 (e.g. fat) are an increasing proportion. When such component traits are expressed as proportions, the allometric coefficients change from b to b−1, so that early maturing parts which are declining as a proportion, have a negative allometric coefficient, and late maturing parts which are increasing as a proportion have a positive coefficient. The allometric coefficients used for pigs, sheep and cattle were, respectively, 0.50, 0.54 and 0.90 for fat, −0.01, +0.02 and −0.07 for muscle and −0.20, −0.28 and −0.29 for bone. At early stages pigs are closer to their mature body composition than are sheep, and much closer than are cattle especially for fat.

Perhaps the simplest way to study changes in body composition during growth is to present the curves of the allometric square in tabular form, as done in Table 7 for carcass muscle, fat and bone. The degree of maturity of each component part is given for each degree of maturity in body weight. Compositional maturity is the ratio of the degree of maturity in the part to that in the whole.

Table 7. Degree of maturity in dissected muscle, fat and bone in relation to degree of maturity in body weight (Butterfield *et al.* 1983).

Trait	Degree of maturity (proportion)								
Whole body	0.2	0.3	0.4	0.5	0.6	0.7	0.8	0.9	1.0
Muscle	0.24	0.35	0.46	0.56	0.66	0.75	0.84	0.92	1.0
Fat	0.05	0.11	0.18	0.27	0.38	0.51	0.65	0.82	1.0
Bone	0.26	0.39	0.50	0.60	0.70	0.79	0.86	0.94	1.0

Chemical composition during growth

Mammals change from being about 98% water at conception to about 60% in the adult. During the first half of gestation, body composition on a dry-matter basis is about 10% lipid, 70% protein and 20% ash (Salmon-Legagneur 1968).

On a fat-free basis, body composition is chemically almost mature when body weight is roughly 0.3 mature, although ash does increase slightly. Blaxter (1967) showed that percentage of ash, nitrogen and water in cattle aged 5 months differed little from percentage at 20–50 months. There are striking species differences in proportion of lipid at birth. For example, the rat and pig both have low proportions (1.1%) compared with an overall average of about 2.5%. The guinea-pig and man both have high fat (10%). However, it does not follow that the pig, for example, is a lean species.

The mean mammalian equations given by Taylor (1985) provide expected values for lipid at different stages of maturity in body weight against which pigs can be compared (Table 8). Pigs are twice as fat as the average for other species at birth (degree of maturity 0.006) and at maturity, while they are five to ten times fatter when 0.05 mature.

Table 8. Body fat concentration (g /kg) in pigs compared with the average for mammals at three stages of body weight maturity (Walstra 1980).

Degree of maturity	Pigs	Mammalian average
0.006	11	5
0.05	100 to 200	20
1.00	300 to 400	150

Distribution of protein

In well-fed cattle, protein is distributed throughout the body in roughly the same way throughout postnatal life : blood (5%), organs (8%), hair and hide (17%), skeleton (slightly less than 20%), fat and lean tissue (just over 50%). Mean values for many species are not available.

Cell size and cell number in adipose tissue

Mammals begin to lay down fat cells about halfway through foetal life, that is at about 25 to 30 metabolic days of age (Lawrence 1980). By the end of gestation there are fat cells in most depots, with intermuscular fat being the last to be laid down. Final adipocytes are formed at about 200 metabolic days of age when mammals are about 0.80 mature in body weight, but species vary greatly. In sheep, Broad *et al.* (1980) have shown that, from 40 metabolic days of age until full maturity in body weight, adipocyte volume increased 70-fold, the same in all depots. Over the same period, total number of adipocytes increased 6-fold on average although depots differed. There may be 10^8 to 10^9 fat cells /kg mammalian body tissue (Cianzio *et al.* 1985).

If Joubert's (1955) results are typical, the number of muscle fibres increases from 15 to 33 metabolic days of age, and thereafter they only enlarge. Severe postnatal dietary protein restriction makes no difference to the number of fibres, but fibre size is strongly affected (Timsen and Dudenhoffer 1985).

Body size and breed differences in body composition during growth
Allometric growth has been studied most extensively in sheep and breed differences were reviewed by Taylor (1987a). Significant differences in allometric coefficients have been reported for individual bones, total bone, individual muscles, omental fat and (chemical) protein. Although breed differences undoubtedly exist, relative growth is nevertheless remarkably similar in most breeds, necessarily so if all breeds are to grow like sheep.

The most commonly found differences in allometric coefficients are those between the sexes, with males having heavier shoulder and neck muscles, and lighter muscles in the hind leg.

Genetic selection within breeds
Genetic selection experiments, mainly in mice, have shown that selection for large and small body size in early life usually results in increased body fat in large selected lines at all ages. Whether the increase is compatible with an increase only in proportion of adult fat has not been established. There is no reason to believe that these size-dependent correlated responses to genetic selection are inevitable. On the contrary, selection for a high body weight at a late age preceeded by selection for a low body weight at an early age has resulted in lean animals at all stages (Riccard 1978).

Decades of genetic selection in poultry have increased the proportion of breast muscle. At present great care is being taken that selection programmes do not increase body fat.

ENERGY METABOLISM AND BODY SIZE IN MATURE MAMMALS

Energy intake
Evans and Miller (1968) have shown that the genetic differences in voluntary energy intake of different species are proportional to their mature metabolic body weight ($A^{0.73}$). Energy intake is about $0.60A^{0.73}$ MJ ME /d. Farlow (1976) gave an estimate for herbivores of $0.97A^{0.73}$ MJ GE /d. During lactation energy intake in all species shows a remarkable two or three fold increase above the non-lactating level. There is considerable genetic variation in the extent of this increase but it is not related to body size (Kleiber 1961).

Heat production
The inter-species relationship of basal heat production in mature mammals to their mature body weight was one of the first to be examined. Kleiber's (1932) estimate was proportionality to body weight to the power 0.74. Brody and Procter (1932) produced their famous mouse-to-horse (later mouse-to-elephant) curve with an exponent of 0.73, a value subsequently adopted as an international standard in 1935 (Brody 1945), and amply confirmed (Hemmingsen 1980; Peters 1983).

Basal heat production is $0.3A^{0.73}$ MJ /d, about half the metabolisable energy normally ingested. Normal adult heat production is therefore about twice basal, and in terms of watts, $H_{normal} = 2 \times H_{basal} = 7A^{0.73}$ watts.

In humans for example, $A^{0.73} = 20$, and $H_{normal} = 100$w, so that a human when quiescent is the equivalent of a 100 watt electric bulb. Maximum (for a trained athlete) is 20 times basal, a 2 kilowatt fire. After a good meal with wine, metabolic rate could be 5 times basal; then each couple is equivalent to a 1 kilowatt fire. The total heat produced in a normally active mammal's adult lifetime of 1500 metabolic days will be 0.6 x 1500 MJ = 900 MJ on average from each kilogram of adult body tissue.

Protein metabolism and body size in mature mammals

Evans and Millar (1968) examined the inter-species relationship for protein intake and Brody (1945) the relationship for protein excretion. Both are proportional to mature metabolic body weight, $A^{0.73}$. Mean protein intake is $1.7A^{0.73}$ per day, which must equal normal excretion rate. On a minimal protein diet, endogenous urinary protein excretion is $0.9A^{0.73}$g /d. Now total body protein in an adult mammal is about 0.186Akg. In a mature equilibrium, protein excreted equals protein ingested. Normal protein excretion is therefore $1.7A^{0.73}/0.186A = 0.0085A^{-0.27}$ of total body protein per day, or about 0.0085 of total body protein per metabolic day.

Protein turnover. The mean rate of whole-body protein turnover rate in various mammals is $18.6A^{0.73}$g /d (Young 1985). Protein turnover, as a proportion of total body protein, is therefore $0.1A^{-0.27}$ /d. In other words, whole-body turnover is 0.1 of total body protein per metabolic day.

The ratio of turnover to excretion is therefore 0.1/0.0085 = 12. Thus the protein ingested by a mature mammal is normally recycled twelve times before excretion.

In the course of an adult lifetime of 1500 metabolic days, the total protein ingested, catabolised, re-synthesised and excreted per kilogram adult body weight will be the same for mammals of all body sizes. For example, total protein ingested is expected to be 2.55 kg /kg adult body weight.

Food intake and heat production during normal growth of a typical mammal

The relationship between growth in body weight and the cumulate food intake of a mammal can be reasonably well described by Spillman's equation (Parks 1982). Thus degree of maturity, u, in body weight at a given cumulate food intake, $\tilde{F}$, in MJ ME /kg adult body weight, is:

$$u = 1 - \exp(-\alpha \tilde{F}) \qquad \text{Equation 4}$$

where α has a mean mammalian value of about 0.04 kg body weight /MJ

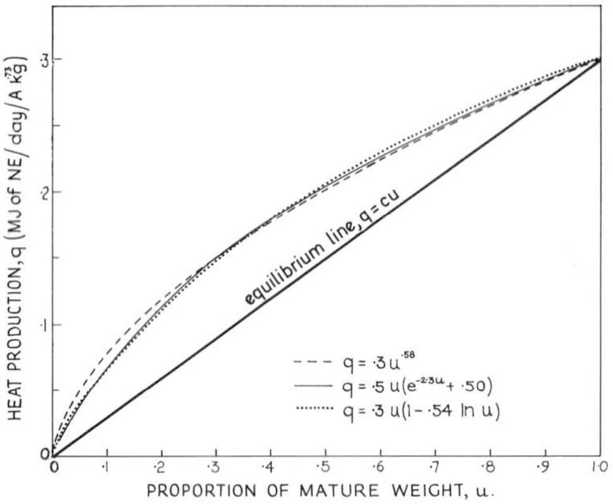

Figure 7. Genetically size-scaled mean mammalian curve for heat production during normal growth, represented by three equations that give almost the same curve. The maintenance equilibrium line gives heat production for a state of zero growth on constant food intake (Taylor 1985).

ME and about 0.025 for ruminants. The mean mammalian curve for daily energy intake during growth depends on the nutrient density of the diet (Taylor 1982). For diets of high nutrient density, maximum intake occurs before animals are fully mature.

The mean mammalian curve for heat produced by a growing mammal is reproduced in Figure 7. The associated relationship for the heat produced in an equilibrium state by an immature mammal was given by Taylor *et al.* (1981) and is represented by the straight line. The extra heat produced by being in a growing as opposed to an equilibrium state follows a curve that closely resembles the shape of the mean growth-rate curve.

Efficiency of food utilisation during normal growth

Food conversion efficiency, that is, the ratio of body weight gain to food consumed over a given maturity interval, declines rapidly with increasing maturity. If growth rate as a function of degree of maturity in body weight is $g(u)\ A^{0.73}$ kg /d and food intake is $f(u)\ A^{0.73}$ kg /d, then food conversion efficiency = $g(u)A^{0.73}/g(a)A^{0.73} = g(u)/f(u)$, which is independent of body size. A comparison of food conversion efficiency in meat-producing animals (Holmes 1977) showed that the main differences were those within species due to a different degree of maturity at slaughter (e.g. veal *vs* beef, and early *vs* late lamb), and there was no obvious trend with body size of species.

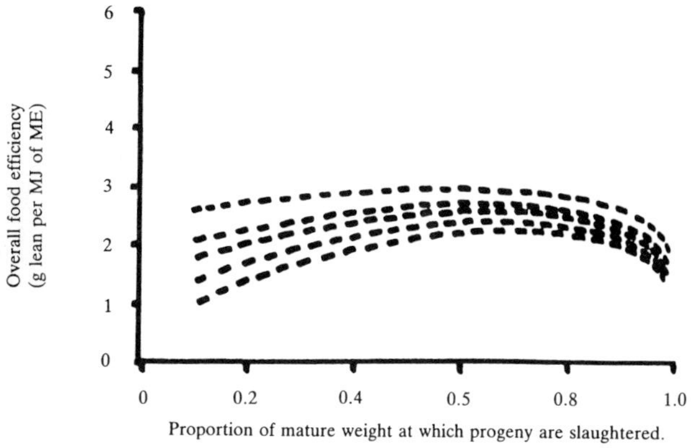

Figure 8. Overall food efficiency curves showing how efficiency depends upon the degree of maturity of progeny at slaughter. Individual curves show efficiency decreasing as the number of progeny per dam increases from 2 (top), 3, 4, 6 to 10 (bottom) (Taylor *et al.* 1985).

Overall food conversion efficiency in a maternal family

The food conversion ratio is only one of many variables that determine the overall food efficiency of a meat production system. A full tally of the meat produced and food consumed in a commercial herd can be achieved if a single (average) maternal family is taken as the production unit. All the male and female offspring grown to slaughter (excluding one replacement breeding female), and the breeding female herself until culled, consume food and produce meat. The optimum degree of maturity of offspring at slaughter is apparent in Figure 8 (see also Taylor 1987b).

Within-species difference in food intake in relation to body size

Ad libitum food intake of both liquid and dry diets by dairy calves of the Friesian, Ayrshire and Jersey breeds were compared by Roy (1967) and Roy *et al.* (1971). Over the period studied, calves of the same age were about equally mature in body weight. The between-breed regression of log *ad lib.* milk intake on log body weight averaged 0.71 $\pm$ 0.04 up to 12 weeks. Equal dry matter intakes from liquid and dry food were achieved at body weights of 36, 47 and 52 kg for Jerseys, Ayrshires and Friesians respectively, yet these body weights as a percentage of mature body weight were the same for all three breeds, being 8.8, 8.9 and 8.4 respectively. Similarly, equivalent net energy intakes were obtained from liquid and dry diets at very different body weights but at 17.2, 17.6 and 17.7% of mature weight. The intake behaviour of the three breeds was either remarkably similar or quite different depending on whether breeds were or were not compared at the same degree of maturity. Monteiro (1972) gave mean *ad libitum* intakes between calvings for lactating Friesian and Jersey cows as 2.9% and 3.2% of body weight per day with the Jersey value 11% greater than the Friesian. In terms of metabolic body weight, the corresponding values are 15.9% and 15.5%, the difference now being reduced to 4%.

Taylor *et al.* (1986) in their study of 25 breeds found that, in relation to expectation, large breeds of cattle tended to eat relatively more than small breeds. M.L. Thonney, St C.S. Taylor and T.H. McClelland (unpublished) found the same in equally mature breeds of sheep. The equivalent regressions from Southgate *et al.* (1982) were 1.19 ± 0.23 and 1.24 ± 0.30, both greater than expected, although not significantly so. Thus large breeds may well have a larger than average early appetite.

Within-species differences in food conversion efficiency in relation to body size

Whether food conversion efficiency is dependent or independent of body size depends on the interval chosen. Genetic comparisons at constant body weight show a strong correlation between breed size and food efficiency. Over the same maturity interval, food efficiency and body size are genetically uncorrelated (Taylor 1987a).

Within-breed genetic correlations

Within breeds quite high genetic correlations have been found between food intake and body weight. Hooven *et al.* (1968), for example, gave a genetic correlation of 0.80 ± 0.09. Even so, intake in mature dairy cattle is not usually entirely *ad libitum*, and variation in intake associated with variation in yield complicates the situation. Genetic regressions are not usually quoted and it remains to be seen whether or not the within-breed genetic regression of log voluntary food intake on log mature body weight has a coefficient of 0.73. Everything points to it being so, but the hypothesis awaits experimental confirmation.

Mice genetically selected for large and small body size showed correlated changes in mature food intake (Roberts 1981). The observed difference between the large and small selected lines was only about half that expected. Large selected mice maintained themselves more efficiently than expected.

Within-breed difference in maintenance efficiency

Taylor *et al.* (1981) found that each individual had its own innate maintenance efficiency. For maintenance equilibria in the same animal at different degrees of maturity, the repeatability was 0.76. Their results were compatible with maintenance efficiency increasing with $A^{0.27}$, but this relationship with body size could not be established with any accuracy.

Genetic selection within breeds for efficient milk production
Several within-breed investigations have found that cows with higher gross food efficiency tend to be lighter (Dickinson *et al.* 1969). The combined result from three different breeds showed a significant decline in food efficiency with body weight and several measures of body size. An interesting feature was that this decline was largely absent in one of the three breeds, namely the Ayrshire, where there was no decline with body weight but a slight decline with increasing height at withers. Hooven *et al.* (1968) also found small to moderate negative correlations between efficiency and body weight. On closer examination they found the actual association was decidedly curvilinear with an intermediate optimum. A curvilinear relationship between milk yield and body weight, particularly at lower levels of feeding, is discussed later in this review. In the case of Hooven *et al.* (1968), however, feeding was at a high level, so that within a breed the case for optimum efficiency occurring at an intermediate body weight seems even stronger.

A further aspect to be considered is whether body weight should be taken into account when selecting animals for dairy merit. Selecting for milk yield alone has been advocated because of the high genetic correlations reported between milk yield and efficiency, for which Syrstad (1966), for example, gives values of 0.81 to 0.88. Milk yield per unit body weight, however, has been advocated as an alternative or even better indirect measure of efficiency with even higher correlations. Nedava (1970) found the correlation of efficiency with yield per unit body weight was in general higher than that for yield itself (0.70 *vs* 0.54, 0.91 *vs* 0.61 and 0.31 *vs* 0.38 in milk, milk-meat and meat type cows respectively). Both criteria will thus improve efficiency, but the correlated responses will be different. In the case of yield, body size will probably be increased, whereas in the case of yield per unit body weight, body size is very likely to be decreased. Neither criterion is likely to result in the extremes of size within a breed being selected. When breed comparisons are involved, however, the problem becomes more acute, since these two criteria could lead to opposite extremes of breed size. The conclusion reached in a previous section would suggest that 305-day milk yield per unit metabolic body weight could be used both within and between breeds without unduly favouring either large or small cows or breeds. This, of course, will produce exactly the same result as selecting for gross efficiency estimated by Brody's (1945) formula, since the metabolic efficiency of producing milk and maintaining body weight can only be presumed equal in the absence of recorded food intake.

Body size and food composition within species
Body size, body shape and body composition, both during growth and at maturity, are affected by food composition, sometimes to an extreme degree when essential dietary components are lacking (Lewis and D'Mello 1967). The main effect, however, is often largely a consequence of food intake being altered by food composition. Parks

(1970) has investigated the way in which different amounts of protein and energy affect growth curve parameters with some interesting results. Milk yield and milk composition are well known to be likewise affected by food composition (Storry *et al.* 1968). In genetic terms, differences in food composition can result in specific genotype–nutrition interactions, the classical example being the differential reaction to roughage of ruminants as compared with non–ruminants. Specific breeds are often claimed to perform better than others on certain diets. Despite this a characteristic of many experimental investigations is that no interaction has been found. Even when they have, there has been no systematic association with body size.

Here, the overall conclusion is that in general there is no association between body size and the effect of differences in food composition, so long as the amount of food is unlimited and animals are fed *ad libitum*. A major exception associated with poor pasture is discussed in the next section.

BODY SIZE AND FOOD SUPPLY

Inter–species relationships

An excellent review of the functional significance of variation in body size among mammals in relation to food availability was given by Clutton–Brock and Harvey (1983). They deal with abundance and quality of food ingested, mouth size and feeding rate, survival time during acute food shortage, time spent eating, and direct competition for food. They conclude that among herbivores, frugivores and omnivores, large–sized species usually spend more time feeding than do smaller species and feed on foods of lower nutritional quality or on foods which are more widely and erratically distributed; the less abundant the food supply the more pronounced become these feeding differences associated with body size.

Inter–species relationships of food supply and body size

Many of the conclusions concerning food quality and body size, especially in herbivores, involve two doubtful premises. Demment and Van Soest (1983, 1985) in particular have stressed extra gut capacity in large species in relation to their metabolic requirements. They give gut contents as directly proportional to adult body weight and equal to 0.123A, in acceptable agreement with the unpublished estimates of Thonney *et al.* of 0.133 ± 0.004 and 0.128 ± 0.004 for seven breeds of sheep fed *ad libitum*. Demment and Van Soest (1983) therefore believe that an animal's ability to digest food increases with increasing body size at a faster rate than its metabolic requirements. They gave the retention time of digesta in the gut as $T_r = 0.60DA^{0.27}$ hours, where D is the digestibility of the diet. Rate of passage of food through the gut is therefore slower in large species. The rate at which ingested food is digested must also be slower, being proportional to gut capacity divided by retention time, that is, to:

$$0.123A/0.60DA^{0.27} = 0.50D^{-1}A^{0.73} \qquad \text{Equation 5}$$

and therefore proportional to metabolic rate.

Equation 5 shows that rate of digestion is slower for less digestible foods, and there will be a correlation between digestion rate and body size only to the extent that larger species eat less digestible foods, a correlation which is likely to vary with abundance of food supply.

A second doubtful premise is that, within a given territory, large species need more food in total than small species and, to find enough, have to eat the less digestible food sufficiently abundant for their needs. The amount of food required by a species in a given territory is found by multiplying the number of animals belonging to that species by the average requirement of an individual. The relationship of animal density to body size (Calder 1983; Peters 1983) has not yet been accurately established, but the allometric coefficient appears to lie between 0.73 and 1.25. Whatever the exact value, the number of animals per hectare declines rapidly with increasing body size. The total food required from any one hectare is $0.6A^{0.73}NA^{-(0.73+e)} = 0.6NA^{-e}$, so that large species do not require more food from each hectare than small species but possibly less. Hence large species are not obliged to eat less digestible food because of its greater abundance. Nor do small species enjoy food that is less abundant but of higher quality because they need less of it. Quite the contrary.

A primary reason for larger species eating less digestible material, which may well be a sufficient reason in itself, is that larger species have bigger mouths which limit the ability of large ungulates to select the most nutritious parts of plants (Jarman 1974) and which make grazing difficult for large herbivores on short-cropped swards (Allden and Whittaker 1970; A.W. Illius and I.J. Gordon unpublished). Conversely, on swards that have been grazed until short, small herbivores with small mouths find further grazing less difficult and more rewarding than do bigger herbivores with large mouths. This most important interaction between pasture condition and body size has been modelled by A.W. Illius and I.J. Gordon (unpublished). Their model is based on incisor arcade breadth which they give as equal to 8.6 $A^{0.36}$mm, the power 0.36 not being significantly different from the 0.33 given by Clutton-Brock and Harvey (1983). When there is an abundant supply of only poor quality food, the argument is quite different. Dinius and Baumgardt (1970) showed that dry matter intake increased but DE remained constant in sheep as the DE of diet decreased from a high value of 15 MJ /g down to about 10.5 MJ /g; below this value, both dry matter and DE intake decreased. Species of all body sizes are then equally affected.

Intra-species relationship of grazed intake to body size

Allden and Whittaker (1970) have shown that on good pasture year-old sheep had a higher rate of intake than had lambs, whereas under sparse pasture conditions lambs were able to consume at a significantly greater rate than the year-old sheep. Differences in rate of intake on the same pasture by animals of different body size correspond to those found for the same animals on pastures of varying sward length. Rate of intake of dry matter increased from 1 to 7 g /min when tiller length increased from 3.7 to 7.7 cm (Allden and Whittaker 1970). Grazing time of 400 min /day on good pasture almost doubled on poor pasture in an attempt to maintain the same total daily intake. Allden and Whittaker (1970) also showed that biting rate steadily increased as intake per bite declined so that intake per minute did not decline until after maximum biting rate was reached.

The unpublished results of A.W. Illius and I.J. Gordon seem to correspond closely to the observed differences in growth rate of large and small breeds as pasture conditions change. An experimental comparison of body weight and growth rate of cattle breeds in intensive, semi-intensive and hill-grazing conditions, showed that the difference between large and small breeds progressively diminished as the availability of food decreased (MLC 1971). Intake per unit metabolic body weight was no longer independent of body weight but was genetically negatively correlated with body weight. A comparison of two (or several) breeds of sheep growing in seasonally changing pasture conditions also illustrates the increasing importance of genetic body size as food availability declines. When food is plentiful in summer, genetically larger animals grow more rapidly, but when food is scarce in winter, they lose body weight more rapidly. Results similar to these for breeds differing in body size, have been observed for the two sexes by A.W. Illius and I.J. Gordon (unpublished).

Within dairy herds where a high quality food is used to supplement relatively poor grazing conditions, another important exception can be found to the absence of any association between food composition and body size. Genetically large animals may graze enough only for maintenance and produce very little milk whereas genetically smaller animals may consume enough for maintenance and a reasonable amount of milk. Feeding concentrates according to yield may then result in small or medium cows actually receiving more concentrates than very large cows, and the concentrate-roughage ratio will be lower for larger cows. A genetic correlation between body size and food composition can thus be created. The previously discussed negative correlation between body size and energy intake then becomes even more exaggerated. The negative association created between body size and milk yield becomes self-perpetuating.

Body size and endurance without food

Large and small mammalian species can all survive for about 10 to 15 metabolic days without any food. In relation to their lifespan and metabolic tempo, all species thus have the same capacity for enduring food shortage. If, however, food supply is seasonal and is acutely scarce for a constant number of days, a large species, having to survive for fewer metabolic days, will suffer less. This ability to endure longer helps large mammals to compete with smaller mammals when food is scarce.

Body size, maximum eating rate and eating time

Feeding times quoted by Demment and Van Soest (1983) for East African herbivores were on average 10 hours per day, 11 to 12 hours for females and 8 to 9 for males, plus, for the ruminants, a further 4 to 5 hours spent ruminating. Feeding time for a species was independent of its body size. Females grazing 0.35 times longer than males yet with a body weight only 0.7 times that of males, is inconsistent with larger species having to eat for a longer time in order to achieve an equivalent intake when nutrition is limited. The simplest explanation is the maternal reproductive load of breeding females, who require more food than males both for pregnancy and lactation and to replace reserves after lactation.

The proportion of its time a mammal spends eating should be independent of its body size. Mice, cats, seals and humans can all consume their daily requirement of high-quality food in 5 to 10 minutes. The impression from Edinburgh Zoo keepers is of great variability but no immediately obvious association with body size (Elaine Leonard unpublished). The great range of adaptability of mammals in time spent eating, which can vary from 10 minutes a day up to 10 hours or more a day, is clearly of major importance to survival in a world of fluctuating food supplies.

In an unpublished genetic experiment on time-controlled feeding, we fed five breeds of cattle differing in body size six meals a day, each of duration 4, 5 or 6 minutes. Maximum eating rate was achieved on all three meal lengths at all stages of growth. Intake per minute increased as animals increased in body weight and age. At all stages eating rate was closely correlated with breed size. Adult eating rate was estimated as $31A^{0.73}$ kJ ME /min for this complete pelleted ruminant diet containing 30% chopped straw.

For sheep grazing on ryegrass, Allden and Whittaker (1970) found that maximum eating rate was about $4.0A^{0.73}$ kJ ME /min, which is about one eighth of the cattle value discussed above. The cattle maintained their adult body weight by eating for 20 minutes per day. Similarly sheep offered the same diet as the cattle twice daily *ad libitum* also spent about half an hour per day eating and ate about 8 times more quickly than the grazing sheep (M.L. Thonney, St C.S. Taylor and T.H. McClelland, unpublished). Given this slow eating rate, grazing time for the sheep would be just under 3 hours per day. Actual grazing time reported by Allden and Whittaker (1970) was about 6 hours per day, so that maximum eating rate on that ryegrass pasture was not maintained beyond the first hour.

Bines and Davy (1970) studied intake and eating behaviour in dairy cows fed *ad libitum* for five hours daily. Eating time increased from 154 min /d for 0% roughage up to 198 min /d for 60% roughage with associated ruminating times of 0 and 468 min /d. Increasing access to food from 5 to 24 hours increased dry matter intake to 148% for 0% roughage but only 105% for 50% roughage. Given 24 hours access the daily intake of dry matter became the same for both diets.

The result for breeds appears to be the same as those for species. Feeding time is independent of breed size, except where a large mouth size is at a disadvantage as is most apparent on short grazed pasture. In such situations of food shortage, not only do small breeds have an advantage over large breeds, but growing young are favoured at the expense of their larger parents.

REPRODUCTION, ABUNDANCE AND SURVIVAL

Calder (1983) suggested that the ecological survival of a population could most simply be examined in terms of time, space, energy and number of offspring. Peters (1983) covers these aspects and many more in his excellent book on the ecological implications of body size.

Time

Millar and Zammuto (1983) examined 29 mammalian species and found that age at maturity, generation interval, life expectancy at birth and at maturity were all highly intercorrelated, with correlations between 0.92 and 0.99, and that all increased in proportion to $A^{0.27}$. For these traits mean values in metabolic years were 0.92, 1.74, 1.13 and 1.63 respectively (the last two both having been adjusted to $A^{0.27}$ from $A^{0.24}$ and $A^{0.23}$ respectively by multiplying the constant of proportionality by the overall mean adult body weight in their data raised to the power -0.03 and -0.04 respectively). The value of 336 metabolic days is somewhat earlier than the 440 metabolic days given by Taylor's (1965) equation for age when 0.98 mature in body weight and is closer to his 350 days for 0.95 mature. Taylor (1985) gives lifespan as 2000 metabolic days (5.5 metabolic years), so that life expectancy at birth and at maturity are only about 0.2 and 0.3 of lifespan. Millar and Zammuto (1983) have clearly shown that genetically larger mammals known to take longer to reach maturity, also had a longer generation interval and a longer life expectancy, all in accordance with genetic size-scaling for time.

Space

Peters (1983) collated a vast amount of data and found that the number of herbivorous mammals per square kilometre was about 100/A. In other words, the total mammalian biomass per square kilometre was on average 100 kg irrespective of body size. Calder (1983), however, gave animal density as proportional to $A^{-0.75}$, and home range as proportional to A. Both Peters (1983) and Calder (1983) agree that the number of animals per home range in square kilometres probably increases in proportion roughly to $A^{0.27}$. In other words, larger herbivores are possibly more gregarious and sociable than smaller herbivores.

Locomotion

Walking speed and maximum speed at which mammals can travel might be expected to be independent of body size (Maynard-Smith 1968). The average of the four estimates given by Peters (1983) was an allometric coefficient of 0.24 which apparently agrees with results for birds flying and fish swimming. In walking homeotherms, normal activity about doubles basal metabolic rate (Peters 1983).

Speed (a) is the product of stride length (b) and stride frequency (c). Schmidt-Nielsen (1984) derived the empirical relationship $aA^{0.17} = bA^{0.33}\ cA^{-0.17}$ (Schmidt-Nielsen 1984) which can be treated either as suspect or as difficult to explain. It has been suggested by Bonner (Schmidt-Nielsen 1984) that speed of locomotion increases most with body size when effort is at its maximum. The suggestion implies that large mammals can increase their normal speed of movement by a greater amount than can small mammals. The longer distances migrated non-stop by larger species can be explained in terms of transport costs being roughly proportional to $A^{0.73}$ while reserve energy stored in the body as fat is proportional to A. In other words, endurance time is longer in large animals. At maximum speed, however, larger mammals would travel further in the same length of time; but they would then use up their reserves relatively more rapidly and endurance times would be about the same.

Energy

The energy costs of reproduction include lactation and litter weight, both of which are discussed elsewhere in this review. Peters (1983) and Calder (1983) conclude that reproductive effort is greatest among small mammals. On critical examination, this result could be attributed either to litter weight per unit of dam weight being larger in smaller mammals (Leitch *et al.* 1959) or to protein concentration in milk decreasing with increasing genetic body size (Payne and Wheeler 1968).

Reproductive rate

Litter size. As discussed in the section on maternal performance, litter size declined as genetic body size increased. Expected litter size is 1.15 for cattle with an adult body weight of 500 kg and 6.0 in mice weighing 0.03 kg.

Lifetime reproduction. Calder (1983) concludes that total number of offspring produced in a lifetime is independent of body size, with a mean value of 5.6 offspring per breeding female. The implication is that fewer litters per lifetime are produced by small mammals. This may be the case in nature. Within their normal lifespans, breeding females from large and small species can both produce a total of about eight litters (Roberts 1961).

Reproductive value. From life tables gathered from 29 species, Millar and Zammuto (1983) derived measures of reproductive value at maturity and a residual reproductive value in the following time interval, which were highly negatively correlated (−0.71). Litter size and reproductive value at maturity were positively correlated (0.71). Thus species with larger litter size were expected to have a higher lifetime total. Reproductive value at maturity was positively correlated with litter size (0.71) and negatively correlated with generation time, age at maturity, and life expectancy at birth and maturity, and declined as genetic body size increased (expected value 4.6 $A^{-0.08}$ female offspring). Residual reproductive value was independent of body size (expected value = 2.5 female offspring).

Population cycles

The number of animals in a population fluctuates with a cycle duration given by Calder (1983) as 6.5 $A^{0.27}$ years. The corresponding estimate from the results of Peterson *et al.* (1984) was 8.2 metabolic years. Cycles of about three years are expected in mice and about 30 years in large ungulates.

Table 9. Perinatal deaths in calves from large, medium and small breeds (Donald 1963).

	Number of calves		Mortality (%)		Average, first 4 parities
	1st parity	2nd parity or later	1st parity	2nd parity or later	
Friesian	73	88	26	1	7.4
Ayrshire	69	113	4	2	2.4
Jersey	55	87	13	5	6.6

Breed variation in mortality

Donald (1963) examined perinatal deaths of calves in three dairy cattle breeds and found that the intermediate breed (Ayrshires) had a much lower mortality rate than either the large Friesian or the small Jersey breeds (Table 9). When he examined all deaths in relation to the ratio of birth weight to dam weight, calves both very large and very small relative to their dam had a high mortality. The relatively small calves were too weak to survive; the relatively large calves were prone to calving difficulties. Monteiro (1969) examined calving difficulties, not necessarily fatal, in the same data and showed that they increased rapidly with breed size (Table 10).

Genetic variation within breeds

Within a breed, body size is also clearly associated with probability of survival. The type of association, however, is not always the same. In general, mortality is least at an intermediate body size, although small size appears to be favoured in some cases and large in others.

Table 10. Calving difficulties in relation to body weight (kg) of dam in three breeds, meaned over first two parities (Monteiro 1969).

Breed	Number of calvings	†Index of calving difficulty	Calving difficulties(%)
Friesian	70	70.9	40.0
Ayrshire	82	62.6	8.5
Jersey	42	51.7	0

†Index = (log (birth weight) −0.4 log (dam weight)) x 100, adjusted for parity.

In terms of embryo survival, Moore *et al.* (1970) found that small size was favoured in mice. In lines of mice selected for large and small body size, significantly more eggs survived to birth when transferred both from (8% more) and to (20% more) the small selected dams than when transferred from or to the large selected mice.

Purser and Young (1959) had earlier found an intermediate optimum in two breeds of sheep. Perinatal mortality was greatest at both extremes of birth weight, and they noted that the same was true in humans. Carpenter *et al.* (1971) studied lifetime reproductive performance and found a significant tendency for breeding cows that were smaller and earlier maturing than average to produce more calves and more regularly. Roberts (1961) found similar results in mice. Mice selected for small size had smaller litters than mice selected for large size, but the small mice reproduced more frequently and more often, so that their lifetime performance in terms of total young born exceeded that of the large line.

The conclusion appears to be that in relation to body size there is in general an intermediate optimum for maternal characteristics connected with survival. Very large calves at birth are prone to calving difficulties; very small calves at birth are often too weak to survive. If milk yield is regarded as a maternal characteristic then, as we saw earlier, there is some evidence pointing to intermediate body size being optimal here also (Hooven *et al.* 1968).

BIRTH WEIGHT, LITTER SIZE AND MATERNAL PERFORMANCE

Mammals produce offspring in litters varying in number (N) from 1 up to about 20, and with individual birth weight ($W_b$kg) varying from about 0.25 times down to less than 0.005 times their mother's adult body weight (A kg). The total weight of offspring ($N\overline{W}_b$) delivered at any one parturition is largely determined by the genetic body size of the mother. This close relationship was first published by Leitch *et al.* as:

$$N\overline{W}_b = 0.165A^{0.83} \qquad \text{Equation 6}$$

Peters (1983) lists several authors who have reported much the same relationship.

Robbins and Robbins (1979) related total birth weight to maternal weight in ungulates and subungulates and obtained:

$$N\overline{W}_b = 0.214A^{0.79} \qquad \text{Equation 7}$$

Assuming that the exponent 0.79 is not significantly different from 0.83, their relationship based on an average maternal weight of about 200kg can be transformed to give:

$$N\overline{W}_b = 0.214\ (200)^{-0.04} = 0.173A^{0.83} \qquad \text{Equation 8}$$

close to Leitch *et al*'s (1959) inter-species relationship of 0.165 $A^{0.83}$. We have obtained constants of proportionality of 0.197 for cattle, 0.183 for sheep and 0.185 for dogs.

The components of litter weight are also related to maternal body weight. Millar and Zammuto (1983) found a systematic decrease in litter size (N) as maternal body weight (A) increased with:

$$N = 3.43A^{-0.16} \qquad \text{Equation 9}$$

The present authors examined over 100 mammalian species and obtained a closely similar result

$$N = 3.3A^{-0.17} \qquad \text{Equation 10}$$

In addition, they examined the relationship for individual birth weight (W_b) and found:

$$W_b = 0.05A \qquad \text{Equation 11}$$

The implication of equation 11 is that individual birth weight is directly proportional to maternal body weight. Degree of maturity in body weight at birth (u_b) must therefore have an expected value of $u_b = W_b/A = 0.05$, irrespective of litter size or gestation length and irrespective of whether the species is large or small.

Equations 10 and 11 combine to give Leitch *et al*'s (1959) relationship between maternal weight and total litter weight, and this derivation demonstrates that the decrease in litter weight is due not to any decrease in the average birth weight of an individual but is associated with the decrease in litter size (Fig. 9). They are in general agreement with other published results, although these show considerable variation (Peters 1983).

Maternal weight accounted for 0.982 of the variation among mammalian species in total litter weight, and this relationship gave a more accurate prediction than either equation 10 or 11. Deviations from expected litter size and expected individual birth weight were

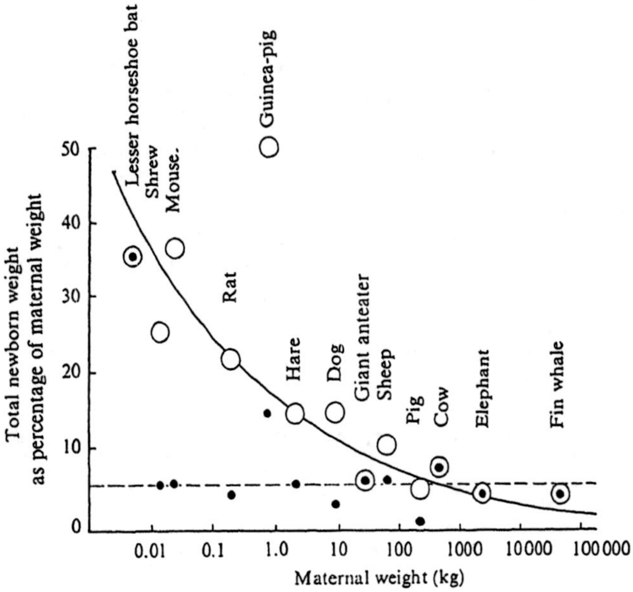

Figure 9. The relationship for eutherian mammals of total newborn liveweight (○ ——), and individual birthweight (● ----) to mature maternal body weight (Land 1985).

often large. Larger than average litters, however, usually had smaller than average birth weight. This negative correlation reduced the magnitude of the deviations in total litter weight (although they were still quite large). Maternal capacity might be defined as total litter weight at birth as a percentage of maternal weight to the power 0.83, and then maternal capacity would be predicted as 0.165 for any species in the range investigated. Total litter weight can be thought of as strongly controlled by maternal weight, most of the gross control being affected by variation in litter size and thus making it possible for individual birth weight to remain about 0.05 times adult maternal weight.

The large deviations in individual birth weight and litter size were then examined in relation to deviations in gestation length as well as maternal weight. The influence of maternal weight was large, accounting for 0.98 of the variation in birth weight, 0.64 in litter size and 0.85 in gestation length; it was eliminated by using size-scaled variables, u_b for degree of maturity in body weight at birth and θ^b for metabolic age at birth, to which were added litter size at birth adjusted for maternal body weight, namely $n_b = NA^{0.17}$. These variables were then examined as deviations from their overall mean values of $\bar{u}_b$

= 0.05, $\bar{\theta}_b$ = 50 and $\bar{n}_b$ = 3.3. The deviations for these three traits were quite highly inter-correlated. Large litter size deviations were associated with less mature individuals at birth (r = −0.7) and a lower metabolic age at birth (r = -0.7). In addition species whose young were immature at birth tended to be born at a younger metabolic age, although this genetic correlation was slightly lower (+0.6). Interestingly, the correlation between adjusted total litter weight and adjusted gestation length was zero, so that gestation length was affected by litter size but not by total litter weight. In other words, total litter weight in general attained the same expected value whether gestation length was short or long. These intercorrelations strongly suggested a factor operating to reduce both maturity at birth and gestation length when litter size increased. This factor possibly operated through the total amounts of various hormones produced *in utero* and depended more on how many foetuses were present than on their size.

The proportion of error variation in these new variables was relatively very much greater than that in the original variables, and a single common factor plus specific errors was all that could be ascertained with any degree of certainty.

This factor accounted for all the residual co-variation (there were only 3 variables), and for 66%, 90% and 62% of the residual variation in birth weight, litter size, and gestation length respectively. Specific error variances corresponded to errors of the order of $\pm$30% for all three variables. Although, altogether, about 99%, 84% and 91% of the total variation in the original variables was accounted for, the amount of variation still unaccounted for was large when measured as a coefficient of variation.

Since the factor accounted for most of the variation in adjusted litter size, it might be thought of as a "litter size factor" with a concomitant effect on individual birth weight and on gestation length.

In summary, of the variation in individual birth weight, 98% was accounted for by maternal weight and a further 1% by the "litter size", factor. Of the variation in gestation length, 85% was accounted for by maternal weight and a further 6% by the "litter size" factor. Of the variation in litter size only 64% was accounted for by maternal weight, and a further 20% by the "litter size" factor.

A striking example of these correlated deviations is the pig with a gestation length half that expected from its adult body size, with an average litter of about ten piglets instead of the expected 1.3, and with individual piglets that are an average 0.006 mature in body weight at birth compared with an expected value twelve times greater; yet total litter weight is close to expectation. Small bats are also exceptional in that their expected litter size is about seven but they give birth to a litter of one that is about 0.25 mature in body weight compared with an expected total litter weight of 0.36; this enormous offspring is produced in a gestation length three times longer than expected.

A simple hypothesis would be to assume that (i) maternal capacity varies genetically with $A^{0.83}$, that the full capacity of 0.165 $A^{0.83}$ is generally attained irrespective of maturity at birth, litter size and gestation length and (ii) that the growth of individuals along the mean mammalian growth curve is largely independent of their being *in utero* or not and independent of number of litter mates, provided these are at their genetically determined mean values.

This hypothesis was applied to pigs. Observed and expected values for total litter weight, individual weight and litter size seriously disagreed until an adjustment for short gestation length led to excellent agreement (Table 11).

Table 11. Observed and expected values for litter weight, individual weight and litter size at birth in pigs (mature maternal weight A=280kg).

	Observed	Expected	
Total litter wt (kg)	16.8	17.7	
Degree of maturity at birth	0.006	0.0063a	0.05b
Litter size	10.0	10.0a	1.3b

(a) from equations 12 or 13
(b) from equations 10 or 11

Equation 11 gives an adult body weight of about 1200 kg for a species with a expected litter size of 1.0. If the hypothesis is to hold, species larger than this would have to shorten their gestation length until individual birth weight was equal to maternal capacity, that is, until $u_b = 0.165A^{-0.17}$. Degree of maturity at birth is therefore expected to fall below 0.05 and decline as species become larger. A predicted value of 0.042 for a 3000 kg elephant agrees well with an observed value of 0.036. Unfortunately gestation length is longer than expected, not shorter. Foetal growth has therefore been slowed down. For an 80,000 kg whale, an expected degree of maturity at birth of 0.024 is in reasonable agreement with the observed value of 0.030; in this case gestation length is unnecessarily short, so that foetal growth has to be faster than expected. Total litter weight, which is also degree of maturity at birth, thus remains predictable, while foetal growth rate and gestation length can apparently be adjusted in various ways to achieve this.

The overall conclusion is that the hypothesis has some predictive value and accounts for much of the observed variation but further factors are clearly operating.

Body size and maternal performance of breeds

The clear genetic association between litter weight and maternal weight among species has also been examined among breeds within a species. The same genetic relationship appears to hold, so that proportionality to maternal weight to the power 0.83 can be uniformly applied to both breeds and species.

We studied the relationship of birth weight to dam weight in over 100 breeds of cattle. The conclusion was fairly convincing that birth weight as a proportion of dam weight declined with breed size but the exact rate was less clear. It was consistent with proportionality to $A^{0.83}$, the inter-species relationship of Leitch *et al.* (1959). Further evidence that the inter-breed and inter-species relationships are the same comes from cattle and dogs. Cundiff *et al.* (1986) gave birth weights for 19 breed crosses in cattle. The inter-breed (log-log) regression of birth weight on cow weight was 0.76 ± 0.13 for crossbred calves out of the same dam breeds and 0.86 ± 0.16 for 3-way-cross calves out of F_1 cows by the same set of sire breeds. The expected regression for dams and calves of the same genotype would be slightly greater than either of these. Dogs, because of their wide range in size of breeds, are an excellent species in which to study inter-breed relationships. Kirkwood's (1985) study yielded two regressions, one for litter size (0.32) and one for individual birth weight (0.56) which can be combined to give a regression for litter weight on maternal body weight of 0.88, with a standard error of about 0.08. The evidence thus strongly favours a uniform $A^{0.83}$ inter-breed and inter-species relationship.

We studied Friesian, Ayrshire and Jersey dairy cattle which have mean maternal body weights in the ratio 1.15 to 1.0 to 0.81 and are expected to have mean birth weights in the ratio 0.977 to 1.0 to 1.036. Observed values given by Monteiro (1969) showed that Friesians had calves 6% larger and Jerseys 16% smaller than expected. In line with these deviations, Friesians had greater calving difficulties and above-average calf mortality while Jerseys had easy calvings and above-average mortality among their small, weak calves.

The constant maternal capacity hypothesis

The uniformity of the inter-breed and inter-species relationship implies that the hypothesis discussed above for species may also hold for breeds. Gestation length does not vary within a species, and maternal capacity must therefore be achieved by altering individual birth weight with an associated change in litter number when possible. Thus the following two equations can be solved for N and u_b, given t_b, k and A:

$$N = k/u_b A^{0.17} \qquad \text{Equation 12}$$

$$\text{and } u_b = \emptyset(\theta_b) = \emptyset((t_b - 3.5)/A^{0.27}) \qquad \text{Equation 13}$$

where k is the appropriate constant of proportionality (either the mammalian mean of 0.165 or a species-specific value) and where gestation length in days, t_b, is constant.

The result is an explanation of the well-known increase in

litter size as breed size increases. Because gestation length is constant, metabolic age at birth is less in large breeds where individual offspring are much less mature than average, with the opposite occurring in small breeds. If maternal capacity is to be maintained, litter size must increase with increasing breed size to compensate for young that are becoming less mature at birth. Kirkwood's (1985) equations give small (2 kg) dogs as about 0.05 mature at birth, medium (15 kg) dogs as about 0.02 mature and large (70 kg) dogs as only about 0.01 mature, with litter sizes of 3.2, 5.6 and 8.2 respectively. This explanation and these equations agree with observed litter size and birth weight in breeds of dogs with mean maternal body weights in the range 2 to 15 kg. For the range 15 to 70 kg, the observed increase in litter size falls short of that required. Instead, foetal growth rate increases. Young are in fact more mature in larger breeds than suggested by the mean growth curve. The large St Bernard breed was calculated to have a foetal growth rate ten times greater than expectation.

The conclusions are that maternal capacity is usually achieved and that to achieve it when gestation length is constant, litter size increases until the individual young become dangerously immature. If necessary, foetal growth rate increases also. The implication of Kirkwood's inter-breed regression is that there is no transition point between litter size increasing and foetal growth rate increasing, but a steady rate of change over the whole range.

Table 12. Observed and expected mean litter weight of mice and degree of maturity at birth before and after genetic selection for body size.

		Strain of selected mice		
		large	control	small
Total litter weight (g)	obs	11.2	9.4	6.5
	exp	11.5	9.3	6.3
Degree of maturity at birth	obs	0.035	0.040	0.056
	exp(a)	0.030	0.038	0.050
	exp(b)	0.050	0.050	0.050

(a) from equations 12 and 13
(b) from equation 11

The hypothesis was also applied to three strains of mice, one unselected and two genetically selected for large and small body size by Falconer (1973). Total litter weight in the large line was almost twice that in the small line. All three were predicted reasonably accurately by equation 6 for maternal capacity (Table 12). Predicted and observed gestation lengths did not agree, since all three lines had the same gestation length. Individual birth weights, predicted as being 0.05A at a metabolic age of 50 days, were then adjusted along the mean mammalian growth curve of Moore (1985) from their predicted to observed age at birth. In terms of degree of maturity at birth, predicted and observed values agreed well after adjustment but not before (cf. Table 11). Lastly litter size, calculated as the ratio of total litter weight to individual weight, is predicted reasonably accurately only after adjustment for gestation length.

The conclusion is that correlated responses in both litter weight and litter size to genetic selection for large and small body size can be predicted from maternal weight if allowance is made for gestation length remaining unchanged.

BODY SIZE AND LACTATION IN A TYPICAL MAMMAL

Inter-species relationships

That the daily milk yield of mammals is proportional to mature body weight to the 0.73 power has been well established by Brody (1945), Payne and Wheeler (1968), Linzell (1972), Taylor (1973) and Hanwell and Peaker (1977). Lactation length has also been shown to behave like other time variables and lactation yield to be directly proportional to mature body weight (Taylor 1973). Other relationships of interest are that milk yield in kg /d is 1.7 times the weight of mammary gland (Hanwell and Peaker 1977), that peak milk yield in MJ /d equals 34 times maximum growth rate of young in kg /d (Robbins and Robbins 1979), and that milk appears to be higher in protein concentration in small species than in large (Payne and Wheeler 1968).

After birth at 50 metabolic days new born mammals suckle until weaned at about 90 to 100 metabolic days of age, when they are about 0.3 mature in body weight. The total amount of milk consumed is directly proportional to the maternal body weight of the species. In fact, lactation yield can be predicted as 16A MJ of milk energy with an average energy concentration of 4.0 MJ /kg. Lactation length is the same as time spent suckling and has an expected value of 50 metabolic days. Average daily yield is therefore expected to be $16A/50A^{0.27} = 0.32A^{0.73}$MJ /d. Peak yield is $0.50A^{0.73}$MJ /d, about 1.5 times average daily yield, and occurs about a third of the way through lactation.

The above formulae yield expected values that can be compared with the observed values reported by Anderson *et al.* (1984) if it is assumed that guinea-pigs have an adult body weight (A) of 1kg. The observed and expected values are: lactation length of 21 days (less than half the expected 50 days), lactation yield of 0.5 kg (expected yield 4 kg), peak yield of 0.04 kg /d (expected yield 0.13 kg /d, or 0.09 kg /d according to Hanwell and Peaker 1977) and an average daily yield only about half peak yield.

The conclusion is clear. Guinea-pigs, with their short lactation and low daily and lactation yields, are not representative of mammalian milk production. Their milk yield may well be in keeping with their newborn being unusually mature in body weight (0.14 *vs* 0.05 expected) and with an unusually high proportion of body fat (0.10 *vs* 0.02 expected). With a litter size of four, a guinea-pig after gestating 65 days produces about half her own weight of offspring, a remarkable feat. Guinea pigs are weaned at about 65 + 21 = 86 days, not very different from the expected 90 days. Moreover, guinea-pigs mature postnatally at a rate well below average (Taylor 1965, 1980b). This example illustrates the power of genetic size-scaling in assessing the validity of using the guinea-pig as a model for dairy cattle or a typical mammal, which it certainly is not.

The expectation for a 600 kg dairy cow is a lactation length of 281 days, with a lactation yield of 2400 kg, close to the observed value for suckled dairy cows (Cundiff 1986). Cows' milk contains an above-average amount of water and 3 MJ /kg energy compared with the expected value of 4 MJ /kg.

Inter-breed relationships

Holmes (1973) and Taylor (1973) both found daily peak milk yield to be proportional to maternal metabolic body weight. Lactation length had an inter-breed regression intermediate between 0 and 0.27, the problem being to obtain data where management practices and the annual cycle do not influence lactation length (Taylor 1973). An interesting result was that peak yield appeared to occur earlier in large than in small dairy breeds.

Within-breed genetic selection

The within-breed relationship between milk yield and the biological efficiency of milk production has been reviewed by Taylor (1973) and Morris and Wilton (1976). Genetically larger animals produce more milk but only up to a certain size beyond which the trend reverses. The same applies to food conversion efficiency but optimum yield occurs at a lower body weight (Fig. 10). The results are compatible with proportionality to metabolic body weight but with additional factors coming into operation as both extremes of body weight are approached; possibly body condition is different at the extremes (Smith *et al.* 1976) or genetic homeostasis breaks down (Lerner 1954).

Body size and milk composition

In a study involving 158 different breeds of cattle, fat percentage was found to decrease very slightly with mature weight of breed but not significantly (Taylor 1971). Since environmental conditions were very different for different breeds, this slight trend may have been brought about by the association between fat percentage and level of feeding. Between breeds (as between species; Payne and Wheeler 1968) some genetic association between body size and milk composition may exist, but it is a very erratic association.

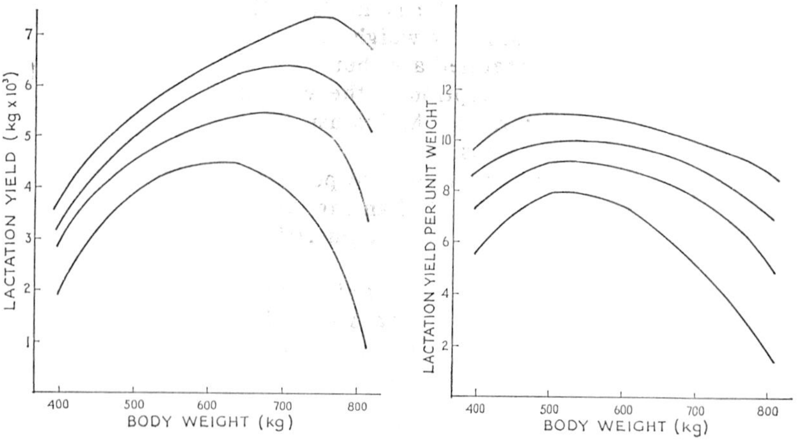

Figure 10. Lactation yield in relation to genetic differences in mature body weight of cows for different planes of nutrition (Taylor 1973).

Within breeds, many correlations have been reported between body weight and both fat and solids-not-fat percentages (Lefebvre and Ricordeau 1966). The general finding is that cows heavier after calving tend to yield somewhat more and have a slightly lower fat percentage although in several studies no significant correlation have been found (Nazarec 1957).

SUMMARY

All sizes of mammal have virtually the same biochemistry, physiology and anatomy. Stemming from this underlying similarity is a body of theory called genetic size-scaling, which depends on each genotype having its own genetic size factor operating throughout growth from embryo to adult. This genetic size factor expresses itself in many traits at every stage of growth.

The traits of interest to ecologists and nutritionists, and of considerable relevance to animal production, are the amount of food ingested and metabolised, the growth of muscle, fat and bone, the yield of milk and wool, reproductive rate, survival rate, and whether the generation interval is short or long. All these traits are quantitative and their relationship to body size is surprisingly uniform over the whole mammalian range. These relationships have been encapsulated in the two formal rules of genetic size-scaling which have then been used to furnish a quantitative description of the growth of a typical mammal.

At present, the best conceptual position to adopt is undoubtedly that, with certain exceptions, genetic size-scaling applies uniformly across species, breeds and strains within species, and sexes and genotypes within breeds. The main exceptions across species are liver and kidney weight and litter weight at birth, all with inter-species regression coefficients intermediate between 1.0 and 0.73. Within species, the outstanding exception is the virtually constant gestation length of all breeds and strains. Males usually have heavier neck and shoulder muscles than females.

The descriptive and predictive power of genetic size-scaling needs to be more widely propagated among students of animal science, especially those involved in genetics, nutrition, and ecology.

This Butler Memorial monograph would never have been completed without the patience and word-processing skill of Bobby Johnston. Figures 5 and 10 are reprinted with permission from the Br. Soc. Anim. Prod.; and 7 from J. Anim. Sci..

Allden, W.G., and Whittaker, I.A. McD. (1970). *Aust. J. Agric. Res.* **21**, 755-66.

Anderson, R.R., Salah, M.S., Sheffield, L.G., and McKenzie, W.N. (1984). *J. Dairy Sci.* **67**, 185-93.

Berg, R.T., and Butterfield, R.M. (1976). "New Concepts of Cattle Growth." (Sydney Univ.: Sydney.)

Bines, J.A., and Davey, A.W.F. (1970). *Brit. J. Nutr.* **24**, 1013-28.

Blaxter, K.L. (1967). "The Energy Metabolism of Ruminants." (2nd ed). (Hutchison: London.)

Broad, T.E., Davies, A.S., and Tan, G.Y. (1980). *Anim. Prod.* **31**, 73-9.

Brody, S. (1945). "Bioenergetics and Growth." (Reinhold: New York).

Brody, S., and Proctor, R.C. (1932). Univ. Missouri Agric. Exp. Sta. Res. Bull. **166**, 89-101.

Butterfield, R.M., Griffiths, D.A., Thompson, J.M., Zamora, J., and James, A.M. (1983). *Anim. Prod.* **36**, 29-37.

Calder, W.A. (1983). *Ann. Rev. Ecol. Syst.* **14**, 213-30.

Calder, W.A. (1984). "Size and Function and Life History." (Harvard Univ.: Harvard.)

Carpenter, J.A., Fitzhugh, H.A., Brown, J.E., and Crouch, E.K. (1971). *Beef Cattle Res. Texas* pp. 41-3.

Cianzio, D.S., Topel, D.G., Whitehurst, G.B.. Beitz, D.C., and Self, H.L. (1985). *J. Anim. Sci.* **60**, 970-6.

Clark, A.J (1927). "The Comparative Physiology of the Heart." (Cambridge Univ.: Cambridge.)

Clutton-Brock, T.H., and Harvey, P.H. (1983). In "Mammalian Behavior and Ecology." (Eds J.F. Eisenberg and D.G. Kleiman.) pp. 632-63. (Smithsonian: Washington.)

Cundiff, L.V., Gregory, K.E., Koch, R.M., and Dickerson, G.E. (1986). Proc. 3rd Wld Congr. Genetics Applied Livestock Prod. (Nebraska) **9**, 271.

Demment, M.W. (1983). *Afr. J. Ecol.* **21**, 219-33.

Demment, M.W., and Soest, P.J. Van. (1983). "Body Size, Digestive Capacity and Feeding Strategies of Herbivores." (Winrock: Morrilton, Arkansas.)
Demment, M.W., and Soest, P.J. Van. (1985). *Amer. Nat.* **125**, 641–72.
Dickinson, F.N., McDaniel, B.T., and McDowell, R.E. (1969). *J. Dairy Sci.* **52**, 489–97.
Dinius, D.A., and Baumgardt, B.R. (1970). *J. Dairy Sci.* **53**, 311–6.
Donald, H.P. (1963). *Anim. Prod.* **5**, 87–95.
Evans, E., and Miller, D.S. (1968). *Proc. Nutr. Soc.* **27**, 121–9.
Falconer, D.S. (1973). *Genet. Res.* **22**, 291–321.
Falconer, D.S., Gauld, I.K., and Roberts, R.C. (1978). *Genet. Res.* **31**, 287–301.
Farlow, J.O. (1976). Ecology **57**, 841–57.
Ferrell, C.L., and Jenkins, T.G. (1985). *J. Anim. Sci.* **61**, 725–41.
Fitzhugh, H.A., and Taylor, St C.S. (1971). *J. Anim. Sci.* **33**, 717–25.
Günter, B., and Guerra, (1955). *Acta. Physiol. Latinaom* **5**, 169.
Hammer, R.E., Palmiter, R.D., and Brinster, R.L. (1984) *Nature*, London **311**, 65–7.
Hansett, R. (1986). In "Exploiting New Technologies in Animal Breeding : Genetic Developments." (Eds C. Smith, J.W.B. King, and J.C. McKay.) pp. 71–80. (Oxford Univ.: Oxford.)
Hanwell, A., and Peaker, M. (1977). In "Comparative Aspects of Lactation." (Ed. M. Peaker) *Symp. Zool. Soc. Lond.* **41**, 297–312.
Hemmingsen, A.M (1980). Reports of the Memorial Hospital and Nordinsk Insulin Laboratorium **9**, 6–110.
Hinks, C.J.M. (1966). *Anim. Prod.* **8**, 467–88.
Holmes, W. (1973). *Proc. Brit. Soc. Anim. Prod.* **2**, 27.
Holmes, W. (1977). *Phil. Trans. R. Soc. Lond. B.* **281**, 121–37.
Hooven, N.W., Miller, R.H., and Plowman, R.D. (1968). *J. Dairy Sci.* **51**, 1409–19.
Huxley, J.S. (1932). "Problems of Relative Growth." (Methuen: London.)
Jarman, P.J. (1974). *Behaviour* **48**, 215–6.
Joubert, D.M. (1955). *Nature*, London. **175**, 936–7.
Keyser, C., and Heusner, A. (1964). *J. Physiol.* Paris, **56**, 489.
Kidwell, J.F., Gregory, P.W., and Guilbert, H.R. (1952). *Genetics* **37**, 158–74.
Kirkwood, J.K. (1985). *J. Small. Anim. Pract.* **26**, 97–110.
Kleiber, M. (1932). *Hilgardia* **6**, 315–53.
Kleiber, M. (1961). "The Fire of Life." (Wiley and Sons: New York.)
Krebs, H.A. (1950). *Biochem. Biophys. Acta.* **4**, 249–69.
Land, R.B. (1985). In "Reproduction in Mammals. 4. Reproductive Fitness." (Eds C.R. Austin and R.V. Short.) pp. 62–102. (Cambridge Univ.: Cambridge.)
Lang, B.J., and Legates, J.E. (1969). *Theoret. Appl. Gen.* **39**, 306–14.
Lawrence, T.L.J. (1980). "Growth in Animals." (Butterworths: London.)
Lefebvre, J., and Ricordeau, G. (1966). 9th Int. Congr. Anim. Prod., Edinburgh. Scient. Progm. Abstr., Eng. ed, **16**.

Leitch, I., Hytten, F.E., and Billewicz, W.Z. (1959). *Proc. Zool. Soc. Lond.* **133**, 11-28.
Lerner. I.M. (1954). "Genetic Homeostasis." (Oliver and Boyd: Edinburgh.)
Lewis, D., and D'Mello, J.P.F. (1967). In "Growth and Development of Mammals." (Eds C.A. Lodge and G.E. Lamming.) pp. 345-67. (Butterworths: London.)
Lindstedt, S.L., and Calder, W.A. (1981). *Quart. Rev. Biol.* **56**, 1-16.
Linzell, J.L. (1972). *Dairy Sci. Abstr.* **34**, 351-60.
Lobley, G.E., Milne, V., Lovie, J.M., Reeds, P.J., and Pennie, K. (1980). *Brit. J. Nutr.* **43**, 491-502.
McCarthy, J.C. (1980). In "Selection Experiments in Laboratory and Domestic Animals." (Ed. A. Robertson.) pp.100-9. (CAB: Farnham Royal, UK.)
MLC (1971). Beef Improvement Scientific Study Group Report. (Meat and Livestock Commission: Bletchley, England.)
Millar, J.S., and Zammuto, R.M. (1983). *Ecology* **64**, 631-5.
Mitchell, H.H. (1962). "Comparative Nutrition of Man and Domestic Animals." (Academic: New York.)
Monteiro, L.S. (1969). *Anim. Prod.* **11**, 293-306.
Monteiro, L.S. (1972). *Anim. Prod.* **14**, 263-81.
Moore, A.J. (1985). *Anim. Prod.* **40**, 441-53.
Moore, R.W., Eisen, E.J., and Ulberg, L.C. (1970). *J. Reprod. Fert.* **23**, 271-5.
Morris, C.A., and Wilton, J.W. (1976). *Can. J. Anim. Sci.* **56**, 613-47.
Nazarec, P.M. (1958). *Anim. Breed. Abst.* **26**, 36-7.
Nedava, V.E. (1970). *Tsitol. Genet.* **4**, 447-50.
Parks, J.R. (1970). *Amer. J. Physiol.* **219**, 840-3.
Parks, J.R. (1982). "A Theory of Feeding and Growth of Animals." (Springer-Verlag: New York.)
Payne, P.R., and Wheeler, E.F. (1968). *Proc. Nutr. Soc.* **27**, 129-38.
Peters, R.H. (1983). "The Ecological Implications of Body Size." (Cambridge Univ.: Cambridge.)
Peterson, R.O., Page, R.E., and Dodge, K.M. (1984). *Science* **224**, 1350-2.
Pitts, G.C., and Bullard, T.R. (1968). In "Body Composition in Animals and Man." pp. 45-79. (Nat. Acad. Sci.: Washington, DC.)
Prange, H.D., Anderson, J.F., and Rahn, H. (1979). *Amer. Nat.* **113**, 103-22.
Priestley, G.C., and Robertson, M.S.M. (1973). *Genetic. Res.* **22**, 255-78.
Purser, A.F., and Young, G.B. (1959). *Anim. Prod.* **1**, 85-91.
Ricard, F.H. (1978). Proc. XVI. Wld Poul. Cong., Rio de Janeiro. **10**, 1786-93.
Roberts, R.C. (1961). *Heredity* **16**, 369-81
Roberts, R.C. (1981). *Genet. Res.* **38**, 9-24.
Robbins, C.T., and Robbins, B.L. (1979). *Amer. Nat.* **114**, 101-16.
Roy, J.H.B. (1967). *Sb. Vys. Sk. Zemed. Brne* **36**, 325-42.
Roy, J.H.B., Stobo, I.J.F., Gaston, H.J., Ganderton, P., and Shotton, S.M. (1971). *Br. J. Nutr.* **26**, 363.

Russell, W.S. (1975). *Anim. Prod.* **21**, 217-26.

Salmon-Legagneur, E. (1967). In "Growth and Development of Mammals." (Eds G.A. Lodge and G.E. Lamming.) pp. 158-91. (Butterworths: London.)

Schmidt-Nielsen, K. (1984). "Scaling. Why is Animal Size so Important?" (Cambridge Univ.: Cambridge.)

Smith, G.M., Fitzhugh, H.A., Cundiff, L.V., Cartwright, T.C., and Gregory, K.E. (1976). *J. Anim. Sci.* **43**, 380-8.

Smith, J.M. (1968). "Mathematical Ideas in Biology." (Cambridge Univ.: Cambridge.)

Smith, R.E. (1956). *Ann. N.Y. Acad. Sci.* **62**, 403-21.

Southgate, J.R,, Cook, G.L., and Kempster, A.J. (1982). *Anim. Prod.* **34**, 155-66.

Stobart, R.H., Bassett, J.W., Cartwright, T.C., and Blackwell, R.L. (1986). *J. Anim. Sci.* **63**, 729-40.

Storry J.E., Hall, A.J., and Johnson, V.W. (1968). *Br. J. Nutr.* **22**, 609-14.

Syrstad, O. (1966). *Acta. Ag. Scand.* **16**, 79-96.

Taylor, St C.S. (1965). *Anim. Prod.* **7**, 203-20.

Taylor, St C.S. (1970). In "The Use of Models in Agricultural and Biological Research." (Ed. J.G.W. Jones.) pp. 107-17. (Grassl. Res. Inst.: Hurley, UK.)

Taylor, St C.S. (1971). *Ann. Genet. Sel. Anim.* **3**, 85-98.

Taylor, St C.S. (1973). *Proc. Br. Soc. Anim. Prod.* **2**, 15-26.

Taylor, St C.S. (1980a). *Anim. Prod.* **30**, 161-5.

Taylor, St C.S. (1980b). *Anim. Prod.* **31**, 223-35.

Taylor, St C.S. (1982). Proc. 2nd Wld. Congr. Genetics Applied to Livestock Production. pp. 218-30.

Taylor, St C.S. (1985). *J. Anim. Sci.* **61**. (Suppl. 2), 118-43.

Taylor, St C.S. (1986). Proc. XIII Internat. Congress. Nutr. pp. 683-5.

Taylor, St C.S. (1987a). *Proc. Aust. Assoc. Anim. Breed. and Genet.* (in press).

Taylor, St C.S. (1987b). *Proc. Aust. Assoc. Anim. Breed. and Genet.* (in press).

Taylor, St C.S., and Craig, J. (1965). *Anim. Prod.* **7**, 82-102.

Taylor, St C.S., Turner, H.G., and Young, G.B. (1981). *Anim. Prod.* **33**, 179-94.

Taylor, St C.S., Moore, A.J., and Thiessen, R.B. (1986). *Anim. Prod.* **42**, 11-8.

Thiessen, R.B. (1985). *Anim. Prod.* **40**, 23-32.

Thiessen, R.B., Hnizdo, E., Maxwell, D.A.G., Gibson, D., and Taylor, St C.S. (1984). *Anim. Prod.* **38**, 323-40.

Thompson, J.M., Butterfield, R.M., and Perry, D. (1985). *Anim. Prod.* **40**, 71-84.

Timson, B.F., and Dudenhoeffer, G.A. (1985). *J. Anim. Sci.* **61**, 416-22.

Tulloh, N.M. (1964). In "Carcase Composition and Appraisal of Meat Animals." (Ed. D.E. Tribe.) pp. 5.1-5.16 (CSIRO: Melbourne.)

Walstra, P. (1980). Meded. Landbhogesch. Wageningen (Nederland.) 80–4. pp. 1–206.
Young, V.R. (1985). *J. Anim. Sci.* **61** (Suppl. 2), 39–56.

INDEX

7 8 9 0 1 2 3 4 5 6
A B C D E F G H I J